LEGAL ASPECTS OF IMPLEMENTING THE CARTAGENA PROTOCOL ON BIOSAFETY

This book, the first in a new series that focuses on treaty implementation for sustainable development, examines key legal aspects of implementing the Cartagena Protocol on Biosafety to the Convention on Biological Diversity (CBD) at national and international levels. The book provides a serious contribution to the current legal and political academic debates on biosafety by discussing key issues under the Cartagena Protocol on Biosafety that affect the further design of national and international law on biosafety and analysing recent progress in development of domestic regulatory regimes for biosafety. This text also examines the legal, political, economic, and practical challenges and solutions encountered in recent efforts to develop and implement domestic biosafety regulations, with a focus on developing countries. Coming after the fifth Conference of the Parties serving as the meeting of the Parties (COP-MOP) to the Cartagena Protocol on Biosafety (COP-MOP 5), where the Parties adopted a new Nagoya – Kuala Lumpur Protocol on Liability and Redress, this timely book examines recent developments in biosafety law and policy.

Marie-Claire Cordonier Segger is the Head of Economic Growth and Trade at the International Development Law Organization (IDLO) in Rome, Italy. She co-edits the Treaty Implementation for Sustainable Development series with Cambridge University Press, and has authored or edited more than eighty publications, including fourteen books in three languages: Sustainable Development in World Investment Law (2010) and Sustainable Development in World Trade Law (2005) with Dr. M. Gehring; Sustainable Development Law: Principles, Practices and Prospects (2004) with A. Khalfan; and Sustainable Justice: Reconciling Economic, Social and Environmental Law (2004) with H. E. Judge C. G. Weeramantry. In an academic capacity, she also serves as Senior Director of the Centre for International Sustainable Development Law (CISDL) in Montreal, Canada; as Affiliated Fellow of the Lauterpacht Centre for International Law (LCIL) at Cambridge University, and as a Visiting Professor at the University of Chile Faculty of Law.

Frederic Perron-Welch is Managing Director of Bionomos Ltd and Programme Coordinator of the CISDL Biodiversity and Biosafety Law Research Programme. He has represented the CISDL at a number of meetings of the Convention on Biological Diversity (CBD), including the 10th Convention of the Parties, and interned at the Biosafety Division of the Secretariat of the CBD in 2009. He is called to the Bar of Ontario, Canada, and articled with the Canadian Environmental Law Association (CELA).

Christine Frison is a Legal Research Fellow with the CISDL and currently conducts joint Ph.D. research as an affiliated junior researcher at the Centre for Intellectual Property Rights (CIR, at the Katholieke Universiteit Leuven, Belgium) and at the Centre for Philosophy of Law (CPDR) at the Université catholique de Louvain, Belgium. She has served as a legal adviser to the Belgian Federal Ministry of Environment, where she remains a member of the Belgian Access and Benefit-Sharing (ABS) Contact Group. She regularly served on the Belgian delegation in ABS-related international meetings between 2006 and 2009.

TREATY IMPLEMENTATION FOR SUSTAINABLE DEVELOPMENT

Over the past three decades, a series of international treaties have entered into force to address pressing global concerns – social and economic development and environmental protection. On climate change, biodiversity and biosafety, desertification, agriculture and seeds, and trade and investment liberalisation, new regimes have been established to implement global commitments related to sustainable development, many with nearly universal membership.

Successful domestic implementation of these international treaty regimes is one of the most significant challenges facing international law today. Although much has been written on the content and form of treaty law, there is relatively little that examines the transition from international legal theory and treaty texts to domestic regulation and practice.

This series of books addresses this need and provides a serious contribution to ongoing global debates by conducting a detailed analysis of how myriad new treaty regimes that cover the future's most pressing concerns can be made to work in practice. It is dedicated to our mentor and first Chairman of the Board of Governors of the Centre for International Sustainable Development Law (CISDL), The Honourable Mr. Justice Charles Doherty, Puisne judge on the Supreme Court of Canada from February 1, 1989, to August 1, 2003.

Series Editors:

Marie-Claire Cordonier Segger

Markus Wilheim Gehring

Legal Aspects of Implementing the Cartagena Protocol on Biosafety

Edited by

MARIE-CLAIRE CORDONIER SEGGER

FREDERIC PERRON-WELCH

CHRISTINE FRISON

Centre for International Sustainable Development Law
International Development Law Organisation
Lauterpacht Centre for International Law

CAMBRIDGE
UNIVERSITY PRESS

CAMBRIDGE UNIVERSITY PRESS
Cambridge, New York, Melbourne, Madrid, Cape Town,
Singapore, São Paulo, Delhi, Mexico City

Cambridge University Press
32 Avenue of the Americas, New York, NY 10013-2473, USA

www.cambridge.org
Information on this title: www.cambridge.org/9781107004382

First published 2013

Printed in the United States of America

A catalog record for this publication is available from the British Library.

Library of Congress Cataloging in Publication Data

Legal aspects of implementing the Cartagena Protocol on Biosafety / [edited by] Marie-Claire Cordonier Segger, Frederic Perron-Welch, Christine Frison.
p. cm – (Treaty implementation for sustainable development)
Includes bibliographical references and index.
ISBN 978-1-107-00438-2 (hardback)
1. Convention on Biological Diversity (1992). Protocols, etc., 2000 Jan. 29. 2. Biotechnology industries – Law and legislation. 3. Biotechnology – Safety measures. 4. Transgenic organisms industry – Law and legislation. 5. Transgenic organisms industry – Safety regulations. I. Cordonier Segger, Marie-Claire, 1973– II. Perron-Welch, Frederic, 1983– III. Frison, Christine.
K3925.B56L45 2012
343′0786606–dc23 2011033039

ISBN 978-1-107-00438-2 Hardback

Contents

Contributors

EDITORS

Marie-Claire Cordonier Segger is the Head of Economic Growth and Trade at the International Development Law Organization (IDLO) in Rome, Italy. She co-edits the Treaty Implementation for Sustainable Development series with Cambridge University Press and has authored or edited more than eighty publications, including fourteen books in three languages: Sustainable Development in World Investment Law (2010) and Sustainable Development in World Trade Law (2005) with Dr. M. Gehring; Sustainable Development Law: Principles, Practices and Prospects (2004) with A. Khalfan; and Sustainable Justice: Reconciling Economic, Social and Environmental Law (2004) with H. E. Judge C. G. Weeramantry. In an academic capacity, she also serves as Senior Director of the Centre for International Sustainable Development Law (CISDL) in Montreal, Canada; as Affiliated Fellow of the Lauterpacht Centre for International Law (LCIL) at Cambridge University, and as a Visiting Professor at the University of Chile Faculty of Law.

Christine Frison is a Legal Research Fellow with the CISDL and currently conducts joint Ph.D. research as an affiliated junior researcher at the Centre for Intellectual Property Rights (University of Leuven [KULeuven], Belgium) and at the Centre for Philosophy of Law (University of Louvain [UCLouvain], Belgium). Her Ph.D. dissertation is titled "Towards Redesigning the Plant Commons: A Critical Assessment of the Multilateral System of Access and Benefit-Sharing of the International Treaty on Plant Genetic Resources for Food and Agriculture." She has served as a legal adviser to the Belgian Federal Ministry of Environment, where she remains a member of the Belgian Access and Benefit-Sharing (ABS) Contact Group. She regularly served on the Belgian ABS delegation in various ABS-related international meetings between 2006 and 2009. In 2006, she performed a national survey on the use of genetic resources by Belgian users and on the degree of knowledge and application of international biodiversity obligations by these users (CBD and Bonn Guidelines) for the Belgian Federal Ministry of Environment in collaboration with the Centre for Philosophy of Law. She has also worked as a consultant for international organizations such as Bioversity International (formerly the International Plant Genetic Resources Institute (IPGRI), Rome) and the International Development Law Organization (IDLO, Rome). At the CISDL, she conducts research within the Biodiversity and Biosafety Law Programme and the Natural Resources Law Programme (with a focus on the International Treaty on Plant Genetic Resources for Food and Agriculture). She worked on the Biosafety Law Project of CISDL in collaboration with UNEP-GEF to design biosafety regulatory assessment tools and assist francophone countries in developing national laws to implement the Cartagena Protocol on Biosafety. Mrs. Frison reviewed more than 15 African countries' National Biosafety Frameworks (NBFs) and served as an independent legal expert to many NBFs' Validation Workshops. Mrs. Frison has represented CISDL in meetings of the Convention on Biological Diversity (CBD), the Cartagena Protocol, and the Food and Agriculture Organization (FAO).

She is the author of several academic and nonacademic publications in the area of agricultural biodiversity, biodiversity, and biosafety international law and policy. Plant Genetic Resources and Food Security: Stakeholder Perspectives on the International Treaty on Plant Genetic Resources for Food and Agriculture was published in 2011 by Routledge. She holds an LL.B. (University of Montpellier I, France), an LL.M. in Public International Law (Free University of Brussels [ULB], Belgium) and in International and Trade Law (University of Lyon I, France), and a B.A. in English and Italian (University of Montpellier III, France). Her working languages are French, English, and Italian; she speaks fair Dutch and has a notion of German and Spanish.

Frederic Perron-Welch was a Contributing Author to the Collaborative Partnership on Forests (CPF) Global Forest Experts Panel on the International Forest Regime and provides technical legal support to the ABS Capacity Development Initiative. He is the Francophone representative on the Board of the Canadian Environmental Network (RCEN) and a member of the RCEN Biodiversity Caucus. He has also worked on legal projects for the Canada Parks and Wilderness Society (CPAWS), Citizens Against Strip Mining (CASM), the East Coast Environmental Law Association (ECELAW), and the Living Oceans Society. He has an LL.B., Environmental Law (Dalhousie), M.A. in History (Toronto), and B.A. *cum laude* in History (Catholic University of America). He is fluent in English and French and understands some Spanish.

CONTRIBUTING AUTHORS

Paige Ainslie is a Solicitor at Canadian Pacific Railway and former legal researcher at the Centre for International Sustainable Development Law.

Jorge Cabrera Medaglia is Lead Counsel for International Sustainable Biodiversity Law with the CISDL. He is Professor of Environmental Law in the Master's programme on environmental law at the University of Costa Rica and was formerly a tutor for World Intellectual Property Organisation's distance courses on intellectual property. Prof. Cabrera also acts as a legal advisor to Costa Rica's National Biodiversity Institute (INBio) and is an international consultant in the area of intellectual property, biodiversity, biotechnology, biosafety, access to genetic resources, and benefit sharing for national and international institutions such as UNCTAD, CBD, ECLAC, IICA, SICA, CCAD, IPGRI, CYMMIT, REMERFI, CATIE, UNITAR, University of California, IUCN Environmental Law Center in Bonn, COSUDE, EU projects, IISD, CAF, USAID, TNC, IDB, ICTSD, PNUMA, and the Institute of Economic Development of the World Bank, among others. In addition, he has served as co-chair of the CBD's Expert Panel on Access and Benefit Sharing, as Chairman of the Subworking Group on IPR and Capacity Building during the 2nd meeting of the Working Group on Access and Benefit Sharing in Bonn. He is a member of the Technical Expert Group on Certificates of Origin–Legal Provenance–Source established by the CBD in the context of the International Regime Negotiations on ABS and of the TEG on terms, working definitions, and sectoral approaches. He also served as a member of the delegation to the WIPO Committee on Genetic Resources and Traditional Knowledge and the Group of Like-minded Megadiverse Countries. Prof. Cabrera has published in the areas of intellectual property rights, access to benefit sharing, biosafety, and trade and environment, and he has participated in the drafting of biodiversity and ABS laws, negotiated ABS contracts, and developed environmental legislation in countries including Costa Rica, Nicaragua, Panama, El Salvador, Chile, Dominica, Bhutan, Seychelles, Honduras, and Paraguay. He has also been a trainer on ABS for several African countries, including Benin, Kenya, Tanzania, Botswana, and Nigeria. He holds a B.C.L. (University of Costa Rica), LL.M. (University of Costa Rica), and M.B.A. (National University of Costa Rica).

Dr Papa Meïssa Dieng is Professor of International Law at Universite Gaston Berger in Saint Louis, Senegal, where he is Senegal Director of all McCormack/Gaston Berger projects, including the innovative Saint Louis Region Multi-Function Community Resource Center. A specialist in environmental law, Professor Dieng is deeply involved in local development policy and ecological policy.

David Duthie is an ecologist with more than 20 years of experience working on environmental issues. He studied Environmental Biology at Oxford Polytechnic from 1974 to 1977 before completing, in 1983, a research degree on migration of insects, also at Oxford Polytechnic. Between 1983 and 2000, he worked as a freelance biology teacher and as an environmental consultant, specializing in Environmental Impact Assessment (EIA) of development projects. He moved to Kenya to join UNEP to manage the GEF-funded Biodiversity Planning Support Programme, providing technical support to developing countries for their National Biodiversity Strategy and Action Plan (NBSAP) planning process. In 2005, he moved to the UNEP Regional Office for Europe in Geneva to join UNEP's Biosafety Unit, working with developing countries to put in place regulatory systems that ensure that novel applications of modern biotechnology do not have negative impacts on biodiversity. In October 2010, he joined the Secretariat of the CBD in Montreal, Canada, as a part of the team created under the Japan Presidency of the Convention to provide technical support for the implementation of the outcomes of the CBD Conference of the Parties held in Nagoya, Japan.

Liina Eek is by education a plant ecologist (Ph.D., Tartu University). Currently she is an adviser of nature conservation (especially issues connected to alien species) and biosafety to the Estonian Ministry of the Environment. During 2003–2005 she worked in the UNEP Biosafety Unit in Geneva as task manager of biosafety projects for Central and Eastern European countries. She has worked for different projects in Tallinn Technical University (UNEP/GEF biosafety project, projects on soil metagenomics). She is a member of the Advisory Committee for Genetic Modifications (advisory body to the government; chairperson 2000–2003). She is vice chairperson of the Cartagena Protocol Compliance Committee. Currently her main task is drafting NBSAP, a new umbrella law on the environment and working on consultancy bases for UNEP, assisting developing countries in preparing biosafety projects. She also gives lectures on biosafety in Tallinn Technical University. Additionally, she is currently finalizing her master's thesis in theology at the Institute of Theology, with the main topic being bioethics.

Martin Endicott (LL.B. (Lond), BCL (Oxon), LL.M. (Penn)) is a former research Fellow with the CISDL. He specializes in the resolution of international business disputes. He has acted as counsel in numerous international commercial and investment arbitrations, as well as in litigation and mediation proceedings and proceedings before the World Bank Administrative Tribunal. He has worked on NAFTA cases, as well as bilateral investment treaty cases involving claims for expropriation and breaches of international fairness standards and other treaty commitments. He has published several articles on investor-state arbitration. He was admitted as a barrister (England & Wales) in 1999 and is a Fellow of the Chartered Institute of Arbitrators.

Mai Fujii is a Ph.D. candidate in international law at the Graduate School for International Cooperation Studies, Kobe University, Japan. Since April 2010, she has been a research Fellow at the Japan Society for the Promotion of Science (JSPS). Her principal field of research is environmental liability in international law. The topic on which she currently focuses is the issue of liability and redress under the Cartagena Protocol on Biosafety. She worked as an intern at the Biosafety Division of the Secretariat of the Convention on Biological Diversity (SCBD) in Montreal, Canada, from October 2009 to February 2010. She has considerable experience observing several MEA meetings including the 4th Meeting of the Parties to the Cartagena Protocol on Biosafety (MOP-4) and "liability and redress" meetings under the Protocol, as a member of the Kobe University Research Institute on MEAs (KURIM), an academic NGO. She holds an LL.B. and LL.M. (Kobe University, Japan).

Kathryn Garforth is the Legal Officer for the Cartagena Protocol on Biosafety with the Secretariat of the Convention on Biological Diversity. She works, in particular, on the handling, transport, packaging, and identification of living modified organisms and socioeconomic considerations in biosafety decision making as well as liability and redress and compliance. Prior to joining the Secretariat, Ms. Garforth worked with the Centre for International Sustainable Development Law as a researcher and consultant in the areas of access to genetic resources and benefit-sharing, biosafety, and health. In this capacity, she led and participated in projects with Environment Canada, GTZ, the International Development Research Centre, the International Development Law Organization, and the United Nations Environment Programme, amongst others. Ms. Garforth holds an LL.B. (Osgoode Hall) and M.E.S. (York).

Prof. Markus Gehring is Tutor in Sustainable Development Law at the Law Faculty, University of Cambridge; Deputy Director of the Centre for European Legal Studies (CELS); and Fellow in Law at Robinson and at the Lauterpacht Centre of International Law. He holds an ad personam Jean Monnet Chair in Sustainable Development Law at the University of Ottawa in Canada and serves as Lead Counsel for Sustainable Trade, Investment and Finance Law at the CISDL. He holds an MA from Cambridge, an LL.M from Yale and a Dr jur from Hamburg.

Ryan Hill is an applied ecologist with more than 15 years of experience in Canada and abroad, as a consultant and as a staff member of inter-governmental organizations including the United Nations. He specializes in risk assessment and management of contaminated sites, design and implementation of environmental monitoring programs, and environmental impact assessment.

Gregory Jaffe is the Director of the Project on Biotechnology for the Center for Science in the Public Interest (CSPI), an advocacy and educational organization that focuses on nutrition and health, food safety, alcohol policy, and sound science. Mr. Jaffe came to CSPI after a long and distinguished career in government service. He first worked as a trial attorney for the U.S. Department of Justice's Environmental and Natural Resources Division for seven years. He then moved on to become Senior Counsel with the U.S. EPA, Air Enforcement Division, before joining CSPI to direct the Biotechnology project. Over the last decade, he has been a strong advocate for federal positions in federal court and frequently has spoken publicly on behalf of the EPA. At the EPA he was awarded a bronze medal for commendable service, a special achievement award, and a gold medal for performance. His interest in biotechnology began early in his career when he wrote a law review article on regulatory issues surrounding biotechnology and genetically modified organisms. In the early 1990s, while at the Department of Justice, he advised the Assistant Attorney General on biotechnology issues and worked with a federal interagency committee addressing biotechnology policy. He is currently a member of the Pew Initiative on Food and Biotechnology's Stakeholders Forum and was a member of the University of Pennsylvania Bioethics Center's GMO Consumer Values Panel. He has published articles on agricultural biotechnology in the *Christian Science Monitor*, the Food and Drug Law Institute's *Update* magazine, and the Environmental Law Institute's *Environmental Forum* magazine. He also has spoken at more than a dozen conferences addressing agricultural biotechnology issues, both in the United States and abroad. He is a recognized expert on the U.S. regulatory structure for agricultural biotechnology as well as consumer issues pertaining to agricultural biotechnology. Mr. Jaffe earned his B.A. in Biology with High Honors from Wesleyan University and then received a degree from Harvard Law School.

Paula Honorato Marin is a lawyer focusing on civil and family law. She holds an LL.B (Hons) in New Criminal Procedure Law (Universidad Andrés Bello, Chile), and an LL.M. in International Law (University of Montreal). During her Masters, she developed her interest in the protection of indigenous culture of the Mapuche of Chile, her native country, as well as the influence of the environment on culture.

Prof. Julian Kinderlerer is Professor of Intellectual Property Law at the University of Cape Town and Professor of Biotechnology & Society at the University of Technology, Delft, The Netherlands. He is a Member of the European Group on Ethics; Chair of the Task Group on Public Perception of Biotechnology, European Federation of Biotechnology; and Former Director of the Sheffield Institute of Biotechnology Law and Ethics Education. He was a member of the United Kingdom's Advisory Committee on Genetic Modification from 1983 to 2003, Advisory Committee on Releases to the Environment 1990–1999, and UK Biotechnology and Biological Science Research Council Bioscience for Society Panel 2005– (strategic input on societal issues surrounding the conduct and outcomes of research supported by BBSRC). He holds a B.Sc. (with distinction in Chemistry) and B.Sc.

(Hons) (with First Class Honours) from the University of Cape Town, and a Ph.D. in Biochemistry from Cambridge University. He was also Specialist Adviser to the House of Lords Select Committee on the European Union during its investigation into the Regulation of Biotechnology in Europe (1998–99); seconded to the United Nations Environment Programme during 2000 to devise programmes to assist developing countries in implementing the Cartagena Protocol on Biosafety; and assisted in writing the initial strategy on Biosafety of the Global Environment Facility. One hundred twenty-five countries are involved in the Development Project, which helps to devise a National Biosafety Framework and a further 12 in implementation projects implementing the Frameworks. He has been adviser to these projects since 2001; involved in many projects supported by the Wellcome Trust and the European Union in relation to bioethics and BioLaw; and worked with UNEP, UNIDO, FAO, and ICGEB in relation to biotechnology law and ethics.

Prof. Veit Koester is an external professor at Roskilde University Centre in Denmark; visiting professor at United Nations University, Institute of Advanced Studies, Yokohama; and a guest lecturer in international environmental law at Copenhagen University, Faculty of Law. He was chairperson of the Compliance Committee of the Aarhus Convention, as well as chairperson of the Compliance Committee of the Cartagena Protocol on Biosafety until 1 January 2009, and he chaired the Working Group of the Parties to the CBD that negotiated the Protocol. He was one of the two vice-chairs of the CBD negotiating body and chaired its Working Group II throughout the negotiations of the Convention. Veit Koester is a member of the IUCN Commission on Environmental Law and a member of the European Council of Environmental Law. He has received, among others, the UNEP – Global 500 Award (1996) and the Elisbeth Haub Award (ICEL and Pace University) for Environmental Diplomacy for the Year 2000.

Dr Konstantia Koutouki is Lead Counsel for Natural Resources with the Centre for International Sustainable Development Law and teaches in the graduate Common Law program as well as advanced courses in international law. Her research examines the links between international trade, intellectual property, and environmental protection. She has extensive experience working on issues concerning international sustainable development law especially as it relates to the social, economic and cultural development of Indigenous and local communities, as well as the preservation of natural spaces and traditional knowledge, particularly traditional medicine and agriculture. She has first-hand experience with Indigenous and local communities around the world, where she has spent much time as a guest and researcher.

Sylvestre-José-Tidiane Manga is a law specialist and a bioscientist with vast experience in bio-industrial innovation and the green economy. He has a doctorate in Agricultural Biotechnology Law, a doctorate in Hydro-Agricultural Planning and Economics/Mathematical modelling, and a Masters in International Law. He has several peer-reviewed publications. Mr. Manga is a Legal Research Fellow at the McGill University–affiliated Centre of International Sustainable Development Law (CISDL). He is also an International Consultant to the Convention on Biological Diversity Cartagena Protocol on Biosafety Parties (UNEP-GEF Biosafety Unit).

Nizar Mohamed is an environment and development consultant with 35 years' experience with technical assistance and capacity-building projects in many parts of the world, particularly in Asia and the Pacific, as well as working with the New Zealand government. The regional, national, and local projects he has worked on were funded by bilateral (e.g. DFID, NZAID) and multilateral agencies (UNDP, UNEP-GEF, FAO, IFAD, and ADB). From 2001 to 2006 he was the Regional Coordinator for Asia for the UNEP-GEF global project on "Development of National Biosafety Frameworks" and coordinated national projects in 23 countries in Asia (East, South, Central, and West Asia) to help them develop national biosafety frameworks (NBFs).

Thomas P. Redick represents clients in the high-technology and agricultural biotechnology industry sectors with issues relating to regulatory approval, liability avoidance, and compliance with industry standards addressing socioeconomic and environmental impacts – particularly "sustainability" initiatives in agriculture and high technology. Before establishing a solo international environmental consulting practice (Global Environmental Ethics Counsel) in 2005 in St. Louis, Missouri, he was a partner chairing the technology risk management practice for Gallop, Johnson & Neuman L.C. He has held many leadership positions in the American Bar Association Section on Environment, Energy & Resources (ABA-SEER), including Committees on Agricultural Management, International Environmental Law and Climate Change, and Sustainable Development and Ecosystems. In 2008, Mr. Redick was appointed to represent ABA SEER (Agricultural Management) on the Council for Agricultural Science & Technology (CAST). In 2009, he was the first attorney to be elected president of CAST. As a grower association representative to the Global Industry Coalition, he attends meetings of the Convention on Biological Diversity and the Cartagena Protocol on Biosafety. He has more than 24 years experience practicing environmental and intellectual property law. He has a law degree and B.A. with high honors from the University of Michigan.

Olivier Rukundo is Programme Officer for ABS Capacity Building at the Secretariat to the Convention on Biological Diversity and former Senior Legal Research Fellow at the CISDL, where he worked on a joint appointment with the biodiversity and international trade programs. His broader legal research interests include intellectual property law, international trade law, and international environmental law. He holds a joint LL.B./B.C.L. (McGill) and LL.M. (WIPO World Academy).

Christine Toczeck Skarlatakis is an Associate Fellow with the CISDL. She has worked with human rights issues in nongovernmental organisations and as a Legal Consultant in the private sector. Her past experiences include research in innovative technologies issues, such as biosafety and biotechnology, at the United Nations Environment Programme (Biosafety Unit), and nanotechnology at the Canadian Institute of Environmental Law and Policy. Ms. Toczeck Skarlatakis is also assigned as an Analyst Team Member of the Oxford G8 Research Group (2007–2008). She is proficient in English, Spanish, and Portuguese and is learning French. She has been involved with the CISDL since March 2007. Her legal interests comprise international biodiversity and biosafety law, health issues, human rights, climate change, and international governance. She holds an M.A. in International Studies (Barcelona) and an LL.B. (PUC-PR) and LL.M. (University of Barcelona).

Worku Damena Yifru is an environmental lawyer currently working on biosafety issues. He is a Programme Officer for biosafety policy and law at the Secretariat of the Convention on Biological Diversity under the United Nations Environment Programme. He, among other things, coordinates activities of the Secretariat in support of the work of the Compliance Committee under the Cartagena Protocol on Biosafety. He served as the secretary of the working group and its successors established over the years since 2004 for conducting the negotiations on the international rules and procedures on liability and redress for damage resulting from the transboundary movements of genetically modified organisms.

Prior to joining the Secretariat in February 2001, he worked for 12 years on environment and development issues at different government positions in his native country, Ethiopia. He was the head of Policy and Legislation Department at the Environmental Protection Agency of Ethiopia. He was once the Director of Ethiopian Conservation Strategy Initiative, which resulted in the development and adoption of the Environmental Policy of Ethiopia. He coordinated the drafting of the basic environmental laws of Ethiopia, which were passed by the government in 2002. He was a member and a legal advisor of the Ethiopian delegation in a number of international environmental negotiations, including the negotiations for the Cartagena Protocol on Biosafety and the International Treaty on Plant Genetic Resources for Food and Agriculture. Mr. Yifru holds a master's degree (LL.M.) in Environmental Law from the London School of Economics and Political Science (1995) and a bachelor's degree in law (LL.B.) from the Law Faculty of Addis Ababa University (1988).

Tomme Rosanne Young is a Senior Policy and Legal Advisor and Consultant with the International Research Institute for Sustainability currently based in Bonn, Germany. Throughout her 25 years as a lawyer, she has developed a specialized expertise in many areas of environmental law and policy. Internationally, Ms. Young has served as a special advisor on environmental and sustainable development issues to foreign governments, under the auspices of several UN agencies, global NGOs, and other institutions. She has worked in many countries, published many articles, and provided advice to global and national bodies on a variety of key environmental legal/policy issues, including GMOs, environmental certification, incentives, and conservation finance. She has advised 35 countries, in Europe, Africa, Asia, Oceania, and the Americas, regarding legal and legislative issues within the natural resource (forest and fisheries) sectors, as well as environmental sectors and energy development.

With regard to genetic resource issues, she has participated in negotiation of international and regional agreements and prepared advisory white papers on all sides of the issues, including the application of biodiversity, sustainable development, environmental protection, commercial, resource development and conservation, and coastal zone management law in the context of LMOs and genetic resources for the United Nations. She has been named by the Global Environment Facility as the legal expert on the international expert team tapped to perform the external evaluation of UNEP-GEF's implementation of the Cartagena Protocol and has led an international expert team tasked by the UN Food and Agriculture Organisation to undertake the external evaluation of international instruments and processes within FAO's mandate. Ms. Young also serves as Editor of Environmental Policy and Law (IOS Press) and has served as an adjunct professor in several contexts, including at the University of San Francisco School of Law. She is a graduate of Hastings College of the Law (1981) and the University of Southern California (1978). At Hastings, she served as an editor of the *Hastings International and Comparative Law Review*.

Foreword

Emerging from the outcomes of the 1992 United Nations Conference on Environment and Development, the Cartagena Protocol on Biosafety to the Convention on Biological Diversity and its Nagoya – Kuala Lumpur Supplementary Protocol on Liability and Redress respond to the realities of biotechnology development across the world. Advancements in biotechnology continue to push the boundaries of science in search of solutions to global problems and there is no question that the application of biotechnology continues to gain ground worldwide. Nonetheless, this field is relatively new, and questions remain about the short- and long-term biological and social effects of living modified organisms. Countries, concerned about their ability to manage these risks in a world of increasingly liberalized trade, negotiated the Cartagena Protocol and the Nagoya – Kuala Lumpur Supplementary Protocol to enable more sovereign control over the use of biotechnology.

The Cartagena Protocol and the Nagoya – Kuala Lumpur Supplementary Protocol are important instruments for implementing the Convention. They establish how Parties can regulate, manage, and control risks associated with the transfer, handling, and use of living modified organisms that may have adverse effects on the conservation and sustainable use of biodiversity, thus contributing to the first two objectives of the Convention.

Sound implementation of the Protocols encourages both sustainable development and the careful use of biotechnology. It does this by not only empowering Parties to exercise precaution in the regulation of living modified organisms, but also by requiring a higher level of transparency in the production and movement of such organisms.

Implementation of the Protocols is a continuing process and, in many cases, a challenge. The development of appropriate laws and administrative measures is ongoing for many countries. For implementation to be successful, stakeholders and participants at all levels need to share their knowledge and experiences in a spirit of co-operation, engagement, and innovation. This book makes an important contribution to such a dialogue. By bringing together authors with a variety of views, this book explores the range of issues and perspectives that need to be considered in the implementation of the Cartagena Protocol.

The insights of experts who contributed to this book will serve as valuable tools for governments, researchers, and lawmakers everywhere, all of whom play important roles in building the capacity of Parties to implement the sustainable development commitments imbedded in the Cartagena Protocol and the Nagoya – Kuala Lumpur Supplementary Protocol.

This book is a valuable contribution to the body of shared knowledge, and crucial debates, on biosafety.

Braulio Ferreira de Souza Dias
UN Executive Secretary of the Secretariat of
the Convention on Biological Diversity
September 2012

Preface

Emerging from the outcomes of the 1992 United Nations Conference on Environment and Development, the Cartagena Protocol on Biosafety to the Convention on Biological Diversity was adopted by the Conference of the Parties to the Convention on 29 January 2000. The Protocol has its roots in Article 19.3 of the Convention, which obliged Parties to consider the need for and modalities of a protocol setting out appropriate procedures for the safe handling and use of living modified organisms (LMOs) that may have an adverse effect on biodiversity. It entered into force on 11 September 2003 and currently has 163 Parties. On 15 October 2010 at the Fifth Conference of the Parties serving as the meeting of the Parties to the Protocol in 2010, the Parties adopted the Nagoya – Kuala Lumpur Supplementary Protocol to the Cartagena Protocol on Biosafety. The Supplementary Protocol was opened for signature on 7 March 2011 and will enter into force nintey days after the deposit of the 40th instrument of ratification, acceptance, approval, or accession.

These Protocols respond to the reality of biotechnology development across the world. Some of these technologies promise to address malnutrition or other medical problems, or to overcome the challenges of conventional agriculture. There is no question that they are gaining ground worldwide. Nonetheless, these technologies are relatively new, and questions remain about their short- and long-term biological and social effects. Parties to the Convention, concerned about their ability to manage these risks in a world of increasingly liberalized trade, negotiated the Cartagena Protocol and Nagoya – Kuala Lumpur Supplementary Protocol to enable more sovereign control over the sustainable use of biotechnology. The Cartagena Protocol and Nagoya – Kuala Lumpur Supplementary Protocol are important tools for implementing the Convention. They establish how Parties can regulate, manage, and control risks associated with the transfer, handling, and use of LMOs that may have adverse effects on the conservation and sustainable use of biodiversity – the first two objectives of the Convention.

Sound implementation of the Protocols encourages both sustainable development and the careful use of biotechnology. It does this by not only by empowering Parties to exercise precaution in the regulation of such technologies within their borders, but also, as a consequence, by requiring a higher level of transparency in the production and movement of the products of biotechnology.

Implementation will be an on-going process that requires the development of appropriate laws and administrative measures. Many Parties, especially developing countries, have faced challenges in implementing the full scope of their commitments. The UNEP-GEF Biosafety Project has helped countries develop national biosafety frameworks (NBFs). Many of the authors in this book participated in the project in various capacities and have used this opportunity to share their experiences in the national implementation of the Protocol.

These legal expert authors' insights will serve as valuable tools for government departments, experts, researchers, and lawmakers everywhere, all of whom play important roles in building the capacity of all Parties to implement the sustainable development commitments in this important Convention and its Protocols effectively. For this process to be successful, stakeholders and participants at all levels need to share their knowledge and experiences in a spirit of co-operation, engagement, and innovation. This book is a valuable contribution to that body of shared knowledge, and these crucial debates.

Dedication and Acknowledgements

This is the first book in a new series of volumes that focuses on the implementation of international treaty commitments on sustainable development, edited by Marie-Claire Cordonier Segger and Markus Wilheim Gehring, and published by Cambridge University Press. The editors respectfully dedicate this series to the memory of the Honourable Justice Charles Doherty Gonthier, who served for many years on the Supreme Court of Canada and was the founding Chairman of the Board of Governors of the Centre for International Sustainable Development Law (CISDL).

We would like to express our gratitude to those who contributed in the legal experiences and practice, collaborative research, drafting, and editing that were essential tasks to the preparation of this volume. First, we are deeply indebted to every contributor for taking the time to share their expertise and knowledge. We also sincerely thank the dedicated and experienced senior government officials from more than 80 countries, the United Nations Environment Programme (UNEP), and the Global Environment Fund (GEF) for their collaboration on the National Biosafety Framework Development Project for the reviews of National Biosafety Frameworks between 2002 and 2008.

This book would not have happened without the research and support of many brilliant and dedicated members of the CISDL and the McGill University Faculty of Law, who assisted in the research, drafting, development, and review of the manuscript, including Patrick Reynaud, Jennifer S. Bond, Dob Feldman, Kathryn Garforth, Olivier Rukundo, Christine Toczeck Skarlatakis, Alexandra Keenan, Freedom-Kai Phillips, Giselle Davidan, and Aleksandra Bojovic. The collaboration and kindness of knowledgeable professors and practitioners from the Lauterpacht Centre for International Law (LCIL) at Cambridge University Faculty of Law, particularly Sir Eli Lauterpacht, Professors James Crawford and Marc Weller, and Dr Joanna Gomula and Anita Rutherford, are acknowledged. We further recognize and appreciate the engagement and support of the dedicated rule of law experts and legal specialists at the International Development Law Organization, including the

Director-General Irene Khan and Giulio Zanetti, Ilaria Bottigliero, Judit Arenas, Charles E Tucker, and previously Tom McInerney and Robert Buergenthal, as well as the entire team of economic and sustainable development lawyers. We would also like to recognize the support and guidance of the Centre de Philosophie du Droit, Université catholique de Louvain, and of the Centre for Intellectual Property Rights, Katholieke Universiteït Leuven.

Last but not least, we were encouraged all through the process with the love, support, and faith of our families, in particular when we were under tight time pressure and constraint to work late nights. To everyone, we sincerely thank you.

Introduction

Marie-Claire Cordonier Segger, Frederic Perron-Welch, and Christine Frison

The exponential growth in the use and trade of living modified organisms (LMOs) has made the safe use of biotechnology (i.e., biosafety) an issue of global relevance.[1] Modern biotechnology has the potential to further development and improve human welfare, but the possible impact of environmental releases of LMOs on other species and varied ecosystems also implies significant and unexplored risks. The adoption of the Cartagena Protocol on Biosafety[2] (Cartagena Protocol) contributes to the development of an international regulatory framework to enable international trade in environmentally sound applications of biotechnology.[3]

This Protocol, as a legal instrument that aims to promote sustainable development and use of biological resources, is part of the emerging body of sustainable development law, integrating economic and social development and environmental protection by setting an international regime in place to govern the transboundary movements of LMOs.[4] Globalisation has greatly broadened the need for biosafety and has complicated its pursuit. In spite of multilateral initiatives for the implementation of national biosafety frameworks, the implementation of international

[1] Global use of biotech crops has grown rapidly – approximately 94-fold from 1996 to 2011 – to 160 million hectares of cropland cultivated by millions of farmers in 29 different countries. Most biotech crops are grown in the USA, but developing countries are rapid adopters and are expected to soon grow the majority of biotech crops. Successful crops (e.g. soybeans, cotton, corn, and canola) have seen high levels of market penetration. See Clive James, *Global Status of Commercialized Biotech/GM Crops: 2011*, International Service for the Acquisition of Agri-biotech Applications (ISAAA) Brief No 43 (Ithaca, NY: ISAAA, 2011).

[2] *Cartagena Protocol on Biosafety*, 2226 UNTS 208; 39 ILM 1027, 29 January, 2000 (entered into force 11 September 2003) [*Cartagena Protocol*].

[3] Secretariat of the Convention on Biological Diversity (SCBD), *Cartagena Protocol on Biosafety to the Convention on Biological Diversity: Text and Annexes* (Montreal, Canada: SCBD, 2000) at Introduction.

[4] Marie-Claire Cordonier Segger, "The Role of International Forums in the Advancement of Sustainable Development" (2009) 10(1) *Sustainable Development Law & Policy* 4 at 8; Marie-Claire Cordonier Segger and Ashfaq Khalfan, eds, *Sustainable Development Law: Principles, Practices and Prospects* (Oxford, UK: Oxford University Press, 2004) at 2 [Sustainable Development Law].

biosafety obligations remains a challenge for developing countries, especially for those with limited scientific, human, technical, and financial resources.

This book analyses the legal aspects of implementing the Cartagena Protocol and provides a serious contribution to current legal and academic debates on biosafety by reviewing key issues under the Cartagena Protocol that affect the design and implementation of national biosafety regulatory regimes. The book takes into account the principles of sustainable development law, and informs future evolution of the international biosafety regime. The text also examines recent experiences with domestic laws and regulations on biosafety, canvassing the practical, legal, political, and economic challenges encountered in the design and implementation of these regulatory schemes, while placing special emphasis on diverse law and policy approaches taken in developing countries.

Overall, this book addresses the legal avenues that are available to implement international law on biosafety by focusing on three particular objectives. First, the book identifies the aspects of international law on biosafety that are pertinent to and reflective of the treaty objective of sustainable development and use of biotechnology and genetic resources, as well as related principles of international law. Second, the book analyses the national implementation of international law on biosafety, focusing on the design and enforcement of biosafety regulations from a sustainable development law perspective, and examines the practical, legal, political, and economic challenges and achievements encountered in implementing international biosafety obligations. Last, the book discusses the future legal practice and research agenda in this field, providing both recommendations to encourage successful implementation of biosafety regulations and insights into international institutions that oversee and further develop international law on biosafety.

DEVELOPMENT OF THE CARTAGENA PROTOCOL

Due to emerging debates on biotechnology and its potential impact on the environment, the topic of biosafety was raised in the 1992 Earth Summit in Rio de Janeiro. Discussions did not result in a treaty on the subject, but negotiators decided that biosafety should be addressed under the aegis of the Convention on Biological Diversity (CBD).[5] The CBD requires Parties to establish or maintain means to regulate risks arising from biotechnology, taking into account those associated with the use and release of LMOs, which are likely to have adverse environmental impacts that could affect conservation and sustainable use of biotechnology, or present risks to human health.[6] Parties also committed to consider the need for a Protocol setting out procedures for the safe transfer, handling, and use of LMOs that might have adverse effect on the conservation and sustainable use of biological diversity, including

5 *Convention on Biological Diversity*, 5 June 1992, 31 ILM 818 (entered into force 29 December 1993) [CBD].

6 *Ibid* at art 8(g).

arrangements for advance informed agreement (AIA) prior to the import/export of LMOs.[7]

The first Conference of the Parties (COP) of the CBD in 1994 initiated consideration of a Protocol, focusing on possible objectives, definitions, scope, application of the AIA procedures, relation to agreements other than the protocol, aspects of risk, relevant national authorities, capacity building, illegal traffic, liability and redress, and the financial mechanisms and resources.[8] In 1995, the second COP to the CBD adopted Decision II/5, launching the Open-ended Ad-Hoc Working Group on Biosafety. This Working Group met six times between 1996 and 1999, concluding with the submission of a draft Protocol to be considered at the first Extraordinary Meeting of the COP (convened with the purpose of adopting a Protocol on biosafety to the CBD).

The Extraordinary Meeting, which was held in two sessions in 1999 and 2000, resulted in the adoption of the Cartagena Protocol and the establishment of the Ad-Hoc Intergovernmental Committee for the Cartagena Protocol (ICCP) to undertake the preparations necessary for the first meeting of the Parties.[9] The Protocol entered into force on 11 September 2003 and by 31 January 2012 engaged 161 Parties. The CBD Conference of the Parties serving as the meeting of the Parties to the Protocol (COP-MOP) acts as the governing body.[10] By 2010, the COP-MOP had met five times. These meetings made significant contributions to the elaboration of international law on biosafety and sustainable development.

The first meeting of the COP-MOP was held in Kuala Lumpur, Malaysia, from 23 to 27 February 2004. Fundamentally, it succeeded in setting up the operational framework required for the long-term effective implementation of the Protocol.[11] The most notable developments included the creation of procedures and mechanisms that would facilitate decision making by Parties of import, a group that includes developing countries, economies in transition, and centres of origin and centres of genetic diversity,[12] and the establishment of an information-sharing mechanism: the Biosafety Clearing-House (BCH).[13] In addition, a roster of experts was created

7 *Ibid* at Art. 19(3).

8 Patricia Birnie, Alan Boyle, and Catherine Redgwell, *International Law & the Environment*, 3rd ed. (Oxford, UK: Oxford University Press, 2009) at 629 [International Law & Environment].

9 *Report of the Extraordinary Meeting of the Conference of the Parties for the Adoption of the Protocol on Biosafety*, UN Doc UNEP/CBD/ExCOP/1/3, Dec EM-I/3.

10 *Cartagena Protocol*, *supra* note 2 at art 29(1). The list of the Parties to the Protocol is available online: http://bch.cbd.int/protocol/parties.

11 International Institute for Sustainable Development (IISD), Earth Negotiations Bulletin (ENB), Vol 9, No 289, *Summary of the First Meeting of the Conference of the Parties to the Convention on Biological Diversity Serving as the Meeting of the Parties to the Cartagena Protocol on Biosafety*, online: http://www.iisd.ca/vol09/enb09289e.html.

12 SCBD, *Global Biosafety: From Concepts to Action: Decisions from the First meeting of the Conference of the Parties to the Convention on Biological Diversity Serving as the Meeting of the Parties to the Cartagena Protocol on Biosafety* (Montreal, Canada: SCBD, 2004), Dec BS-I/2.

13 *Ibid*, Dec BS-I/3.

for the purpose of capacity building in developing countries;[14] the Action Plan for Building Capacities for the Effective Implementation of the Cartagena Protocol was launched;[15] and the group made first steps toward establishing identification requirements for LMOs intended for direct use as food or feed, or for processing.[16] Finally, that first meeting resulted in the establishment of procedures and mechanisms on compliance,[17] provision of guidance on transboundary movement of LMOs with nonparties,[18] the creation of a medium-term programme of work for the COP-MOP,[19] and the funding of capacity building and implementation measures through the Global Environment Facility.[20]

The following year, the second meeting of the COP-MOP took place in Montreal, Canada, from 30 May to 3 June 2005. Noticeable progress was made on three key issues under the Protocol. First, discussions on risk assessment and management led to the creation of the Ad Hoc Technical Expert Group (AHTEG) on Risk Assessment.[21] Second, major decisions were made relating to the operations and activities of the Biosafety Clearing House, capacity-building activities, and the financial mechanism/resources.[22] Third, the Parties recognized and emphasized the importance of public awareness and participation.[23] The main point of contention during the negotiations, however, was documentation for the transboundary movement of LMOs.[24] Discussions between the Parties did result in the articulation of measures to elaborate documentation for the contained use and intentional introduction of LMOs into the environment,[25] but no decision was made regarding documentation for LMOs meant for food, feed, or processing (LMO-FFP) as required by Article 18.2(a) of the Protocol.[26]

COP-MOP 3 was held in Curitiba, Brazil, from 13 to 17 March 2006. The main result of that meeting was the adoption of a compromise package on LMO-FFP documentation requirements.[27] Other major decisions taken included the

[14] *Ibid*, Dec BS-I/4.
[15] *Ibid*, Dec BS-I/5.
[16] *Ibid*, Dec BS-I/6.
[17] *Ibid*, Dec BS-I/7.
[18] *Ibid*, Dec BS-I/11, Annex.
[19] *Ibid*, Dec BS-I/12.
[20] *Ibid*, Dec. BS-I/15.
[21] SCBD, *Facing the Biosafety Challenge: Towards Effective Implementation of the Protocol: Decisions of the Second Meeting of the Conference of the Parties to the Convention on Biological Diversity Serving as the Meeting of the Parties to the Cartagena Protocol on Biosafety* (Montreal, Canada: SCBD, 2005), Dec BS-II/9 at para 4.
[22] *Ibid*, Dec BS-II/2, BS-II/3, BS-II/4 and BS-II/5.
[23] *Ibid*, Dec BS-II/13.
[24] *Ibid*, at v.
[25] *Ibid*, Dec BS-II/10.
[26] IISD, ENB, Vol 9, No 320, *Summary of the First Meeting of the Ad Hoc Group on Liability and Redress and the Second Meeting of the Parties to the Cartagena Protocol on Biosafety*. Online: <http://www.iisd.ca/vol09/enb09320e.html>.
[27] IISD, ENB, Vol 9, No 351, *Summary of the Third Meeting of the Parties to the Cartagena Protocol on Biosafety*, online: <http://www.iisd.ca/vol09/enb09351e.html>; SCBD, *Biosafety: Building Further*

adoption of an updated version of the Action Plan for Building Capacities for the Effective Implementation of the Cartagena Protocol on Biosafety,[28] directions relating to the financial mechanism,[29] further discussion on guidance and capacity building for risk assessment and risk management,[30] and the adoption of a monitoring and reporting format to help fulfill obligations under Article 33 of the Protocol.[31]

At COP-MOP 3 the Parties decided to meet every two years rather than every year; thus COP-MOP 4 was held in Bonn, Germany from 12 to 16 May 2008. One major achievement was the decision to further elaborate international rules and procedures for liability and redress resulting from the transboundary movement of LMOs.[32] Furthermore, the Parties issued a comprehensive decision on risk assessment and risk management, agreeing to the development of training activities and the establishment of an Ad Hoc Technical Expert Group on Risk Assessment and Risk Management.[33] These initiatives were in addition to the adoption of measures to promote long-term biosafety education and training and to strengthen the Coordination Mechanism to support those measures.[34]

The fifth meeting of the COP-MOP took place from 11 to 15 October 2010 in Nagoya, Japan. The Parties focused on adopting rules and procedures pertaining to liability and redress, and concluded the meeting with the adoption of the Nagoya – Kuala Lumpur Supplementary Protocol on Liability and Redress to the Cartagena Protocol.[35]

At COP-MOP 5, the Parties also adopted a ten-year strategic plan for the implementation of the Protocol from 2011 to 2020 with a focus on five main areas: 1) facilitating the establishment and further development of effective biosafety systems for the implementation of the Protocol; 2) capacity-building; 3) compliance and review; 4) information sharing; and 5) outreach and cooperation.[36] Due to the work done at COP-MOP 5, the Parties, in cooperation with relevant organizations,

Consensus for Action: Decisions of the Third Meeting of the Conference of the Parties to the Convention on Biological Diversity Serving as the Meeting of the Parties to the Cartagena Protocol on Biosafety (Montreal, Canada: SCBD, 2007), Dec BS-III/10.

28 *Ibid*, Dec BS III/3.

29 *Ibid*, Dec BS-III/5.

30 *Ibid*, Dec BS-III/11.

31 *Ibid*, Dec BS-III/14.

32 SCBD, *Biosafety: Taking Further Steps Towards Effective Implementation of the Protocol. Decisions of the Fourth Meeting of the Conference of the Parties to the Convention on Biological Diversity Serving as the Meeting of the Parties to the Cartagena Protocol on Biosafety* (Montreal, Canada: SCBD, 2008), Dec BS-IV/12.

33 *Ibid*, Dec BS-IV/11.

34 *Ibid*, Dec BS-IV/3.

35 *Nagoya – Kuala Lumpur Supplementary Protocol on Liability and Redress to the Cartagena Protocol on Biosafety*, Annex to Dec BS-V/11, online: http://bch.cbd.int/protocol/NKL_text.shtml.

36 SCBD, *Biosafety: Setting a New Agenda: Decisions of the Fifth Meeting of the Conference of the Parties to the Convention on Biological Diversity Serving as the Meeting of the Parties to the Cartagena Protocol on Biodiversity* (Montreal, Canada: SCBD, 2011), Dec BS-V/16, at Annex I, at para 7.

are also approaching the adoption of a roadmap for risk assessment, as well as guidelines on the elements and procedures of risk assessments for different LMO types, to help countries make informed decisions about the development, handling and use of LMOs.[37]

SUSTAINABLE DEVELOPMENT LAW AND THE CARTAGENA PROTOCOL

Sustainable development law refers to a set of legal instruments and related principles that include, among their objectives, the realization of sustainable development.[38] To date, most sustainable development law has been adopted through "hard law" treaty regimes, although there is also a body of emerging customary principles of law on sustainable development. The process of identifying and promoting respect for these principles of law has been reasonably complex and continues to be the subject of debates in 2010.[39] In 2002, however, at the 70th Conference of the International Law Association (ILA), a resolution by the Committee on the Legal Aspects of Sustainable Development, the ILA New Delhi Declaration of Principles of International Law Relating to Sustainable Development, which was annexed to the outcomes of the 2002 World Summit on Sustainable Development (WSSD), provided an important benchmark that is used in much of the relevant academic and legal literature today.[40]

The New Delhi Declaration elaborates on seven central principles that are common to most international treaties related to sustainable development, many of which were recognized and reaffirmed in the 2002 Johannesburg Plan of Implementation from the World Summit.[41] In this book, these principles provide a benchmark against which the Cartagena Protocol can be analyzed to assess its commitment to the realization of sustainable use of biological resources, and ultimately, to sustainable development. These principles include the duty of States to ensure sustainable use of resources; the principle of equity and the eradication of poverty; the principle of common but differentiated responsibilities; the principle of the precautionary approach to human health, natural resources, and ecosystems; the principle of public participation and access to information and justice; the principle of good governance; and the principle of integration and interrelationship, in particular as these relate to human rights and social, economic, and environmental objectives.

37 *Ibid*, Dec BS-V/12.

38 *Sustainable Development Law*, *supra* note 4 at 103.

39 Cordonier Segger, "The Role of International Forums in the Advancement of Sustainable Development," *supra* note 4 at 10.

40 *ILA New Delhi Declaration of Principles of International Law Relating to Sustainable Development*, UN Doc A/CONF.199/8, Annex [*New Delhi Declaration*].

41 Cordonier Segger, "The Role of International Forums in the Advancement of Sustainable Development," *supra* note 4 at 10.

Principle of Integration and Interrelationship

The principle of integration and interrelationship provides the context for international law on sustainable development by emphasizing the interdependence among economic development, social and human rights, and environmental priorities in international law.[42] The 1992 Rio Declaration from the United Nations Conference on Environment and Development (UNCED) states that "[i]n order to achieve sustainable development, environmental protection shall constitute an integral part of the development process and cannot be considered in isolation from it."[43] In the WSSD, this principle was arguably broadened to recognize that human rights and social development priorities also constitute an integral part of this balance.[44] It is essential that sustainable development be implemented at all sectors of society and governance.[45] To respect this principle, States may seek to resolve overlaps or perceived conflicts between economic, social, and environmental concerns either through the activation of existing institutions or the establishment of new ones that can balance the competing goals.[46]

The Cartagena Protocol is a prime example of the principle of integration and interrelationship because it is a treaty regime that attempts to address social, economic, and environmental themes in a balanced and holistic fashion. This recognized, it is clear that the Cartagena Protocol permits and even encourages the economic development of biotechnology, attempting to promote safe, equitable, and environmentally sound uses. The negotiation of the Protocol served as an attempt to reconcile competing narratives on modern biotechnology through the creation of a regime that promoted the environmentally sound application of biotechnology. The Protocol's preamble makes this objective clear, since it recognizes the great potential of modern biotechnology while also expressing awareness of its rapid expansion and the resulting growing public concern over its potential adverse effects on biodiversity and human health. It also recognizes that trade and environment agreements should be mutually supportive, with the aim of achieving sustainable development.[47] Additionally, several provisions in the Protocol allow for economic considerations to receive priority, such as the exemption of LMOs in transit from the advance informed agreement procedure,[48] and LMOs intended for contained use in accordance with the importing Party's standards.[49] Similarly,

42 *New Delhi Declaration, supra* note 40 at Principle 7.1; *Sustainable Development Law, supra* note 4 at 102.

43 *Rio Declaration on Environment and Development*, UN Doc A/CONF.151/26 (Vol I), 31 ILM 874 (1992) at Principle 4.

44 *Sustainable Development Law, supra* note 4 at 103.

45 *New Delhi Declaration, supra* note 40 at Principle 7.2.

46 *Ibid* at Principle 7.3.

47 *Cartagena Protocol, supra* note 2 at Preamble.

48 *Ibid* at art 6(1).

49 *Ibid* at art 6(2).

a distinct procedure for commodities (LMOs–FFP) is provided for, rather than the standard advance informed agreement procedure.[50] The Protocol also reflects the principle of integration and interrelationship through its integration of social considerations by the application of socioeconomic considerations in the risk assessment process.[51] Although the Protocol recognizes and attempts to integrate environmental and social concerns into the process of developing and using biotechnology, and even activates economic instruments to achieve social and environmental aims, it remains far from clear whether the current interrelationships among economic, social, and environmental nuances in the treaty could be described as adequately balanced.

Principle of Sustainable Use of Natural Resources

In principle, States have a duty to ensure sustainable use of natural resources, particularly in a transboundary context. This duty stems from the recognition that the sovereign right of States to manage their own natural resources comes with the responsibility to ensure that activities within their jurisdiction or control do not cause significant environmental damage elsewhere.[52] This principle encourages States to manage their natural resources sustainably to contribute to the development of their peoples, with particular regard for the rights of indigenous peoples, and to the conservation and sustainable use of natural resources and the protection of the environment, including ecosystems.[53] The principle also emphasizes that the protection, preservation, and enhancement of the natural environment, and biological diversity in particular, is a common concern of humankind.[54] This principle is reflected in many international treaties and legal instruments in the field of sustainable development.[55]

Fundamentally, the Cartagena Protocol was conceived with this duty in mind. This is clear from its objective, which holds that the Protocol's aim is to contribute to ensuring an adequate level of protection in the safe transfer, handling, and use of LMOs that could have adverse effects on the conservation and sustainable use of biological diversity.[56] The Parties themselves are also obliged to ensure that the development, handling, transport, use, transfer, and release of LMOs is undertaken in a manner that prevents or reduces the risks to biological diversity.[57] In essence, the Cartagena Protocol is seeking to contribute to the sustainable use of biotechnology and biological resources, in particular LMOs.

50 *Ibid* at art 11.
51 *Ibid* at art 26.
52 *New Delhi Declaration, supra* note 40 at Principle 1.1.
53 *Ibid* at Principle 1.2.
54 *Ibid* at Principle 1.3.
55 *Sustainable Development Law, supra* note 4 at 120.
56 *Cartagena Protocol, supra* note 2 at art 1.
57 *Ibid* at art 2(2).

Principle of Equity and the Eradication of Poverty

The principle of equity – both inter- and intragenerational – is central to the attainment of sustainable development.[58] Although the present generation has a right to use and enjoy the resources of the Earth, this generation is also under an obligation to take into account the long-term impact of its activities and to sustain the resource base and the global environment for the benefit – in its broadest meaning – of future generations.[59] The right to development must be implemented to meet developmental and environmental needs of present and future generations sustainably and equitably. This includes exercising the duty to cooperate for the eradication of poverty in accordance with Chapter IX on International Economic and Social Co-operation of the Charter of the United Nations and the Rio Declaration on Environment and Development, as well as the duty to cooperate for global sustainable development and the attainment of equity in the development opportunities of developed and developing countries.[60] The principle of equity is reflected in different ways in many international treaties and legal instruments in the field of sustainable development.[61] Although it is the primary responsibility of the State to aim for conditions of equity within its own population, and to ensure, at a minimum, the eradication of poverty, both the UN Charter and the 2000 U.N. Millennium Declaration recognize that all States that are in a position to do so have a further responsibility, to assist in the global achievement of this objective.[62] In attempting to protect the rights of future generations to safe food and a sound environment, the Cartagena Protocol reflects a certain respect for the principle of equity. By making possible the safer import and use of drought-resistant varieties of plants, high-yield food crops, and other beneficial LMOs, it could be argued that the Cartagena Protocol is also contributing in certain ways to reducing poverty. Furthermore, by attempting to balance the need of exporters of LMOs for a stable regulatory climate and the need of importers to be able to manage the risks and potential impacts of biotechnology on their populations, especially in poor or rural areas, the Protocol also has the potential to address imbalances, thereby securing greater equity.

Principle of Common but Differentiated Responsibilities

The principle of common but differentiated responsibilities is a manifestation of general principles of equity. States and other relevant actors have a common responsibility for the achievement of global sustainable development and protection of the environment, but each stakeholder's differing circumstances must be taken into

58 *New Delhi Declaration*, *supra* note 40 at Principle 2.1.
59 *Ibid* at Principle 2.2.
60 *Ibid* at Principle 2.3.
61 *Sustainable Development Law*, *supra* note 4 at 122.
62 *New Delhi Declaration*, *supra* note 40 at Principle 2.4.

account when examining their contribution toward those goals.[63] All States have a duty to cooperate in the achievement of global sustainable development and the protection of the environment, and international organizations, corporations (including in particular transnational corporations), nongovernmental organizations, and civil society should also be a part of this global partnership. Corporations have a role pursuant to the "polluter-pays" principle.[64]

Differentiation of responsibilities, although principally based on the contribution that a State has made to the emergence of environmental problems, must also take into account the economic and developmental situation of the State, in recognition of the special needs and interests of developing countries and of countries with economies in transition, particularly least-developed countries and those affected adversely by environmental, social, and developmental considerations.[65] Developed countries bear a special burden of responsibility in reducing and eliminating unsustainable patterns of production and consumption and in contributing to capacity building in developing countries by providing financial assistance and access to environmentally sound technology. In particular, developed countries should play a leading role and assume primary responsibility in matters of relevance to sustainable development.[66]

The Protocol recognizes that centres of origin and centres of genetic diversity are of crucial importance to humankind and takes into account the limited capacities of many countries, particularly developing countries, to cope with the nature and scale of known and potential risks associated with LMOs.[67] The Protocol's capacity-building measures, resources, and financial mechanism may also contribute to the realization of this principle in the treaty.[68]

Principle of the Precautionary Approach

The precautionary approach to human health, natural resources, and ecosystems requires that where there are threats of serious or irreversible damage, lack of full scientific certainty shall not be used as a reason for postponing cost-effective measures to prevent degradation.[69] This requires States, international organizations, and nongovernment actors to avoid activities, in light of scientific uncertainty, that could cause significant harm.[70] The approach includes accountability for harm caused, planning based on clear criteria and well-defined goals, consideration of

63 *Sustainable Development Law*, *supra* note 4 at 132–3.
64 *New Delhi Declaration*, *supra* note 40 at Principle 3.1.
65 *Ibid* at Principle 3.2 and 3.3.
66 *Ibid* at Principle 3.4.
67 *Cartagena Protocol*, *supra* note 2 at Preamble.
68 *Ibid* at arts 22 and 28.
69 Cordonier Segger, "The Role of International Forums in the Advancement of Sustainable Development," *supra* note 4 at 12.
70 *New Delhi Declaration*, *supra* note 40 at Principle 4.1.

all possible means to achieve an objective when completing an environmental impact assessment, and establishing an appropriate burden of proof on the person or persons carrying out, or intending to carry out, activities that could cause serious long-term or irreversible harm.[71] Decision-making processes should adopt a precautionary approach to risk management and appropriate precautionary measures in particular.[72] Last, precautionary measures should be based on up-to-date and independent scientific judgment and be transparent, and not used as a cover for economic protectionism. Transparent structures that involve all interested parties, including nonstate actors, in the consultation process should be established, and appropriate review by a judicial or administrative body made available.[73]

The Cartagena Protocol is founded on the precautionary approach, with express reference made in the preamble and article 1, and implicit reference found in the conditional "may have an adverse effect" on biodiversity and/or human health.[74] The AIA procedure is the primary mechanism used to ensure that Parties are provided with the information necessary to make informed decisions before agreeing to the import of LMOs.[75] It must be considered with the Protocol's risk assessment and risk-management provisions for safe use, handling, and transboundary movement; emergency measures in the event of unintentional release; and provisions on handling, transport, packaging, and identification.[76]

The flexibility to maintain a multidisciplinary case-by-case approach to LMOs for the purpose of risk assessment is provided for by the Protocol, but decision making must still be grounded in sound science; nonscientific factors alone will not provide unchallengeable grounds for refusal to import LMOs.[77] In general, the Protocol requires Parties to ensure that the development, handling, transport, use, transfer, and release of LMOs are undertaken in a manner that prevents or reduces the risks to biodiversity, taking also into account risks to human health.[78] It does not prevent a Party from taking action that is more protective of the conservation and sustainable use of biodiversity than called for, provided that the action is consistent with the objective and provisions of the Protocol and is in accordance with that Party's obligations under international law.[79] The rules contained in the Protocol are a minimum standard rather than a ceiling, meaning that Parties can adopt a more precautionary approach if they deem fit.[80]

71 *Ibid* at Principle 4.2.
72 *Ibid* at Principle 4.3.
73 *Ibid* at Principle 4.4.
74 International Law & Environment, *supra* note 8 at 640.
75 *Ibid* at 640; See *Cartagena Protocol, supra* note 2 at art 7.
76 *Ibid* at 641; See *Cartagena Protocol, supra* note 2 at arts 15, 16, 17, and 18.
77 *Ibid* at 644–5.
78 *Cartagena Protocol, supra* note 2 at art 2(2).
79 *Ibid* at art 2(4).
80 Ruth Mackenzie et al, *An Explanatory Guide to the Cartagena Protocol on Biosafety* (Gland, Switzerland and Cambridge, UK: IUCN and FIELD, 2003) at para 184 [IUCN Guide].

Principle of Public Participation and Access to Information and Justice

The principle of public participation and access to information and justice requires States to ensure that people have proper access to appropriate, comprehensive, and timely information concerning sustainable development that is held by public authorities and the opportunity to participate in decision-making processes, as well as effective access to judicial and administrative proceedings, including redress and remedy.[81] It is a condition for responsive, transparent, and accountable governments; the active engagement of civil society organizations, including industrial concerns and trade unions; and ensuring that women can play their vital role in sustainable development in every country.[82]

Public participation is based on the effective protection of the human right to hold and express opinions and to seek, receive, and impart ideas. This is predicated on access to appropriate, comprehensible, and timely information held by governments and industrial concerns on economic and social policies regarding the sustainable use of natural resources and the protection of the environment, access that does not impose undue financial burdens on the applicants and that gives due consideration for privacy and adequate protection of business confidentiality.[83] The empowerment of peoples in the context of sustainable development also depends on access to effective judicial or administrative procedures in the State where the measure has been taken to challenge such measure and to claim compensation. States should ensure that where transboundary harm has been or is likely to be caused, individuals and groups affected have nondiscriminatory access to the same judicial and administrative procedures as would individuals and groups of the State in which the harm is caused.[84]

Public awareness and participation play a prominent role in the Protocol. There are likely to be a wide range of stakeholders consulted in the field of biosafety, and Parties are required to ensure that the public is actively consulted on LMOs and biosafety, with the objective of promoting transparency and informed decision making, in accordance with their respective laws and regulations.[85] The Biosafety Clearing-House (BCH) is a publicly accessible information- sharing mechanism to facilitate the exchange of scientific, technical, environmental, and legal information.[86] The Protocol's BCH and capacity-building provisions are designed to facilitate developing states' participation. The Protocol also allows for the protection of business confidentiality, balancing social with economic considerations.[87] Last,

[81] Cordonier Segger, "The Role of International Forums in the Advancement of Sustainable Development," *supra* note 4 at 12.
[82] *New Delhi Declaration, supra* note 40 at Principle 5.1.
[83] *Ibid* at Principle 5.2.
[84] *Ibid* at Principle 5.3.
[85] *Cartagena Protocol, supra* note 2 at art 23; International Law & Environment, *supra* note 8 at 642.
[86] *Ibid* at art 20.
[87] *Ibid* at art 21.

it mandates measures for the adequate handling, packaging, transport, and identification of LMOs.[88]

Principle of Good Governance

The principle of good governance commits States and international organizations to adopt democratic and transparent decision-making procedures and financial accountability, take effective measures to combat official or other corruption, respect the principle of due process in their procedures, and observe the rule of law and human rights.[89] Citizens can expect good governance by States and international organizations, whereas nonstate actors should be subject to internal democratic governance and to effective accountability.[90] The application of the principle can evoke the need for corporate social responsibility and a global market aimed at a fair distribution of wealth among and within communities.[91]

The Cartagena Protocol contains numerous aspects that reflect the principle of good governance at both the domestic and international level. Its first general provision explicitly requires each Party to take the necessary and appropriate legal, administrative, and other measures to implement its obligations.[92] Second, it establishes a transparent and procedurally sound decision-making process to govern the first intentional transboundary movement of LMOs for intentional introduction into the importing Party's environment – the advance informed agreement (AIA) procedure.[93] Third, the Protocol establishes notification and consultation requirements on Parties relating to unintentional transboundary movements of LMOs,[94] as well as requiring appropriate domestic measures aimed at preventing and penalizing illegal transboundary movements.[95] Fourth, each Party is required to monitor the implementation of its obligations and report to the COP-MOP on its implementation measures.[96] Fifth, the COP-MOP is required to consider and approve cooperative procedures and institutional mechanisms to promote compliance with the Protocol and to address cases of noncompliance, including provisions to offer advice or assistance when appropriate.[97] Last, the COP-MOP is required to undertake an evaluation of the effectiveness of the Protocol, including an assessment of its procedures and annexes, at least every five years.[98] The COP-MOP shall also

88 *Ibid* at art 18.

89 *New Delhi Declaration*, *supra* note 40 at Principle 6.1; Cordonier Segger, "The Role of International Forums in the Advancement of Sustainable Development," *supra* note 4 at 12.

90 *Ibid* at Principle 6.2.

91 *Ibid* at Principle 6.3.

92 *Cartagena Protocol*, *supra* note 2 at art 2(1).

93 *Ibid* at art 7(1).

94 *Ibid* at art 17.

95 *Ibid* at art 25(1).

96 *Ibid* at art 33.

97 *Ibid* at art 34.

98 *Ibid* at art 35.

keep under regular review the implementation of the Protocol and shall make decisions necessary to promote its effective implementation, including considering and adopting amendments to the Protocol and annexes that are deemed necessary for its implementation.[99] A broad list of observers is allowed, including any body or agency, whether national or international, governmental or nongovernmental, that is qualified in matters covered by the Protocol.[100]

Conclusion

In summary, the Cartagena Protocol on Biosafety shares an important objective with the CBD – the conservation of biological diversity and sustainable use of its components. These components include genetic resources and technologies, and in particular, the Protocol seeks to secure the safe and sustainable use of LMOs, which form a significant part of the growing field of biotechnology. The principles of international law on sustainable development, as enshrined in soft law instruments such as the 1992 Rio Declaration and analyzed in the work of scholarly bodies such as the ILA 2002 New Delhi Declaration, are both recognized and made operational in the Cartagena Protocol in certain ways, often through the establishment of specific mechanisms or instruments. The key challenge, at this point, is to analyze and monitor outcomes to give effect and implement this key Protocol. This endeavour may be able to improve the abilities of the Parties to comply with their obligations within their diverse territories and broaden understanding of the overall regime, contributing to its eventual refinement and evolution internationally.

BIOSAFETY BECOMES BINDING

This volume is presented in five parts. Part I (Essentials of Biosafety and Sustainable Development Law) lays the foundations for the book. It begins with Chapter 1 by Kathryn Garforth, Worku Damena Yifru, and Mai Fujii. This chapter establishes the historical and legal context of international biosafety law, defines biotechnology, and explains why modern biotechnology is a common concern. This chapter also provides a general explanation of the key obligations in the Cartagena Protocol, with emphasis on its main decision-making provisions. The next chapter, by Christine Frison, Sylvestre-José-Tidiane Manga, and Marie-Claire Cordonier Segger, outlines certain challenges faced in the development of National Biosafety Frameworks (NBF) and related regulatory regimes and comments on the role of such national policies and laws in the promotion of sustainable development. Part I concludes with a chapter by Greg Jaffe that details the components and characteristics of a functional and protective biosafety regulatory system.

99 *Ibid* at art 29(4)(d).
100 *Ibid* at art 29(8).

Part II (Sustainable Development Law and Policy on Biosafety) expands on the constituent elements of the Cartagena Protocol and in many cases discusses how these provisions both reflect important principles of international law on sustainable development and seek to contribute to the safe and sustainable use of genetic resources. It begins with Chapter 4, by Ryan Hill, which addresses issues and challenges related to risk assessment and risk management in the context of LMOs. Chapter 5, by Worku Damena Yifru, Mai Fujii, and Kathryn Garforth, examines the relevance and validity of the AIA procedure and related procedure for LMOs intended for direct use as food or feed or for processing. Chapter 6, by Thomas Redick, examines key issues regarding the handling, transport, packaging, and identification of LMOs bound for use as seed or in food, feed, or processing, and the ways in which these issues affect trade. Chapter 7, by Julian Kinderlerer and Christine Toczeck, explores the question of public participation in the debate over the safe use of biotechnology, explaining the importance of meaningful public participation and transparency in decision-making processes. Chapter 8, by Frederic Perron-Welch, discusses the role and importance of the Biosafety Clearing-House. Chapter 9, by Tomme Rosanne Young, considers progress in the implementation of the Biosafety Clearing-House, drawing on a careful analysis of recent domestic legislative experiences to constructively highlight important legal and policy challenges for the regime. Chapter 10, by Frederic Perron-Welch, examines socioeconomic considerations that relate to biosafety in the context of the Cartagena Protocol and how they can play a role in decision-making to help promote sustainable development. Chapter 11, by Veit Koester, discusses the implications and effectiveness of compliance procedures and mechanisms under the Cartagena Protocol. Finally, Chapter 12, by Frederic Perron-Welch and Olivier Rukundo, examines issues of liability and redress in the context of biotechnology and the *Nagoya – Kuala Lumpur Protocol on Liability and Redress to the Cartagena Protocol*.

Part III (Domestic Implementation of Biosafety Regulatory Aspects) conducts a detailed analysis of the implementation of the Cartagena Protocol on Biosafety through the development of national biosafety frameworks. It opens with a discussion of legislative options for national implementation in Chapter 13, by Tomme Rosanne Young. Through the analysis of the experiences of various countries and regions in Chapter 14, by Liina Eek and David Duthie, on Central and Eastern Europe; Chapter 15, by Gregory Jaffe and Papa Miessa Dieng, on West Africa; Chapter 16, by Gregory Jaffe, on East Africa; Chapter 17, by Nizar Mohamed, on Asia; and Chapter 18, by Jorge Cabrera Medaglia, on the Americas, this third part identifies specific problems related to the development of national biosafety regulatory regimes and seeks to outline key lessons for future capacity-building activities in biosafety.

Part IV (Case Studies of Domestic Regimes and Sustainable Development) builds on the analysis in Part III with case studies of specific legislative approaches to the Cartagena Protocol. This part opens with an overview of various responses to key aspects of the Protocol in Chapter 19, by Tomme Rosanne Young. Chapter 20, by

Jorge Cabrera Medaglia, explains Costa Rica's legal framework concerning GMOs in the agricultural sector. Chapter 21, by Frederic Perron-Welch, describes New Zealand's biosafety framework in the context of that country's unique history and geography. In Chapter 22, by Kathryn Garforth and Paige Ainslie, the issue of liability and redress in Canadian case law is investigated by following the case of a group of organic farmers seeking to bring a class action suit against Monsanto and Bayer for damages allegedly caused by the companies' genetically modified varieties of canola. Last, Chapter 23, by Konstantia Koutouki and Paula Honorato Marin, discusses the implications of GMOs in Chile for the protection of Indigenous culture.

Part V (Global Policy Trends in Biosafety) explores global policy trends in the implementation of the Cartagena Protocol on Biosafety. Chapter 24, by Frederic Perron-Welch, Marie-Claire Cordonier Segger, Christine Frison, and Jorge Cabrera Medaglia, draws on the experience of the authors in reviewing the design of numerous national biosafety frameworks and regulatory regimes to draw certain conclusions about the diverse approaches being adopted by Parties to the Protocol. The chapter identifies key future legal research questions and priorities, and provides preliminary recommendations regarding the future design and implementation of biosafety rules in the context of sustainable development law. Chapter 25, by Marie-Claire Cordonier Segger and Markus Gehring, examines the findings and corresponding implications for sustainable development of the *European Communities Biotech Dispute* submitted to the World Trade Organization (WTO). Chapter 26, by Martin Endicott, explores the fraught relationship between food aid and biosafety.

This book concludes with a final scholarly and practical legal analysis of implementation of the Cartagena Protocol, from a sustainable development law perspective, summarizing key conclusions on the strengths and weaknesses of implementation efforts across the world, and offering advice to the over 160 countries seeking to set laws and policies in place related to biosafety, and to further refine their international regime for biosafety.

PART I

Essentials of Biosafety and Sustainable Development Law

1

Biosafety, the Cartagena Protocol, and Sustainable Development

Kathryn Garforth, Worku Damena Yifru, and Mai Fujii

As one of a growing number of multilateral instruments that seek to promote sustainable development, the *Cartagena Protocol on Biosafety*[1] (Cartagena Protocol) represents, in many respects, an interesting development in the law of sustainable development.[2] The Protocol is not solely economic, environmental, or social but incorporates elements of each of these pillars of sustainable development.[3] The Cartagena Protocol has strong links to environmental law through its foundation in the precautionary principle and the development of sound science; links to social and human rights law through its elements of public awareness, public participation, and the public right to know, as well as by permitting consideration of the socioeconomic impacts of living modified organisms (LMOs); and links to trade law because of its focus on the transboundary movement of LMOs.

The sections that follow examine the context in which the Cartagena Protocol was developed and describe some of the different aspects of the Protocol through the lens of sustainable development.

CONTEXT

Biosafety "refers to the need to protect human health and the environment from the possible adverse effects of the products of modern biotechnology."[4] As this definition

1 *Cartagena Protocol on Biodiversity*, 29 January 2000, 2226 UNTS 208 (entered into force 11 September 2003) [*Cartagena Protocol*].

2 UP Thomas, "The CBD, the WTO, and the FAO: The Emergence of Phytogenetic Governance" in Philippe G LePrestre, ed, *Governing Global Biodiversity: The Evolution and Implementation of the Convention on Biological Diversity* (Burlington, VT: Ashgate Publishing Co, 2002) 177 at 200.

3 Marie-Claire Cordonier Segger and Ashfaq Khalfan, eds, *Sustainable Development Law: Principles, Practices and Prospects* (Oxford, UK: Oxford University Press, 2004) at 50.

4 Secretariat of the Convention on Biological Diversity (SCBD), *Cartagena Protocol on Biosafety to the Convention on Biological Diversity: Text and Annexes* (Montreal, Canada: SCBD, 2000) at 1.

illustrates, biosafety is intricately bound to biotechnology. In turn, *biotechnology* is a broad term that can encompass a range of activities in a variety of different fields. The scope of the Cartagena Protocol is limited, however, to "modern biotechnology," which is defined as the application of:

a. In vitro nucleic acid techniques, including recombinant deoxyribonucleic acid (DNA) and direct injection of nucleic acid into cells or organelles, or
b. Fusion of cells beyond the taxonomic family, that overcome natural physiological reproductive or recombination barriers and that are not techniques used in traditional breeding and selection.[5]

The field of biosafety stems from the development of these modern techniques of biotechnology. In the early 1970s, American scientists Herbert Boyer and Stanley Cohen collaborated on a series of experiments that resulted in the creation of the first molecules of recombinant DNA and the first genetically engineered bacteria.[6] The early transformations of bacteria and viruses in the laboratory raised concerns among scientists, leading to the development of guidelines for recombinant DNA research focusing particularly on containment measures for laboratory experiments.[7]

By the end of the 1970s, the field of biotechnology began to shift from experiments in the laboratory to industrial applications in the field. The first environmental releases of genetically modified bacteria began in the 1980s, as did field trials for genetically modified plants.[8] As the science advanced, the risks became more complex, leading to concerns about the possible hazards posed by recombinant DNA.

Until this point, biotechnology was primarily the domain of developed countries. Recombinant DNA experiments were conducted in the laboratories of industrialized countries, and genetically modified organisms (GMOs) were being created to meet needs in developed countries. These industrialized countries also implemented some of the first regulatory regimes for GMOs, such as Canada's National Biotechnology Strategy, created in 1982, or the Coordinated Framework for the Regulation of Biotechnology, published in the United States in 1986.[9]

In this context began the negotiations on the *Convention on Biological Diversity*[10] (CBD) in 1987. The emerging role and potential value of genetic resources in the world economy were becoming increasingly obvious. The United States was

5 *Cartagena Protocol*, *supra* note 1 at art 3(i).

6 Susan R Barnum, *Biotechnology: An Introduction* (Scarborough, ON: Scarborough, Canada: Wadsworth Publishing Company, 1998) at 21.

7 See, for example, the outcomes from the Asilomar Conference on Recombinant DNA Molecules held in February 1975 and the "Guidelines for research involving recombinant DNA molecules," (1976) 41 Federal Register 131 at 27911–27943 developed by the U.S. National Institutes of Health; Barnum, *ibid* at 22–5.

8 Barnum, *ibid* at 25–6.

9 Barnum, *ibid* at 192.

10 *Convention on Biological Diversity*, 5 June 1992, 31 ILM 818 (entered into force 29 December 1993 [CBD].

beginning to grant patents on genetically modified life forms,[11] and the debate over Plant Breeders' Rights versus Farmers' Rights had been raging at the Food and Agriculture Organization of the United Nations since 1980. From the very beginning of the discussions that led to the CBD, developing countries insisted that biotechnology be one of the key issues addressed in the Convention.[12]

Developing countries' insistence about this issue was successful and led to the inclusion of Article 19 in the CBD, which discusses the "handling of biotechnology and distribution of its benefits." The article provides that Parties to the CBD are to take measures to include countries that provide genetic resources in biotechnology research activities, share access to the results and benefits of biotechnology, and provide information about LMOs to Parties where they are to be introduced. Paragraph 3 of Article 19 obliges the Parties to the Convention to consider the need for a protocol on LMOs resulting from biotechnology that could have adverse effects on the conservation and sustainable use of biodiversity. This paragraph was somewhat controversial during the negotiations because some countries wanted to make the development of a protocol mandatory and not simply something for further consideration.[13]

At the same 1992 United Nations Conference on Environment and Development (UNCED), where the CBD was opened for signature, more than 140 governments agreed to *Agenda 21*.[14] Section II of *Agenda 21* is devoted to the "conservation and management of resources for development" and includes chapter 16 on the "environmentally sound management of biotechnology." The chapter focuses on five areas within biotechnology: increasing the availability of food, feed, and renewable raw materials; improving human health; enhancing protection of the environment; enhancing safety and developing international mechanisms for cooperation; and establishing enabling mechanisms for the development and environmentally sound application of biotechnology. The fourth of these areas is of particular relevance here because it calls for

> further development of internationally agreed principles on risk assessment and management of all aspects of biotechnology, which should build upon those developed at the national level. Only when adequate and transparent safety and border-control procedures are in place will the community at large be able to derive

[11] The U.S. Supreme Court decision in *Diamond v Chakrabarty*, 206 USPQ 193 (USSC 1980) granted a patent on a genetically modified microorganism and found that whether something is living is not relevant in determining whether it is patentable. In 1987, the US Patent and Trademark Office issued a statement declaring that "nonnaturally occurring non-human multicellular living organisms" are patentable subject matter, US PTO 1077 Official Gazette 24 (21 April 1987).

[12] Désirée M McGraw, "The Story of the Biodiversity Convention: From Negotiation to Implementation" in LePrestre, *supra* note 2, 7 at 34.

[13] Ruth Mackenzie et al, *An Explanatory Guide to the Cartagena Protocol on Biosafety*, IUCN Environmental Policy and Law Paper No 46 (Gland, Switzerland: IUCN, 2003) at para 11 [IUCN Guide].

[14] *Report of the United Nations Conference on Environment and Development*, UN Doc A/CONF.151/26/Rev.1 (Vol 1), 31 ILM 874 (1992) [UNCED Report].

maximum benefit from, and be in a much better position to accept the potential benefits and risks of, biotechnology.[15]

Three years after UNCED, the United Nations Environment Programme released the International Technical Guidelines for Safety in Biotechnology.[16] According to the Preface, these guidelines are intended as a contribution to the implementation of *Agenda 21*. They were developed "on the basis of common elements and principles derived from relevant existing regional and international instruments and national regulations and guidelines, drawing upon experience already gained through their preparation and implementation."[17] The guidelines "provide technical guidance on evaluating biosafety, identifying measures to manage foreseeable risks and to facilitate processes such as monitoring, research and information exchange."[18] The same year, the Parties to the CBD agreed to establish an Ad Hoc Working Group on Biosafety to elaborate a protocol on biosafety.[19] The Working Group was to endeavour to complete its work in 1998.

In the end, it took until the year 2000 for the negotiating countries to reach agreement on the text of a protocol. The negotiations were rocky and reached the brink of failure when no agreement had been reached by the February 1999 meeting in Cartagena, Colombia. The meeting was adjourned, and a resumed session of the first extraordinary meeting of the Conference of the Parties was called for January 2000 in Montreal, at which consensus finally was achieved. Three and a half years later, in September 2003, the Cartagena Protocol on Biosafety entered into force, having received the fifty necessary ratifications. To date, of the 193 Parties to the CBD, 163 also have become Parties to the Protocol.[20]

As other chapters in this book explore, some of the most contentious issues during the negotiations included the scope of the advance informed agreement procedure under the Protocol and the types of organisms it would cover; inclusion of the precautionary principle in the operational text of the Protocol, the allowance for socioeconomic considerations in decision making under the Protocol, and the relationship between the Protocol and other international agreements, particularly those of the World Trade Organization (WTO).

Although the text of an international treaty might appear to be static, in fact the understanding and application of the words in any agreement are constantly

[15] *Ibid* at para 16.29.

[16] United Nations Environment Programme, "International Technical Guidelines for Safety in Biotechnology" (1995), online: United Nations Environment Programme, http://www.unep.org/biosafety/Documents/Techguidelines.pdf.

[17] *Ibid* at ix.

[18] IUCN Guide, *supra* note 13 at para 119.

[19] CBD Dec II/5, "Consideration of the need for and modalities of a protocol for the safe transfer, handling and use of living modified organisms".

[20] As of 11 July 2012, online: Biosafety Clearing-House, "List of Parties" http://bch.cbd.int/protocol/parties. Only Parties to the CBD may become Parties to the Protocol, see art 32(1) of the CBD.

evolving. Much work takes place to flesh out the meaning of a treaty. In the case of the Cartagena Protocol, this work is led by its governing body – the Conference of the Parties serving as the meeting of the Parties to the Protocol (COP-MOP). The COP-MOP has met five times to date: Kuala Lumpur, Malaysia, 23 to 27 February 2004; Montreal, Canada, 30 May to 3 June 2005; Curitiba, Brazil, 13 to 17 March 2006; Bonn, Germany, 12 to 16 May 2008; and Nagoya, Japan, 11 to 15 October 2010.

The COP-MOP adopts decisions on different aspects and issues of the Protocol. In addition to exploring the relationship between the Protocol and sustainable development, the discussion that follows also examines the role of the COP-MOP in elaborating on the provisions of the Protocol.

THE CARTAGENA PROTOCOL AND THE THREE PILLARS OF SUSTAINABLE DEVELOPMENT

Much of this text is devoted to detailed explorations of the different facets of the Cartagena Protocol. Our intention in this chapter is to provide a brief overview and introduction to these facets and to describe what the Protocol does – and does not – do. The book explores the different aspects of the Cartagena Protocol through the lens of sustainable development to demonstrate the treaty's integrative nature.

The concept of sustainable development[21] has been incorporated into various international instruments, including multilateral environmental agreements, since the 1980s. Although sustainable development may be critiqued as having become an "umbrella term for a host of disparate issues,"[22] or a "rhetorical mantra, often repeated but seldom truly understood,"[23] the notion clearly has obtained broad support in the international community. Since the World Summit on Sustainable Development in Johannesburg in 2002, environmental protection and economic and social development have been considered "interdependent and mutually reinforcing" pillars of sustainable development.[24] It is a common understanding that attaining all three of these pillars concurrently is essential to achieve genuine sustainable development.[25]

21 The notion of sustainable development was originally defined as "development that meets the needs of the present without compromising the ability of future generations to meet their own needs," World Commission on Environment and Development, *Our Common Future* (Oxford, UK: Oxford University Press, 1987) at 43.

22 International Law Association, "International Law on Sustainable Development" in *Report of the Seventy-First Conference: Berlin* (London: International Law Association, 2004) 566 at 576.

23 International Law Association, "International Law on Sustainable Development" in *Report of the Seventy-Third Conference: Rio de Janeiro* (London: International Law Association, 2008) 895 at 900.

24 *Plan of Implementation of the World Summit on Sustainable Development*, WSSD Res 2, 17th plenary meeting, (4 September 2002) in United Nations, *Report of the World Summit on Sustainable Development* (New York, UN, 2002) at para 2, online: United Nations: Johannesburg Summit 2002, http://www.johannesburgsummit.org.

25 International Law Association, "International Law on Sustainable Development" in *Report of the Seventy-Second Conference: Toronto* (London: International Law Association, 2006) 467 at 472.

The Cartagena Protocol contains all three pillars in its provisions. Implementation of the Protocol moves from rhetorical mantra to actual integration of the three components and promotion of sustainable development.

The scope and limitations of the Protocol are largely set out in its first five articles. Article 1 defines the objective of the Cartagena Protocol as follows:

> In accordance with the precautionary approach contained in Principle 15 of the Rio Declaration on Environment and Development, the objective of this Protocol is to contribute to ensuring an adequate level of protection in the field of the safe transfer, handling and use of living modified organisms resulting from modern biotechnology that may have adverse effects on the conservation and sustainable use of biological diversity, taking also into account risks to human health, and specifically focusing on transboundary movements.

There are a few key things to note from this objective. First, the Protocol concerns LMOs rather than the more commonly used term *genetically modified organisms*. This is because the primary concern of countries was the possible environmental impact of the release of new organisms into the environment. The ability of living organisms to reproduce and spread independently led to their being the focus of the Protocol. A second aspect to note is that the objective includes the phrase "taking also into account risks to human health." The Protocol includes human health within its ambit, though the meaning and effect of this phrase has not yet been well defined.

A final aspect to note is that the Protocol focuses specifically on transboundary movements. Although some provisions in the Protocol concern domestic development of LMOs (e.g., Article 16 on risk management), the Protocol mostly addresses what should or could happen when an LMO is moved from one country to another.

These parameters are also reflected in the scope of the Protocol as set out in Article 4: "This Protocol shall apply to the transboundary movement, transit, handling and use of all living modified organisms that may have adverse effects on the conservation and sustainable use of biological diversity, taking also into account risks to human health." Although the scope states that the Protocol applies to *all* LMOs that might have adverse effects on the conservation and sustainable use of biodiversity, Article 5 on pharmaceuticals may limit this somewhat. The article states that the Protocol does not apply to the transboundary movement of LMOs that are pharmaceuticals for humans when these LMOs are addressed by other relevant international agreements or organizations.

With the basic framework of the Protocol set, we now turn to examining the interlinkages among the provisions of the Protocol and the three pillars of sustainable development. By necessity, this exploration is somewhat simplified, because several individual aspects of the Protocol can themselves be understood as cutting across economic, social, and environmental concerns. Nonetheless, it is a useful exercise for understanding the Cartagena Protocol as an integrative instrument that has environmental protection as its fundamental goal but also recognizes the social and economic dimensions necessary for achieving this goal.

Environment

The environmental dimension of sustainable development can be found most particularly in the risk assessment requirements of the Cartagena Protocol. The Protocol sets out provisions on risk assessment (Article 15 and Annex III) and risk management (Article 16) to identify, evaluate, regulate, manage, and control possible adverse effects and risks of LMOs. The risk assessment provisions are tied most closely – although not exclusively – to the advance informed agreement (AIA) procedure of the Protocol. This procedure is set out in Articles 7 through 10 and 12 and applies "prior to the first intentional transboundary movement of living modified organisms for intentional introduction into the environment of the Party of import."[26] It requires that decision making be based on a risk assessment[27] that is "carried out in a scientifically sound manner . . . taking into account recognised risk assessment techniques."[28] Annex III sets out the risk assessment requirements in more detail, providing the objective, general principles, methodology, and points to consider in the conduct of risk assessment.

Article 16 requires Parties to, among other things, adopt measures and strategies for preventing adverse effects and for managing and controlling risks identified by risk assessments. When scientific certainty about the extent of the potential adverse effects of an LMO is lacking, a Party may take precautionary action to avoid or minimize the potential adverse effects.[29]

Discussions on risk assessment under the Protocol have led to agreement that there is a need for additional guidance on specific aspects of risk assessment and risk management, and to this end, an Ad Hoc Technical Expert Group (AHTEG) on Risk Assessment and Risk Management was established at COP-MOP 4. The AHTEG developed "Guidance on Risk Assessment of Living Modified Organisms," which includes a roadmap on the steps for conducting a risk assessment as well as guidance on specific types of LMOs and traits.[30] An open-ended online forum on specific aspects of risk assessment was also established, and several real-time online conferences were conducted to provide input to the AHTEG.

At COP-MOP 5, the Parties welcomed the guidance developed by the AHTEG. They agreed to undertake a scientific review of the guidance and also to extend the open-ended online forum and the AHTEG. In 2011 and 2012, the AHTEG

26 *Cartagena Protocol*, *supra* note 1 at art 7(1).

27 *Ibid* at art 10(1).

28 *Ibid* at art 15(1).

29 *Ibid* at art 10(6).

30 Executive Secretary of the Convention on Biological Diversity, "Risk Assessment and Risk Management (Articles 15 and 16)," UN Doc UNEP/CBD/BS/COP-MOP/5/12 (30 July 2010) at Annex III.

contributes to the scientific review and testing of the guidance and also develops further guidance on new specific topics of risk assessment.[31]

The risk assessment and risk management requirements of the Protocol concern the intentional transboundary movements of LMOs. Article 17 addresses the situation in which a transboundary movement is unintentional. The article requires a Party to take notification measures as soon as it knows of an occurrence under its jurisdiction resulting in a release that leads or might lead to an unintentional transboundary movement of an LMO that is likely to have significant adverse effects on the conservation and sustainable use of biodiversity.[32] It also sets out specific types of information that should be included in the notification.[33] The article closes by asking each Party under whose jurisdiction a release occurs to consult the affected or potentially affected States immediately "to enable them to determine appropriate responses and initiate necessary action, including emergency measures."[34]

The ultimate purpose of Article 17 is to reduce the impact of unintentional transboundary movements of LMOs on biodiversity. Also evident in the language of the article are echoes of Principles 18 and 19 of the *Rio Declaration on Environment and Development*,[35] which similarly call for states to notify each other in cases of "natural disasters or other emergencies that are likely to produce sudden harmful effects on the environment of [other] States" (Principle 18) and to notify and consult with each other "on activities that may have a significant adverse transboundary environmental effect" (Principle 19).

Society

Public awareness and public participation are essential elements to achieve sustainable development. In international environmental forums, transparency and broad public participation in governmental decision making are required to ensure not only social development itself but also the integration of environmental, social, and economic considerations.[36] Public participation in the context of sustainable development is a broad concept that also includes elements of access to information.

The Protocol includes provisions on information sharing, confidential information, and public awareness and participation. Article 20 of the Protocol establishes

31 Cartagena Protocol, COP-MOP 5, Dec BS-V/12, "Risk assessment and risk management (Articles 15 and 16)" (2010).

32 *Cartagena Protocol*, *supra* note 1 at art 17(1).

33 *Ibid* at art 17(3).

34 *Ibid* at art 17(4).

35 UNCED Report, *supra* note 14.

36 An Expert Group Workshop on International Environmental Law aiming at Sustainable Development convened by the United Nations Environment Programme in 1996 recognized "that the integration of environmental, social and economic policies requires transparency and broad public participation in governmental decision-making." See Final Report of the Expert Group Workshop on International Environmental Law Aiming at Sustainable Development, UN Doc UNEP/IEL/WS/3/2 (1996) at para 35(d).

a Biosafety Clearing-House (BCH) as a tool for information sharing and exchange among Parties and other stakeholders. Parties are required to make certain information available through the BCH, such as laws, regulations, and guidelines for the implementation of the Protocol,[37] decisions taken under the AIA procedure,[38] and decisions on domestic use of LMOs intended for direct use as food or feed, or for processing.[39] The BCH also includes a Biosafety Information Resource Centre where users can post biosafety-related materials.

The BCH is unique because it was the first Internet-based information exchange mechanism mandated by a multilateral environmental agreement. Furthermore, as the power of the Internet grows, so does the functionality of the BCH. In recent years, the BCH has served as the platform for online discussion forums and real-time online conferences, complementing the more traditional meeting formats used in the United Nations.

Balancing the access to information facilitated by the BCH is Article 21 on confidential information. This article requires, among other things, a Party of import to allow a notifier to identify information that is to be treated as confidential when it is submitted to the Party of import.[40] The Party of import must protect confidential information received under the Protocol.[41] The Party of import must also consult the notifier if it decides that information identified by the notifier as confidential does not qualify as such, and, prior to any disclosure, the Party of import must inform the notifier of its decision, providing reasons on request as well as an opportunity for consultation and for an internal review of the decision prior to disclosure.[42]

Article 23 of the Protocol requires Parties to promote and facilitate public awareness and education, including access to information, about the safe transfer, handling, and use of LMOs. It also requires Parties to consult the public in the decision-making process, to make public the final decision taken, and to inform the public about the means of access to the BCH. This obligation to involve the public in decision making on LMOs is qualified by a reference to national laws and regulations.

Overall implementation of Parties' obligation to promote and facilitate public awareness, education, and participation faces major challenges, however.[43] COP-MOP 5 thus adopted a programme of work on public awareness, education, and participation concerning the safe transfer, handling, and use of LMOs for the period

37 *Cartagena Protocol, supra* note 1 at art 20(3).
38 *Ibid* at art 10(3).
39 *Ibid* at art 11(1).
40 *Ibid* at art 21(1).
41 *Ibid* at art 21(3).
42 *Ibid* at art 21(2).
43 Executive Secretary of the Convention on Biological Diversity, "Public Awareness and Participation: Interim report on the status of implementation of public awareness, education and participation concerning the safe transfer, handling and use of living modified organisms (Article 23, paragraph 1(a))," UN Doc UNEP/CBD/BS/COP-MOP/4/16 (3 March 2008) at para 25.

2011–2015.[44] The programme of work includes several elements and suggested activities directed toward Parties, civil society, and educational institutions.

It is also worthwhile mentioning the *Aarhus Convention on Access to Information, Public Participation in Decision-making and Access to Justice in Environmental Matters*[45] (i.e., the Aarhus Convention). In 2005, the Parties to the Aarhus Convention adopted an amendment "on public participation in decisions on deliberate release into the environment and placing on the market of genetically modified organisms."[46] Once the amendment enters into force,[47] it requires Parties to provide for "early and effective information and public participation prior to making decisions on whether to permit the deliberate release into the environment and placing on the market of genetically modified organisms."[48] A new annex to the Convention would set out the modalities for Parties to follow in providing early and effective information and public participation. The Secretariats of the Aarhus Convention and the CBD are already collaborating and have undertaken joint activities such as an international workshop on access to information, public participation, and access to justice regarding GMOs that was held in Cologne, Germany, in May 2008.

A final and somewhat separate aspect of the Cartagena Protocol that supports the society pillar of sustainable development is Article 22, on capacity building. Similar to several other multilateral environmental agreements, the Cartagena Protocol requires Parties to cooperate in developing or strengthening human resources and institutional capacities in biosafety in developing country Parties and Parties with economies in transition.[49] According to paragraph 2 of Article 22, capacity building must be needs driven and should include scientific and technical training in the proper and safe management of biotechnology, the use of risk assessment and risk management for biosafety, and the enhancement of technological and institutional capacities in biosafety.

The Parties have adopted numerous extensive decisions on capacity building. These include an Action Plan for Building Capacities for the Effective Implementation of the Protocol, which was initially adopted at COP-MOP 1 and then revised at COP-MOP 3. Even prior to the Protocol's entry into force, the Global Environment Facility – the financial mechanism of both the CBD and the Cartagena

44 Cartagena Protocol, COP-MOP 5, Decision BS-V/13, "Public awareness, education and participation" (2010).

45 *Aarhus Convention on Access to Information, Public Participation in Decision-making and Access to Justice in Environmental Matters*, 25 June 1998, 38 ILM. 517 (entered into force 30 October 2001) [*Aarhus Convention*].

46 Aarhus Convention, MOP 2, Decision II/1, "Genetically modified organisms" (2005). UN Doc ECE/MP.PP/2005/2/Add.2.

47 Article 14(4) of the Aarhus Convention requires that three-quarters of the Parties to the Aarhus Convention ratify, approve, or accept the amendment in order for it to enter into force. As of 20 June 2012, there were 46 Parties to the Convention and 27 Parties (59%) to the amendment on public participation in decisions on the deliberate release into the environment and placing on the market of GMOs.

48 Dec II/1, *supra* note 46 at Annex, art 6bis(1).

49 *Cartagena Protocol, supra* note 1 at art 22(1).

Protocol – undertook an extensive capacity-building project for the development and implementation of national biosafety frameworks to facilitate national compliance with the Cartagena Protocol. More than 120 countries were involved in the project, which formed the basis of many of these countries' national approach to biosafety.

Economy

The Cartagena Protocol permits the inclusion of socioeconomic considerations arising from the impact of LMOs on the conservation and sustainable use of biodiversity in deciding whether to allow the import of these organisms.[50] The inclusion of socioeconomic considerations in Parties' decision making must be consistent with their other international obligations.

The provision's wording reflects a compromise among the different economic interests at stake. During the negotiations, countries that were developers and exporters of LMOs were concerned that socioeconomic considerations could be used as a trade barrier by other countries to block the import of LMOs. Conversely, other countries wanted to assess whether the import of LMOs would be in their social and economic best interests before agreeing to adopt the technology. As a result, the language in paragraph 1 of Article 26 tries to balance the right of Parties to include socioeconomic considerations in decision making with their other obligations under international (trade) law.

The provision on socioeconomic considerations is one of the least elaborated aspects of the Protocol. To date, the COP-MOP has largely restricted its consideration of the issue to discussions on cooperation on research and information exchange on socioeconomic impacts of LMOs as required by paragraph 2 of Article 26. At COP-MOP 5, however, the Parties agreed to a number of activities on socioeconomic considerations in the context of capacity building.[51] These include regional online conferences and a workshop – activities to be carried out in 2011 and 2012. The Parties also included an operational objective on socioeconomic considerations in the Strategic Plan for the Protocol.[52]

The Cartagena Protocol is not unique, however, in allowing for socioeconomic considerations in decision making. Paragraph 3 of Article 5 of the WTO *Agreement on the Application of Sanitary and Phytosanitary Measures*[53] requires WTO Members to take certain economic factors into account when assessing a risk and determining the measure to protect against the risk. The International Plant Protection Convention's International Standard for Phytosanitary Measures (ISPM)

50 *Ibid* at art 26(1).

51 Cartagena Protocol, COP-MOP 5, Decision BS-V/3, "Status of capacity-building activities" (2010), see, in particular, section IV of the decision, "cooperation on identification of capacity-building needs for research and information exchange on socio-economic considerations."

52 Cartagena Protocol, COP-MOP 5, Decision BS-V/16, "Strategic plan for the Cartagena Protocol on Biosafety for the period 2011–2020" (2010), see operational objective 1.7.

53 *Agreement on the Application of Sanitary and Phytosanitary Measures*, 15 April 1994, being part of Annex IA to the *Agreement Establishing the World Trade Organization*, 15 April 1994, 33 ILM 1144.

No. 11 on "Pest Risk Analysis for Quarantine Pests Including Analysis of Environmental Risks and Living Modified Organisms" includes provisions on the assessment of potential economic consequences as part of the pest risk assessment.[54] Furthermore, the World Organisation for Animal Health includes consideration of both biological and economic consequences in its definitions of risk and risk assessment.[55]

Although the allowance for socioeconomic considerations might be the most obvious way in which the Cartagena Protocol integrates the economic pillar of sustainable development, at least three other aspects of the Protocol demonstrate this as well: the procedure for LMOs intended for direct use as food or feed, or for processing (LMOs-FFP); the handling, transport, packaging, and identification of LMOs; and liability and redress.

Article 11 of the Protocol establishes the procedure for LMOs-FFP. It provides that a "Party that makes a final decision regarding domestic use, including placing on the market, of a living modified organism that may be subject to transboundary movement for direct use as food or feed, or for processing shall, within fifteen days of making that decision, inform the Parties through the Biosafety Clearing-House" (Article 11(1)). Annex II to the Protocol sets out the minimum information that must be provided to the BCH. Parties may then take decisions on the import of LMOs-FFP under their domestic regulatory framework that is consistent with the objective of the Protocol (Article 11(4)).

Developing country Parties and Parties with economies in transition also have the option to use the decision-making process provided for in paragraph 6 of Article 11 if they do not have a domestic regulatory framework. The process requires a Party to declare (through the BCH), prior to the first import of an LMO-FFP, that its decision will be made according to a risk assessment undertaken in accordance with Annex III of the Protocol and made within a predictable timeframe, not exceeding 270 days. To date, only two Parties – Barbados and Saint Lucia – have declared through the BCH that they will follow the process in paragraph 6 of Article 11. Paragraph 8 of Article 11 allows Parties to take precautionary decisions when faced with scientific uncertainty because of insufficient information.

The distinction between LMOs for intentional introduction into the environment and LMOs-FFP emerged only late in the Protocol's negotiation. At its core, the distinction is economic. Developers of LMOs, the shipping industry, and exporting countries did not want to hamper bulk trade in commodities (e.g., corn, canola, and soy) by onerous and lengthy decision-making processes that required a risk assessment for every import of every LMO. The procedure for LMOs-FFP is intended to

[54] See, for example, sections 2.1.1.5 and 2.3 in ISPM No. 11, "Pest Risk Analysis for Quarantine Pests Including Analysis of Environmental Risks and Living Modified Organisms" (2004).

[55] *Aquatic Animal Health Code* (2011) online: http://www.oie.int/international-standard-setting/terrestrial-code/access-online/ and *Terrestrial Animal Health Code* (2011) online: http://www.oie.int/international-standard-setting/terrestrial-code/access-online.

be "relatively speedy while allowing an importing country to exercise some degree of sovereignty and control over the regulation of imports of LMO-FFP commodities."[56] Creating a legal distinction between LMOs-FFP and LMOs for intentional introduction into the environment was not without controversy, however, because some countries argued that ensuring that LMOs-FFP would not be introduced into the environment was impossible.

One of the most difficult aspects of the Protocol both during its negotiation and following its entry into force was its provision on the handling, transport, packaging, and identification of LMOs – Article 18. Paragraph 1 of Article 18 requires each Party to "take necessary measures to require that living modified organisms that are subject to intentional transboundary movement within the scope of this Protocol are handled, packaged, and transported under conditions of safety, taking into consideration relevant international rules and standards." Paragraph 2 of the article goes on to set more specific documentation requirements for three categories of LMOs: LMOs-FFP,[57] LMOs for contained use,[58] and LMOs for intentional introduction into the environment.[59] Finally, paragraph 3 requires the COP-MOP to "consider the need for and modalities of developing standards with regard to identification, handling, packaging and transport practices, in consultation with other relevant international bodies." Several economic issues are bound up in this article; we shall focus on those associated with LMOs-FFP.

Paragraph 2(a) of Article 18 states that Parties are to take measures to require that documentation accompanying LMOs-FFP clearly identifies that they "may contain" LMOs and are not intended for intentional introduction into the environment, as well as a point of contact for further information. The paragraph also specifies that the COP-MOP is to take a decision on the detailed requirements for this purpose, including specification of the identity of the LMOs and any unique identification, no later than two years after the date of entry into force of the Protocol. This was the last issue to be resolved during the negotiation of the Protocol and agreement was possible only by punting the details of the issue to COP-MOP for them to resolve.

Although the details of the debate around documentation for LMOs-FFP are quite complex, at its most basic the issue encompasses two economic concerns: first, the current structure of the international trade in bulk commodities, and second, the desire to label genetically modified foods. To understand this, imagine the modern international trade in bulk commodities as an hourglass shape. At the very top are the farmers who grow crops on their individual farms. Their harvests are gradually aggregated through on-farm storage, grain elevators, and transportation terminals, where they are loaded onto ships, trains, or trucks, leaving the country

56 Francois Pythoud & Urs P Thomas, "The Cartagena Protocol on Biosafety" in LePrestre, *supra* note 2 at 48.

57 Cartagena Protocol, *supra* note 1 at art 18(2)(a).

58 *Ibid* at art 18(2)(b).

59 *Ibid* at art 18(2)(c).

of origin to travel to the importer country, which is the narrowest point in the hourglass. Once the goods arrive in the importing country, they are distributed to processing companies that produce and package new goods, which end up in stores for consumers – the very bottom of the hourglass – to purchase. Throughout this chain, the different consignments of the same commodity generally are considered interchangeable. The effect is that if one farmer grows genetically modified corn, for example, but his neighbour does not, the two harvests are treated the same and mixed together after they leave the farm. The system is not set up to segregate genetically modified crops from non-genetically modified crops. Part of the opposition to strict documentation requirements for LMOs-FFP was due to the added costs that would have been required to create separate streams of products.

On the other side were countries that wanted to be able to label genetically modified foods. The desire to label such foods is partly economic – facilitating consumer choice – but also goes beyond this , because consumers wanting to avoid genetically modified foods often do so for social or environmental reasons. Although foods per se are outside the scope of the Protocol, labeling them requires knowledge of the ingredients and therefore documentation to this effect must accompany the shipment of the commodity.

As described, paragraph 2(a) of Article 18 left it to COP-MOP to sort out the details of the documentation requirements for shipments of LMOs-FFP no later than two years after the date of entry into force of the Protocol, which effectively meant by COP-MOP 2. The Parties were, however, unable to reach a compromise at that meeting. The debate continued at COP-MOP 3, at which a deal was finally struck – in the form of decision BS-III/10 – after nearly 48 hours of continuous negotiations.

The text of decision BS-III/10 does include more details on the documentation to accompany LMOs-FFP. For example, documentation should clearly state "[i]n cases where the identity of the living modified organisms is known through means such as identity preservation systems, that the shipment contains living modified organisms that are intended for direct use as food or feed, or for processing."[60] When the identity is not known, however, the documentation should clearly state "that the shipment may contain one or more living modified organisms that are intended for direct use as food or feed, or for processing."[61] Paragraph 7 of the decision states that Parties are to consider a decision at COP-MOP 6 "to ensure that documentation accompanying living modified organisms intended for direct use as food or feed, or for processing covered by paragraph 4 clearly states that the shipment contains living modified organisms that are intended for direct use as food or feed, or for processing." At COP-MOP 5, however, the Parties noted the limited experience in the implementation of decision BS-III/10 and decided to postpone the decision

[60] Cartagena Protocol, COP-MOP 3, Dec BS-III/10 at para 4(a).

[61] *Ibid* at para 4(b).

referred to in paragraph 7 to COP-MOP 7.[62] In essence, the issue has been punted forward again.

At their most basic, questions of liability and redress concern who is responsible for damage and what they must do to remedy the problem. In the case of the Cartagena Protocol, there was insufficient time during the negotiation of the Protocol itself to elaborate the rules on liability and redress. Instead, the Protocol includes an enabling clause in Article 27 that required the Parties at their first meeting to "adopt a process with respect to the appropriate elaboration of international rules and procedures in the field of liability and redress for damage resulting from transboundary movements of living modified organisms, and . . . [to] endeavour to complete this process within four years." Six years of negotiations drew to a close at the COP-MOP 5 in Nagoya with the adoption of the Nagoya – Kuala Lumpur Supplementary Protocol on Liability and Redress. The Supplementary Protocol provides for a so-called administrative approach and its key provision is Article 5 on response measures. It largely addresses actions to be taken by operators and competent authorities to respond to damage caused by LMOs. The Supplementary Protocol also includes one provision on civil liability, namely Article 12.[63] The Supplementary Protocol was open for signature from 7 March 2011 to 6 March 2012 and will enter into force ninety days after the deposit of the fortieth instrument of ratification, acceptance, approval, or accession.

It might also be noted that six of the largest biotechnology developers – Monsanto, Syngenta, BASF, Bayer CropScience, Dow AgroSciences, and DuPont/Pioneer – have banded together to back their products financially. They have developed a private contractual compensation mechanism concerning recourse in the event of damage to biological diversity caused by LMOs – known as *the Compact* – that would provide redress (primarily remediation rather than monetary compensation) to states in cases in which the release of one of the companies' products causes damage to biodiversity.[64]

CONCLUSION

Much progress has been made in the implementation of the Cartagena Protocol, but much work remains to be done. As with all international agreements, issues and points of contention evolve. Research in the field of biotechnology is expanding. Most commercial applications of the technology are still for agricultural crops, but increasing work is being done on organisms such as insects and fish. Furthermore,

[62] Cartagena Protocol, COP-MOP 5, Dec BS-V/8, "Handling, transport, packaging and identification of living modified organisms: paragraph 2(a) of Article 18" (2010) at para 6.

[63] Cartagena Protocol, COP-MOP 5, Dec BS-V/11, "International rules and procedures in the field of liability and redress for damage resulting from transboundary movements of living modified organisms" (2010).

[64] For more details, see online: Crop Life International, http://www.croplife.org/the_compact.

countries' priorities and needs change as they gain experience with biotechnology and their associated regulatory regimes. The key is for the Protocol to remain a flexible instrument that can address new challenges as they arise. With its provisions that integrate social, economic, and environmental considerations, the Protocol is well suited to the task.

2

Implementing Sustainable Development through National Biosafety Frameworks

Christine Frison, Sylvestre-José-Tidiane Manga, and Marie-Claire Cordonier Segger

The international community is facing significant economic, social, and environmental challenges. In 1992, Agenda 21 of the Rio Conference on Environment and Development[1] emphasized the need for a coherent understanding of biodiversity, biotechnology, and sustainable development. Almost twenty years later, scholars still identify an urgent need for support to countries so that they can be helped in materializing the "ambitious prognosis in Agenda 21 of the benefits of sustainable development of biotechnology."[2] Addressing this need through the Cartagena Protocol on Biosafety (Cartagena Protocol) requires structured implementation plans, monitoring, and enforcement.

Implementing the Cartagena Protocol, taking into account the various and often conflicting environmental, social, and economic interests at stake, is not an easy task, especially for developing countries. To help Parties to pursue implementation with a sustainable development approach, the Global Environment Facility (GEF)[3] adopted an Initial Strategy on Biosafety in November

[1] Agenda 21, Report of the United Nations Conference on Environment and Development, Rio de Janeiro, 3–14 June 1992, UN Doc. A/CONF.151/26, Vols. I-III, 12 Aug 1992.

[2] Catherine Redgwell, "Biotechnology, Biodiversity and Sustainable Development," in Francesco Francioni and Tullio Scovazzi, eds, *Biotechnology and International Law* (Hart Publishing, 2006) at 78.

[3] The Global Environment Facility (GEF) unites 182 member governments – in partnership with international institutions, nongovernmental organizations, and the private sector – to address global environmental issues. As an independent financial organization, the GEF provides grants to developing countries and countries with economies in transition for projects related to biodiversity, climate change, international waters, land degradation, the ozone layer, and persistent organic pollutants. These projects benefit the global environment, linking local, national, and global environmental challenges and promoting sustainable livelihoods.

This chapter builds on the experiences of the CISDL in the development of national biosafety frameworks through the UNEP-GEF Biosafety Initiative, expanding on Martin Endicott, Christine Frison, and Kathryn Garforth with Marie-Claire Cordonier Segger, Jorge Cabrera, and Sylvestre-José-Tidiane, *Innovations in Biosafety Law: A CISDL Working Paper* (Montreal: Centre for International Sustainable Development Law, 2005). Thanks are due to Chris Briggs and the other members of the UNEP-GEF Biosafety Initiative, as well as many national stakeholders and decision makers whose sincere and courageous efforts inspired this chapter.

2000.[4] This strategy aims to assist countries preparing for the implementation of the Cartagena Protocol through the establishment of National Biosafety Frameworks (NBFs). The objectives of the GEF Initial Strategy on Biosafety Are to assist in the development of NBFs through capacity-building initiatives; to promote information sharing and collaboration, especially at the regional and subregional level; and to promote collaboration with other organizations to assist capacity building for the Protocol. Various activities have been proposed to implement these strategies, including assisting developing Parties in NBF development and implementation, supporting countries to participate in the Biosafety Clearing-House (BCH), coordinating with other organizations to provide biosafety-related assistance, and enhancing scientific and technical advice to GEF on biosafety issues.[5]

The United Nations Environment Programme (UNEP) joined the GEF to run three activities and to implement the GEF Strategy under the mandate of the UNEP-GEF Biosafety Unit. The UNEP-GEF Biosafety Unit Projects aim to implement the strategy by assisting countries in establishing their NBFs; promoting information sharing and collaboration, especially at the regional and subregional level; and promoting collaboration with other organizations to assist capacity building for the Protocol. These activities include assisting 123 countries to develop their own NBFs, to implement those NBFs in some cases, and to allow 139 countries to participate in and benefit from the BCH. As of July 2009, 111 countries had completed their draft NBFs, 8 countries had completed the implementation of their NBFs, 11 countries began the implementation phase, and 122 countries are currently setting up their participation in the BCH.[6]

The Centre for International Sustainable Development Law (CISDL) partnered with the UNEP-GEF on the Development of National Biosafety Frameworks project. This global project was designed to help countries comply with the Cartagena Protocol and aims to enable countries to make technical, political, and administrative decisions on the safe transfer, handling, and use of living modified organisms (LMOs), and to meet the requirements of the Cartagena Protocol.

Indeed, legal and regulatory structures are required to implement the Cartagena Protocol. NBFs seek to build and implement an effective normative structure at a national level for research and trade in products containing LMOs. Although NBFs vary from country to country, they contain several common components. These include a government policy on biosafety, which is usually part of a broader policy such as policy on biotechnology in general, agricultural production, health care, or environmental protection; a regulatory regime for biosafety, which often is a combination of an act or decree, complemented by implementing technical regulations and guidelines; and a system to handle notifications or requests

[4] GEF Council, *Initial Strategy for Assisting Countries to Prepare for the Entry into Force of the Cartagena Protocol on Biosafety*, November 1–3, 2000, UN Doc. GEF/C.16/4/Rev.1, online: http://www.unep.org/biosafety/Documents/GEF_strategy.pdf.

[5] *Ibid* at paras 20–21.

[6] See UNEP, "About Biosafety Unit," online: http://www.unep.org/biosafety/About.aspx.

for authorizations for certain activities, such as releases of GMOs into the environment. Such systems typically include administrative functions, risk assessment, decision making, and public participation; and systems for follow-up, such as enforcement and monitoring for environmental effects. *Monitoring* is a term used for evaluating actual impacts on the environment and human health, whereas *enforcement* typically focuses on compliance with the regulatory regime and approaches for public information and public participation, that is, informing and involving stakeholders in the development and implementation of the national biosafety framework.[7]

CISDL participated closely in the preparatory phase of the project by contributing to the design of the technical material proposed to contracting parties to help in implementing the Protocol. This technical material, designed as a toolkit for countries, contained three phases, titled the Biosafety Project:

Phase 0 – Starting the project
Phase 1 – Taking stock
Phase 2 – Consultation and analysis
Phase 3 – Drafting the NBF
  Part I: Formulation of the regulatory regime
  Part II: Designing the administrative systems for handling applications and notification

The aim of the toolkit is to provide a practical "how-to" guide for countries to assist them in developing and implementing NBFs, in particular new regulatory regimes. CISDL assisted more than thirty developing countries participating in the UNEP-GEF project, including in Francophone Africa, Latin America, Asia, and the Middle East, by providing an independent, technical, and legal review of the draft NBFs, and in many cases, draft legislation. The first part of this chapter discusses best practices and lessons learned in the design and implementation of NBFs. The second part comments on principal challenges faced by many of the Parties, taking into consideration the basic requirements of the provisions of the Cartagena Protocol. This commentary includes a section examining aspects of the biosafety partnership and a section on the commercial, sanitary, and phytosanitary aspects of emerging national frameworks on biosafety from a global perspective.

THE UNEP-GEF BIOSAFETY PROJECT

The UNEP-GEF Biosafety Project has supported developing countries in designing NBFs since 2001. This support to developing countries was a priority for the Parties

7 UNEP-GEF Biosafety Unit, *Biosafety Project: Phase 3 – Drafting the NBF – Part I: Formulation of the regulatory regime* at 56–7, online: http://www.unep.org/biosafety/Toolkit.aspx. See Annex I to this book for a figure on the components of a National Biosafety Framework (NBF) [Biosafety Toolkit – Phase 3 – Part I].

in the early days of the Protocol, because many developed and developing countries did not have fully developed and implemented policies or laws on biosafety. Furthermore, through this process, the international community aimed to ensure developing countries of its support and assistance in addressing the challenge of NBF development. The lack of human, financial, and structural resources, among others, in many developing countries is an obvious obstacle to the development and implementation of any effective national biosafety policy. Indeed, the development of a national biosafety regime, as well as national legislation on modern biotechnology and biosafety, in accordance with the general requirements of the Cartagena Protocol and biosafety principles, is a difficult and time-consuming process that requires substantial and concrete support to developing countries. The NBFs developed by Parties to the Cartagena Protocol through the UNEP-GEF Biosafety Project, particularly in Francophone Africa, offer important lessons that can raise awareness among others seeking to implement the Protocol. A careful study of these experiences can serve as a source of ideas to guide the development and implementation of national policies on biosafety.

According to the UNEP-GEF Biosafety Unit, a NBF is "a combination of policy, legal, administrative and technical instruments that are set in place to address safety for the environment and human health in relation to modern biotechnology."[8] Given this definition, the NBF must fulfil these imperatives and build a real biosafety system that includes all existing and potential stakeholders in the process.

In the light of experiences from several developed countries that have biosafety legislation and guidelines, the UNEP-GEF Biosafety Unit was able to derive key components to structure a model NBF, despite the variability of different NBFs. The first component is a national policy on biosafety. This is generally part of broader policies on biotechnology, agriculture, health, or environmental protection. The second component is a regulatory regime on biosafety. Usually, this consists of a legislative text supplemented with practical guides on implementation. The third component is a system for handling notifications and applications for authorization of activities, such as the release of LMOs into the environment. These systems require risk assessment, and administrative and decision-making mechanisms. The fourth component is environmental monitoring and inspections. It generally consists of the implementation of a mechanism to monitor the impact of LMOs on the environment and human health. The last component is a requirement for access to information and public participation.[9]

Although these elements are generally present in the NBFs that were developed through the UNEP-GEF Biosafety Project, significant gaps exist in several of the NBFs, and countries continue to improve and strengthen those frameworks from

[8] UNEP-GEF Biosafety Unit, *Guide for the implementation of national biosafety frameworks* (Geneva: UNEP-GEF, 2003) at 2, online: http://www.unep.org/biosafety/Documents/Impl.Guide-RegReg.pdf.

[9] *Ibid.*

the very beginning of the process. As countries design their NBFs and draft their legislation on biosafety, a continuing engagement with both national stakeholders and international partners is of great value. This continued engagement can secure support for developing, implementing, and reviewing regulatory regimes for biosafety. The regulatory regimes in particular can benefit from efforts to secure clarity, transparency, consistency, workability, enforceability, and adaptability.[10] To secure these conditions of good governance, countries can provide a clear and precise general structure for NBFs and draft legislation, implement a systematic approach in the design of the NBF, and complement their NBF with a bibliography, tables, and other graphic information.

PROVIDING A CLEAR AND PRECISE STRUCTURE FOR NBFs AND DRAFT LEGISLATION

Most of the ideas and concepts of the Cartagena Protocol are present in national biosafety policies but are not always sufficiently brought forth in a clear and well-structured manner. Thus, restructuring the presentation and logical arrangement of these concepts can be beneficial. For example, including all draft legislation within a NBF is important. This legislation is part of the NBF but is a separate, independent document. Instead of having national legislation on biosafety incorporated into their NBF, several African countries developed a strong policy that mentioned the objective as well as the main sections of their national law. In proceeding this way, those countries met the requirements to the second component of a NBF, namely the regulatory regime, and were able to develop and adopt suitable and efficient legislation on a smoother schedule. Such a procedure allowed Parties to avoid adopting a NBF while their laws were still under development, given the reality that it is often easier to adopt a policy document to guide the work of government and other officials prior to developing, promulgating, and enforcing binding new legislation.

Implementing a Systematic Approach in Designing NBFs

To meet all the obligations under the Cartagena Protocol, NBFs should contain a description of all tasks, activities, skills, and status of the different actors involved in the national biosafety policy. Furthermore, it is helpful to be as precise and rigorous as possible, to avoid accidentally omitting mechanisms to comply with particular obligations under the Protocol . Hence, after the creation of a subsection within the NBF or of an article in the draft legislative text, a systematic review of each corresponding article in the Protocol can help to ensure fulfillment of each obligation.

10 *Ibid* at 6.

To respond to these proposals effectively, the following systematic approach can be useful. National consultants designing NBFs can examine each and every obligation under the Protocol; make a list of the requirements necessary for compliance with the Protocol and its annexes (this can be done using the UNEP-GEF Toolkit); divide the stated obligations into subgroups and then organize them in a legally sound manner; and only then create an outline of the NBF and accompanying draft legislation to form a clear and detailed structure without omitting any obligation.

Complementing the NBF with a Bibliography, Tables, and Diagrams

Tables and explanatory diagrams are valuable aids for understanding and illustrating a NBF project, and they are welcome as long as they are used wisely. For example, a clearly annotated table, briefly explaining its meaning, can be very useful in quickly understanding a procedure or the structure of an institution. A complete bibliography is also crucial in enabling stakeholders to access he information referred to in the NBF.

The UNEP-GEF Biosafety Unit Project on NBF development offered guidance oriented not only toward compliance with the minimum requirements of the Cartagena Protocol, but also with an eye to continual and adaptive development. To accomplish this, the Parties need to continue being as active during implementation of the Protocol as they were during negotiation. Furthermore, it is expected that the NBFs and corresponding legislations will create provisions allowing for a continual interaction between the Parties, including major stakeholders such as industry and NGOs, in particular through BCH activities and regular meetings of the Parties to the Cartagena Protocol.

SUSTAINABLE DEVELOPMENT ASPECTS OF NBFs

A Partnership for the Safe Use of Modern Biotechnology

The establishment of a national biosafety framework is not simple. Modern biotechnology is a technological innovation whose adoption by southern countries calls for the establishment of an institutional framework and various mechanisms that contribute to the safe use of LMOs in pursuit of their national development. Such a challenge can seem overwhelming and insurmountable for developing countries that can suffer from a chronic lack of skilled labour, investment, and adequate equipment and materials.

Fortunately, the development of national biosafety regimes is a partnership process in which developing countries meet this challenge at the national level with specialized UN agencies and their experts' services. The reality is that developing countries are not left on their own in this process of designing, developing, and implementing a national biosafety framework. Developing countries are sure to benefit in this process from the full participation and continued and constructive support of the UNEP, and of the Secretariat of the Convention on Biological Diversity (SCBD) and Cartagena Protocol. UNEP-GEF financial assistance and expertise and the activities of the BCH are good examples of such support.

Developing countries can thus be assured of participating in the safe use of LMOs on the basis of effective future national biosafety schemes, designed according to international requirements and national expectations, realities, and conditions. Moreover, developing countries are expected to put in place a national structure whose provisions will take full advantage of intended participation in the implementation activities of the Cartagena Protocol. Analyzing the implications of the Protocol for the development of NBFs comes back to exploring the ways and means that characterize or ought to characterize the contribution of the Secretariat of the CBD and the Biosafety Clearing House to the implementation (notably alongside developing countries) in developing strategies and administrative structures for the safe use of modern biotechnology. Such a contribution can be addressed from two angles. The first are the *obligations* of developing countries with regard to the Protocol, and the second are the *expectations* of developing countries with regard to the Cartagena Protocol and the BCH. Both angles are convergent and complementary.

Obligations of Developing Countries Relating to the Cartagena Protocol

The obligation of the Parties with respect to the Protocol were identified and expanded at the Open-Ended Meeting of Experts on Capacity-Building for the Implementation of the Cartagena Protocol, which was held in Havana, Cuba, in July 2001. At this meeting, an action plan for capacity building for effective implementation was discussed and included for consideration by the Second Meeting of the Intergovernmental Committee for the Cartagena Protocol on Biosafety (ICCP 2), held in Nairobi, Kenya, from the 1–5 October 2001. Following this meeting, a concrete list was drafted that contained all obligations for which Parties to the Protocol are bound. This list was further elaborated by an implementation toolkit, adopted at COP-MOP 1.[11] This implementation toolkit provides a compilation, as a checklist, of obligations found in the Cartagena Protocol. These obligations are organized into three categories, which are administrative tasks (initial and future), legal requirements and/or undertakings, and procedural requirements such as advanced informed agreement (AIA).[12]

[11] Cartagena Protocol, COP-MOP 1, Dec BS-I/5 "Capacity-building" at Annex III, "Implementation Tool Kit." Reproduced as Annex II to this book.

[12] Second Intergovernmental Committee for the Cartagena Protocol on Biosafety (ICCP 2), "Capacity-building (Article 22, Article 28)," UN Doc UNEP/CBD/ICCP/2/10, Annex II at 33–37.

Parties have obligations in regard to the Cartagena Protocol, but they also have expectations. The latter differ depending on the country. Developed countries are currently in the consolidation phase of their biosafety laws, which in most cases were established well before the era of the Protocol. Conversely, developing countries need assistance to reach a level that allows them to participate in the safe trade of biotechnology products.

Developing Countries' Expectations of the Cartagena Protocol and the Biosafety Clearing-House

The obligations discussed in the previous section can be considered accomplished by most developed countries; however, these can be a challenge for a developing country to meet. Indeed, in most developing countries, everything is still to be designed, developed, and implemented for the safe use of LMOs in pursuit of national development. A roadmap or dossier on Parties' expectations regarding the Protocol and the BCH could therefore be particularly pertinent for developing countries. Parties can be encouraged, in the spirit of encouraging compliance under Article 34 of the Protocol, to pursue capacity-building efforts already adopted by the Protocol and COP-MOP that support developing countries in this new global challenge. In such circumstances, developing countries can find partnership to help navigate the thorny issues of notification, authorization applications, monitoring of capacity-building efforts, and mechanisms for public participation.

In fact, Articles 33 and 34 of the Protocol offer an interactive forum for dialogue between the Protocol and the Parties, especially developing countries.[13] The raison d'être of the BCH is made obvious in this situation. Through this dynamic and interactive forum, developing countries can improve and develop their NBFs through dialogue, experience, and awareness of innovations in modern biotechnology, according to their national priorities, their particularities, and their obligations as Parties to the Protocol.[14]

How Countries Developed Their NBFs to Meet SD Policy Goals

The following comments relate to the six items that are common to the structure of many NBFs. These six points are supplemented with some examples taken from the NBFs that were developed by Parties during the course of the UNEP-GEF Biosafety Project.

[13] The reports will be an effective means of communication through which developing countries can draw attention to the difficulties encountered in the implementation of the Protocol and the desire of these countries to find solutions to their problems. Article 33 states: "Each Party shall monitor the implementation of its obligations under this Protocol, and shall, at intervals to be determined by the Conference of the Parties serving as the meeting of the Parties to this Protocol, report to the Conference of the Parties serving as the meeting of the Parties to this Protocol on measures that it has taken to implement the Protocol."

[14] See Chapters 8 and 9 for further discussion of the Biosafety Clearing-House.

A Comprehensive National Policy on Biosafety

The fundamental starting point for the creation and implementation of any NBF is a comprehensive national policy on biosafety. The objective is multifaceted: to offer stakeholders the possibility to choose whether to import and use LMOs; ensure safety; provide input, support capacity building, ensure stability in the use of LMOs; and realize the provisions of the Cartagena Protocol. All NBFs reviewed provided such a national policy to implement the Protocol. Sometimes it was based on existing legislation, but sometimes it was developed *ab initio*. Many countries expanded the scope of application of the CPB from LMOs to genetically modified organisms (GMOs). In many countries, where there is no policy or law on either LMOs or GMOs, it can be simply a question of efficiency and common sense to design the policy and regulatory framework to cover the larger set of issues, rather than drafting a second policy and law within a short period. Countries explain this expansion with reference to sustainable development principles, and/or for the national (social and economic) interest. For example, Togo[15] and Algeria[16] explicitly stressed that enlarging the scope of their NBF to include all GMOs implements sustainable development perspectives and promotes the national interest. By doing so, many countries have also provided ways to implement the precautionary principle in a strict manner.

A System or Framework to Ensure the Safe Use of Modern Biotechnology

This framework constitutes the legislative implementation of a national biosafety policy. For the most part, developing countries have provided a detailed inventory of their existing regulatory framework. The Protocol requires Parties to create a legislative framework based on the objectives of the Protocol, and to regulate any situation related to the provisions of the Protocol, a step that has been described in the toolkit.[17] Comprehensive legislation, in addition to a detailed inventory of

[15] See National Biosafety Framework of Togo at vii, s 2.2, which states: "Le cadre juridique de gestion de la biosécurité que le Togo s'engage à élaborer, vise à *préserver la santé humaine, l'environnement et les tissus socio-économiques* des risques liés à l'utilisation de la biotechnologie moderne. Il s'articule autour des composantes essentielles ci-après: (i) *le champ d'application* qui ne se limite pas au concept réducteur d'organisme vivant modifié, mais *couvre tout organisme génétiquement modifié et produit dérivé*, et compris ceux utilisés dans la production pharmaceutique destinés à l'utilisation humaine." (Emphasis added.)

[16] For instance, see Algeria, *Projet de loi relative à la circulation des ressources biologiques, au contrôle des organismes génétiquement modifiés et a la prise en charge des risques liés a l'usage des biotechnologies modernes*, which states that "*Dans la perspective d'un développement durable et bénéfique pour l'intérêt national*, la présente loi a pour objet de déterminer: – les conditions de collecte, de circulation et d'utilisation des ressources biologiques et des connaissances qui leur sont associées – les conditions de détention et/ou d'usage *d'organisme génétiquement modifié*, ainsi que les règles de sécurité liées à la prise en charge des risques découlant de l'utilisation des biotechnologies modernes, *aux fins de protection de l'ensemble des écosystèmes en général, et des ressources biologiques en particulier*" (emphasis added).

[17] See Biosafety Toolkit – Phase 3 – Part I, *supra* note 7.

existing national regulatory frameworks, can be included in the NBF. These make up two distinct but complementary parts of a NBF.

Centralising Document Management Relating to the Use of LMOs

The importance of an institutional framework cannot be overstated. Indeed, the effectiveness of institutions and of management procedures is fundamental in the implementation phase of NBFs. An effective administrative framework leads to a more efficient exchange of information and more effective handling of procedures at domestic borders and in international cooperation. All countries proposed such an administrative framework to implement and manage the use of GMOs. Many different options have been adopted in the context of the UNEP-GEF Biosafety Project.

Decision-Making Process for Risk Evaluation and Risk Management

An effective decision-making process helps to prevent most accidents related to the use and trade of LMOs. To be as efficient as possible, this process consists of risk evaluation and management mechanisms that must necessarily conform to the requirements of the Protocol, especially the provisions of Annexes I and III. The precautionary principle was often discussed and analyzed during the national workshops for the validation of the NBFs throughout the UNEP-GEF Biosafety Project. Often, these discussions stemmed from a general discomfort with the use of LMOs because of a lack of knowledge and appropriate, reliable, and independent scientific information as to risks. This has been translated in the NBFs in very different manners, in particular through risk assessment and risk management procedures, as well as liability and redress mechanisms. A good example is given by the Togo NBF, which has decided to base its liability and redress procedure on the polluter-pays concept.[18] In the Togo NBF, the three sustainable development components (i.e., environmental, economic, and social components) are directly discussed when specifying the chosen redress procedures.[19]

18 See National Biosafety Framework of Togo at 64, s 5.2.2.3: "Les régimes de responsabilité et de réparation en matière d'environnement sont généralement fondés sur le principe du pollueur payeur. Le principe du pollueur payeur vise à imputer au pollueur les coûts écologiques, économiques et sociaux de la pollution qu'il engendre. Ce qui traduit un mécanisme de responsabilité pour dommage écologique couvrant tous les effets d'une pollution non seulement sur les biens et les personnes, mais aussi sur la nature elle-même. En terme économique, il s'agit de l'internalisation des coûts externes ou de la théorie des externalités qui permet d'imposer des taxes ou redevances de dépollution au pollueur sans faire supporter la dépollution par l'ensemble des citoyens. Le principe du pollueur payeur s'applique aux personnes physiques et morales tant publiques que privées. Plusieurs dispositions de la loi n° 88–14 du 3 novembre 1988 instituant code de l'environnement sont fondées sur ce principe."

19 See National Biosafety Framework of Togo at 70, s. 6.7: "Le responsable de tout dommage sera tenu de réparer les préjudices directs ou indirects induits par le ou les produits en cause, sur la biodiversité, les tissus socio-économiques et les normes culturelles."

Public Participation and Awareness Mechanisms

Public participation and awareness are essential pillars of decision making and constitute a key feature in the implementation of sustainable development principles. Five reasons influence countries to develop such effective mechanisms: (1) promoting sustainability by encouraging a sense of ownership within the public as a stakeholder in decision making; (2) improving understanding of the subject and establishing relationships of trust and respect between groups of stakeholders with different points of view; (3) ensuring an inclusive process that involves all stakeholders; (4) promoting better decision making based on accurate information and the use of all national resources and expertise available; and (5) promoting transparency and accountability of governments and decision-making processes relating to LMOs. In the process of developing NBFs during the UNEP-GEF Project, particularly those in Francophone Africa, many developing countries explained that public participation and awareness mechanisms should depend on the social, cultural, and economic environment,[20] which differs significantly even within one country. For this reason, the various NBFs have established very different types of procedures that have been identified in the UNEP-GEF Phase 2 Toolkit as empowering, interactive, consultative, functional, and passive.[21]

Monitoring Institutions and Systems

It is important for developing countries to create monitoring institutions and systems to adapt quickly and effectively to developments in modern biotechnology. Indeed, this field is in constant flux and requires a constant adaptability of stakeholders. The BCH appears to be a key partner in this regard. Conversely, however, institutional systems for monitoring aim to set up efficient traceability mechanisms to make the right decisions at the right time when unexpected events occur from released LMOs and GMOs. This component thus encompasses some of the most important biosafety policies for sustainable development. The NBFs developed by countries such as Gabon, the Democratic Republic of the Congo, Chad, and Djibouti, among others, opted to involve all competent national institutions under their monitoring and control mechanism. For example, in the case of Djibouti's NBF, several institutions take part in the monitoring and control system for the release of GMOs and products thereof: the Customs Service, the Standards and the Quality Control Service of the Ministry of Trade and Industry, the Phytosanitary Service of the Ministry of Agriculture, the Public Health Service of the Ministry of Health, and the Study and Research Centre of Djibouti (CERD). Such a mechanism is run by a national committee known as Biotechnology Watch. The Competent National Authority will have

[20] See National Biosafety Framework of Sénégal at 12, s 3.8: "l'approche adoptée pour informer, sensibiliser, éduquer le public sur les questions de biotechnologies et de biosécurité, en vue de sa participation dans le processus décisionnel diffère en fonction du contexte socioculturel."

[21] UNEP-GEF Biosafety Unit, *Biosafety Project: Phase 2 – Consultation and analysis* at 45, online: http://www.unep.org/biosafety/Toolkit.aspx.

to rely on these institutions to carry out the monitoring and control aspects of the NBF, especially the information and monitoring of projects in progress authorized by the regulatory procedure that follows the application of the Advance Informed Agreement (AIA) procedure. When unexpected difficulties occur from released LMOs or GMOs, the Monitoring and Control Policy has set up safeguard measures consistent with the National Biosafety Law, elaborating actions to be taken according to different scenarios and circumstances. Those decisions can be to make modifications to the initial terms of the authorization; to simply withdraw the GMO from trade; to suspend the activity while waiting for additional information favourable to its continuation; to impose on the author of the notification new terms of release for the GMO; to order the destruction of the GMO in the event of an acknowledged breach and obvious risk; or to sanction by appropriate measures if necessary.[22]

Commercial, Sanitary, and Phytosanitary Aspects of National Biosafety Frameworks

The Cartagena Protocol is a dynamic regulatory framework. The current text provides a foundation that allows Parties to begin the implementation and continual updating of national biosafety systems as innovations in modern biotechnology continue to evolve. The Protocol and its COP-MOP, through the activities of the BCH, monitor these innovations and new data as well as their repercussions in national regimes. The main tools used by the Protocol to contribute to the implementation of national biosafety regimes are risk assessment and management, labelling, traceability, liability, redress, and trade dispute resolution. Regarding this last point, international trade jurisprudence on agricultural products containing LMOs points to the possibility for mutual support between the recent dispute settlement mechanism of the CBD and the powerful dispute settlement body (DSB) of the WTO. As noted in the Preamble of the Cartagena Protocol, the Parties agree that trade and environment agreements should be mutually supportive with a view to achieving sustainable development, that the Protocol does not change the rights and obligations of a Party under any existing international agreements, and that no hierarchy exists between the Protocol and other international agreements. These issues are further discussed in later chapters of this book.

Regarding the evaluation and management of biotechnology risks, the intensity of activities concerning biosafety within international organizations, States, NGOs, and industry suggests convergence toward harmonization of methods and techniques for biotechnology risk prevention. Through the goodwill and good faith of the various stakeholders, the BCH is given the means to standardize biosafety risk assessment and gene detection methods, as well as traceability techniques. Traceability techniques

[22] UNEP-GEF Biosafety Unit, Djibouti National Biosafety Framework, English Translation at 39–40 (Geneva, Switzerland: UNEP-GEF, 2006), online: http://www.unep.org/biosafety/files/Djibouti NBF Eng.pdf.

should allow for the removal of products from commercial and food channels. Harmonization standards regarding toxicity and innocuousness levels or concerning the labelling of LMO presence in food (to respect the consumer's freedom to choose LMO-free food) should also allow the removal of these products from the commercial chain. Consideration is also given to provide legal recourse assistance and alternative dispute resolution mechanisms.

CONCLUSION

Through the UNEP-GEF Biosafety Project, many countries invested a great deal of effort in drafting efficient, sound, and complete NBFs and draft legislation. Although biosafety, biodiversity, and sustainable development are clearly intertwined concepts, their coherent implementation at the national level remains a challenge for developing countries. Stakeholders gained experience in activating the NBFs and securing the implementation of the CPB in their countries by focusing on specific issues (e.g., risk assessment, public participation, and advanced informed agreement), reporting on regional experiences in implementing the Protocol (e.g., East and West African countries, Near East, Central and Eastern Europe), and learning from their own domestic experiences with biosafety (e.g., Costa Rica, New Zealand, and Canada).

3

Crafting National Biosafety Regulatory Systems

Gregory Jaffe

The purposes of a national biosafety regulatory system are to scientifically assess the safety of genetically engineered (GE) organisms to humans and the environment, manage any potential risks, and authorise the development and marketing of safe GE organisms and their products. To develop such a regulatory system, governments can use existing laws or develop new laws. During their deliberations, they can access the deep pool of knowledge and experience that exists in countries with already established biosafety regulatory systems and seek advice from experts throughout the international academic community. Any national biosafety regulatory system that is proposed, however, must be functional and protective, and also comply with international trade standards that are evolving in recognition of the growing importance of GE organisms in world affairs.

This chapter sets forth some of the important building blocks needed in a national biosafety regulatory system to make it functional, protective, and compliant with international obligations. These fundamental elements, which might be directly relevant to the principle of good governance that is discussed in the introduction of this book,[1] can be effectively realised in domestic policy and law on biosafety. First, the chapter discusses many of the common attributes, components, and characteristics that are found throughout the world in existing biosafety regulatory systems that are both functional and protective. Second, a brief overview provides some of the more important international legal regimes that guide and constrain national biosafety law. Together, the analysis provides a roadmap of the important issues that governments must address in the development and adoption of a national biosafety regulatory systems to manage and assess living modified organisms (LMOs).

[1] See Introduction.

COMPONENTS AND CHARACTERISTICS OF A FUNCTIONAL AND PROTECTIVE SYSTEM

A national biosafety regulatory system is a regulatory regime responsible for assessing and managing the full range of potential risks that might be posed by a living modified organism (LMO) and its products. The system addresses potential risks to the environment and biologic diversity as well as any food/feed risks or other safety-related issues involving LMOs (e.g., worker health or drug safety). A protective biosafety regulatory system ensures that any risks from genetically engineered organisms are managed and allows safe LMOs to be developed, marketed, and used for their intended purpose. Such a system, however, must also be functional, which means that it should be understandable, workable, equitable, fair, adaptive, and enforceable.[2]

Existing biosafety regulatory systems from around the world reflect, among other things, the type of government and politics of the country, the country's view on the relative safety of GE organisms, and the country's regulation of food, agriculture, and environmental issues. Establishing those systems required balancing numerous goals and trading off different interests. Through an analysis and comparison of different existing biosafety regulatory systems, however, one can identify key characteristics and components that are generally important to a functional and protective biosafety regulatory system.[3] Incorporating each of those characteristics and components in a functional and protective biosafety regulatory system involves problem solving, because tensions between the different characteristics can exist. Those key characteristics and components are set forth here.

Comprehensive

A biosafety regulatory system should be comprehensive.[4] First, the system should cover the different stages of development for a GE organism,[5] such as releases into the environment as confined and unconfined field trials, releases of commercial products, and consumption of GE organisms by humans and/or animals. Second, a comprehensive regulatory system analyses the range of potential safety issues associated with GE organisms. It addresses not only the environmental and biodiversity issues highlighted in the Cartagena Protocol on Biosafety (Cartagena Protocol) but

[2] UNEP-GEF, *Project on Development of National Biosafety Frameworks: Phase 3 Toolkit Module Part (i): Developing the Regulatory Regime*, online: United Nations Environment Programme – Global Environment Facility http://www.unep.ch/biosafety/old_site/devdocuments.htm. [*Developing the Regulatory Regime*].

[3] Gregory Jaffe, "Regulating Transgenic Crops: A Comparative Analysis of Different Regulatory Processes" (2004), 13 Transgenic Research 5 at 19 [Regulating Transgenic Crops].

[4] *Ibid.*

[5] *Developing the Regulatory Regime*, *supra* note 2.

also food safety issues and any other potential safety questions (e.g., worker safety).[6] Finally, the regulatory system's scope includes all plants and animals that could be engineered and the different products that they might produce. Comprehensive regulatory systems cover not just engineered plants used for food or feed, but also plants engineered to nonfood substances such as pharmaceuticals, nonfood crops like trees, and engineered animals.

Adequate Legal Authority

The biosafety regulatory system needs sufficient legal authority to subject each GE organism to a food-safety and environmental risk assessment and approval process before any unconfined release into the environment or any GE organism is placed into commerce.[7] Such mandatory authority over GE organisms helps to ensure protection of the environment or human health. "Clear responsibility and legal authority is [*sic*] important not only for ensuring the protection of health and the environment, but also for providing the public and technology developers with a clear understanding of the regulatory pathway to market."[8]

To ensure adequate legal authority for a biosafety regulatory system, countries must decide whether they can establish a system using existing laws or whether they need to pass new biosafety-specific legislation. An example of a country that used existing laws for its biosafety regulations is the United States, which regulates GE organisms using a coordinated framework involving numerous laws and several agencies.[9] The European Union, South Africa, and New Zealand have established new laws to address GE organisms specifically.[10] Whether a country passes a new law or uses existing laws, the legal authority for the biosafety regulatory system still must be exercised within that country's broader legal system, including its constitution, its judicial system, and its other laws and regulations.

A Clear Safety Standard

Biosafety regulatory systems usually establish safety standards for their approval processes.[11] The safety standard sets forth which level of protection must be satisfied

6 Klaus Von Grebmer and Steven Were Omamo, "Lessons and Recommendations" in Steven Were Omamo and Klaus von Grebmer, eds. *Biotechnology, Agriculture, and Food Security in Southern Africa* (Washington, D.C.: Food, Agriculture, and Natural Resources Policy Network, 2005).

7 Jaffe, "Regulating Transgenic Crops," *supra* note 3 at 19; J. Cohen, J. Komen, and P. Zambrano, "Public GM Innovations, Safety Regulation, and ... Results?" (Presentation delivered to The Globalization of Agricultural Biotechnology: Multi-disciplinary Views from the South, University of Warwick, Coventry, UK. 11–13 March 2005).

8 Pew Initiative on Food and Biotechnology, *Issues in the regulation of genetically engineered plants and animals* (Washington, D.C.: Pew Initiative on Food and Biotechnology, 2004).

9 Office of Science and Technology Policy, "Coordinated framework for regulation of biotechnology products" (1986) 51 *Federal Register* 23302; "Regulating Transgenic Crops," *supra* note 3 at 19.

10 "Regulating Transgenic Crops," *supra* note 3.

11 *Ibid.*

to approve an application and which factors the government will consider before making an approval decision, including the baseline for any risk analysis.[12] The safety standard guides the government on how to use its legal authority to regulate GE organisms. The standard also identifies whether the benefits from the GE organism or the opportunity costs of not introducing the organism will be considered.[13] In a functional and protective system, all interested parties know and understand the safety standard beforehand, and government decisions apply that standard uniformly and fairly.

Proportional Risk-Based Reviews

Biosafety regulatory systems look at each application individually and assess any potential risks to human health and the environment through a scientific risk-based analysis. The system usually has the flexibility to treat products differently depending on the potential risks and concerns raised.[14] It prioritises applications that it reviews based on the potential risk and gives the most scrutiny to products with the most relative risk while allocating fewer resources and less time to products that raise less concern.[15] Regardless of the review procedure, all GE organisms must meet the applicable safety standards. The procedures and the data needed to meet that standard, however, should vary depending on the nature of the product and its potential risks, so that the potential risks match the regulatory procedure.

The key to establishing a proportional risk-based review process is providing the biosafety regulatory system with the flexibility to address particular products differently depending on the nature of the product (i.e., the organism and the added gene), and the use of the product (i.e., a confined laboratory experiment, a field trial, or a commercial release).[16] For example, a confined field trial does not require the same detailed risk assessment and governmental review as a commercial release of that same product; the confined trial is released under specific conditions, limited in duration, and designed to have minimal impact on the environment, whereas the commercial release might not be controlled and will remain in the environment. Thus, if a biosafety regulatory system allows for proportionate risk-based reviews, it seeks to minimise the regulatory costs for products with minimal risks.

12 *Developing the Regulatory Regime*, *supra* note 2; "Regulating Transgenic Crops," *supra* note 3 at 19; Gregory Jaffe, "Implementing the Cartagena Protocol through National Biosafety Regulatory Systems: An Analysis of Key Unresolved Issues" (2005) 5 Journal of Public Affairs 299 at 311 [Implementing the Cartagena Protocol].

13 Deborah P Delmer, "Agriculture in the Developing World: Connecting Innovations in Plant Research to Downstream Applications" (2005) 103(44) Proceedings of the National Academy of Sciences 15739 at 15745.

14 *Ibid.*

15 "Implementing the Cartagena Cartagena Protocol," *supra* note 12 at 311.

16 Julian Kinderlerer, "Regulation of Biotechnology: Needs and Burdens for Developing Countries" (UNEP-GEF Biosafety Projects: 2002), online: UNEP, http://www.unep.org/biosafety/Documents/BTregulationJK.pdf.

Transparent and Understandable

A common component of many biosafety regulatory systems is transparency.[17] Public access to information about the regulatory system and the organisms that go through it can lead to greater public confidence in regulatory decisions.[18]

Biosafety regulatory systems that are transparent and understandable usually provide information to the public about five key issues: the regulatory process, including the steps, data requirements, and time lines for the applicant (a roadmap of the process and what is expected of the applicant); how the agency will conduct its review, which criteria and standards it will use, and who will be the accountable public officials; where, when, and how the public can be involved in the regulatory process; particular applications that seek approval from the government (the application without the confidential business information); and the agency decision on a particular application, including the analysis of a particular application and the reasoning behind its decision.[19]

With that information, interested persons can understand how the regulatory system processes applications, which information the public will have access to, which opportunities for public participation exist in the process, and what is the basis for any decisions.

While making information available to the public, however, the regulatory system must balance the competing interests of the applicant, who may want to keep some information confidential for business purposes. A regulatory system with no protection for trade secrets and proprietary information might not receive any applications, because private enterprises would not be able to market a product successfully if certain information is not kept confidential. Thus, a good regulatory system must also protect from disclosure the confidential business information of applicants.

Participatory

Public participation is an important component in biosafety regulatory systems in democratic societies.[20] Public participation can include the opportunity to provide information and comments to regulators on regulations, guidance documents, and specific applications before a regulatory decision has been made.[21] Public participation also may include the opportunity to provide oral and/or written testimony at

[17] *Developing the Regulatory Regime, supra* note 2; UNEP-GEF Biosafety Unit, "Draft Guide for the implementation of national biosafety frameworks", (2003) UNEP-GEF Projects on Implementation of National Biosafety Frameworks, UNEP, http://www.unep.ch/biosafety/old_site/Implementation/Impdocs/Impl.Guide-RegReg.pdf. Pew Initiative on Food and Biotechnology, *supra* note 8.

[18] Pew Initiative on Food and Biotechnology, *supra* note 8.

[19] M McLean et al, *A Conceptual Framework for Implementing Biosafety: Linking Policy, Capacity, and Regulation* (Netherlands: ISNAR, 2002).

[20] *Developing the Regulatory Regime, supra* note 2; M. McLean et al, *supra* note 19.

[21] Pew Initiative on Food and Biotechnology, *supra* note 8.

public hearings. In most instances, the regulatory system responds to relevant comments in its decision-making documents to improve its overall decision and assure the public that any relevant concerns were seriously considered. Thus, although public participation helps to inform the decision-making process, the ultimate decisions remain with the regulatory agencies and the designated leaders.

Postapproval Oversight

A biosafety regulatory system does not stop its oversight once a GE organism has been approved for a confined field trial or for a commercial release. The system continues to ensure adequate protection of humans and the environment after the product is released into the environment or enters the marketplace. Postapproval activities can include monitoring for adverse environmental or health effects and monitoring for compliance with any risk management conditions imposed on the GE organism.[22]

Flexible and Adaptable

Because biotechnology is a rapidly changing discipline, it is impossible to anticipate the range of future applications fully. Thus, if a country is setting up a biosafety regulatory system to address currently unknown applications of genetic engineering, flexibility to adapt to new evidence on risks and benefits and new types of products can be important.[23]

There are several ways to build flexibility and adaptability into a biosafety regulatory system. First, laws, regulations, and guidance must be forward looking and be written broadly to accommodate not only the products being proposed today but also products that might be developed ten or twenty years from now. Nonflexible systems either do not have the authority to regulate new products or put in place moratoriums while they implement new laws, regulations, and procedures to address them. Second, the regulatory system should learn from its experiences regulating products and adapt accordingly.[24] As the system regulates more products, it should become familiar with the benefits and risks of particular applications, allowing some applications with low risk to obtain a streamlined review process while increasing regulatory scrutiny for products that are similar to previous high-risk applications.

22 J Cohen, J Komen, & P Zambrano, *supra* note 7.

23 Klaus Von Grebmer & Steven Were Omamo, *supra* note 5; Draft Guide for the Implementation of National Biosafety Frameworks, *supra* note 17.

24 J Cohen, J Komen & P Zambrano, *supra* note 7.

Efficient, Workable, and Fair

Although the primary responsibility of a biosafety regulatory system is to ensure that GE organisms are safe for humans and the environment, it is also important that the system is efficient, workable, and fair. Efficient regulatory systems minimise costs to applicants and the government agencies that implement them. Functional systems make decisions promptly in a reasonable amount of time, understanding that if the decision-making process takes too long or never is completed, the nondecision is in effect a rejection. Finally, fair systems treat similar products similarly, and decisions on those similar products are consistent with one another.[25] The system should be fair and equitable to developers, researchers, and their products.

INTERNATIONAL OBLIGATIONS RELEVANT TO BIOSAFETY

Countries do not have complete discretion when deciding how to set up their biosafety regulatory system. There are several international treaties and agreements that relate to biosafety. If a country is bound by those international agreements, then the biosafety regulatory system must be compliant with those obligations. The international obligations that most directly affect biosafety are discussed in the next section.

Cartagena Protocol on Biosafety

The Cartagena Protocol, which came into force on September 11, 2003, is a binding international agreement related to its parent treaty, the Convention on Biological Diversity.[26] The Protocol is currently a primary driving force behind the establishment of national biosafety regulatory systems in countries that have ratified or acceded to the Protocol. The Protocol empowers countries to establish biosafety procedures and provides the scientific and legal boundaries under which such systems should operate. It has the potential to effectuate a common set of processes and procedures for biosafety that safeguard the environment and the public while allowing for international commerce and product innovation.[27]

The Protocol's scope is the "transboundary movement, transit, handling, and use of all living modified organisms that may have adverse effects on the conservation and sustainable use of biological diversity, taking also into account risks to human health."[28] To ensure the safe transfer, handling, and use of LMOs, the Protocol sets up two separate procedures. When LMOs are intentionally introduced into the

[25] *Ibid.*

[26] Secretariat of the Convention on Biological Diversity, *Cartagena Protocol on Biosafety to the Convention on Biological Diversity: Text and annexes* (Montreal: Secretariat of the Convention on Biological Diversity, 2000) [*Cartagena Protocol*].

[27] *Ibid*; "Implementing the Cartagena Protocol," *supra* note 12 at 311.

[28] *Cartagena Protocol, supra* note 26 at art 4.

environment for the first time, the Protocol sets up an Advance Informed Agreement (AIA) procedure.[29] This procedure requires that an exporter of an LMO provide a notice with detailed information about the LMO to the importing country.[30] The importing country then reviews the information, conducts a risk assessment, and decides, based on the risk assessment results, whether to approve or reject the LMO.[31]

The second procedure set up by the Protocol is for LMOs for food, feed, or for processing (e.g., corn, soybeans, wheat, or other grains to be fed directly to humans or animals, or used for processing).[32] For those LMOs, the AIA procedure is not required. Instead, the Protocol establishes a simpler system that reflects the decreased likelihood that those LMOs will affect the exporting country's biodiversity. Before the LMO can be exported to another country, the only requirement is that the safety decision in the exporting country is communicated to other countries through the Biosafety Clearing-House (BCH). For LMOs used for other purposes, such as LMOs used in the laboratory, the Protocol leaves any regulation to the discretion of the individual country. The Protocol also does not cover products derived from LMOs, such as processed foods that have ingredients that came from LMOs.

If countries look to the Protocol for guidance when establishing their own domestic regulatory systems, the Protocol is helpful on several important issues. First, the Protocol establishes a system that gives proportional treatment to an LMO based on its proposed use. For LMOs intentionally released into the environment – an activity that poses, on a relative scale, greater potential risk to biodiversity – the Protocol establishes an informed consent process requiring a detailed risk assessment, risk management, and consent by the importing country. For an LMO used for food, feed, or processing – an activity in which the risk of harm to biodiversity is significantly less – the Protocol does not require advanced informed consent but allows parties to make decisions based on the safety decision from the exporting country or to conduct their own risk assessment. For the contained use of an LMO in a laboratory or greenhouse – an activity having less potential risk than a deliberate release – the Protocol has no required procedures. Thus, depending on the LMO's use, the Protocol establishes different procedures that correspond to the relative risk of that activity. Second, the Protocol provides for differential treatment of LMOs based on particular risk characteristics. Article 13 sets up a "simplified procedure" that allows certain LMOs that would normally qualify for the AIA procedures to have a streamlined process or complete exemption from AIA if that LMO can be released safely. Article 7 also allows the Parties to exempt certain LMOs collectively from the AIA procedures if those LMOs are not likely to have adverse effects. Finally, the Protocol provides useful details about the information needed for a risk assessment

29 *Ibid* at art 7. See Chapter 5 for further discussion of the advance informed agreement (AIA) procedure.
30 *Ibid* at art 8.
31 *Ibid* at arts 10 and 15.
32 *Ibid* at art 11.

(Annex II) and an explanation of what a scientific risk assessment of an LMO should entail (Annex III).

Although the Protocol comprehensively covers many issues, it leaves unresolved issues that each country must address when establishing its biosafety regulatory regime. First, the Protocol has no discussion about which level of protection is adequate before an LMO is approved or how much potential risk must be identified to justify withholding consent. The Protocol is silent on what should happen after a risk assessment is conducted and potential risks are identified (as invariably happens, because most activity has some potential risk). Instead, each individual country decides which safety standard it believes must be satisfied before consenting to an LMO. Second, although Article 26 of the Protocol allows parties to take into account, consistent with other international obligations, socioeconomic considerations that arise from LMOs, the Protocol defines socioeconomic considerations only vaguely and does not explain how they might be factored into the procedures set forth in the Protocol.[33] Individual countries have to decide which socioeconomic considerations can or should be included in their decision-making process and when and how those issues will be analysed.[34] Finally, the Cartagena Protocol does not comprehensively address all of the major risk issues associated with GE organisms. Although the Protocol provides detailed legal and scientific procedures to ensure that LMOs do not adversely affect biological diversity, it does not substantively address human health or food safety concerns surrounding LMOs.[35] Those concerns actually might be more important biosafety concerns to the world's population than biodiversity risks. Thus, a biosafety regulatory system implementing only the Protocol might miss relevant food safety issues.

World Trade Organization Agreements and Other Relevant Economic Treaties

There are several other international agreements that may have a direct impact on countries establishing national biosafety regulatory systems. First, the Agreement on the Application of Sanitary and Phytosanitary Measures (SPS Agreement) provides countries with the sovereign right to establish appropriate levels of sanitary and phytosanitary protection in international trade, which includes food and agriculture.[36] Under the SPS Agreement, those protections must do so in a way that minimises

33 "Implementing the Cartagena Protocol," *supra* note 12 at 311.

34 L Fransen et al, *Integrating Socio-Economic Considerations into Biosafety Decisions* (Washington, D: World Resources Institute, 2005).

35 F Kalibwani et al, "Governing Biotechnology in Africa: Toward Consensus on Key Issues in Biosafety" (Johannesburg, South Africa; Harare, Zimbabwe; and Washington, DC: NEPAD; FANRPAN; and IFPRI, 2004). Prepared for "African Policy Dialogues on Biotechnology-Southern Africa" held 20–21 September 2004 in Harare, Zimbabwe.

36 *Agreement on the Application of Sanitary and Phytosanitary Measures*, being part of Annex IA to the *Agreement Establishing the World Trade Organization*, 15 April 1994, 33 ILM 1144 [*SPS Agreement*].

negative trade effects (Article 5.4). The SPS Agreement also requires in Article 5.5 that countries "avoid arbitrary or unjustifiable distinctions in the level of protection they consider to be appropriate for different situations, if such distinctions result in discrimination or disguised restrictions on international trade."[37] The SPS Agreement does allow countries to adopt precautionary measures when relevant scientific evidence is insufficient (similar to the Protocol) but allows that decision to remain for a reasonable period only while additional scientific evidence is actively gathered.[38] Finally, the SPS Agreement also sets forth risk assessment procedures that include the use of both scientific and socioeconomic considerations, but it provides a fairly narrow definition of which socioeconomic considerations can be legitimately used in the decision-making process. Thus, the SPS Agreement may restrict the scope of a biosafety regulatory system.

Two other agreements under the World Trade Organization (WTO) also could impact biosafety. The General Agreement on Tariffs and Trade (GATT) could impact national biosafety regulations systems because it requires that "like products" be treated in the same manner, whether produced domestically or imported.[39] Under GATT, it is unclear whether GE products can legitimately be distinguished solely by their process of production.[40] In addition, the Technical Barriers to Trade Agreement (TBT) requires that countries' technical regulations may not be more trade restrictive than necessary to fulfil a legitimate objective.[41] The TBT may be relevant to provisions on the labelling and tracing of GE organisms and their products, because various regimes might not meet the "no more trade-restrictive than necessary" requirement.

The SPS and TBT Agreements encourage the use of international scientific standards. The SPS Agreement recognizes the standards developed by three relevant organizations: the FAO/WHO Codex Alimentarius Commission, the Office of International des Epizooties (OIE – the World Organisation for Animal Health), and the International Plant Protection Convention (IPPC). Those standard-setting bodies all have their working groups on safety aspects of GMOs and GM foods, and the resulting standards, recommendations, and guidelines could become the basis for WTO members' sanitary and phytosanitary measures or technical regulations.

37 S Zarrilli, *International Trade in GMOs and GM Products: National and Multilateral Legal Frameworks* (New York, USA: United Nations Conference on Trade and Development, 2005).

38 *SPS Agreement*, *supra* note 36; J Kinderlerer and M Adcock, "Agricultural Biotechnology, Politics, Ethics, and Policy" in Steven W Omamo and Klaus von Grebmer, eds, *Biotechnology, Agriculture, and Food Security in Southern Africa* (Washington, DC: Food, Agriculture, and Natural Resources Policy Network, 2005).

39 *General Agreement on Tariffs and Trade*, being part of Annex IA to the *Agreement Establishing the World Trade Organization*, 15 April 1994, 1867 UNTS 187, 33 ILM 1153.

40 Zarrilli, *supra* note 37.

41 *Agreement on Technical Barriers to Trade*, being part of Annex IA to the *Agreement Establishing the World Trade Organization*, 15 April 1994, 33 ILM 1144.

Of note, the relationship between the Cartagena Protocol and the WTO and its agreements currently is unclear. The WTO might or might not find that the Cartagena Protocol's obligations are consistent with WTO obligations, and it may not recognise the Protocol in disputes between members, if one member is not a party to the Protocol. Furthermore, many of the obligations found in the WTO Agreements are included, often with deeper degrees of liberalization, in regional and bilateral trade and investment treaties to which Parties to the Cartagena Protocol may also be signatories. The complexities of these particular relationships, although beyond the scope of this chapter, are worthy of further legal research and analysis in the case of the specific rules that bind each Party to the Protocol, in that country's economic, social, and environmental circumstances.[42]

The Codex Alimentarius Commission

Under the auspices of the United Nations Food and Agriculture Organization (FAO) and the World Health Organization (WHO), the Codex Alimentarius Commission was established to develop internationally acceptable standards for use in the areas of food quality and food safety. The SPS Agreement specifically cites Codex standards, guidelines, and recommendations as the preferred international measure for facilitating international trade in food. The Codex Commission has been working on food safety issues surrounding GE organisms and their food products for several years and has generated consensus documents that discuss how to conduct food safety risk assessments for GE organisms and their food products.[43] Thus, those documents should be considered in biosafety decisions involving food safety issues.

International Plant Protection Convention

The International Plant Protection Convention[44] (IPPC) establishes the sovereign authority to regulate the entry of plants or plant products and other regulated articles with the use of phytosanitary measures. Those measures, however, must be

[42] See M Gehring and MC Cordonier Segger, eds., *Sustainable Development in World Trade Law* (The Hague, Netherlands: Kluwer Law International, 2005); Gilbert R Winham, "International regime conflict in trade and environment: the Biosafety Protocol and the WTO" (2003) 2 World Trade Review 131; Sebastian Oberthür and Thomas Gehring, "Institutional interaction in global environmental governance: The case of the Cartagena Protocol and the World Trade Organization" (2006) 6(2) Global Environmental Politics 1, for a more detailed discussion of the relationship between trade agreements and the Cartagena Protocol.

[43] These documents include "Principles for the Risk Analysis of Foods Derived from Modern Biotechnology," "Guideline for the Conduct of Food Safety Assessment of Food Produced Using Recombinant-DNA Microorganisms," "Guidelines for the Conduct of Food Safety Assessment of Food Produced From Recombinant-DNA Plants," and "Guidelines for the Conduct of Food Safety Assessment of Food Produced from Recombinant-DNA Animals." See online: Codex Alimentarius Commission, www.codexalimentarius.net/web/standard_list.do?lang=en.

[44] *International Plant Protection Convention*, 6 December 1951, 150 UNTS 67.

technically justified, transparent, and not applied in a way that constitutes either a means of arbitrary or unjustified discrimination or a disguised trade barrier. Similar to the Codex Commission, the IPPC currently has working groups addressing issues involving GE organisms that fall within the scope of the international agreement. Those working groups provide documents for countries addressing plant pest issues and LMOs.[45]

CONCLUSION

Although the needs and resources of each nation are different, there are universal qualities of a functional and protective biosafety regulatory regime. Such a system should be comprehensive, fair, participatory, transparent, and provide for risk assessment and post-approval oversight while operating with sufficient legal authority. These attributes can work synergistically – for example, a system that is transparent and understandable encourages greater participation. Implementing those attributes also can require policy choices and balancing of interests. For example, increasing the transparency of a regulatory system could reduce its ability to be flexible or as efficient as it might be with less transparency.

A regulatory system must not only assess potential risk but also decide how much risk is tolerable and include an analysis of the potential risks to human health, the environment, and biodiversity. The procedures for assessing risk and authorizing the transfer and/or release of genetically engineered organisms are informed by the Cartagena Protocol and other international agreements. Those agreements and international obligations seek to prevent unnecessary and inconsistent barriers to trade, promote research and development, and ensure that harmful or untested organisms do not enter the environment. They are thus bound and constrain the national biosafety regulatory systems that countries can develop and implement. Although those international agreements bear on different aspects of biosafety regulation, those agreements, as well as the numerous functional and protective national biosafety regulatory systems in existence in many countries, provide guidance and a robust wealth of knowledge to lawmakers intending to craft a national biosafety regulatory system

45 See FAO, ISPM No. 11, "Pest risk analysis for quarantine pests including analysis of environmental risks and living modified organisms", online: FAO, http://www.fao.org/docrep/008/y5874e/y5874e00.htm.

PART II

Sustainable Development Law and Policy on Biosafety

4

Risk Assessment and Risk Management

Ryan Hill

Risk assessment and risk management are integral components of the Cartagena Protocol and of any biosafety law and policy framework. Risk assessment is the process of estimating the potential consequences of the use and/or release of living modified organisms (LMOs), as well as the likelihood of such consequences arising. Risk assessment, like other forms of environmental assessment, aims to ensure that decisions about development policies and activities are taken with a full understanding of the potential adverse impacts to the environment. In addition, the Cartagena Protocol allows Parties to apply the precautionary approach so that uncertain risks that could have adverse impacts on the environment can be considered as a part of the risk assessment process. This approach, discussed at greater length in the Introduction to this book, is often mentioned in international treaties in the field of sustainable development.

In the Cartagena Protocol, risk assessment is a key element of decision making regarding the use and release of LMOs. Article 10.1 requires that decisions regarding the import of LMOs intended for intentional introduction into the environment be taken in accordance with risk assessment.[1] For LMO-FFPs, Parties make decisions on import in accordance with their domestic regulatory frameworks.[2] Most Parties that have established specific domestic frameworks for LMOs have included risk assessment and management as part of the requirements set out for the import of LMOs. The Protocol specifically establishes that in the absence of such frameworks, developing country Parties and Parties with economies in transition may base specific import decisions on risk assessment.

Risk management under the Cartagena Protocol comprises both measures taken to address potential risks identified through specific risk assessments, and broader measures to manage and mitigate risks associated with LMOs. Unlike in some

[1] *Cartagena Protocol on Biosafety*, 29 January 2000, 2226 UNTS 208 (entered into force 11 September 2003) at art 10(1) [*Cartagena Protocol*].

[2] *Ibid* at art 11(4).

national frameworks, risk management and decision making regarding the import of LMOs are addressed separately in the Cartagena Protocol. This chapter aims to analyse the provisions of the Protocol regarding risk assessment and risk management; compare the provisions of the Cartagena Protocol to other frameworks for risk assessment and risk management, focusing particularly on key principles and methodology for risk assessment; and explore challenges for the implementation of the risk assessment provisions under the Cartagena Protocol.

THE PROTOCOL'S PROVISIONS ON RISK ASSESSMENT AND RISK MANAGEMENT

Risk assessment under the Cartagena Protocol is provided for primarily by Article 15 and Annex III. Article 15 describes the basic context for risk assessment under the Protocol as follows:

1. Risk assessments undertaken pursuant to this Protocol shall be carried out in a scientifically sound manner, in accordance with Annex III and taking into account recognized risk assessment techniques. Such risk assessments shall be based, at a minimum, on information provided in accordance with Article 8 and other available scientific evidence in order to identify and evaluate the possible adverse effects of living modified organisms on the conservation and sustainable use of biological diversity, taking also into account risks to human health.
2. The Party of import shall ensure that risk assessments are carried out for decisions taken under Article 10. It may require the exporter to carry out the risk assessment.
3. The cost of risk assessment shall be borne by the notifier if the Party of import so requires.

Paragraph 1 of Article 15 lays out basic requirements for risk assessment under the Protocol, whereas paragraphs 2 and 3 ensure, to the extent possible, that risk assessment is actually implemented. The Protocol is flexible as to who carries out the assessment, recognizing that in some cases national authorities conduct the risk assessment, whereas in other cases national authorities audit a risk assessment that has been completed by the exporter.[3]

Actual guidance on risk assessment is not contained in Article 15 but rather in Annex III to the Protocol. Of note, Annex III includes the following selection of key principles for risk assessment:

3. Risk assessment should be carried out in a scientifically sound and transparent manner, and can take into account expert advice of, and guidelines developed by, relevant international organizations.
4. Lack of scientific knowledge or scientific consensus should not necessarily be interpreted as indicating a particular level of risk, an absence of risk, or an acceptable risk.

[3] For further discussion, see Ruth Mackenzie, et al, *An Explanatory Guide to the Cartagena Protocol on Biosafety*, IUCN Environmental Law and Policy Paper Series No 46 (Gland, Switzerland and Cambridge, UK: IUCN and FIELD, 2003).

5. Risks associated with living modified organisms or products thereof, namely, processed materials that are of living modified organism origin, containing detectable novel combinations of replicable genetic material obtained through the use of modern biotechnology, should be considered in the context of the risks posed by the non-modified recipients or parental organisms in the likely potential receiving environment.
6. Risk assessment should be carried out on a case-by-case basis. The required information may vary in nature and level of detail from case to case, depending on the living modified organism concerned, its intended use and the likely potential receiving environment.

This is not an exhaustive set of risk assessment principles but does cover some of the most important principles common to many risk assessment frameworks for LMOs (see discussion of frameworks later in this chapter)[4] and for other types of stressors. Paragraph 4, it has been suggested in other chapters of this book, provides a tailored and highly context-specific reflection of the precautionary approach because lack of scientific evidence or consensus is not to be interpreted as indicating a particular risk, absence of risk, or acceptable risk, so the burden of addressing situations of scientific uncertainty stays with the LMO proponent.[5] Paragraphs 4 and 5 of Annex III are important in establishing the context of risk assessment – that is, risk assessment under the Protocol is used to support decision making but does not itself result in a decision. Other inputs besides risk assessment are also relevant to decision making (e.g., socioeconomic considerations, as provided for in Article 26.1).

Annex III also outlines the basic elements of methodology for risk assessment in paragraph 8:

8. To fulfil its objective, risk assessment entails, as appropriate, the following steps:
   (a) An identification of any novel genotypic and phenotypic characteristics associated with the living modified organism that may have adverse effects on biological diversity in the likely potential receiving environment, taking also into account risks to human health;
   (b) An evaluation of the likelihood of these adverse effects being realized, taking into account the level and kind of exposure of the likely potential receiving environment to the living modified organism;
   (c) An evaluation of the consequences should these adverse effects be realized;
   (d) An estimation of the overall risk posed by the living modified organism based on the evaluation of the likelihood and consequences of the identified adverse effects being realized;
   (e) A recommendation as to whether or not the risks are acceptable or manageable, including, where necessary, identification of strategies to manage these risks; and

[4] Ryan A Hill and Cyrie Sendashonga, "General Principles for Risk Assessment of Living Modified Organisms: Lessons from Chemical Risk Assessment" (2003) 2 Environmental Biosafety Research 81.

[5] See also the discussion of the precautionary approach in the Introduction to this book.

(f) Where there is uncertainty regarding the level of risk, it may be addressed by requesting further information on the specific issues of concern or by implementing appropriate risk management strategies and/or monitoring the living modified organism in the receiving environment.

This methodology, particularly subparagraphs (a) to (d), is consistent with existing risk assessment frameworks for LMOs[6] and is similar to the general paradigm for risk assessment of chemicals and other stressors that has evolved over a period of more than two decades.[7] The key element is that risks are characterised through the evaluation of the likelihood of a potential adverse effect (or, in cases in which there is uncertainty, the likelihood of different potential magnitudes of effect) and its consequences (see Annex III, Figure 1).

Subparagraphs (e) and (f) of paragraph 8 are part of risk assessment methodology in the Protocol but are closely linked to risk management. Subparagraph (e) refers to the "identification of strategies to manage . . . risks." Such strategies are then implemented in accordance with paragraphs 1 and 2 of Article 16, which deals with risk management. Subparagraph (f) refers to the implementation of risk management strategies and monitoring, and therefore directly overlaps with Article 16. These elements of risk management are sometimes but not always integrated into frameworks for risk assessment.[8]

Interestingly, subparagraph (e) also refers to a recommendation from the risk assessor as to whether risks are "acceptable." Typically, risk assessment frameworks do not explicitly ask risk assessors for value judgments as to acceptability but assume instead that such judgments are the realm of decision makers and stakeholders. In theory, this is an important element of distinction between risk assessment and decision making. In practice, however, risk assessors can often provide decision makers with information about acceptability based on their knowledge of what has been considered as "acceptable" in past cases and/or based on an understanding of the risks associated with alternatives.

Although risk management is touched on in Annex III of the Protocol, the main provisions for risk management under the Protocol are contained in Article 16:

1. The Parties shall, taking into account Article 8 (g) of the Convention, establish and maintain appropriate mechanisms, measures and strategies to regulate, manage and control risks identified in the risk assessment provisions of this

6 Ryan A Hill, "Conceptualizing Risk Assessment Methodology for Genetically Modified Organisms" (2005) 4 Environmental Biosafety Research 67.

7 Ryan A Hill and Cyrie Sendashonga, *Guidelines for Ecological Risk Assessment* (United States Environmental Protection Agency, EPA/630/R-95/002F, April 1998), online: http://www.epa.gov/lawsregs/guidance/bytopic-e.html#ecosystems.

8 *Ibid.*

Protocol associated with the use, handling and transboundary movement of living modified organisms.

2. Measures based on risk assessment shall be imposed to the extent necessary to prevent adverse effects of the living modified organism on the conservation and sustainable use of biological diversity, taking also into account risks to human health, within the territory of the Party of import.
3. Each Party shall take appropriate measures to prevent unintentional transboundary movements of living modified organisms, including such measures as requiring a risk assessment to be carried out prior to the first release of a living modified organism.
4. Without prejudice to paragraph 2 above, each Party shall endeavour to ensure that any living modified organism, whether imported or locally developed, has undergone an appropriate period of observation that is commensurate with its life-cycle or generation time before it is put to its intended use.
5. Parties shall cooperate with a view to:
   (a) Identifying living modified organisms or specific traits of living modified organisms that may have adverse effects on the conservation and sustainable use of biological diversity, taking also into account risks to human health; and
   (b) Taking appropriate measures regarding the treatment of such living modified organisms or specific traits.

Subparagraphs 1 and 2 relate mainly to measures for addressing risks identified by one or more risk assessments. On the other hand, subparagraphs 3 to 5 are more general and do not imply measures to be taken in response to risk assessments. Many existing frameworks for risk management are not so broad as to contain these latter elements.

COMPARISON OF THE PROTOCOL TO OTHER LMO RISK ASSESSMENT FRAMEWORKS

There are several national and international frameworks for risk assessment and/or risk management of LMOs. These range from legally binding international instruments to national guidance documents. This section reviews one well-known international framework and two frameworks developed by Parties to the Protocol. The elements of these frameworks are compared to the risk assessment and risk management provisions of the Cartagena Protocol (see Annex III, Table 1 of this book). This comparison is important for demonstrating the areas of consistency between the Protocol's provisions on risk assessment and risk management and the frameworks that have been developed by other bodies and governments.

Prior to examining the risk assessment frameworks previously mentioned, it is useful to understand differences in terminology. First, although not used by the Protocol, the term *risk analysis* is commonly used internationally. Many guidance documents, in particular those from some international organizations, use this term to refer to risk assessment, risk management, and in some cases risk communication. Parties to the CBD also use the term in relation to Article 8(h) on invasive alien species. Second, the initial step in risk assessment, provided for in paragraph 8(a) of Annex III of the Protocol, is often referred to as *hazard identification* in many frameworks, or in some cases this step is a broader process called *problem formulation*. Problem formulation is considered in more detail later in this chapter. Third, the Protocol refers to the "likelihood" of potential adverse effects in paragraph 8(b) of Annex III. Some frameworks use this term, whereas others refer to this step more generically as *exposure assessment*. Fourth, the Protocol refers to the evaluation of "consequences" in paragraph 8(c) of Annex III. Once again, some frameworks use this same term, whereas others use the more generic expression *effects assessment*.[9]

The following risk assessment frameworks are considered: the 2004 IPPC International Standard for Phytosanitary Measures No. 11: Pest risk analysis for quarantine pests including analysis of environmental risks and living modified organisms;[10] the 2005 Australian Government's Risk Analysis Framework for License Applications to the Office of the Gene Technology Regulator;[11] and the EC's 2002 Decision establishing guidance notes supplementing Annex II to Directive 2001/18/EC of the European Parliament and of the Council on the deliberate release into the environment and repealing Council Directive 90/220/EEC.[12]

These represent a very limited selection of risk assessment frameworks. There are many other frameworks specific to LMOs (or GMOs), including some that focus on potential risks to human health rather than biodiversity.[13] More important, it should be noted that many bodies and governments have not developed specific legislation or guidelines for LMOs but rely on existing legislation and guidelines.

9 *Ibid.*

10 IPPC Secretariat, "ISPM No. 11: Pest Risk Analysis for Quarantine Pests Including Analysis of Environmental Risks and Living Modified Organisms" (International Standard for Phytosanitary Measures, 2004), online: https://www.ippc.int/file_uploaded/1146658377367_ISPM11.pdf.

11 Australia, Commonwealth, Office of the Gene Technology Regulator, *Risk Analysis Framework for License Applications to the Office of the Gene Technology Regulator* (Handbook, 2005).

12 Commission Decision (EC) 2002/624 of 24 July 2002 Establishing Guidance Notes Supplementing Annex II to Directive 2001/18/EC of the European Parliament and of the Council on the Deliberate Release into the Environment of GMOs and Repealing Council Directive 90/220/EEC [2002] OJ L200/22.

13 For example, various documents developed by the Codex Alimentarius Commission. For a specific list, see Annex to the COP-MOP 2 document "Risk Assessment and Risk Management (Articles 15 and 16)," UN Doc. UNEP/CBD/BS/COP-MOP/2/9, online: http://www.cbd.int/doc/meetings/bs/mop-02/official/mop-02-09-en.pdf.

These frameworks were reviewed to determine if they contained some of the same elements as the Protocol, especially in terms of principles and methodological steps in risk assessment. This chapter focuses on the more operational principles of the Protocol and does not consider the more broadly worded elements, such as paragraph 3 of Annex III, which calls for risk assessment to be carried out in a scientifically sound and transparent manner. Because rigorous use of science is the basis for risk assessment, it is not expected that other frameworks would necessarily repeat such an obvious point. Similarly, paragraph 4 of Annex III was not considered because it appears to be related not to risk assessment itself but rather to the use and interpretation of risk assessment in decision making.

The analysis shows that some of the key elements of risk assessment in the Cartagena Protocol are found in other risk assessment frameworks. Methodology in particular is quite similar among frameworks and follows the conventional paradigm for risk assessment,[14] and more variability occurs among principles. Even when principles address common issues, the details differ. For example, a principle related to scientific uncertainty is included in the last row of Table 1, but in reality how this principle is addressed in the frameworks in hand varies greatly, and the Protocol's treatment of it is rather weak and indirect.

Interestingly, one relevant principle that is increasingly explicit in other fields of risk assessment, particularly chemical risk assessment, is missing from most LMO risk assessment frameworks, with the exception of the Australian framework. This principle establishes that risk assessment should consider the "weight of evidence" from multiple types of information.[15] In fact, there is often more than one type of information to support a risk assessment. For example, a risk assessment in one country could simultaneously consider field trials in that country, experience with releases in other countries, and basic biological information about the species in hand (e.g., does it have characteristics that are typical of weedy species?). None of these three lines of evidence is ideal – field trials are limited in scope, experience in other countries is usually linked to a different receiving environment, and basic biological arguments constitute generalizations that might not be applicable to a

14 For further discussion, see Hill, *supra* note 6.

15 R Hill et al, "A Weight-of-Evidence Approach for Evaluating Ecological Risks: Report of the Massachusetts Weight-of-Evidence Workgroup" (1996) 2 Journal of Human and Ecological Risk Assessment 277; TE Nickson and MJ McGee, "Ecological Assessment of Crops Derived through Biotechnology" in John A Thomas and Roy L Fuchs, eds, *Biotechnology and Safety Assessment*, 3rd ed. (Oxford, UK: Elsevier, 2002); RN Hull and S Swanson, "Sequential Analysis of Lines of Evidence – an Advanced Weight-of-Evidence Approach for Ecological Risk Assessment" (2006) 2 Integrated Environmental Assessment and Management 302; PM Chapman, BG Macdonald, and GS Lawrence, "Weight-of-Evidence Issues and Frameworks for Sediment Quality (and Other) Assessments" (2002) 8 Journal of Human and Ecological Risk Assessment 1489; GA Burton, PM Chapman, and EP Smith, "Weight-of-Evidence Approaches for Assessing Ecosystem Impairment" (2002) 8 Journal of Human and Ecological Risk Assessment 1657.

particular case. Nevertheless, a good risk assessment considers each of these lines of evidence carefully. One line of evidence might be given more weight than others, but that does not mean that the others should be ignored completely. In some cases, two lines of evidence can be completely contradictory. Although this creates challenges for the risk assessor, ignoring one line of evidence implies ignoring uncertainty rather than confronting it.

The Cartagena Protocol's provisions on risk assessment and risk management are consistent with the conventional paradigm for risk assessment; however, although Annex III outlines a few key principles and generic methodologic steps, the level of detail is inadequate for providing practical guidance on implementation.

Fortunately, a considerable amount of guidance materials and case studies on risk assessment and risk management for living modified organisms has been developed. These documents vary widely in their scope and application, and include:

- Guidance documents that are specific to a particular living modified organism or group of organisms. For example, at international level, the Organisation for Economic Co-operation and Development (OECD) has been particularly active, through its Working Group on Harmonization of Regulatory Oversight in Biotechnology, in developing consensus documents on the biology of certain organisms, including maize, oilseed rape, potato, soybean, sugar beet, and many other species for which modified varieties have been commercialized.[16]
- Guidance documents that are specific to a particular trait or group of traits. Again, at the international level, the OECD has been active in this regard, publishing and developing consensus documents related to herbicide tolerance and other traits.
- Guidance documents that are specific to one or more risk pathways or mechanisms (e.g., weediness, effects on non-target organisms, and development of insect resistance). There is a large body of journal articles and other guidelines on this topic.
- Guidance documents that focus on a particular issue in the practice of risk assessment or risk management for living modified organisms, such as particular elements of methodology (e.g., problem formulation and monitoring) or methodological issues (e.g., characterizing uncertainties). This type of guidance is found mostly in journal articles and books, although some work has been done by international bodies.[17]

[16] H Bergmans, "Basic Framework for Risk Assessment for Transgenic Plants Developed by the OECD: 20 Years After the OECD Blue Book" (2006) 5 Environmental Biosafety Research 213; P Macdonald and S Yarrow, "Current Harmonization Activities Related to Risk/Safety Assessment by the OECD Working Group on Harmonization of Regulatory Oversight in Biotechnology" (2006) 5 Environmental Biosafety Research 223.

[17] For example, OECD, Environmental Directorate, *Guidance Document on the Use of Taxonomy in Risk Assessment of Micro-organisms: Bacteria*, Doc. No. ENV/JM/MONO(2003)13 (2003), online: http://bch.cbd.int/database/record-v4.shtml?documentid=41900.

Guidance documents on how to submit risk assessments for regulatory consideration. This is perhaps the most practical type of guidance relevant to the implementation of the Protocol. The Public Research and Regulation Initiative has drafted an extended manual on how to submit a risk assessment that meets the administrative and technical information requirements of a country of import and/or the Cartagena Protocol.[18] Another example of the information requirements for the risk assessment of a particular LMO is provided by AGBIOS for the maize line MON810.[19]

A review of available guidance documents on risk assessment and risk management was conducted for COP-MOP 2 and is available on the CBD Web site.[20] A database of guidance materials is also available through the Biosafety Information Resource Centre, a component of the BCH.[21] This database is expected to expand and an overview of the scope and applicability of each guidance material is also to be provided.[22]

CHALLENGES IN IMPLEMENTING RISK ASSESSMENT UNDER THE PROTOCOL

This section focuses on a few particular issues and challenges related to the current practice and implementation of risk assessments for LMOs. Many of the challenges in risk assessment and risk management result from scientific uncertainty. Indeed, uncertainty is inherent in the concept of risk. This section does not focus on the broad issue of limitations in capacity, which is arguably the most significant challenge to the implementation of the Protocol.

Problem Formulation

As mentioned earlier, the conventional paradigm for risk assessment begins with a step sometimes called *hazard identification* or *problem formulation*. The latter usually implies a broader scope of work that goes beyond the identification of a potential hazard. Paragraph 8(a) of Annex III consists of only one sentence but constitutes

18 Public Research and Regulation Initiative, "Guide for Notifications and Risk Assessments for Releases into the Environment of Genetically Modified Organisms" (Draft, 9 November 2005), online: http://bch.cbd.int/database/record-v4.shtml?documentid=48124.

19 Monsanto Company / Agriculture & Biotechnology Strategies (AGBIOS), "MON 810 Environmental Risk Assessment Case Study," online: http://bch.cbd.int/database/record-v4.shtml?documentid=42184.

20 "Risk Assessment and Risk Management (Articles 15 and 16)," *supra* note 13.

21 "Biosafety Information Resource Centre" (Database of Guidance Materials), online: http://bch.cbd.int/database/resources/.

22 In accordance with paragraphs 2–6 of Dec BS-III/11 *Risk Assessment and Risk Management*, U Doc. UNEP/CBD/BS/COP-MOP/3/11, online: http://bch.cbd.int/protocol/decisions/decision.shtml?decisionID=11067.

perhaps the most important component of risk assessment. Problem formulation is relatively standard for chemical risk assessment and typically involves setting the stage for a risk assessment by clarifying goals (e.g., maintaining a functioning ecosystem and/or preventing adverse effects), selecting assessment end points (e.g., toxicity to nontarget species and/or survival of related native species), and developing a conceptual model that links the hazard (e.g., a chemical or an LMO) to potential adverse effects by one or more mechanisms or pathways.[23] Problem formulation also often includes a plan for further analysis and/or collection of data and is a critical component of risk assessment. If a risk assessment is framed incorrectly, it will not provide the appropriate information to support decision making.

Some recent case studies regarding risk assessment of LMOs explicitly include problem formulation components involving conceptual models. These case studies have been described by Wolt and Peterson and have focused on particular risks to a particular nontarget species.[24] Problem formulation would necessarily be much broader in scope if a risk assessment were to consider all possible risks associated with a class of LMOs. A couple of recent efforts to explore risk assessment methodology have focused on hypothetical case studies evaluating the introduction of Bt cotton in Brazil[25] and the introduction of Bt maize in Kenya.[26] In both cases, the range of types of risks considered was necessarily broad (e.g., effects on nontarget organisms, potential for gene flow and associated consequences, and potential for development of insect resistance).

An interesting issue in problem formulation is the degree of consultation and consideration of external factors that is appropriate. The Bt maize and Bt cotton case studies mentioned here put significant emphasis on defining the issues and using a consultative process to frame the risk assessment, and recommend the consideration of social and economic issues. This is consistent with a deliberative approach to risk assessment[27] and might be appropriate where assessments are specifically intended to take into account broader socioeconomic considerations. It should be noted, however, that the Cartagena Protocol, and many national frameworks, address broader

23 United States Environmental Protection Agency, *supra* note 7; Alan Raybould, "Problem Formulation and Hypothesis Testing for Environmental Risk Assessments of Genetically Modified Crops" (2006) 5 Environmental Biosafety Research 119.

24 Jeffrey D Wolt, Chris A Conlan, and Kei Majima, "An Ecological Risk Assessment of Cry1F Maize Pollen Impact to Pale Grass Blue Butterfly" (2005) 4 Environmental Biosafety Research 243; RKD Peterson et al, "Genetically Engineered Plants, Endangered Species, and Risk: A Temporal and Spatial Exposure Assessment for Karner Blue Butterfly Larvae and Bt Maize Pollen" (2006) 26 Risk Analysis 845.

25 A Hilbeck, DA Andow, and EMG Fontes, eds, *Environmental Risk Assessment of Genetically Modified Organisms Vol 2: A Case Study of Bt Cotton in Brazil* (Wallingford, UK: CABI Publishing, 2006).

26 A Hilbeck and DA Andow, eds, *Environmental Risk Assessment of Genetically Modified Organisms, Vol 1: A Case Study of Bt Maize in Kenya* (Wallingford, UK: CABI Publishing, 2004).

27 National Research Council, Committee on Risk Assessment Methodology, *Understanding Risk – Informing Decisions in a Democratic Society* (Washington, DC: National Academy Press, 1996).

socioeconomic inputs to decision making separately from risk assessment.[28] Indeed, most risk assessors consider socioeconomic considerations to be important for decision making but not part of risk assessment. Chapters 7 and 10 of this book focus specifically on the issues of public involvement and socioeconomic considerations. The problem formulations for the Bt case studies are focused heavily on clarifying the overall decision-making context and identifying general management goals, but they do not clearly define assessment end points or derive a conceptual model of the various risk pathways. These latter elements are the subject of subsequent chapters on the development of the methodology for those case studies.

Problem formulations for risk assessment of LMOs are often weak.[29] Fortunately, this component of risk assessment of LMOs has recently started to attract more focused interest. The International Life Sciences Institute (ILSI) is currently developing a broad framework for problem formulation for application to genetically modified crops,[30] based on input from a diverse group of scientists. The output of this effort will be useful in providing guidance to risk assessors on the important first phase of risk assessment.

Determining the Appropriate Level of Detail for Risk Assessment

The Cartagena Protocol (in paragraph 6 of Annex III) and many risk assessment frameworks explicitly state that the level of detail in risk assessment varies on a case-by-case basis. There are several reasons for this. First, the appropriate level of detail is somewhat proportional to the estimated risks. For example, if risk estimates are extremely low, even when several conservative assumptions are employed, a conclusion of negligible risk may be easily reached without need for further assessment. Conversely, if risk estimates are ambiguous or appear to be highly based on conservative assumptions, a more detailed assessment is usually warranted. For this reason, many risk assessment frameworks are explicitly "tiered," beginning with screening-level assessments and progressing to more complex assessments only if warranted.[31]

28 In the case of the Protocol, Article 26.1 states that Parties "in reaching a decision on import under this Protocol or under its domestic measures implementing the Protocol, may take into account, consistent with their international obligations, socio-economic considerations."

29 Raybould, *supra* note 23.

30 JD Wolt and S Olin, "Problem Formulation for Environmental Risk Assessments of Genetically Modified Crops" (Abstract of a Presentation for the Annual Meeting of the Society for Risk Analysis in San Antonio, December 2007), online: Society for Risk Analysis, http://birenheide.com/sra/2007AM/program/singlesession.php3?sessid=W4-B.

31 A Raybould and I Cooper, "Tiered Tests to Assess the Environmental Risk of Fitness Changes in Hybrids Between Transgenic Crops and Wild Relatives: the Example of Virus Resistant Brassica Napus" (2005) 4 Environmental Biosafety Research 127; R Hill et al, "Level of Detail in Ecological Risk Assessment" (2000) 40 Marine Pollution Bulletin 471; M Garcia-Alonso et al, "A Tiered System for Assessing the Risk of Genetically Modified Plants to Non-target Organisms" (2006) 5 Environmental Biosafety Research 57; TE Nickson and MJ McGee, *supra* note 15.

A second reason that the level of detail varies among risk assessments is that the quality and quantity of available information varies greatly. For cases in which information is more limited, decision makers must consider the advantages and disadvantages of collecting more information. The advantages of additional data collection could include reducing uncertainty, which supports a more informed decision. The disadvantages could be the economic costs of collecting more information and delaying the approval of a cost-effective LMO, as well as the ecological costs associated with continuing to use a more harmful alternative.[32] The availability of information does not necessarily mean that it should be included in a risk assessment. Assessments should focus only on what is needed to support a decision.[33]

A third reason that the level of detail varies among risk assessments is the need to predict risks under alternative management scenarios. Once risks are predicted in an initial assessment, it might be necessary to assess risks under other scenarios, in which one or more mitigation measures are implemented. Under any such scenario, certain exposure pathways (i.e., mechanisms of risk) might need to be considered in more or less detail.

A fourth and critical reason that the level of detail varies among risk assessments is the standard of proof expected by assessors, decision makers, and stakeholders. For example, if decision makers have a low tolerance for uncertainty and think that the benefits of introducing an LMO are marginal to begin with, they might require proportionally more evidence that an LMO is safe. In contrast, if the potential benefits of an LMO are thought to be large, and if a decision maker has more tolerance for uncertainty, the standard of proof regarding safety might be less stringent.

Differences in the standard of evidence required in risk assessment are well illustrated by the consideration of potential toxic effects of insect-resistant crops on nontarget organisms. This type of risk is analogous to the assessment of the potential toxicity of chemicals in the environment, and even after decades of work, risk assessors and regulators struggle with the following questions:

Which functional groups should be considered as being potentially affected by a chemical?

Within a functional group, which species should be selected as being representative, or should there be more than one species?

How should species that are protected by national legislation be addressed in the risk assessment?

[32] J Romeis, M Meissle, and F Bigler, "Reply to Ecological Risk Assessment for Bt Crops" (2006) 24 Nature Biotechnology 751.

[33] Australia, Office of the Gene Technology Regulator, *supra* note 11; Hill et al, *supra* note 31; Hill and Sendashonga, *supra* note 4.

Which type of measurement end points should be considered – mortality, growth, or reproduction?

Which duration of toxicity tests is adequate?

Normally, chemical risk assessments focus on the functional groups that are known to be sensitive to the chemical in hand or that are most directly affected, and within those groups, assessors try to select relatively sensitive species and measurement end points. In this regard, certain species and end points are often used repeatedly in laboratory bioassays as surrogates because of their relative sensitivity to the stressor and their performance in bioassays. It is simply neither feasible nor necessary to test every end point for every species. An understanding of ecosystems and ecosystem functions helps to narrow choices and facilitates extrapolation to species for which data do not exist. Proposals regarding important functional groups in the case of nontarget effects have been made.[34] The potential impacts of Bt maize on butterflies have been well studied,[35] and questions have (expectedly) arisen regarding which species and end points should be considered,[36] what is the appropriate duration of bioassay,[37] and how data should be extrapolated to species for which specific data do not exist.[38]

Ultimately, the level of detail in a risk assessment will vary for several reasons. For chemicals, some risk assessments are terminated at the problem formulation stage, because for a particular case there are no exposure pathways that warrant further analysis. Similarly, for LMOs, risk assessors should not presume in advance that certain types of data are necessary to support decision making. Careful clarification of management goals and assessment end points, and review of existing data, could reveal that no new data are needed to support a risk-based decision about the use and/or release of an LMO in a particular case.

The Relationship between Risk Assessment and Precaution

An interesting issue, given that this book focuses on the legal aspects of implementing the Cartagena Protocol, is the relationship between precaution and risk assessment in the Protocol. Many developing (and developed) countries struggle with how to "implement" the precautionary principle. The word *precaution* is not evident in the

34 For example, Garcia-Alonso et al, *supra* note 31; Hilbeck and Andow, *supra* note 26.

35 For example, MK Sears et al, "Impact of Bt Corn Pollen on Monarch Butterfly Populations: A Risk Assessment" (2001) 98 Proceedings of the National Academy of Sciences 11937.

36 Romeis, Meissle, and Bigler, *supra* note 32; DA Andow, GL Lovei, and S Arpaia, "Ecological Risk Assessment for Bt crops" (2006) 24 Nature Biotechnology 749.

37 J Romeis, M Meissle, and F Bigler, "Transgenic Crops Expressing Bacillus Thuringiensis Toxins and Biological Control" (2006) 24 Nature Biotechnology 63; J Romeis, M Meissle, and F Bigler, "Reply to Early-tier Tests Insufficient for GMO Risk Assessment" (2007) 25 Nature Biotechnology 36; A Lang, E Lauber, and B Darvas, "Early-tier Tests Insufficient for GMO Risk Assessment" (2007) 25 Nature Biotechnology 35.

38 Wolt, Conlan, and Majima, *supra* note 24.

Protocol, but there is reference in Article 1 to the "precautionary approach contained in Principle 15 of the Rio Declaration on Environment and Development." Principle 15 reads as follows:

> In order to protect the environment, the precautionary approach shall be widely applied by States according to their capabilities. Where there are threats of serious or irreversible damage, lack of full scientific certainty shall not be used as a reason for postponing cost-effective measures to prevent environmental degradation.[39]

Similar language can be found in the operational provisions of the Protocol. Article 10.6 reads as follows (the text contained in Article 11.8 is almost identical but applies to LMO-FFPs):

> Lack of scientific certainty due to insufficient relevant scientific information and knowledge regarding the extent of the potential adverse effects of a living modified organism on the conservation and sustainable use of biological diversity in the Party of import, taking also into account risks to human health, shall not prevent that Party from taking a decision, as appropriate, with regard to the import of the living modified organism in question as referred to in paragraph 3 above, in order to avoid or minimize such potential adverse effects.

These paragraphs acknowledge that current experience might not provide sufficient information on the nature, severity, probability, or scope of an activity's environmental consequences.[40] The difficulties in its implementation arise in determining whether the magnitude, distribution, and probability of environmental damage in a given case is proportional to the cost of taking action to prevent the risk.[41] In the context of the Cartagena Protocol, the costs of employing precaution might include the loss of economic gains that could be realized from the import of LMOs, or legal claims by other states under the international trade regime. Also important is that environmental harms usually are distributed unequally.[42]

Principle 15 and the Cartagena Protocol's Article 10.6 simply recognize that parties have the right to take a precautionary decision in the face of scientific uncertainty. The degree of precaution that is exercised in any particular case is up to the decision maker. Two decision makers might take different decisions based on the same assessment, because one is more cautious (or "risk-averse") than the other. Precaution is an attitude (a risk-averse attitude) applied by decision makers, not by risk assessors.[43]

39 *Rio Declaration on Environment and Development*, 13 June 1992, UN Doc A/CONF.151/26 (vol. I), 31 ILM 874 (1992) at Principle 15.

40 Marie-Claire Cordonier Segger and Ashfaq Khalfan, *Sustainable Development Law: Principles, Practices & Prospects* (Oxford, UK: Oxford University Press, 2004) at 3.

41 *Ibid* at 144.

42 *Ibid.*

43 For further discussion, see: Ryan Hill, Sam Johnston, and Cyrie Sendashonga, "Risk Assessment and Precaution in the Cartagena Protocol" (2004) 13 RECIEL 263.

CONCLUSION

The basic framework for risk assessment and risk management of LMOs is set out in the Cartagena Protocol and is consistent with other frameworks for LMOs and with the conventional paradigm for risk assessment and management. The Protocol does not provide details about the implementation of this framework, however. In general, capacity is still the most critical limiting factor for the implementation of the risk assessment and risk management provisions of the Protocol. Even where capacity exists, there are certain elements of risk assessment that are particularly challenging to implement, including but not limited to those discussed in this chapter.

5

The Decision-Making Procedures of the Protocol

Worku Damena Yifru, Mai Fujii, and Kathryn Garforth

Concerns regarding living modified organisms (LMOs) are associated primarily with the uncertainty about the potential impacts of uncontrolled propagation and gene transfer on both particular species (e.g., similar nongenetically modified organisms) and ecosystems as a whole (e.g., pollinators and other species interacting within an agricultural setting). To promote the safe international transfer, handling, and use of LMOs, the Cartagena Protocol on Biosafety establishes the advance informed agreement (AIA) procedure regarding the first importation of an LMO for intentional introduction into the environment. A separate procedure is provided for LMOs that are intended for direct use as food or feed, or for processing (LMOs-FFP).

THE ADVANCE INFORMED AGREEMENT PROCEDURE

The AIA procedure (Articles 7 to 10 and 12 of the Protocol) is the central mechanism for regulating the transboundary movement of LMOs under the Cartagena Protocol. The procedure covers LMOs for *intentional introduction into the environment* and thus goes directly to the purpose of the Protocol, which is to contribute to the conservation and sustainable use of biodiversity. The Protocol specifies that the AIA procedure does not apply to certain categories of LMOs, namely LMOs that are pharmaceuticals for humans and that are addressed by other agreements or organizations,[1] LMOs in transit,[2] LMOs destined for contained use,[3] and LMOs-FFP.[4] Parties do, however, have the right to regulate the import of these categories of LMOs on the basis of domestic legislation.

The AIA procedure requires that, prior to the first intentional transboundary movement of a specific LMO into its jurisdiction, the Party of import be notified

1 *Cartagena Protocol on Biosafety to the Convention on Biological Diversity*, 29 January 2000, 2226 UNTS 208 (entered into force 11 September 2003) at art 5.

2 *Ibid* at art 6.

3 *Ibid*.

4 *Ibid* at art 7(3).

of the proposed transboundary movement, receive information about the LMO and its proposed use, and have the opportunity to decide whether the import should be allowed and on which conditions. The AIA procedure mainly involves a four-step process: (1) notification by the Party of export or the exporter; (2) acknowledgment of receipt of the notification by the Party of import; (3) the decision procedure; and (4) review of decisions.

The exporting Party or the exporter shall notify the importing Party prior to the first transboundary movement of an LMO for introduction into the environment and provide, at minimum, the information specified in Annex I of the Cartagena Protocol.[5] This information includes things such as identification information of the exporter and importer, the name and identity of the LMO, a description of the modifications made to the organism and the resulting characteristics of the LMO, intended uses of the LMO, a previous risk assessment report, and the regulatory status of the LMO in the country of export. The Party of export must ensure that there is a legal requirement for the accuracy of information provided by the exporter.[6] The Party of import shall acknowledge receipt of the notification within 90 days.[7]

The decision procedure is set out in Article 10 of the Protocol. The key point is that the decisions of the importing Party must be based on risk assessments that are carried out in a scientifically sound manner and take into account recognized risk assessment techniques.[8] The Party of import may require the exporter to carry out the risk assessment and the notifier to cover the cost of the risk assessment. In a case of lack of scientific certainty, the precautionary approach may be applied.[9]

Article 10 also provides that the importing Party must, within 90 days of receipt of the notification, inform the notifier (i.e., the Party of export or the exporter, as the case may be) whether the intentional transboundary movement may proceed only after the written consent of the Party of import or after no less than 90 days without written consent. Furthermore, the importing Party must communicate, in writing, its decision to the notifier and to the Biosafety Clearing-House (BCH) within 270 days of the date of receipt of notification. The Party of import may make a decision to approve the import with or without conditions or prohibit the import. Except in cases of unconditional approval, the reasons underlying the decision are required. In its communication, the importing Party may request additional information or inform the notifier that the period for making a decision has been extended by a defined period. Failure by an importing Party to acknowledge receipt of a notification or communicate its decision regarding import shall not be considered as its implied consent to an intentional transboundary movement.[10]

5 *Ibid* at art 8(1).

6 *Ibid* at art 8(2).

7 *Ibid* at art 9.

8 *Ibid* at arts 10(1) and 15 and Annex III.

9 *Ibid* at art 10(6)

10 *Ibid* at arts 9(4) and 10(5). The Parties to the Protocol decided at their first meeting that "[i]n the case where no acknowledgement of receipt of notification or decisions are [*sic*] communicated by a

A Party may review and change a decision in light of new scientific information. To do so, it must inform any relevant notifiers and the BCH within 30 days and must set out the reasons for its decision. The Party of export or the notifier may also request the Party of import to review the decisions on the basis of a change in circumstances that may affect the risk assessment, or the availability of additional relevant scientific or technical information.[11]

The AIA procedure is considered a flexible mechanism because the Protocol provides the Parties with broad discretion in its application. Any Party may apply its domestic regulatory procedure instead of the AIA decision process in Article 10 of the Protocol.[12] In addition, according to Article 13, the Party of import can apply a simplified procedure to the import of LMOs. Parties that choose to follow the simplified procedure must specify in advance to the BCH cases in which an intentional transboundary movement may take place at the same time as the movement is notified, and imports of LMOs exempted from the AIA procedure. These flexibilities are constrained only by requirements that the domestic regulatory framework be "consistent with this Protocol"[13] or that there be "adequate measures to ensure the safe international transboundary movement of LMOs in accordance with the objective of the Protocol."[14]

Of note, the Protocol includes a provision on socioeconomic considerations.[15] This provision allows the importing Party to take into account socioeconomic considerations that arise from the impact of LMOs on the conservation and sustainable use of biodiversity, in a manner that is consistent with the Party's other international obligations.

THE PROCEDURE FOR LMOs-FFP

Article 11 of the Cartagena Protocol provides a separate decision-making procedure for LMOs-FFP. These LMOs are, in essence, commodities, and they represent a large percentage of the international bulk trade in agricultural products.[16]

Party of import that is a developing country Party or a Party with an economy in transition, under the time frame established under Articles 9 and 10 of the Protocol, and after the Party of export has sought clarification from the Party of import on the reason for lack of response or decision, the Party of export may remind the Party of import of the need for an acknowledgement and, as appropriate, help it financially to obtain expert or other assistance, including through the use of the roster of experts, in order to enable the Party of import to reach a decision." See paragraph 7 of Decision BS-I/2, "Procedures and Mechanisms for Facilitating Decision-Making by Parties of Import (Article 10, paragraph 7)."

11 *Ibid* at art 12.

12 *Ibid* at art 9(2)(c). See also *ibid*. at Art. 14.4.

13 *Ibid* at art 9(3).

14 *Ibid* at art 13(1).

15 *Ibid* at art 26.

16 Secretariat of Convention on Biological Diversity, "Frequently Asked Questions on the Cartagena Protocol," online: at http://bch.cbd.int/protocol/cpb_faq.shtml.

The procedure for LMOs-FFP in the Protocol is the result of lengthy and difficult negotiations. The issue of the scope of the Protocol, that is, whether and how to include various categories of LMOs within the framework of the Protocol, was one of the most controversial issues during the negotiations.[17] Major LMO exporting countries, that is, members of the Miami Group,[18] were concerned that the Protocol would require shipment-by-shipment approvals of agricultural commodities. They argued that these commodities should be excluded from the Protocol given that their purpose is for human, animal, or industrial consumption rather than release into the environment and so are less likely to create adverse impacts on the environment than other LMOs.[19] In contrast, importing countries expressed safety concerns and advocated for the application of the AIA procedure to all LMOs, stressing the possibility of accidental releases into the environment.[20]

After years of debate, delegations ultimately reached an agreement during the final stages of the negotiations.[21] They agreed that LMOs-FFP would be covered under a more straightforward procedure than the AIA. Consequently, the procedure for LMOs-FFP was established as a multilateral information exchange mechanism. As Stoll states, although this procedure accommodates the Miami Group's trade-related concern by exempting LMOs-FFP from the AIA procedure, the importing countries' safety concern is also addressed by providing an opportunity for the timely consideration of the risks associated with new LMOs and for decision making regarding future imports of that LMO.[22]

Under this procedure, a Party is required to inform other Parties through the BCH, within 15 days, of its decision regarding domestic use of LMOs that might be subject

[17] The economic value of the transboundary movement of agricultural commodities is much higher than that of other LMOs. This partly explains the different positions during the negotiations of the Protocol. See Eric Schoonejans, "Advanced Informed Agreement Procedures," in C Bail, R Falkner, and H Marquard, eds, *The Cartagena Protocol on Biosafety: Reconciling Trade in Biotechnology with Environment and Development?* (London, UK: Earthscan, 2002) 299 at 306.

[18] The members of the Miami Group during the negotiations were Argentina, Australia, Canada, Chile, Uruguay, and the United States.

[19] Secretariat of the Convention on Biological Diversity, *The Cartagena Protocol on Biosafety: A Record of the Negotiations* (Montreal, Canada: Secretariat of the Convention on Biological Diversity, 2003) at 30, online: http://bch.cbd.int/protocol/publications/a_record_of_negotiation_en.pdf; Schoonejans, *supra* note 17 at 306.

[20] Tewolde Egziabher, "Ethiopia" in Bail et al, *supra* note 17, 115 at 122; *Record of the Negotiations, supra* note 19 at 30.

[21] Stanley W Burgiel, "The Cartagena Protocol on Biosafety: Taking the Steps from Negotiation to Implementation" (2002) 11(1) Review of European Community and International Environmental Law 53 at 54; Robert Falkner, "Negotiating the Cartagena Protocol: The International Process" in Bail et al, *supra* note 17, 3 at 18–22. See also "Informal Consultations on Biosafety Protocol held in Vienna from 15 to 19 September 1999: Chairman's Summary," UN Doc. UNEP/CBD/ExCOP/1/INF/3 (20 October 1999), online: http://www.cbd.int/doc/meetings/cop/excop-01/information/excop-01-inf-03-en.pdf.

[22] Peter-Tobias Stoll, "Controlling the Risks of Genetically Modified Organisms: The Cartagena Protocol on Biosafety and the SPS Agreement" (1999) 10 Yearbook of International Environmental Law 82 at 92.

to transboundary movement.[23] Any Party may then request additional information from the authority responsible for the decision regarding the LMO-FFP.[24]

A Party may also make a decision on the import of the LMO-FFP under its domestic regulatory framework that is consistent with the objective of the Protocol.[25] In the absence of such a domestic framework, a developing country Party or a Party with an economy in transition can declare, through the BCH, that it will make a decision prior to the first transboundary movement of an LMO-FFP in accordance with the risk assessment provisions of the Protocol and within a predictable and limited time frame not exceeding 270 days. To date, only two countries have made such a declaration, namely Barbados and Saint Lucia.

As with the AIA procedure, the failure by an importing Party to communicate its decision within the prescribed time frame does not imply the Party's consent or refusal to the import of the LMO-FFP.[26] A Party may also take a decision on the import of an LMO-FFP on the basis of the precautionary approach and may take socioeconomic considerations into account. The provisions of Article 12 (review of decisions) also can apply to the procedure for LMOs-FFP, although the article was initially developed in the context of the AIA procedure.

In addition to Article 11, Article 18 is an important provision for transboundary movements of LMOs-FFP.[27] The procedure under Article 11 is supplemented by paragraph 2(a) of Article 18, which sets out requirements for documentation accompanying transboundary movements of LMOs-FFP. Article 18 requires the safe handling, packaging, and transport of LMOs. It stipulates that documentation accompanying LMOs-FFP should clearly identify that they "may contain" LMOs and are not intended for intentional introduction into the environment. This ambiguous wording is a result of the compromise reached in the final hours of the negotiations.[28] Although many delegations supported mandating clearer identification requirements, major exporting countries of LMOs stated that precise determinations of LMOs contained in bulk shipments of commodities was virtually impossible and financially prohibitive, given existing domestic systems for collection of seeds, grains, and cereals.[29] Reflecting the dissatisfaction of many delegates, the second sentence of Article 18(2)(a) provides that the Conference of the Parties serving as the meeting of the Parties to the Protocol (COP-MOP) will make a decision on detailed identification requirements no later than two years after the date of entry into force of the Protocol.

[23] *Cartagena Protocol on Biosafety*, *supra* note 1 at art 11.

[24] *Ibid* at art 11(3).

[25] *Ibid* at art 11(4).

[26] *Ibid* at art 11(7).

[27] See Chapter 6 for a more detailed discussion on Article 18.

[28] Ruth MacKenzie, "The Cartagena Protocol after the First Meeting of the Parties" (2004) 13 Review of European Community and International Environmental Law 270 at 274.

[29] Burgiel, *supra* note 21 at 55.

A decision was indeed made at the third meeting of the COP-MOP in 2006. The decision – BS-III/10 – sets out more specific identification requirements. Parties are urged to take measures ensuring that documentation accompanying LMOs-FFP clearly states that in cases in which the identity of the LMOs is known through means such as identity preservation systems, the shipment may contain one or more LMOs that are intended for direct use as food or feed, or for processing; or in cases in which the identity of the LMOs is not known through means such as identity preservation systems, the shipment may contain one or more LMOs that are intended for direct use as food or feed, or for processing.[30]

Parties are, however, supposed to consider a decision at their sixth meeting to ensure that documentation accompanying LMOs-FFP clearly states that the shipment contains these organisms and includes some of the detailed information elements set out elsewhere in the decision. This was to be based on an assessment and review during COP-MOP 5 of experience gained with the implementation of the relevant parts of decision BS-III/10. The information received for the fifth meeting of the Parties demonstrated there has been little experience to date with the implementation of the COP-MOP 3 decision.[31] The Parties thus agreed to postpone consideration of a decision until their seventh meeting. In addition to this lack of experience, it appears that the debate has shifted in recent years from how LMOs-FFP should be identified in shipping documentation to which scientific methods should be used for the sampling, detection, and identification of LMOs in bulk shipments of commodities like grains and oilseeds.[32]

ADVANCE INFORMED AGREEMENT AND PRIOR INFORMED CONSENT

It is often pointed out that the AIA procedure under the Protocol has similar features to the prior informed consent (PIC) procedures under other multilateral environmental agreements (MEAs)[33], notably the Convention on the Control of

30 Decision BS-III/10, "Handling, Transport, Packaging and Identification of Living Modified Organisms: Paragraph 2(a) of Article 18" (2006) at para 4(a) and (b).

31 "Handling, Transport, Packaging and Identification of Living Modified Organisms – Synthesis of information on experience gained with the implementation of requirements related to paragraph 2 (a) of Article 18." UN Doc UNEP/CBD/BS/COP-MOP/5/8, online: www.cbd.int/doc/meetings/bs/mop-05/official/mop-05-08-en.pdf.

32 For example, see "Handling, Transport, Packaging and Identification of Living Modified Organisms – Synthesis of Information on Experience Gained with the use of Sampling of Living Modified Organisms and Detection Techniques and on the Need for and Modalities of Developing Criteria for Acceptability of, and Harmonizing, Sampling and Detection Techniques (Paragraph 2(a) of Article 18)" (7 April 2008) UN Doc. UNEP/CBD/BS/COP-MOP/4/9, online: www.cbd.int/doc/meetings/bs/mop-04/official/mop-04-09-en.pdf. Work on these issues is also taking place under the Codex Alimentarius Commission, and the International Organization for Standardization and the International Seed Testing Association, among others.

33 David A Wirth, "Hazardous Substances and Activities" in Daniel Bodansky, Jutta Brunnée, and Ellen Hey, *The Oxford Handbook of International Environmental Law* (Oxford, UK: Oxford University Press,

Transboundary Movements of Hazardous Wastes and Their Disposal[34] (Basel Convention) and the Convention on the Prior Informed Consent Procedure for Certain Hazardous Chemicals and Pesticides in International Trade[35] (Rotterdam Convention). The three regimes do share the same core concept: they all require the consent of an importing country prior to the transboundary movement of certain products regulated by each treaty regime. Although there are similarities, the AIA procedure under the Cartagena Protocol differs from the PIC procedures in some important respects.[36]

PIC has been defined in international fora as the procedure requiring an importing country's consent for international shipment of a product that is banned or severely restricted to protect human health or the environment.[37] The Secretariat of the Rotterdam Convention states that "[t]he PIC procedure provides all Parties with an opportunity to make informed decisions as to whether they will consent to future imports.... All Parties are required to ensure that their exports do not take place contrary to an importing Party's import decision."[38] Similarly, Article 6 of the Basel Convention obliges each Party of export to inform the States concerned of the intended transboundary movement of hazardous wastes and to receive written consent for each transfer from the Party of import.

The term *advance informed agreement* (AIA) is contained in Article 19(3) of the *Convention on Biological Diversity* (CBD), which requires Parties to "consider the need for and modalities of a protocol setting out appropriate procedures, including, in particular, *advance informed agreement*, in the field of the safe transfer, handling and use of any living modified organism resulting from biotechnology that may have adverse effect on the conservation and sustainable use of biological diversity" (emphasis added). When the second meeting of the Conference of Parties to the CBD adopted decision II/5, in which they agreed to negotiate a biosafety protocol, they also decided that the protocol was to include an "appropriate procedure" for AIA

2007), 394 at 414–415; Stephen McCaffrey, "Biotechnology: Some Issues of General International Law" (2001) 14 Transnational Lawyer 91 at 96.

34 *Convention on the Control of Transboundary Movements of Hazardous Wastes and Their Disposal*, 22 March 1989 (entered into force 24 May 1992), 28 ILM 657.

35 *Convention on the Prior Informed Consent Procedure for Certain Hazardous Chemicals and Pesticides in International Trade*, 11 September 1998 (entered into force 24 February 2004), 38 ILM 1.

36 P Birnie, A Boyle, and C Redgwell, *International Law and the Environment*, 3d ed, (Oxford: Oxford University Press, 2009) at 640; Catherine Redgwell, "Biotechnology, Biodiverisity and Sustainable Development," in Francesco Francioni and Tullio Scovazzi eds, *Biotechnology and International Law* (Oxford, UK: Hart Publishing Ltd, 2006) 61 at 69.

37 See UNEP, Governing Council Decision 15/30 (25 May 1989); Philippe Sands, *Principles of International Environmental Law*, 2d ed (Cambridge, UK: Cambridge University Press, 2003) at 630; Cyrus Mehri, "Prior Informed Consent: An Emerging Compromise for Hazardous Exports" (1988) 21 Cornell International Law Journal 365 at 366.

38 Secretariat of the Rotterdam Convention, "The Prior Informed Consent (PIC) procedure," online: http://www.pic.int/.

with reference specifically to transboundary movements of LMOs.[39] Schoonejans points out that the AIA procedure became part of the terms of reference for the negotiations partly because developing countries were inspired by international efforts to regulate trade in hazardous substances, which centred on the concept of prior informed consent. He notes, however, that although there is a close resemblance between the AIA and PIC procedures, the latter term was avoided in the biosafety context partly because of concern that it would imply a link between the potential risks of LMOs and the known hazards of certain chemical products.[40]

Both the Basel and Rotterdam Conventions provide, in their annexes, lists of hazardous wastes or chemicals to be regulated. The PIC procedure under the Rotterdam Convention applies to international trade in the Annex III list of severely hazardous pesticides and other chemicals. Under the Basel Convention, "hazardous wastes" are listed in Annex I and II and Parties may consider other wastes as hazardous and regulate them under their national legislation. It is therefore clear that the PIC procedures under the Basel and Rotterdam Conventions apply to categories of substances and materials for which there is international agreement that they are hazardous in nature.

Conversely, the procedures under the Cartagena Protocol are set out and applied on an LMO-by-LMO basis.[41] This is the key difference from the procedures of the other two MEAs. The fact of a substance's being listed triggers the PIC procedures under the Basel and Rotterdam Conventions, whereas there is no listing under the Cartagena Protocol of LMOs that do or do not require AIA prior to transboundary movement.[42] Instead, it is the fact that the organism is genetically

39 Dec II/5, "Consideration of the Need for and Modalities of a Protocol for the Safe Transfer, Handling and Use of Living Modified Organisms," online: http://www.cbd.int/decision/cop/?id=7078.

40 Schoonejans, *supra* note 17 at 299.

41 Stoll, *supra* note 22.

42 Amanda Wolf, "The Emergence and Implementation of the Advance Informed Agreement Procedure" in Philippe G LePrestre, ed, *Governing Global Biodiversity: The Evolution and Implementation of the Convention on Biological Diversity* (Burlington, VT: Ashgate Publishing Co, 2002) 127 at 134. It should be noted that Article 7(4) of the Protocol provides that the AIA procedure shall not apply to the intentional transboundary movement of LMOs identified in a COP-MOP decision as being not likely to have adverse effects on the conservation and sustainable use of biodiversity, taking also into account risks to human health. Thus, the development of lists is possible, but it would create exemptions to the AIA procedure rather than triggering it. The COP-MOP has yet to make such a decision, although it has requested the submission of information in this area. See Dec BS-V/12, "Risk Assessment and Risk Management" at paras 12 and 13. Furthermore, Article 16 on risk management provides that the Parties are to cooperate to identify LMOs or specific traits thereof that could have adverse effects on the conservation and sustainable use of biodiversity, taking also into account risks to human health and to take appropriate measures regarding the treatment of such LMOs or traits. Consideration of this provision is only just beginning, see document UNEP/CBD/BS/COP-MOP/5/12, "Risk Assessment and Risk Management (Articles 15 and 16)" Dec BS-V/12 at paras 10 and 11, and operational objective 1.4 of the Strategic Plan for the Cartagena Protocol on Biosafety, Dec BS-V/16, "Strategic Plan for the Cartagena Protocol on Biosafety for the Period 2011–2020", Annex I.

modified and will be released into the environment that prompts the requirements for AIA. The procedures under the Protocol require establishing national decision-making systems based on risk assessment and risk management, without mandating a particular outcome.[43] Unlike the substances listed under the Basel and Rotterdam Conventions, there is no agreement at the international level as to the nature and extent of any special risks that any specific LMO may pose to the environment.

A second difference between PIC and AIA is the level of customization of information.[44] The Basel and Rotterdam Conventions start from the basis that their listed substances are hazardous to all. Standard information is distributed to all Parties to help them decide whether they consent to the import of the substance. Conversely, under the Cartagena Protocol, "[r]isk assessment should be carried out on a case-by-case basis. The required information may vary in nature and level of detail from case to case, depending on the living modified organism concerned, its intended use and the likely potential receiving environment."[45] The intention is that a risk assessment be tailored to the particularities of each country where the LMO in question is to be released.

CURRENT ISSUES AND CHALLENGES RELATED TO THE DECISION-MAKING PROCEDURES

More than eight years have passed since the Protocol entered into force on 11 September 2003. Many achievements have been made over the past eight years;[46] however, many things still need to be done. A review by the Secretariat in 2009 indicated that several developing countries had reported in their first national reports that they were still developing their national biosafety frameworks.[47] Clearly, it is essential to develop a national or regional biosafety framework addressing legal, administrative, and scientific aspects to ensure effective implementation of the Protocol, under which governments and their competent authorities are ultimately responsible for carrying forward the procedures mentioned previously and managing any risks that may result from the transboundary movement of LMOs.

43 Redgwell, *supra* note 36 at 72.

44 Wolf, *supra* note 42.

45 *Cartagena Protocol on Biosafety*, Annex III at para 6.

46 Executive Secretary of the CBD, Message on the Occasion of the Fifth Anniversary of the Entry Into Force of the Cartagena Protocol on Biosafety, "The Cartagena Protocol on Biosafety: Five Years of Global Cooperation Towards Sustainable Development" online: http://www.cbd.int/doc/external/mop-04/anniversary/statement-es-en.pdf.

47 "Review of General Issues of Compliance: Information-Sharing" UN Doc. UNEP/CBD/BS/CC/6/3 at paras 5–8, online: http://www.cbd.int/doc/meetings/bs/bscc-06/official/bscc-06-03-en.pdf.

There is an urgent need to build the required capacity to enable countries to implement the requirements under the Protocol. This includes building capacity to perform risk assessments, to identify LMOs in shipments and to make information available through the BCH, thus facilitating transparency and the sharing of information, all of which are vital to a dynamic and effective global biosafety system.

Although the amount of information reported to the BCH for all the main categories of data has increased over the years, the Secretariat's review shows that information such as national decisions on the approval of LMOs and associated risk assessment reports are not posted on the BCH in a complete and timely manner.[48] Article 20(3) of the Protocol requires that summaries of risk assessments be made available to the BCH. Currently, however, there is approximately twice the number of AIA decisions on the BCH as risk assessments reports. The Compliance Committee has considered this issue as part of its review of general issues of compliance under the Protocol, but enhancing the quality and completeness of the information in the BCH is a challenge for Parties and developing country Parties in particular.

A further challenge is the willingness of Parties to follow the spirit of the Protocol. The Protocol is concerned with the protection of the environment and biodiversity, hence the creation of the AIA procedure. The provisions of this procedure – most notably the requirements for risk assessment – do not apply to the import of LMOs for contained use. A Party may import an LMO for the purpose of contained use and then decide to release it into the environment. The release into the environment does not involve a transboundary movement, and so the requirements of the AIA procedure can be avoided. This can effectively create a loophole in the Protocol if Parties choose to use importation for contained use as a mechanism to avoid the requirements of the AIA procedure.

Article 20, which established the BCH, requires Parties to make available to the BCH their final decisions on the release of LMOs (Article 20(3)(d)). Thus, all releases of LMOs into the environment should be notified to the BCH regardless of whether the AIA procedure has been applied. Parties have been reminded of this obligation in the decision on the BCH adopted by COP-MOP 5.[49] Furthermore, Parties have other means at their disposal – most notably the simplified procedure in Article 13 – if they wish to exempt LMOs from the AIA procedure. It is clear, however, that Parties that import LMOs for contained use simply for the purpose of expediting release into the environment run against the spirit, if not the letter, of the Protocol.

48 *Ibid.*

49 Dec. BS-V/2, "Operation and Activities of the Biosafety Clearing-House" at para 1(a).

CONCLUSION

The decision-making procedures in the Cartagena Protocol reflect the political process through which they were developed. They are neither perfect nor foolproof, but they were the compromise that was available to meet the various interests represented in the negotiations. The different procedures for decision making on LMOs for intentional introduction into the environment, on the one hand, and LMOs-FFP, on the other, aim to keep the focus on the possible impacts of LMOs on biodiversity to meet the objective of the Protocol and, ultimately, the Convention.

6

Handling, Transport, Packaging, and Information

Thomas P. Redick

This chapter examines key issues involving Article 18 of the Protocol, in particular, the potential complications and trade disruption that can arise from handling, transport, packaging, and identification (HTPI) of living modified organisms (LMOs), with particular emphasis on the issue of identifying genetic events (i.e., the traits giving crops particular qualities). As Chapter 5 notes, little experience has been reported to the Secretariat of the Convention on Biological Diversity (SCBD) to date regarding the implementation of the decision relating to commodities shipped for FFP reached at the third Conference of the Parties serving as the meeting of the Parties to the Cartagena Protocol (COP-MOP 3). Given that limited information was submitted in advance of COP-MOP 5, case studies in HTPI will help to inform Parties that pass laws implementing Article 18.[1]

Experiences in testing for LMOs in the European Union (EU) and Japan provide examples of incidents of trade disruption that raise the potential to inhibit innovation in agricultural biotechnology. In contrast to the policies of the EU and Japan, many developing nations engage in genetic testing for genetic events only intermittently, with little disruption of trade. At the same time, developing nations' passing biosafety laws generally feature a requirement to recall any "new technology" that lacks regulatory approval.[2]

After a review of a sample of Parties' implementing laws, this chapter suggests a model for avoiding such costs in implementation of Article 18. The combined efforts of reasonable regulatory approaches (with reasonable tolerances for stray rDNA) and

[1] Dec UNEP/CBD/BS/COP-MOP/5/8, "Handling, Transport, Packaging and Identification of Living Modified Organisms – Synthesis of information on experience gained with the implementation of requirements related to paragraph 2 (a) of Article 18" (13 May 2010).

[2] Socialist Republic of Vietnam, *Law on Food Safety*, No 55/QH12/2010 at art 48, online: http://www.nafiqad.gov.vn/b-legal-documents?set_language=en.

industry stewardship might be able to prevent most of the unwanted commingling that causes trade disruption.

OVERVIEW OF APPROACHES TO HTPI IMPLEMENTATION

In general, two distinct regulatory approaches have emerged in implementing HPTI legislation: (1) a process-based approach (e.g., the EU regulatory system), which emphasizes the testing and tracing of any "genetic event" that lacks EU authorization, and (2) a product-based approach (e.g., Canada and the United States) that emphasizes risk-based regulatory action with more focus on the end use of the product and its potential risks.[3]

A 2008 United Nations University (UNU) study has observed that in practice, many national regulatory systems and international instruments appear increasingly to reflect a "mixed" approach, subjecting GMOs and GM products to both general and specific safety rules and standards.[4] These issues of implementation can be better understood by looking at a representative sample of nations passing laws implementing Article 18 of the Cartagena Protocol in three regions: (1) Europe (including the European Community [EC] and Turkey, which hopes to join the EC); (2) Asia (including Japan, Malaysia, and Vietnam); and (3) Africa (including Kenya and South Africa) being the focus of discussion. Each of these nations is trying, with varying levels of success, to find the right level of testing and detection of unapproved varieties, while also allowing imports to proceed. With reasonable approaches to the low-level presence (LLP)[5] of unapproved-for-import varieties of biotech crops, nations can promote safe and sustainable innovation in agricultural biotechnology.

The Objectives Underlying Article 18 Requirements

Article 18 of the Protocol provides for handling, transport, packaging, and identification of living modified organisms (LMOs) that move in intentional trade (i.e., transboundary movement).

3 See Claire Dunlop, "GMOs and Regulatory Styles" (2000) 9(2) Environmental Politics 149; Gregory Jaffe, "Regulating Transgenic Crops: A Comparative Analysis of Different Regulatory Processes" (2004) 13 Transgenic Research 5 at 19 [Regulating Transgenic Crops].

4 Sam Johnston et al, "Internationally Funded Training in Biotechnology and Biosafety: Is it Bridging the Biotech Divide?" (Yokohama, Japan: United Nations University Institute of Advanced Studies, 2008).

5 The concept of LLP is parallel to "adventitious presence" (AP), both of which have comparable meanings. Some interpretations distinguish between an LLP that is simply a marketing question and those that relate to regulatory status (some would limit AP to the latter).

As discussed in Chapter 5 of this book, Article 18.2(a) addresses transboundary movement of LMOs bound for food, feed, or processing (LMOs-FFP), whereas paragraph 18.2(b) addresses contained uses and paragraph 18.2(c) covers seeds. This chapter expands on Chapter 5 by discussing the experiences to date in implementing laws relating to HTPI. After offering observations on the intent of Article 18 requirements in this section, the third section discusses implementation of laws.

FFP Commodities under Article 18.2(a)

To control the potential import of unapproved varieties in FFP shipments, Article 18.2(a) requires Parties to "take measures to require that documentation clearly identifies that they 'may contain' living modified organisms and are not intended for intentional introduction into the environment."[6] Because this disclosure does not allow the importer to determine whether all of the genetic events that a shipment may contain are approved, it serves only to set up further testing or to enable GM food labeling. Although a "may contain" declaration does not give specific information about the LMOs present, its generic description of the contents of a shipment can enable routing of that shipment to appropriate uses (e.g., in animal feed and biofuel uses).

In any of the nearly fifty nations that have mandatory GM food labeling laws, this "may contain" designation can enable testing to determine if the GM content exceeds the applicable tolerance in that nation.[7] This disclosure may allow a nation with either a voluntary non-GMO labelling marketplace or a mandatory GM food-labelling law to avoid food recalls that arise from commingling of LMOs and products thereof at levels exceeding applicable tolerances.

Implementing legislation under Article 18 can require more disclosure of genetic events than the "may contain" designation previously mentioned. Developing nations passing laws implementing the Cartagena Protocol may choose to follow the EU traceability and labelling model and impose restrictive documentation or genetic testing of incoming shipments bound for food, feed, or processing without regard for risk-based analysis of the threats posed by low-level presence of an

[6] As noted in Chapter 4, Dec BS-III-10 set up two options for LMOs-FFP. First, identity-preserved LMOs-FFP shipment must state that the shipment "does contain" the listed LMOs-FFP, and second, LMOs-FFP in commercial production in the exporting nation that are not identity preserved need state only that they "may contain" LMOs-FFP on their invoice or other shipping document. See BRIDGES, "Constructive Ambiguity Saves LMO Labeling Discussions at MOP-3;" (2006) 6(5) BRIDGES Trade Bioresearch 1. The first level of shipments identifies the genetic event(s) it contains, whereas the "may contain" designation does not specify – it merely enables GM labeling or testing driven by marketplace concerns (eg, an importer planning to market food using voluntary non-GM labeling will test to determine which level of GM content the shipment "may contain").

[7] CBD Notification 2009-142, "Submission of information on experience gained with respect to identification of living modified organisms intended for direct use as food or feed or for processing (LMOs-FFP) – Decision BS-III/10 paragraph 2(a) of Article 18 of the Cartagena Protocol on Biosafety," online: http://www.cbd.int/doc/notifications/2009/ntf-2009-142-bs-en.pdf.

unapproved variety. They do so at risk of stopping shipments that are necessary to maintain reasonably low-priced supplies of food, feed, and fiber, however. For nations that seek self-sufficiency in meat, eggs, and dairy, as Russia and other nations have declared as their objective,[8] there is a clear need for reasonable, science-based approaches to feed safety. Without such an approach, a nation can encourage the import of conventional or organic corn that might contain higher levels of mycotoxins, while turning away shipments that contain low levels of Bt corn that is not yet approved.[9]

Article 18.2(b) Seed Exports

Seed shipments are addressed by Article 18.2(b).[10] Because seeds are bound for environmental use, there is generally more information required by national implementing laws. A shipping document for seed identifies the LMO by event code, with contact information and other useful guidance.

As of COP-MOP 4 in 2008, no incidents of trade disruption in seed and contained use LMOs had been reported to the SCBD, but disruption of trade in seed (e.g., Canadian canola, discussed later) has taken place. A survey conducted by the Global Industry Coalition and the International Seed Federation (ISF), however, suggests that shipments to and from Parties and non-Parties under Article 18.2(b) and (c), using the existing guidance provided by the Parties are working reasonably well.

Article 18.2(c) Contained Uses

Contained uses, including microorganisms or laboratory supplies, are subject to Article 18.2(c). Although not intended for environmental release, these exports are subject to documentation that is specific to the LMO being shipped.[11] Documentation for LMOs destined for contained use generally includes the identity and characteristics of the LMO, exporter-importer contact data, dates of the transboundary movement, if known, and handling requirements.

[8] Russian Federation, "Poultry and Products Semi-Annual 2010," online: http://www.themeatsite.com/articles/934/russian-federation-poultry-and-products-semiannual-2010. On 30 January 2010, President Medvedev signed Russia's Food Security Doctrine, seeking 85 percent self-sufficiency in meat by 2020.

[9] For example, Mexico's poor populations have a largely corn-based diet, which can cause adverse human health effects from mycotoxins. FDA-CFSCAN, First Mexico-U.S. Food Safety Research Conference (May 2000); Stacey A Missmer et al, "Exposure to Fumonisins and the Occurrence of Neural Tube Defects along the Texas-Mexico Border" (2006) 114 Environ Health Perspect 237, online: ehp.niehs.nih.gov/docs/2005/8221/abstract.html. The legal implications of this are discussed in Drew Kershen, "Health and Food Safety: The Benefits of Bt-corn" (2006) 61 Food Drug LJ 197 [Kershen, Bt].

[10] *Cartagena Protocol on Biosafety to the Convention on Biological Diversity*, 29 January 2000, 2226 UNTS 208 (entered into force 11 September 2003) at art 18.

[11] Croplife International, "Documentation Requirements," online: http://www.croplife.org/public/doc_requirements.

Once an LMO arrives in the importing nation for contained use, national laws specify the conditions for release. For example, the shipment of eggs for salmon that are LMOs intended to be grown in contained facilities might fall within the national laws and regulations relating to contained use.[12]

Article 18.3 Online Review of HTPI Issues

At COP-MOP 4, the Parties requested the Executive Secretary to hold an online conference to (1) identify the relevant standards with regard to handling, transport, packaging, and identification of LMOs; (2) identify where gaps exist; and (3) suggest how to fill these gaps. This conference discussed other international organizations with HTPI expertise to offer, including the Codex Alimentarius Commission for food and feed, the International Plant Protection Convention (IPPC) for plants, and the World Organisation for Animal Health (OIE) for animals. At COP-MOP 5, the Parties prepared a document on the need for and modalities of developing standards regarding HTPI.[13]

COP-MOP 5 Decision Postponing Review of HTPI Issues

At COP-MOP 5, given the limited experience gained to date in the implementation of paragraph 4 of decision BS-III/10, a final decision was made to "postpone the decision-taking referred to in paragraph 7 of decision BS-III/10 until its seventh meeting."[14]

Biosafety Clearing-House and Implementing the COP-MOP 3 Decision

At COP-MOP 3, decision BS-III/2 suggested that the Biosafety Clearing-House (BCH) be used as a reference point, in commercial invoices, for lists of the genetic events that a shipment "may contain."[15] The shipping document would list the "unique identifier code" (UI) and the BCH Web site providing further information. The BCH is used to make information available to Parties on existing rules and standards governing handling, transport, packaging, and identification of LMOs and is discussed in greater detail in Chapters 9 and 10 of this book.

[12] Andrew Pollack, "Genetically Altered Salmon Get Closer to the Table New York Times, 25 June 2010." GE salmon would be grown only in inland tanks or other contained facilities, and not in ocean pens where they might escape into the wild.

[13] "Handling, Transport, Packaging and Identification: the Need for and Modalities of Developing Standards (Paragraph 3 of Article 18)," UN Doc UNEP/CBD/BS/COP-MOP/5/9.

[14] Decision BS-V/8, "Handling, transport, packaging and identification of living modified organisms: paragraph 2(a) of Article 18" at para 61.

[15] Decision BS-III/2, "Operations and Activities of the Biosafety Clearing-House" at para 7.

Because the BCH is not always updated regularly, it can be an unreliable reference for information regarding the genetic events that a shipment of commodities may contain. Governments that are posting regulatory decisions to the BCH can take months or years to get these data onto the system.[16] This decision makes it mandatory for the BCH to keep updated records of UIs. The CBD and its Parties consider assignment of UIs for all biotech crops that a commodity shipment "may contain" to be a matter of importance.

The EC has financially supported the SCBD on regional training workshops for the identification of LMOs. The EC has also supported the SCBD of the CBD's work with the Green Customs Initiative on implementation of the Protocol documentation requirements.[17]

Although the posting of regulatory approval decisions is not always up to date for many nations, current information is generally available about the genetic events commercialized in exporting nations. Modern biotechnology developers maintain a database of products that differentiate among (1) those that have received an authorization but have never been in commercial production, (2) those in commercial production, and (3) the last seed sale date for those products that have been discontinued, to facilitate decisions on detection priorities.[18] This posting enables importing nations to determine, with a reasonable level of certainty, the genetic events that a shipment of FFP might contain. Despite these industry efforts to list events in commercial use, a joint review by the SCBD and seed industry of information posted on the BCH to date indicates many errors in the posting of regulatory approval decisions by parties.[19]

Economists studying the high cost of identifying genetic events that incoming shipments "may" contain have suggested that the BCH provide a "default" position

[16] Industry provides UIs for each genetic event in LMOs when available. The OECD gave biotech seeds unique identifiers. Distinguishing between biotech traits gives each trait a unique identifier signifying each genetic transformation event (e.g., resistance to herbicides or drought, or an ability to repel certain plant pests). Unique identifiers allow for detection and "traceability" of genetic traits – including those lacking approval for import. See Organisation for Economic Co-operation and Development (OECD), "Guidance for the Designation of a Unique Identifier for Transgenic Plants," online: http://bch.cbd.int/database/organisms/uniqueidentifiers/about.shtml.

[17] "Synthesis of information on experience gained with the implementation of requirements related to paragraph 2 (a) of Article 18," UN Doc UNEP/CBD/BS/COP-MOP/5/8.

[18] "Commercial Status of Certain Agricultural Biotechnology Products as of June 25, 2010," online: http://www.biotradestatus.org [Commercial Status].

[19] Croplife International, "The Biosafety Clearing-House Must Be Complete and Accurate," online: http://www.croplife.org/public/biosafety_clearing_house. "While the BCH is fully operational, the absence of information available limits its utility. Few Parties have posted the necessary information on the BCH that outlines how they intend to implement these provisions of the Protocol, or have otherwise clarified the applicable processes for such imports. In addition, much of the posted information is dated as most Parties do not appear to have a system for periodic updating of their postings."

for nations that cannot afford to test each incoming shipment. Gruere, Bouet, and Leroy studied the economic effects of implementing Article 18 to require traceability for genetic events (e.g., testing for an unapproved variety that is listed as "commercialized" by the exporting nation on the BCH) in the specific setting of maize exports.[20] Gruere et al. concluded that this testing would cause "trade distortion, diverting exports from their original destination. The measure would also lead to significant negative welfare effects, for all members of the Protocol and non-member that produce GM maize." As is discussed later, in the context of Kenya's new biosafety law, laws requiring genetic testing of incoming shipments could lead to high costs for testing to determine whether commodities contain an unapproved-for-import genetic event that has been commercialized in the exporting nation.[21]

Cartagena Protocol COP-MOP 5 (Nagoya, Japan)

At their fifth meeting, COP-MOP 5, the Parties postponed the difficult question of whether all LMOs-FFP shipments that "may contain" LMOs should be identified in shipping documentation or via postings to the BCH.[22] The Parties made more progress on issues of the scientific methodology for sampling, detection, and identification of LMOs in bulk shipments of commodities. It can be argued that these issues are progressing toward possible consensus, which could broaden the scope of testing methods beyond modern biotechnology and cover other and more serious threats to human health.[23]

REVIEW OF IMPLEMENTATION LAWS FOR HTPI

Protocol parties may choose to implement the Protocol through legislation or regulations that impose a system for tracing each trait that is in commercial use in the exporting nation. If a trait lacks import approval, this tracing could trigger recalls

20 Antoine Bouët, Guillaume Gruère, and Laetitia Leroy, "From 'May Contain' to 'Does Contain'. The price and trade effects of strict information requirements for GM maize under the Cartagena Protocol on Biosafety," Paper for Agricultural & Applied Economics Association's 2010 Meeting 25–27 July 2010).

21 J Huang et al, "Will the Cartagena Protocol hinder or protect the developing world: Learning from China's experience" (2008) 33(1) Food Policy 1.

22 Decision BS-V/8 at 61.

23 At the Thirty-First Session of the Codex Committee on Methods of Analysis and Sampling, Budapest, Hungary (8–12 March 2010) several delegations noted that there was no need to "place specific emphasis on foods derived from modern biotechnology." See "Report of the 31st Session of the Codex Committee on Methods of Analysis and Sampling" (ALINORM 10/33/23) at para 24, online: www.codexalimentarius.net/download/report/738/al33_23e.pdf.

or remediation of unlawfully imported traits that might be found in the importing countries' food supply or environment.

European Union Traceability and Labelling Directives

The EU Traceability and Labelling (T&L) Directives were passed to implement its obligations under the Cartagena Protocol.[24] The EU applies the "precautionary principle" to regulate a wide range of hypothetical adverse impacts of biotech crops, and it has implemented Cartagena Protocol Articles 11 and 18.2(a) to require a "precautionary approach" to regulatory approval for both seeds and FFP commodities. The EU also requires grain exporters from nations using biotech crops to specify the traits that a shipment of grain commodities "may contain" when they ship commodities to the EU.

The EU's 2004 law implementing the Cartagena Protocol (the Traceability Directive) recognized the commercial impacts that zero tolerance for unapproved traits could cause, and so the EU created a "grandfather" provision allowing some commingling ("adventitious presence") of crops lacking full EU approval. From 2004 to 2007, the EU had a 0.5 percent temporary tolerance for pending-approval genetic events (i.e., only those with approval in the exporting nation and a positive safety finding from the European Food Safety Authority [EFSA]). Although corn-based processed feeds (e.g., the distillers dried grains with solubles [DDGS], which can be a by-product of corn ethanol production) continued to flow to the EU through 2007, by early 2008 the EU officials began testing for new genetic events in corn that had been only partially approved in the EU (by approval for safety by the EFSA) and rejecting shipments of animal feed containing unapproved corn traits.

In the EU, trade disruption can occur even when both the exporting nation's food safety body and EFSA have reviewed the risks and recommended approval. Pending a full regulatory approval in the EU's complex comitology process (which allows EU member states to vote on each event), there is zero tolerance for any trace of biotech DNA or proteins in the corn-based feed. With corn-based feed exports dropping off steeply in early 2008, the EU feed industry woke up to the high cost of EU's zero tolerance, but perhaps too late to change this policy before it causes billions of euros in economic harm to the EU animal feeding industry.[25] This economic harm can arise in the exporting nation as well, leading to litigation over economic impacts in the exporting nation.[26]

[24] See EU Regulation 1830/2003, 2003 OJ (L 268) 24, online: http://ec.europa.eu/environment/biotechnology/pdf/regu1830_2003.pdf. EU Regulation 1829/2003, 2003 OJ (L 268) 1, online: http://ec.europa.eu/environment/biotechnology/pdf/regu1829_2003.pdf [T&L Directives].

[25] Joshua Chaffin, "Farmers face animal feed shortage" Financial Times, September 7, 2009.

[26] Liberty Link Rice for example.

Finding Flax in Seeds Shipped for Food

The EU's strict interpretation of the Protocol demands genetic information from an agricultural system that depends on commingling[27] and builds a certain level of anonymity of seed suppliers into the system.[28]

In late 2008, biotech flax with the Triffid genetic event was detected in 2008, and it began appearing in Canada's export to the EU, on the EU's online Rapid Alert System for Food and Feed (RASFF), which reports violations of the EU food laws (including the traceability directive). The unapproved biotech flax/linseed variety was found in cereal and bakery products in Germany. Because this variety is not authorised for food or feed use in the EU, a food recall is now under way, according to the zero tolerance standards that apply in the EU.

There are no GM varieties of flaxseed registered or commercially produced in Canada; in 1998, Canada's flax industry shunned Triffid, pending EU approval.[29] The Canadian Flax Council estimates that the EU would have taken 500,000 to 700,000 metric tonnes of Canadian flaxseed (about two-thirds of Canada's flax crop) in 2008, but this is doubtful, pending a full assessment of the commingling of the unapproved variety. Estimates of the economic impact vary with price assumptions[30] and the rules of partial equilibrium economic analysis, but the range of economic impact could be more than 100 million Canadian dollars.

In the past, Canada has endured a string of annual economic impacts as the result of Canadian canola seed exports[31] and canola grain commodities export,[32] which

27 For some crops, the commingling of various farmers' deliveries of grain allows moisture levels to be adjusted by commingling wetter with drier deliveries. Moisture control is critically important to the management of mycotoxins in grain. See European Mycotoxin Awareness Network, "Fact Sheets on HACCP – Prevention and Control," online: http://services.leatherheadfood.com/mycotoxins/index.asp.

28 American Seed Trade Association (ASTA), "Seed Purity–Understanding Label Information and Industry Practices" online: http://www.amseed.com/newsDetail.asp?id=118 (federal law allows up to 5 percent impurity from biotech or other varieties, and "pure seed" need only be "95 [percent] of the hybrid named on the label") [ASTA 5% Purity]. Seed companies have made it clear that zero tolerance for commingling of unapproved events is "not realistic." See International Seed Federation, "Motion on Adventitious Presence of GM Material in non-GM Seeds," online: http://www.worldseed.org/Position_papers/GMOmotion2001e.htm.

29 Ian Bell, "GM Flax Variety Is Gone, But Not Forgotten: Transgenic flax development was halted to protect the export market," Western Producer, March 15, 2001, online: http://archives.foodsafety.ksu.edu/agnet/2001/3–2001/ag-03–16–01–01.txt.

30 A recent high price of $810 CAD per metric ton prevailed in late 2008 with September 2009 prices in the $290 to $390 range per ton. At $400 per ton, the range of lost exports at risk from the commingling would be $200 million to $350 million.

31 "Renewed Optimism for Canadian Canola Exports to EU," 17 July 2009, online: http://www.ccga.ca/CCGA_News.aspx ("Canada's canola farmers are encouraged that the Government of Canada has resolved a six year trade dispute preventing genetically modified canola seed from entering the EU").

32 Max Foster, "GM Canola: What are its economics under Australian conditions?" online: http://www.jcci.unimelb.edu.au/GMCanola2007/ABARE%20report%202003.pdf. "The EU restrictions have provided an opportunity for Australian canola because of its non-GM status and has largely locked out Canada from the EU market."

are unapproved-in-EU varieties. This disruption included varieties fully approved in Canada, causing economic loss to growers. Economic loss for non-GMO farmers was litigated and not recognised as compensable to the growers.[33]

Tracing Seed Impurities in Rice

Two recent episodes of trade disruption in rice illustrate the challenges that lie ahead for both developed and developing nations that hope to use agricultural biotechnology to address food security issues. China has had exports disrupted by its Bt rice, which lacked approval in the EU and other Protocol parties, whereas the United States had exports to the EU of commodity rice disrupted by Liberty Link (LL) rice, which Bayer CropScience (Bayer) and researchers at Louisiana State University allegedly allowed to commingle with foundation rice seed in the United States. Both the Bt rice from China and the unapproved LL rice in the United States triggered costly recalls of food and years of genetic testing of food and grain shipments.

In 2007, the EU and Mexico reported finding LL610 in rice shipments, which led to the destruction of some rice shipments to the EU and temporary suspension of Mexico's imports of U.S. rice. At the close of 2007, the EU and the Philippines were both suspending rice imports pending USDA certification that U.S.-origin rice was free of LL rice. After an initial drop in rice futures in the United States, growers sued to seek recovery of financial losses caused by the commingled rice[34] under various theories, including private and public nuisance, asking for compensation for economic loss.[35]

A German importer sued growers (i.e., a grower cooperative, Riceland Foods) in Arkansas under contract law for violating a warranty. Unwittingly shipping

[33] *Hoffman v Monsanto Canada Inc et al*, 2007 SKCA 47, online: at http://www.saskorganic.com/oapf/pdf/2007SKCA047.pdf.

[34] An economist retained by Greenpeace estimated the range of possible economic impact between $741,223 and $1,284,626 over a five-year period. *In re LL Rice 601 Contamination Litigation*, MDL Docket No. MDL-1811 (ED MO), (Honorable Catherine D. Perry, Presiding), pleadings at http://www.llrice601contaminationlitigation.com/caseinformation.html#mdl).

[35] *In re Genetically Modified Rice Litigation*, 2009 US Dist LEXIS 98302. In April 2006, Bayer disclosed to USDA a commingling incident arising from field trials beginning in 1998 near seed-producing areas in Louisiana. Bayer's corporate predecessor, Aventis CropSciences, was responsible for the field trials that led to this seed impurity (the LL 601 rice genetic event). This revelation triggered an outcry in rice export markets, with genetic testing of U.S.-origin rice commodities shipments commencing in many overseas markets. Although the USDA had promptly approved this rice for use in food after it was discovered, this approval did not stop the trade disruption. Bayer's defenses included intervening causes "due to the intervening and/or superseding acts or omissions of parties or non-parties to this action for whose acts or omissions the BCS Defendants and Bayer Corporation are not liable," but in a summary judgment ruling, the court did not hold the handlers (Riceland, a grower cooperative) liable for any degree of comparative fault. LL601 MSJ at 91.

illegal-in-EU rice allegedly violated a regulatory compliance clause. The German company attached the 2003 EU traceability directive as an exhibit to the complaint.[36] Similar traceback litigation has not been reported as having been filed in China to date, but the disruption from Bt rice caused food recalls, which has presumably led to significant costs for EU importers. Claims for breach of contract can be arbitrated and are not necessarily disclosed publicly.

Tracing Corn Dust in Soybean Meal

In 2009, U.S.-origin soybean meal shipments were barred from unloading in Rotterdam because of their soy having tested positive for unapproved-in-EU biotech corn events (e.g., MON880, Syngenta's MIR 604, and two others). Germany had tested for unapproved "foreign material" biotech corn, and Spain soon joined in. As importers began cancelling purchases of U.S. soybean meal pending a resolution of this crisis, the economic costs began to mount. Industry estimates of economic impact of such disruption run into billions of dollars.[37]

The EU animal feeding industry was already operating in the red[38] from past disruption, beginning in 2007, of U.S.-origin corn gluten feed. The EU feed industry estimates that with higher feed costs for the EU, the livestock industry cannot be sustained economically if the zero tolerance regulatory mandate is not relaxed for animal feed and biofuel.[39] This could cost the EU feed industry and animal feeding operations several billion euros in losses owing to higher prices for imported food and feed.[40] This testing for foreign material is a new level of process-based restriction on exports and could prove to be an untenable position, particularly when other factors increase food prices.[41]

Like soybean oil, corn dust cannot reproduce and pose an ecological threat. At low levels, it also could not pose a health threat. Moreover, the potential for this to expand to other Parties to the Protocol makes such testing in the EU potentially more troubling. If other export markers follow the EU's lead, this could exponentially

36 *Rickmers Reismüehle Gmbh v Riceland Foods Inc*, Case No 4–07-CV00000733-JMM (ED. Ark. complaint on file with the author). Warranty is the strictest form of liability, with no intent required and few of the defenses found in tort law.

37 "Europe's Lag on Approvals Hurts Industry" Feed International, November 2009 (on file with author).

38 "GMO Zero Tolerance Devastating for EU feed industry," All About Feed, 19 July 2007, online: http://www.allaboutfeed.net/news/gmo-zero-tolerance-devastating-for-eu-feed-industry-id921.html.

39 Martin Roberts, "U.S. Soya Shipments Resume but More EU Woes feared," Reuters Newswire, 19 April 2010. "Last summer, more than 200,000 tonnes of soya beans and meal were refused entry to EU ports, largely in Spain, because they contained small traces of GMO maize varieties then forbidden in Europe." (On file with author).

40 COCERAL Press Release, "Concern Over Viability of EU Livestock Sector," 13 October 2008.

41 FEFAC Press Release, "EU feed and food price inflation on the rise," 3 August 2010, online: http://www.fefac.org/file.pdf?FileID=30415.

increase the impact of zero tolerance policies, contrary to the emerging Codex "adventitious presence" initiative.[42]

Ironically, EU consumers consume large amounts of imported meat products from meat-exporting nations feeding their animals biotech crops. This places livestock producers in the EU at a disadvantage, denying them access to GM feed sources that could be both safer and cheaper.[43]

Turkey

In late 2009, Turkey adopted a regulation related to the control of food and feed products containing GMOs that would approve only biotech crops that had been grown for three preceding years in the exporting nation (the "wait three years" law). Turkey decided in early May 2010 to stop implementing this overly restrictive approval law. Turkey's implementing directive now lists as approved EU-approved genetic events. Turkey is now allowing specified imports (e.g., soybeans) if they are free from any unapproved-in-EU varieties. Turkey has benefitted from its weaker environmental regulations for livestock, with lower costs than producers in western Europe, becoming the world's fourteenth largest poultry producer. With Russia seeking imports from Turkey, Turkey needs these imports to feed its animals at competitive prices.

Japan

Japan's approach to handling and transport issues lies somewhere between the very restrictive approach taken by the EU and the science-based approach to low-level presence that prevails in North American nations (under the trilateral agreement among Mexico, Canada, and the United States).[44] The Japanese system offers a model for imports using a commercial tolerance for non-GMO shipment (5 percent) that matches the trilateral arrangement in North America. Japan also provides a 0.5 percent tolerance for pending-approval biotech traits (i.e., in animal feed), which

[42] "Report of the Fifth Session of the Codex Ad Hoc Intergovernmental Task Force on Foods Derived from Biotechnology," (July 2006), online: www.codexalimentarius.net/download/report/653/al29_34e.pdf.

[43] Irish producers have stated, in an op-ed article, that 600 percent price increases for feed in Europe could cause cataclysmic impacts on EU feeding operations. Marybelle Ang, "A Slow Motion Tennis Match: GMO debate continues, American Bar Association Section on Environment, Energy & Resources Agricultural Management Committee Newsletter, January 2010, online: http://www.abanet.org/environ/committees/agricult/newsletter/archiveslist.html.

[44] "Documentation Requirements for Living Modified Organisms for Food or Feed, or for Processing," online: http://www.agr.gc.ca/itpd-dpci/to-su/4915-eng.htm [Tripartite NAFTA agreement] See also, "IGTC Notice #5," online: http://www.naega.org/images/IGTC_2005.pdf.

helps that nation to avoid the disruption to animal feed operations that the EU recently suffered.

Past Food Recalls in Japan: Pringle Potato Chips

For food products, there is a zero tolerance in place that has been enforced several times, including the recall of StarLink maize beginning in 2000 and Pringles brand potato chips in 2001.[45] Monsanto announced in 2000 that they were closing down production of Bt potatoes owing to a lack of market acceptance.[46]

In the summer of 2001, while StarLink maize was still being traced and recalled, Japan also ordered recall of 800,000 units of U.S.-origin potato chip products (Pringles) that were found to contain unapproved-in-Japan biotech potatoes. These Bt potatoes were approved in Canada and the United States and grown on 50,000 acres (about 4 percent of North America's potato crop in 1999[47]) but were not yet approved by Japanese authorities.[48]

Environmental Release: Japan's Handling of Unapproved Canadian Canola

Japan's management of biotech canola from Canada pending regulatory approval provides a good example of how to handle the discovery of an unapproved trait without triggering economically devastating trade disruption. Canadians grew Monsanto's RT73 canola event while Monsanto was still seeking regulatory approval in Japan, and this canola was apparently blown from Japanese railcars or other transport vehicles in transit to its destination in Japan in April 2005.

Although activists continue to find stray GM canola,[49] and Greenpeace has called for environmental remediation,[50] Japan found no adverse environmental impact and approved the canola for release.[51] Japan's approach did not depend on the genetic

45 Scott Kilman, "McDonald's, Other Fast-Food Chains Pull Monsanto's Bio-Engineered Potato," Wall Street Journal, 28 April 2000.

46 Ronnie Cummins, "Who's Winning the Frankenfoods Fight? News and Analysis on Genetic Engineering, Factory Farming, & Organics," The Coastal Post, June 2000, online: http://www.inmotionmagazine.com/geff8.html [Cummins, Who's Winning?].

47 Kilman, *supra* note 45. "Two biotech potatoes (NewLeaf Plus™ and NewLeaf Y™) with genes from the soil bacteria, *Bacillus thuriengiensis* (resistant to the Colorado potato beetle) were grown. In 2000, two large potato processors told farmers they would no longer accept gene-altered spuds, and McDonald's, Burger King, Frito-Lay, and Procter & Gamble declared non-GMO purchasing policies for brand-name French fries and potato chips. Monsanto closed down production of Bt potatoes due to lack of market acceptance." See also, Cummins, Who's Winning, *supra* note 46.

48 "Pringles Potato Chips in GMO Recall," Japan Times Online, July 18, 2001, online: http://search.japantimes.co.jp/cgi-bin/nn20010718b3.html.

49 Consumers Union of Japan, "Getting Ready for Nagoya," 15 September 2010, online: http://www.nishoren.org/en/. "Since 2004, groups of volunteers have participated in activities to investigate wild-growing GM canola near roads and harbours in many locations around Japan."

50 Greenpeace, "Genetically Engineered Canola Contamination across Japan," 24 May 2005, online: http://www.greenpeace.or.jp/campaign/gm/imgindex_e_html.

51 Japan is testing to keep the unapproved event out of food and feed. See "Response to Monsanto Petition 98-216-01p for Determination of Nonregulated Status for Glyphosate-Tolerant Canola Line RT73,"

testing of all incoming commodities and demand a food recall or a forcible return of shipments to North America. Instead it relied on a simpler form of postmarket monitoring. Japan controlled this using measures specified under the International Plant Protection Convention (IPPC),[52] which specify mowing along railsides and containing transport shipment.

Japan's policies are also noteworthy for its 5 percent tolerance for approved biotech crops in non-GMO bulk shipments, which allows it to import soybeans from the United States for non-GMO food products without triggering excessive recalls (in contrast to the EU's 0.9 percent tolerance for similar non-GMO shipments).[53] This standard makes it much more cost-effective to meet Japan's demand for non-GMO products.[54] In this author's opinion, other Asian nations should follow Japan's lead.

Malaysia

Malaysia became a signatory to the Cartagena Protocol on May 24, 2000, but Malaysia passed its law for the regulation of biotech crops (LMOs) only in 2009.[55] This law established a National Biosafety Board and Genetic Modification Advisory Committee that regulate imports of LMOs. The function of the advisory committee is to provide scientific, technical, and other relevant advice to the minister or the board.

Under this biosafety law, "No person shall undertake any release activity, or any importation of living modified organisms, or both without the prior approval of the Board" without risking a "fine not exceeding five hundred thousand ringgit and, in the case of a continuing offence, to a further fine not exceeding twenty thousand ringgit for each day." The Biosafety Regulations Advisory Committee (BRAC) is responsible for drafting regulations. Citing EU Directive 2001/18/EC, which excludes "organisms obtained through certain techniques of GM which

BIOTECH BASICS, January 1999, online: http://www.aphis.usda.gov/brs/aphisdocs2/98_21601p_com.pdf.

52 *International Plant Protection Convention*, online: http://www.ippc.int.

53 Policy Planning Division, Department of Food Safety, Pharmaceutical and Food Safety Bureau, MHLW, "Mandatory Labeling of Genetically Modified Foods and Foods Containing Allergens," online: http://www.mhlw.go.jp/english/topics/qa/gm-food/gm4.html. See also, "The Organic and Non-GMO Report, Japan's Legislation on Labeling of Genetically Engineered Foods," online: http://www.non-gmoreport.com/Japan_legislation_labeling_GM_foods.php. "[GM] foods include soy- and corn-based products, including tofu, natto, soymilk, miso, products made with soy ingredients, corn snacks, cornstarch, popcorn, and products made with corn ingredients. If the GM content of these 30 foods exceeds 5 percent they must be labeled as 'GM Ingredients Used' or 'GM Ingredient Not Segregated.' For a product to be labeled as 'Non-GM,' the GM content of the food must fall below 5 percent and the processor must be able to show that all non-GM ingredients were identity preserved from production through processing. The 5 percent tolerance applies only to GM varieties that have been approved in Japan."

54 Colin Carter and Guillaume Gruere, "International Approaches to the Labeling of Genetically Modified Food," 4 March 2003, online: http://www.agmrc.org/media/cms/cartergruere_16D28F0207B6C.pdf.

55 ICTSD Reporting Services, "Malaysian Biosafety Act in Force," Crop Biotech Update, 4 December 2009, online: http://ictsd.org/i/news/biores/65569/.

have conventionally been used in a number of applications and have a long safety record," the Government FAQ suggests that the BRAC may "simplify procedures for approvals" and may recognize automatic approval for import, based on approvals elsewhere in the world.[56]

The FAQ also cites the "precautionary approach" to LMOs, "where there are threats of irreversible damage, lack of full scientific evidence may not be used as a reason not to take action to prevent such damage" and states that Malaysia's obligations under the Cartagena Protocol on Biosafety compel such an approach.[57] One Malaysian commentator suggested that the draft biosafety bill's language, adopting Article 10(6) of the Protocol (in Clause 35), was a double-edged sword. The bill would allow Malaysia to bar an import of LMO commodities with no scientific basis for the refusal when it deems fit, but could also justify accepting LMOs-FFP that are not approved under its biosafety system that are necessary socially and economically. Malaysia appears to have imposed a more strict approval requirement for LMOs-FFP, but its addition of the word *appropriate*, in the eyes of that observer, leave it the option not to raise a barrier to unapproved varieties in LMOs-FFP shipments.[58]

This import law, like many others, links Malaysia's mandatory GM food labeling law to the Cartagena Protocol. Indeed, the generic phrase "may contain LMOs" on incoming FFP shipments serves only to inform importers with (1) a generic GM label or (2) an indication that further inquiry is required to determine the approval status of each LMO that the shipment may contain. Malaysia's FAQ suggests that its mandatory GM food label law "would also push exporting countries to segregate their GM and non-GM crops, shifting the burden to exporter countries, rather than to importing countries like Malaysia, to detect and identify GMO shipments."

Malaysia's approach to testing for traces of unapproved varieties of biotech crops has included periods when genetic testing for suspected unapproved varieties would take place for a period of 90 days. To date, however, Malaysia has not reported detection of any unapproved varieties.

Although its import laws embrace a precautionary approach more reminiscent of the EU than Japan, Malaysia's government is also very supportive of building its own biotech industry, particularly for the bio-prospecting of its rich biodiversity.[59] Its "BioValley" innovation corridor must coexist with Malaysia's rich biodiversity – the Biovalley activities will not be allowed to "exploit or deplete the country's rich pool of natural resources."[60]

56 Malaysia, *The Biosafety Act of Malaysia: Dispelling the Myths* (Kuala Lumpur: Ministry of Natural Resources and Environment, 2008) at 23.

57 *Ibid* at 18.

58 "Chris Kwan," Malaysia's Biosafety Bill – Throwing Precaution to the Wind?" (2007) 5 Current Law Journal, online: http://papers.ssrn.com/sol3/papers.cfm?abstract_id=1016008.

59 Azmi & Associates, "Biotech: Natural Resources, New Economy," Asia Law, October 2003, online: http://www.asialaw.com/Article/1972413/Channel/16961/Biotech-Natural-Resources-New-Economy.html.

60 *Ibid.*

Vietnam

In 2008, Vietnam's Biosafety Decree was issued by the Ministry of Environment and Natural Resources, while the Ministry of Agriculture and Rural Development (MARD) notified the WTO of draft regulations for biotech crops, with the comment period closing November 2009.[61] Vietnam's 2009 law implementing the Cartagena Protocol requires field trials before issuing safety certificates for imports of biotech crops. The MARD and the Ministry of Health (MoH) issues a Biosafety Certification good for 5 years, after which the applicants need to re-register.[62] Anyone handling biotech crops must comply strictly with risk management standards in handling and be prepared to remediate for harm. Anyone who produces, trades, imports, or exports biotech crops must assume responsibility for detection of unwanted releases, to remediate any consequences.[63]

Vietnam's mandatory GM food labeling law sets a 5 percent tolerance but will not be implemented for several years. GM foods are expected to reach Vietnam's market in 2015,[64] and animal feed already contains GMOs. With China launching a pollution-reducing phytase corn for feed use, Vietnam's flourishing animal feeding industry will find uses for biotech feed to control phosphorous runoff from feedlots and application of manure. Vietnam is embracing biotechnology's use in agriculture and as feed for its livestock and aquaculture operations.

Kenya

As the most developed economy in East Africa and home to noted biotech innovators,[65] Kenya uses its geographic location and industrial development to act as the trade hub for the east and central African region. It is also a leader in setting an example for biosafety legislation that balances the need for information with the need for food, feed, and other products from imported commodities that "may contain" LMOs.

Kimani and Gruere reviewed the potential for trade disruption in Kenya's top four imported LMOs-FFP (e.g., maize, soybeans, cottonseed, and rapeseed/canola). Maize is a particularly important food in Kenya; a tiny percentage goes to livestock, whereas soybeans and cottonseed are mainly used in animal feed. Soybean imports

[61] United States Department of Agriculture, Global Agriculture Information Network (GAIN) report VM8005, 17 January 2008, online: http://www.fas.usda.gov/gainfiles/200801/146293488.pdf.

[62] *Ibid* at 7.

[63] *Ibid* at 8.

[64] *Ibid.*

[65] Virginia Kimani and Guillaume Gruere, "Implications of Import Regulations Information Requirements under the Cartagena Protocol on Biosafety for GM Commodities in Kenya," April 2010, Table A4, including "Feathery mottle virus (SPFMV) Sweet potato" being developed with Florence Wambugu [Kimani and Gruere].

hit a peak of 45,000 tons in 2007, mainly from Uganda, before dropping off steeply, triggering shortages.

Production of maize dominates, and no approved biotech maize is grown in Kenya. Two-thirds of the growers grow maize, producing two-thirds of Kenya's needs. The rest has increasingly been imported, quadrupling between 1970 and 1990 and remaining stable since then. An import duty of 25 percent may be waived in times of need. U.S. exports are World Food Program grain that is sent for refugees. Testing has been conducted to confirm aflatoxin and moisture content, but not for LMOs to date, despite imports that probably contained maize, soy, or other crops approved in the exporting nation.[66] More than 90 percent of Kenya's maize imports are from South Africa, the first and largest producer of GM crops in Africa, despite official objections to GM imports. Non-GM maize sells at premium prices in Africa, and importers might be reluctant to pay high premiums.[67]

Kenya's new biosafety law does not mandate GM food labelling, but it may require some testing of imports of commodities, particularly those that disclose that they "may contain LMOs." LMOs-FFP are subject to a two-tiered import system, simplified for East African trading partners, and more complex for other nations. Since 2005, Kenya's KS EAS 2:2005 standard required shipments coming in under the more complex process to state if they contained LMOs, and although apparently this was not enforced, shipments from the United States and Argentina declined.[68]

In 2009, the new standard[69] requires that an import consignment clearly identify the LMO and its intended use,[70] with inspectors enforcing the 2008 Biosafety Act. Import control at border points is bound to be very challenging and costly, especially as it relates to land-based ports of entry. If traders have to pay for testing and validation of permits, the regulation will act as a new barrier to entry that could have significant effects, especially for small traders. In the short run, consignments coming from neighboring countries may bypass testing because of the frequency of imports.

Like many African nations, Kenya's borders are porous to informal trading of grain, and testing at major ports is not a complete containment strategy. As neighbouring nations in the region plant more GM crops, Kenya might have difficulty controlling the entry of such commodities. As such, given its inefficacy in preventing the import of LMOs approved in exporting nations, Kenya's move toward an EU-style testing

66 *Ibid* at 19.

67 *Ibid* at 15–16: "In 2009 and 2010, several news outlets reported that shipments of GM maize had been imported from South Africa despite the preference of the government ... Non-GM maize was sold at a premium of about 80–100 USD/tonne or up to 33% in 2009 (Boit, 2009) and reported in the media as being $300/tons in 2010 compared to $200 for conventional maize."

68 *Ibid* at 19.

69 "Code of Practice for Handling, Transfer and Use of Genetically Modified Organisms and Derived Products" (Kenya Bureau of Standards, 2009).

70 Kimani and Gruere, *supra* note 65 at 17–18. "Each consignment would have to be accompanied by a certificate of analysis clearly stating and verifying the GM events in the material."

and tracing for Kenya-approved genetic events could be a disincentive to wary grain exporters in the United States, Argentina, and South Africa.

As a result, Kimani and Gruere note that Kenya could have used the designation "may contain LMOs," which is accompanied by a reference to the BCH, where the events in commercial production in the exporting nation should be listed[71] in a regularly updated database. Indeed, any nation can follow this line, and if they do, according to Kimani and Gruere, "it will not have to account for accuracy of tests, and will focus its efforts on testing for and rejecting unapproved events (something that needs to be done to enforce import regulation)."[72] They suggest that a "may contain" approach with BCH information (as the COP-MOP 3 decision suggested), would work better for developing nations as opposed to strict information requirements. This approach would likely have limited impact on trade, if any.

In contrast, tracing every single genetic event raises costs significantly at export, costs that nations might not want to incur, and adds additional costs for public agencies. This can lead to price increases without obvious benefits for regulators,[73] which is to be expected given the similarity to the EU's traceability regime. As nations pass laws approximating the EU approach, the EU is slowly coming to grips with the high cost of testing and turning away shipments at zero tolerance for any trace of DNA (e.g., corn dust in soybean meal). As the EU reconsiders its regime as applied to animal feed, other nations follow it by adopting zero tolerance and testing for unapproved events, while Kenya's animal feeding operations suffer from shortages of necessary soy protein. Although Kenya might find identity-preserved production within its borders or with neighbors, and thereby ensure purity of some locally grown commodities, its main exporting neighbors (e.g., Tanzania and Uganda) are leaning toward commercializing biotech crops, including potentially LMOs in locally adapted maize.[74]

These conclusions suggest that the Kenyan government should reconsider supporting the "does contain" option as it has done in the past. Import regulations at the regional level, and a "may contain" approach, are likely to be more effective and certainly less costly for all parties.

Gruere, Bouet, and Leroy conclude that some economic benefits could accrue for nations preserving non-GMO exporter status (a class of nations that is narrowing and not growing, with Latin American nations producing more biotech acres with each passing year), whereas producers and consumers in parts of Africa will pay more for seed and food. They suggest that developing nations and other affected

[71] *Ibid* at 23.

[72] *Ibid.*

[73] *Ibid* at 27.

[74] *Ibid.* See also, Kristi Heim, "Gates Foundation's Agriculture Aid a Hard Sell," Seattle Times, 20 Jan. 2008, online: http://seattletimes.nwsource.com/html/localnews/2004135235_gatesagriculture20o.html.

Parties should reconsider their support for EU-style restrictive implementation of traceability regulations under Article 18.2(a).[75]

South Africa

South Africa exports to Kenya and shares an interest with other exporters who would like to avoid the high cost of testing, disrupted transport, delays in clearance of consignments, storage pending release, data recording, laboratory capacity, and other implications of strict testing of incoming LMOs-FFP shipments.[76]

South Africa's draft biosafety law, "Requirements for receiving, handling, transportation and storage of living modified organisms (LMO) not approved for general release," released in 2010,[77] would ensure that all activities using LMOs limit "possible harmful consequences to human and animal health, the environment as well as potential influences on the South African trade and industry." South Africa growers planted mostly GM maize (78 percent), soybeans (85 percent), and cotton (92–95 percent) in 2009, a 17 percent increase in the area totaling over 2 million hectares (an increase of 17 percent over the previous year), allowing South Africa to retain its position as the eighth largest producer of GM crops in the world. The cumulative benefits (income gain for farmers) from 1998 to 2009 from the sale of GM maize, soybean, and cotton was estimated at US $500 million.[78] No "harmful consequences" have been reported beyond the usual struggles to contain resistance in insects and weeds.

South Africa could have transporter liability in mind as well, because its 2010 draft law states that whenever unapproved "GM" grain is spilled in transit through South Africa, a "professional clean up shall be undertaken, under strict supervision of the owner of the spilled product" and that "insurance facilitators shall be advised of the loss."[79]

As Africa's largest producer of biotech crops, South Africa stands to lose the most from restrictive biosafety laws in other African nations that depend on its exports of maize. Bouet and Gruere estimate that South Africa could lose about $122 million per year, including $56 million for producers.[80]

[75] *Ibid* at 2 and 23.

[76] *Ibid* at 24–5.

[77] LMO imports are regulated under South Africa's *Genetically Modified Organisms Act, 1997 (Act 15 of 1997)*.

[78] Republic of South Africa, Department of Agriculture, Forestry & Fisheries, "Briefing of Parliamentary Committee on GMOs," online: www.pmg.org.za/files/docs/100504gmo.doc.

[79] South African National Standard 910:2009, Ed 1 at 7.1.9. "Requirements for receiving, handling, transportation and storage of living modified organisms (LMO) not approved for general release" online at: https://www.sabs.co.za/content/uploads/files/04–09.pdf. Accessed September 28, 2010.

[80] Bouët, Gruère, and Leroy, *supra* note 20 Table 11. Welfare effects (B3-Base) by country ranked in decreasing order of total surplus effects.

MODELS FOR IMPLEMENTATION WITH MINIMAL TRADE DISRUPTION

As is noted above, COP-MOP 3 established a "may contain LMO" option, with a reference to the BCH, that enables an importing competent authority to determine which LMOs a shipment is allowed contain (i.e., those in commercial use in the exporting nation). Complete risk information is also available, allowing a nation to assess whether the approval process in the exporting nation ignored a risk that the importing nation must assess.

Moreover, several international organizations, such as the IPPC, International Standards Organization (ISO), Codex Alimentarius (Codex), and the Organisation for Economic Co-operation and Development (OECD), have well-established work plans that focus on development and harmonisation of systems and standards for LMOs in commerce. These organisations incorporate applicable scientific expertise and experience to determine the appropriate integrated systems, standards, and specifications to best enable global trade in LMOs. Paired with industry stewardship, these preexisting systems allow Parties to manage trade without undue disruption.

Role of Industry Stewardship

In the United States, commercial launches of biotech crops have sometimes been accompanied by conscientious product stewardship – that is, environmental management – going well beyond compliance requirements. Insect resistance issues were managed first through voluntary programs that became regulatory permit requirements,[81] and the Biotechnology Industry Organization, based in Washington, D.C., has stewardship programs and quality management systems for the full life cycle of biotechnology-derived plant products.[82] Such stewardship programs promote compliance with applicable regulatory requirements, plant product integrity, and prevention of trade disruption.

Given the need to protect export markets for soybeans, a high level of attention is paid to avoiding the commingling of varieties lacking approval in the EU or other overseas markets. The American Soybean Association has a long-standing policy with the grain trade (e.g., North American Export Grain Association) of requiring "major market approval" for all biotech soybeans, so generic biotech soybeans that are not approved in the EU or other major market are prevented from being grown in the U.S. farm belt. A variety fully approved for unrestricted planting in the exporting nation, but lacking overseas import approval, can cause trade disruption.

[81] Stanley H Abramson and J. Thomas Carrato, "Crop Biotechnology: The Case for Product Stewardship" (2001) 20 Virginia Environmental Law Journal 241, online: http://nationalaglawcenter.org/assets/bibarticles/abramsoncarrato_case.pdf.

[82] See Commercial Status, *supra* note 18. Biotechnology Industry Organization, "Excellence Through Stewardship," online: http://excellencethroughstewardship.org/.

Codex "Adventitious Presence" Model

To implement harmonized controls for the incidental low-level presence (or "adventitious presence") of biotech crops that are approved in the exporting nation but not yet approved in importing nations, the Codex Alimentarius Ad Hoc Task Force on Foods Derived from Biotechnology is developing international guidelines for low level presence of GM material in food commodities. In 2008, the Codex Alimentarius Commission and its member countries approved Annexes on Food Safety Assessment in Situations of Low-Level Presence of Recombinant-DNA Plant Material in Food (LLP Annex), Food Safety Assessment of Foods Derived from Recombinant DNA-Plants Modified for Nutritional or Health Benefits, and the Guideline for the Conduct of Food Safety Assessment of Foods Derived from Recombinant-DNA Animals. The guidelines leave it up to national authorities to determine when a recombinant-DNA plant material is present at a level low enough for the Codex provisions to be appropriate and do not preclude national authorities from conducting a safety assessment.[83] If implemented, China might feel more ready to commercialize several products in its pipeline, notably, phytase corn and Bt rice. Although Asia appears likely to use these Codex guidelines, the EU's "precautionary approach" will likely continue to disrupt trade there as long as unreasonable fears overcome scientific risk assessment.[84]

International Plant Protection Convention Model

The IPPC was established to promote appropriate measures to prevent and control the spread and introduction of pests of plants and plant products. Its objectives include the development and application of international standards in international trade to prevent introduction and dissemination of plant pests. The Convention addresses natural flora and plant products, is not solely concerned with transborder transfer, and covers direct and indirect damage by pests, including weeds.

The IPPC establishes international standards for phytosanitary measures (ISPM) that could be relevant to LMOs, such as ISPM No. 11 on pest risk analysis for quarantine pests, including analysis of environmental risks and LMOs. Other codes of conduct developed under the IPPC could be relevant to the transboundary movement of LMOs to some extent, for example, the Code of Conduct for the Import and Release of Exotic Biological Control Agents.

83 ALINORM 08/31/34, Ad Hoc Intergovernmental Task Force on Food Derived from Biotechnology, Joint FAO/WHO Food Standards Program, Codex Alimentarius Commission, 31st Session, 30 June – 5 July 2008, para 106 and Appendix IV, online: http://www.codexalimentarius.net/web/archives.jsp.

84 Andrew Apel, "The Tolerance of Food Contamination in Europe," online: http://www.cropgen.org/european_food.pdf.

CONCLUSION

In evaluating the biosafety legislation of various Parties to the Protocol, there is a trend toward greater acceptance of biotech crops, both in planting and in the food, feed, and fuel supply. In the case of the EU and some of its trading partners (e.g., Turkey), however, there is a trend toward increasingly restrictive regulation of imports, with tests to determine whether any trace of unapproved variety is present. In many nations, the planting of biotech crops remains controversial and strictly regulated at the national level, often in a manner penalising imports of feed and fuel sources, even those that contain small amounts of biotech crop content.

Incidents involving trade disruption or the inhibition of innovation can be very costly for industry actors. Trade disruption can cause significant financial impact while inhibiting innovation in biotechnology (particularly in agriculture and forestry), thereby causing on indirect adverse impact. These ill effects include loss of investment in biotechnology (e.g., a $100 million write-off on a corporate balance sheet for a product not marketed) and the impact on human and animal health that, in the view of this author, comes with stifling such innovation. Finally, trade disruption that limits the launch of new biotech crops can lead to lost opportunity costs from the failure to use the best available technology for agricultural production.

7

The Importance of Public Participation

Christine Toczeck Skarlatakis and Julian Kinderlerer

Continuing advances in an increasingly technological society certainly will have an impact on the capacity of policy makers to make informed decisions about the safe use of biotechnology. There are, however, several potential benefits and risks associated with the use of these technologies that are still unknown, or at least are not in the public domain. All of these latent prospects have affected the uptake of new technologies in agriculture, trade, food supply, human health, social and economic order, environment, and ultimately, public perceptions of biotechnology.

The benefits and risks of modern biotechnology were described relatively early in the development of the technology and its commercialisation. Almost as soon as it was possible to cut and paste DNA within and across species barriers, the potential of the science was realised. Both potential benefits and harm were appreciated. In 1974, Paul Berg and others wrote a letter published in *Science* calling for a self-imposed moratorium until a meeting had been held to discuss the "potential biohazards" among the purely scientific issues. Its purpose would be to "review the progress, opportunities, potential dangers and possible remedies associated with the construction and introduction of new recombinant DNA molecules into living cells."[1] The meeting was held in Asilomar, California, in 1975, with the goal of identifying the measures needed to ensure the safe use of the new technology. "Asilomar remains an important scientific landmark, a rare if not unique instance of scientists independently questioning and successfully regulating their own cutting-edge work."[2] Asilomar could be considered one of the first uses of the precautionary principle. The possible impact of modern biotechnology on public perception was

[1] P Berg, D Baltimore, HW Boyer, et al, "Potential biohazards of recombinant DNA molecules" (1974) 185 Science 303, online: http://www.sciencemag.org/cgi/issue_pdf/toc_pdf/185/4148.pdf.

[2] E Russo, "Reconsidering Asilomar – scientists see a much more complex modern day environment" (2000) 14(7) The Scientist 14, online: http://www.the-scientist.com/article/display/11789/.

foreseen right at the beginning, before much of the science had been identified and the ability to use that science effectively had been achieved.

Public perceptions affect policy decisions, which in turn are also directly influenced by policy decisions. These complex interconnections are better grasped in the context of public participation in decision-making processes, whereas credibility, effectiveness, and legitimacy of any policy makers' decisions depend on the extent to which public and key stakeholders take part in the process. Policy decisions that accommodate diverse public participation, and therefore take into consideration public perceptions, might be more likely to elicit public trust in public institutions.

Public participation is a question of relevance in international sustainable development law as well as in controversies and debates surrounding innovative technologies, such as biotechnology. As noted in the Introduction and Chapter 1 of this book, the commitment to facilitate public participation, supported by access to information and access to justice, is an important principle of many international treaties on sustainable development, including the Cartagena Protocol. Examining how public participation works in different worldwide scenarios, what the mechanisms and challenges are for effective public participation in biosafety policy, and how all of these aspects are tackled by international and national sustainable development regulations reassure everyone that biosafety is a sustainable development issue.

As a component of regulatory frameworks, public participation in the biosafety policy process is an important and crucial element to address and guarantee the safe use of biotechnology. In the international scenario, the Cartagena Protocol, among other international instruments, recognises public participation as an essential element in the decision-making process for living modified organisms (LMOs) as well as of sustainable development and good governance. Governance should involve all stakeholders, with a cross-section of perspectives and aspirations, to implement effectively the role of public participation in the decision-making process for LMOs. The Cartagena Protocol requires Parties to promote and facilitate public awareness, education and participation. The Protocol also requires Parties (within the contest of national law) to consult the public in developing policy about the use of LMOs.

This chapter attempts to highlight the importance of public participation in the decision-making process surrounding the safe use of biotechnology. Essentially, the chapter examines public participation in the context of international sustainable development law, dimensions of the concept of public participation, challenges to be faced in biosafety policy, mechanisms for effective public participation, and global perceptions in the biosafety and biotechnology debate. The discussion concludes with some relevant notes on why public participation is important in the present and the future of biotechnology from the perspective of sustainable development, transparency, and good public governance.

PUBLIC PARTICIPATION AND INTERNATIONAL SUSTAINABLE DEVELOPMENT LAW

Public participation in the decision-making process related to environmental issues has been flagged as an essential component to ensure sustainable development and good governance. Most recent international instruments, such as the Rio Declaration on Environment and Development (Rio Declaration)[3] and the Convention on Access to Information, Public Participation in Decision-Making and Access to Justice in Environmental Matters (Aarhus Convention),[4] have recognised that participatory partnerships play an important role in furthering an equitable society responsive to the sustainable development rationale.

In 1986, the United Nations General Assembly's *Declaration on the Right to Development* first underscored public participation as a central developmental objective. Article 1 of the Declaration recognises that "every human person and all people are entitled to participate in, contribute to, and enjoy economic, social, cultural and political development, in which all human rights and fundamental freedoms can be fully realised."[5]

The relation of public participation and international sustainable development law is identified in the text of the Brundtland Report while defining inherent concepts of sustainable development. According to the report, "Humanity has the ability to make development sustainable to ensure that it meets the needs of the present without compromising the ability of future generations to meet their own needs."[6] To meet essential needs, however, equity issues shall be addressed and supported by "political systems that secure effective citizen participation in decision-making and by greater democracy in international decision-making."[7] The Brundtland Report identified effective public participation as a necessary component of sustainable development.[8]

Public participation and access to information and justice have been recognised as principles of international law relating to sustainable development in the *New Delhi Declaration of Principles of International Law Relating to Sustainable Development*,[9]

[3] *Rio Declaration on Environment and Development: Report of the United Nations Conference on Environment and Development*, UN Doc A/CONF.151/6/Rev.1 (1992) [Rio Declaration].

[4] *Convention on Access to Information, Public Participation in Decision-making and Access to Justice in Environmental Matters*, 25 June 1998, online: http://www.unece.org/env/pp [Aarhus Convention].

[5] *Declaration on the Right to Development*, 4 December 1986, UNGA Res 41/128.

[6] G Brundtland, *Our Common Future: World Commission on Environment and Development* (Oxford, UK: Oxford University Press, 1987) at 8 [Brundtland Report].

[7] *Ibid.*

[8] Marie-Claire Cordonier Segger and Ashfaq Khalfan, *Sustainable Development Law: Principles, Practices and Prospects* (Oxford, UK: Oxford University Press, 2004) at 157.

[9] International Law Association, Resolution 3/2002, *New Delhi Declaration of Principles of International Law Relating to Sustainable Development*, online: http://www.ila-hq.org. See also International Law Association, Committee on the Legal Aspects of Sustainable Development, Searching for the Contours of International Law in the Field of Sustainable Development, Fifth and Final Report (London, UK:

adopted by the International Law Association in 2002. In addition, the *International Covenant on Civil and Political Rights*[10] and the *Universal Declaration of Human Rights*[11] identify, in Articles 25 and 21, respectively, rights of participation at the national levels.

More recent international instruments, in particular key legal texts produced during the United Nations Conference on Environment and Development,[12] the Earth Summit, namely, the *Rio Declaration*, *Rio Programme of Action*, and *Agenda 21*,[13] have recognised effective public participation as a principle and a necessary element of sustainable development.[14]

The Rio Declaration builds on basic perceptions concerning the attitudes of individual people and nations towards the environment and development, first identified at the United Nations Conference on the Human Environment (UNCHE) and the Stockholm Declaration on the Human Environment.[15] The preamble to the Rio Declaration calls for the establishment of a new and equitable global partnership which requires a high level of cooperation among States and with other sectors of society. Nevertheless, it is Principle 10 of the Rio Declaration that clearly refers to the importance of public participation in environmental governance. Principle 10 establishes that

> [e]nvironmental issues are best handled with participation of all concerned citizens, at the relevant level. At the national level, each individual shall have appropriate access to information concerning the environment that is held by public authorities, including information on hazardous materials and activities in their communities, and the opportunity to participate in decision-making processes. States shall facilitate and encourage public awareness and participation by making information widely available. Effective access to judicial and administrative proceedings, including redress and remedy, shall be provided.[16]

During the early 1990s, there was a belief that modern biotechnology could have a substantial beneficial impact on the world. Chapter 16 of Agenda 21, agreed on in 1992 in Rio de Janeiro (by most of the nations of the world), stressed the benefits that could flow from using our ability to manipulate biologic systems:

2002). Also reprinted in UN General Assembly, 57th session, UN Doc A/57/329 (2002; hereinafter New Delhi Principles).

10 *International Covenant on Civil and Political Rights*, 19 December 1966 (entered into force 23 March 1976), 999 UNTS 171, 6 ILM 368.

11 *Universal Declaration of Human Rights* (1948) UNGA Res 217 (III), UN GAOR 3d Session Supp No 13 UN Doc A/810.

12 Rio Declaration, *supra* note 3.

13 *Agenda 21: Report of the UN Conference on Environment and Development* (1992) UN Doc A/CONF.151/26/Rev.1 (1992).

14 Rio Declaration, *supra* note 3, Principle 10.

15 *Stockholm Declaration on the Human Environment*, (1972) UN Doc A/C.48/14 . [Stockholm Declaration].

16 Rio Declaration, *supra* note 3.

> By itself, biotechnology cannot resolve all the fundamental problems of environment and development, so expectations need to be tempered by realism. Nevertheless, it promises to make a significant contribution in enabling the development of, for example, better health care, enhanced food security through sustainable agricultural practices, improved supplies of potable water, more efficient industrial development processes for transforming raw materials, support for sustainable methods of afforestation and reforestation, and detoxification of hazardous wastes. Biotechnology also offers new opportunities for global partnerships, especially between the countries rich in biological resources (which include genetic resources) but lacking the expertise and investments needed to apply such resources through biotechnology and the countries that have developed the technological expertise to transform biological resources so that they serve the needs of sustainable development.[17]
>
> It was thought that biotechnology could assist with
>
> a. Increasing the availability of food, feed and renewable raw materials
> b. Improving human health
> c. Enhancing protection of the environment
> d. Enhancing safety and developing international mechanisms for cooperation
> e. Establishing enabling mechanisms for the development and the environmentally sound application of biotechnology[18]

Indeed, Agenda 21 outlines the importance of broad public participation in the decision-making process to contribute to the attainment of sustainable development goals. The document encourages the participation of different actors with a cross-section of perspectives, such as government, nongovernmental organisations, industry, academia, women, youth, and indigenous people. The participation and inclusion of these groups were also recognised in the 2002 World Summit for Sustainable Development (WSSD) in Johannesburg[19] as a key element to the success of sustainable development.

Although the underlying principles presented at the Earth Summit and in other international instruments represent the foundation of participatory, transparent, equitable, and accountable governance by means to pursue sustainable development goals, they do not refer to public participation as a right.

In 1995, to put this principle into practice, the member States of the United Nations Economic Commission for Europe (UNECE) adopted the *Sofia Guidelines*.[20] This document frames the importance of public information, public participation, and access to justice on environmental issues. In 1998, governments took a further step

[17] Agenda 21, *supra* note 13 at Chapter 16, para 1.

[18] *Ibid.*

[19] *Johannesburg Plan of Implementation, Report of the World Summit on Sustainable Development*, 4 September 2002, UN Doc A/CONF.199/20 (hereinafter Johannesburg Summit 2002).

[20] United Nations Economic Commission for Europe, *Draft Guidelines on Access to Environmental Information and Public Participation in Environmental Decision-Making*, ECE/CEP/24, Ministerial Conference Environment for Europe, 23–25 Oct 1995.

and adopted the Aarhus Convention. This far-reaching international agreement relating to public participation has also made a significant contribution to the development of the principle of public participation and represents a step forward in international law. This treaty defines and provides guidance on how public participation should be reached in the context of environmental governance. Among the provisions of the instrument, it acknowledges the right of every person to "live in an environment adequate to his or her health and well-being, and the duty, both individually and in association with others, to protect and improve the environment for the benefit of present and future generations."[21] The convention is one of the first binding international instruments to recognize the right to a healthy environment.

To assert the right and observe the duty for a healthy environment, the Aarhus Convention recognises that "citizens must have access to information, be entitled to participate in decision-making and have access to justice in environmental matters,"[22] and that Parties to the convention shall guarantee such rights. The provisions of the Convention embody a broad range of requirements, such as adequate, timely, and effective public notification, provision of relevant information, provision of opportunities for public comment, and responses to such comments. Most important, the convention recognises the citizen's rights to participate in the decision-making process for a healthy environment founded on the principles of sustainable development.

The acknowledgement of public participation as a principle and right in international law is embodied in the text of many other international legal documents, mainly related to sustainable development law. Among the international instruments that identify the right to public participation in relevant decision-making processes are the International Labour Organization (ILO) 1989 Convention on Indigenous and Tribal Peoples; 1994 Draft Principles on Human Rights and the Environment; 1994 Draft UN Declaration on Indigenous Peoples; and the International Union for the Conservation of Nature and Natural Resources (IUCN) Draft Covenant on Environment and Development. Other international environmental conventions that include the public participation principle are the 1994 UN Convention to Combat Desertification in those Countries Experiencing Serious Drought and/or Desertification, Particularly in Africa; 1992 UN Convention on Biological Diversity; 1992 UN/ECE Convention on the Transboundary Effects of Industrial Accidents; 1992 UN Framework Convention on Climate Change; and the 1991 UN/ECE Convention on Environmental Impact Assessment in a Transboundary Context.

In particular, the Aarhus Convention addresses public concern related to the deliberate release of genetically modified organisms (GMOs) into the environment and the need for increased transparency and greater public participation in the decision-making process for LMOs. Article 6 (11) of the convention determines

[21] *Aarhus Convention*, *supra* note 4 at Preamble.

[22] *Ibid.*

that the provisions associated with public participation shall apply to "decisions on whether to permit the deliberate release of genetically modified organisms into the environment";[23] however, the implementation of the terms of this provision shall comply with each Party's national regulatory framework. Discussion on how this could effectively be put into practice continues.

The Convention on Biological Diversity,[24] also signed in 1992, places the possibility of developing a protocol to ensure that the use of modern biotechnology poses minimal risk to conservation and the sustainable use of biodiversity within Article 19 titled "Handling of Biotechnology and Distribution of its *Benefits*" (these authors' emphasis). Moreover, Article 14(1) (a) of the Convention encourages public participation in environmental impact assessment of projects that are likely to have an adverse impact on biological diversity.

More specifically, the Cartagena Protocol on Biosafety to the Convention on Biological Diversity[25] requires public participation, awareness, and consultation concerning the "safe transfer, handling and use of living modified organisms in relation to the conservation and sustainable use of biological diversity, taking also into account risks to human health."[26] The Protocol is a strong example of sustainable development law, using economic measures to promote safer use of biological technologies.[27]

Several international treaties in the field of sustainable development have included, among their provisions, the public participation component as well as governments' accountability to ensure public engagement.[28] Conceivably, the involvement of different groups of stakeholders is vital to ensure transparency and effectiveness of any regulatory framework. Neither environmental nor developmental strategies are likely to be sustainable unless all affected actors, both State and nonstate, are involved in decision making.[29]

DIMENSIONS OF THE CONCEPT OF PUBLIC PARTICIPATION

The recognition of the principle and right to public participation in international law and sustainable environmental governance principles reflects the emergence of the three dimensions of the concept of public participation, which are essentially enunciated in Principle 10 of the Rio Declaration: (1) the access to adequate information; (2) the opportunity to participate in environmental decisions that affect the

[23] *Aarhus Convention*, *supra* note 4 at art 6.

[24] *Convention on Biological Diversity*, 5 June 1992, 1760 UNTS 79, 31 ILM 818 (entered into force 29 December 1993) at arts 8 and 10.

[25] *Cartagena Protocol on Biosafety to the Convention on Biological Diversity*, 29 January 2000, 2226 UNTS 208 (entered into force 11 September 2003), online: http://bch.cbd.int/protocol/text/.

[26] *Ibid* at Article 23(1)(a).

[27] *Sustainable Development Law: Principles, Practices and Prospects*, *supra* note 8 at 109.

[28] *Ibid* at 160–3 and 183.

[29] *Ibid* at 164.

public; and (3) people's access to mechanisms of redress and justice when their rights are violated or when they have suffered harm.

Public access to adequate information is a key component of public participation. This point is emphasised in Agenda 21 and Aarhus Convention texts, which outline that public authorities must provide the public with access to all relevant information to the decision-making process. Obtaining information is crucial to public participation in the decision-making process, particularly with environmental impact processes or other permitting procedures.[30] In particular, public access to transparent information as well as tools to understand the information would help to identify the potential risks and benefits of modern biotechnology.

The opportunity for individual people to participate in official decision-making activities that directly affect them and their welfare is an important dimension of public participation. In this context, the Aarhus Convention requires Parties to ensure that public participation in decision making related to environmental impact take place in the early stages of the process. Moreover, public opinion, as gauged through the procedures for public participation, must be taken into consideration. Public participation must surpass pro-forma requirements and be integrally implemented into practice.

Ultimately, any person who considers that his or her rights concerning environmental matters were violated shall have access to judicial or administrative processes within the framework of its national legislations. The preamble to the Aarhus Convention emphasises that "effective judicial mechanisms should be accessible to the public, including organizations, so that their legitimate interests are protected and the law is enforced." Besides, Article 9 of the Convention requires Parties to ensure access to justice, within the provisions of its domestic laws, in three circumstances: (1) Parties must ensure access to a review procedure before a court or another independent and impartial body established by law for any person who considers that his or her information request has been ignored, wrongfully refused, inadequately answered, or not dealt with in accordance with the Convention's access to information provision under article 4; (2) Parties must ensure that "members of the concerned public" have access to a review procedure before a court or another independent and impartial body established by law to challenge "the substantive and procedural legality of any decision, act or omission subject to the provisions of article 6" and other relevant provisions of the Convention; and finally, (3) Parties shall ensure that "members of the public have access to administrative or judicial procedures to challenge acts and omissions by private persons and public authorities which contravene provisions of its national law relating to the environment."

[30] A Kiss and D Shelton, *International Environmental Law*, 2nd ed. (New York: Transnational Publishers, 1994) at 204 [Kiss & Shelton].

Several levels of public participation can be necessary, and different participating countries interpret international agreements differently. In the first instance, the public could be consulted on development of policy in relation to a particular activity. This could relate to the development of legislation (including primary, secondary, and guidance) and defining rules for the use of particular types of activity that might have an adverse effect on the environment or human health. In the second instance, information could be made available to the public in relation to particular specific instances of activities that might have an adverse impact. The public also could be given input into the decision-making process in relation to these individual uses.

CHALLENGING PUBLIC PARTICIPATION IN BIOSAFETY POLICY

The Cartagena Protocol and Biosafety Policy

The debate about the role of public participation in decision-making processes has stressed that participation leads to a more adequate and effective public policy, mainly concerning environmental matters. In the biosafety context, this argument persists. The requirement of public participation and awareness in biosafety policy has been consolidated through several international agreements, such as the Aarhus Convention and the Cartagena Protocol. Indeed, biosafety regulatory systems that are intended to support sustainable development must consider how to fulfill the requirement of public participation.

Thus far, the Cartagena Protocol is the most comprehensive international agreement to promote the safe use of biotechnology,[31] and it is one of the best examples of sustainable development law, because it integrates socioeconomic considerations, environmental and human health concerns, precaution and trade issues, and the public participation principle into its provisions. The challenges of biosafety, particularly in the context of the transboundary movement of LMOs, require an international regime to address common concerns and efficiency of regulatory systems. That is exactly what the Protocol represents and the reason for its negotiation. This international instrument contains important new rights and obligations for its Parties, relating to the transboundary movement, handling, and use of LMOs. It ensures that countries receiving, exporting, and using LMOs can identify eventual risks to human health and the environment resulting from modern biotechnology.

Article 23 of the Cartagena Protocol requires Parties to promote and facilitate public awareness, education, and participation in the context of the safe transfer, handling, and use of LMOs grounded by conservation and sustainable use of biodiversity. The protocol also establishes mandatory consultations with the public

[31] It primarily addresses the transboundary movement of such organisms.

in the decision-making process for LMOs and disclosure of decisions to the public.

Essentially, the article indicates three aspects of public participation:[32] public awareness, education, and participation itself. Public awareness contemplates the use of media and other channels of communication to inform the public; public education comprises broader means of disseminating information to the public, and specific learning programmes through the formal and informal educational system; and public participation itself envisions adequate mechanisms for public consultation in the decision-making process for the transfer, handling, and use of LMOs.

In practice, these three aspects are intermingled and must not be employed separately. Public access to information, awareness, and education on LMO issues enable citizens' participation in the decision-making process. Likewise, the challenges involved in fostering these three aspects of public participation are very similar,[33] which include engaging the general public to participate in the biosafety debate rather than only motivated stakeholders.

In greater detail, Article 23(1) (a) of the Cartagena Protocol requires Parties to "promote and facilitate" public awareness, education, and participation in the context of transfer, handling, and use of LMOs. Conversely, Article 23(1) (b) "endeavours" Parties to ensure that access to information is provided through public awareness and education, specifically in the case of transboundary movements of LMOs. In essence, Article 23 (1) infers that circumstances involving transfer, handling, and use of LMO shall contemplate the three aspects of public participation; and imports of LMOs shall ensure access to information within the extent of public awareness and education mechanisms. Nevertheless, the levels of which Parties have agreed to be bound are quite softer[34] and are reflected by vocabulary.

The provision regarding public consultation appears in Article 23(2) of the Protocol and establishes that Parties shall consult the public in the LMO decision-making process and make the result of such decisions available to the public. Public consultation shall be in accordance with national laws and respect confidential information (Article 21).

The question of confidential information has been argued during the negotiations of the Protocol[35] because confidential business information was already protected

[32] Ruth Mackenzie et al, *An Explanatory Guide to the Cartagena Protocol on Biosafety*, IUCN Environmental Policy and Law Paper No 46 (Gland: IUCN, 2003) at 150.

[33] Dominic Glover et al, *Public Participation and the Cartagena Protocol on Biosafety, Part I: The Main Report*, Institute of Development Studies at 30, online: http://www.ids.ac.uk/UserFiles/File/knots_team/NBFreport_Main.pdf.

[34] *An Explanatory Guide to the Cartagena Protocol on Biosafety*, *supra* note 32 at 150.

[35] Chee Yoke Ling and Lim Li Lin, *Public Participation in the Implementation of the Cartagena Protocol* (Third World Network, 2005) at 2, online: http://www.biosafety-info.net/pubart.php?pid=41.

under international trade laws.[36] The provision about confidential information remained in the text of the Protocol. To this point, Parties shall define the conditions and extent of confidentiality claims in their national laws coherently with the provisions of the Article 23 of the Protocol.

Finally, Article 23(3) of the Cartagena Protocol requires Parties to try earnestly to inform the public about the means of public access to the Biosafety Clearing-House (BCH). As an information exchange mechanism, the BCH assists Parties in implementing the Protocol by receiving information submitted by the Parties and providing access to them, as well as to other information exchange mechanisms. The availability of and accessibility to an extensive source of information plays a crucial role in terms of implementation of the Cartagena Protocol, in particular, Article 23.

Challenges of Public Participation in Biosafety Policy

Although meaningful public participation in the design and implementation of multilateral environmental agreements and national laws is widely accepted and desired, there are still many hindrances to motivate public engagement, particularly regarding biosafety issues. Legal processes and legal literacy[37] are among the obstacles that limit public participation in the biosafety debate. Legal and policy issues generally require a minimal knowledge of decision-making processes, legal terminology, and background information.

Scientific language associated with the use of biotechnologic products and processes also challenges public participation. Indeed, the language of science and technology can be difficult for general public comprehension; however, scientific information that affects human health and the environment ought to be made accessible, useful, and intelligible for the general public to ensure expressive and informed public participation. Increasing levels of awareness of legal and scientific understanding contribute to effective public participation engagement.

Sustained authentic public engagement is another challenge in the biosafety-biotech debate. Pragmatically, "environmental threats rarely capture the attention of the public and policymakers unless or until they are linked to human health."[38] Engaging the general public in the biosafety debate, and not simply the motivated people of society, can be difficult. The key is to motivate the citizenry to hold governments accountable for any decisions that could adversely affect their welfare and environment, not just in the near future but long term. It is essential to go

36 We refer to the World Trade Organization (WTO) Agreement on Trade-related Aspects of Intellectual Property Rights (TRIPS) and the Agreement on Technical Barriers to Trade (hereinafter TBT).

37 Dominic Glover et al, *supra* note 33 at 36.

38 Lisa Heinzerling, "Climate Change, Human Health, and the Post-Cautionary Principle" (2007). Georgetown Law Journal Research Paper No 4/2007 at 9.

beyond polarized opinions, establish a process with opportunities for all public and multisector stakeholders to share all views and relevant information, and work together toward a consensus.

The question of confidential information is also part of the biosafety debate. Modern biotechnology requires confidentiality; however, this is often used to avoid disclosure of information.[39] LMO products and processes are subject to stringent legal protection, for example, under the terms of intellectual property rights. As such, confidentiality issues affect the attempt to inform the public about eventual risks and benefits resulting from biotechnology, which often leads to conflict of interest. In this respect, Parties to the Cartagena Protocol will be challenged to balance demands on confidentiality and public inquiry. National laws shall define the scope and limits of confidentiality consistent with the terms of the Protocol, particularly regarding public participation and awareness issues.

Another factor that challenges public participation rests on framing national biosafety regulations according to the provisions of international instruments. For example, there are different approaches between national and international regulations, in the regulation of certain aspects of risk assessment and management, and international trade agreements and disputes. If people perceive that governments are restricted to incorporating public concerns into regulations because of international obligations, this could negatively influence public engagement. The spur to address the challenge of international influence shall focus on the recognition of biosafety as a sustainable development issue that requires a response to needs and priorities at national levels. Ensuring that relevant national sustainable elements and priorities are taken into consideration in regulatory biosafety frameworks promotes public and political support. An inclusive approach is vital to guarantee the involvement of all stakeholders and promote the acceptance of national biosafety regulations in the domestic context.[40]

Mechanisms for Public Participation in Biosafety Policy

Although public participation is recognised under the terms of the Cartagena Protocol, specific guidance on how to disseminate information, raise awareness, and promote public consultations is not provided. Other existing international agreements, for example, the Aarhus Convention, provides some guidelines on how to implement public participation, but it is ultimately entrusted to the Parties, at the national level, to decide how public participation is to be effectively implemented.

39 Tomme Young, *Genetically Modified Organisms and Biosafety: a background paper for decision-makers and others to assist in consideration of GMO issues* (Gland: Switzerland, 2004) at 22.

40 UNEP-GEF Biosafety Unit "A Comparative Analysis of Experiences and Lessons" (December 2006) at 45.

In this regard, formal and informal mechanisms to implement Article 23 of the Cartagena Protocol might be used in different stages of the decision-making process for LMOs and by different stakeholders, such as government or the public. Among these approaches are consultations with the public, deliberative inclusionary processes (DIP), and public participation in committees.

The formal right to access information and participate in the biosafety decision-making process has been consolidated through numerous international and national laws. Some countries have encouraged the submission of oral and written comments from the public on decisions and policies; others have required public hearings before decisions regarding the introduction of LMOs.[41] Moreover, judicial systems have also been used to contest biosafety laws and their transparency and legitimacy. In fact, access to national courts is an essential tool for those whose rights have been affected by decisions on LMOs. Because legal systems vary considerably among countries, however, this mechanism might not be an option in some jurisdictions.

At the international level, several mechanisms ensure the right to access information. For instance, the BCH, the exchange mechanism of the Cartagena Protocol, assists Parties to implement the provisions of the Protocol and facilitates information and experience sharing. The FAO Electronic Forum on Biotechnology in Food and Agriculture provides and exchanges information and experiences on agricultural biotechnology in developing countries. Among other mechanisms are the Biosafety Information Network and Advisory Service (BINAS), the International Centre for Genetic Engineering and Biotechnology (ICGEB), and several other knowledge centres.

In the national realm, members of the concerned public, interest groups, or government authorities can trigger mechanisms for public participation. Members of the general public or community and nongovernmental organisations have the right to request information from public authorities. Authorities shall provide the information in a timely and accurate manner; however, disclosure of information and its procedures is sometimes restricted or not completely covered under national laws. Conversely, governments have the proactive obligation to collect and disseminate information to the public and raise public awareness concerning environmental matters.[42] In this case, governments can use a variety of outreach activities, such as public databases containing information about LMO products, processes, and research and development projects; national media channels; informative materials on the risks and benefits of biotechnology; LMO product labeling; workshops and discussion forums for specific stakeholders and the general public; informal meetings in the community to encourage public engagement and awareness about

[41] Dominic Glover et al, *supra* note 33 at 41.

[42] The obligation to disseminate information is underscored under Article 5 of the Aarhus Convention.

biosafety and biotechnology issues; and support for local civil society groups and nongovernmental organisations. All of this encompasses broader interactions between the general public and other specific groups. "These mechanisms can be effective if they are accurate, specific, clearly expressed in understandable language, unbiased, and based on full disclosure of the relevant facts."[43]

The right to public participation must also be endorsed by informal approaches to participation, not depending merely on government initiatives. These approaches aim to allow a broader public deliberation in decision making through the inclusion of all stakeholders. Most recently, deliberative and inclusionary processes have been employed in policy dialogues in both developing and industrialised countries. Deliberative and inclusionary processes intend to elicit democratic values through an inclusive and participatory dialogue among stakeholders having different or similar perspectives and other interested parties. Through the interaction among diverse actors of the society, these processes capture value through the creation of small public spaces and generate a wide range of information and sharing among participants. The deliberative approach can guarantee that information shared will be used in relevant authorities' decisions to ensure a responsive approach to local sustainable elements and priorities. Several methods and techniques are used to include multiple stakeholders in the deliberative and inclusionary processes. These methods include public meetings, deliberative polling, focus groups, citizen's juries and panels, workshops, conferences, and seminars.

The importance of broader participation from all sectors of the society and inclusionary approaches in the debate of the safe use of biotechnology is vital to increase and ensure legitimacy, adequacy, and efficiency of policy formulation, as well as the quality of decision making for LMOs. Although governments still must provide the public with information, civil society organisations, nongovernmental organisations, and other stakeholders are openly responsible to the growing public interest, mainly concerning environmental and health matters.

GLOBAL PUBLIC PERCEPTIONS ON BIOTECHNOLOGY

Public perceptions on environmental matters have changed swiftly as people appreciate them as being potential risks to human health and the environment itself.[44] Public perceptions of biotechnologies are connected to these potential risks and vary considerably depending on individual values, experiences, and interests. Assessing benefits and risks of biotechnologic products and processes is still difficult for the public if they are not provided with relevant and trustworthy information.

43 *Genetically Modified Organisms and Biosafety*, *supra* note 39 at 22.

44 Julian Kinderlerer and Mike Adcock, "Agricultural Biotechnology, Politics, Ethics, and Policy," Working Paper No. 3, (FANRPAN/IFPRI Regional Policy Dialogue on Biotechnology, Agriculture, and Food, 2003) at 18.

Several surveys and studies[45] about public perceptions of biotechnology and related sciences are conducted periodically at all levels of governance and in different parts of the globe. In fact, legal, political, and cultural aspects range vastly among countries and affect the way in which the public perceives modern biotechnology. People might not be aware of particular scientific or technical features when assessing potential risks and benefits resulting from biotechnology, but they are capable of forming judgments on their own about principles of fairness.

For example, surveys show that European citizens are more optimistic, informed, and trustful about technologies. In greater detail, "there is widespread support for medical (red) and industrial (white) biotechnologies, but general opposition to agricultural (green) biotechnologies in all but a few countries."[46] The European public is willing to accept innovative technologies that promise tangible benefits, whereas in the particular case of genetically modified (GM) foods, there is still an ethical unacceptability issue.

In North America, particularly in the United States, awareness of biotechnology remains an abstract concept,[47] mainly concerning GM foods. Despite unfamiliarity with scientific concepts and terminology, there is considerable approval of GM foods among Americans, even more in terms of potential benefits, such as decreased use of pesticides;[48] however, most Americans show moderate resilience to certain applications of modern biotechnology, such as GM animals.

Public opinion in North America and Europe differs significantly. In Europe, public perception seems to address the question, "why?" "Why should this product be introduced onto the market; do we need it?" In contrast, in North America, the question appears to be, "why not?" Let the market decide between that which is currently on the market and that which is now being introduced. In Europe, LMOs are assessed in relation to the risk that they pose to either human health or to the environment. In North America, the assessment is essentially a comparison of the benefits and risks, including the change in practice that introduction of LMOs might cause. The rest of the world has polarised behind one or the other approach, which has resulted in many countries choosing to consider LMOs as truly dangerous and arguing that their import into their country should not be permitted. Others have

[45] See Philipp Aerni, *Assessing stakeholder attitudes to agricultural biotechnology in developing countries* (Biotechnology and Development Monitor, 2001), online http://www.biotech-monitor.nl/4702.htm; Claire Marris et al, *Report on Public Perceptions of Agricultural Biotechnologies in Europe* (Centre for the Study of Environmental Change, Lancaster University, 2001); William K Hallman et al, *Public Perceptions of Genetically Modified Foods: Americans Know Not What They Eat* (Food Policy Institute, Rutgers, The State University of New Jersey, RR-0302–001, 2002), online http://www.foodpolicyinstitute.org.

[46] See George Gaskell et al, "Europeans and Biotechnology in 2005: Patterns and Trends" (2006) 64(3) Eurobarometer 3.

[47] W. K. Hallman et al, *Public Perceptions of Genetically Modified Foods: A National Study of American Knowledge and Opinion* (Publication number RR-1003–004; Food Policy Institute, 2003) at 11.

[48] *Ibid* at 15.

accepted the technology with little consideration of the possibility of harm, perhaps arguing that the benefits outweigh any harm that might occur.

Public perceptions in developing countries have begun to be examined only recently. The extent to which innovative technologies, in particular biotechnology, are perceived in developing countries is uneven. Although some countries agree that biotechnology could assist in agriculture and that GM food does not represent a serious risk to human health, others are concerned about the safety of GM food and the effect of biotechnology on agriculture.[49] In prominent countries such as China, policies encourage the development and plantation of GM crops; in others, the planting of GM crops is not yet officially approved.[50] Thus far, the evidence of acceptance of biotechnology in developing countries varies.

Regardless of polarised public opinions on biotechnology and its uses worldwide, some regions share certain concerns. Labelling for GM food is one of these areas. Consumers insist on having the right to choose between non-GM and GM food, which is a consumer's choice in relation to sustainable development issues. Acceptability of GM food depends on how consumers perceive risks and benefits associated with biotechnology and their priorities. In particular, policy makers and stakeholders play a crucial role in communicating the pros and cons of modern biotechnology to the public, because many people do not think about modern biotechnology until the concepts are presented directly to them.

Policy makers have an "opportunity to design a credible regulatory process for assuring consumers that products reaching the market are safe."[51] Consumers informed of tangible benefits and potential risks have more tools to make informed decisions as consumers and therefore could be more or less willing to accept the risks connected with products and processes resulting from modern biotechnology.

Public Perceptions and Trust

Certainly, the inclination to promote or prohibit modern biotechnology is influenced by a country's policy choices in several areas, such as trade and environmental issues. As far as public perception is concerned, this inclination drives policy decisions and triggers food production industry responses. Conversely, public perceptions are directly affected by policy decisions. These complex interconnections are better

49 "Many developing countries are fearful of the impact of agricultural biotechnology. Zimbabwe and Zambia, for example, have been wary of permitting food-aid which contains transgenic maize into the country, even though many of its people are starving." *Agricultural Biotechnology, Politics, Ethics, and Policy*, *supra* note 44 at 24–5.

50 Robert L Paarlberg, *Governing the GM Crop Revolution: Policy Choices for Developing Countries*, Food Agriculture and the Environment Discussion Paper 33 (IFPGRI, 2000) at 251.

51 GA Baker and TA Burnham, "Consumer response to genetically modified foods: market segment analysis and implications for producers and policy makers" (2001) 26 J Agricultural Resource Economics 387 at 400.

realised in the context of public participation in decision-making processes, whereas credibility, effectiveness, and legitimacy of any policy decision depend on the extent to which public and key stakeholders participate in the process. Policy decisions that accommodate diverse public participation and therefore take into account public perceptions are more likely to elicit public confidence in public institutions.

The level of trust in public institutions is an important aspect that contributes to build the public's confidence in products and processes resulting from modern biotechnology. For example, in the United States the public tends to be confident about public institutions, such as the Food and Drug Administration and the Environmental Protection Agency,[52] and thus policy decisions tend to have a greater influence in Americans' public perceptions of biotechnology. Conversely, in Europe, the public is more resilient to trust in regulatory institutions;[53] therefore, it tends to be less influenced by or at least more sceptical about policy decisions. Risk perception and trust in risk regulation still contribute moderately but significantly to the explanation of acceptability.[54]

In addition, diverse public perceptions of the benefits and risks of modern biotechnology are a result of cultural, ethical, societal, political, and economical aspects, which vary considerably in different societies. In some developing countries, for instance, food insecurity and shortage of food are aspects that affect consumers' acceptance of GM food, whereas food insecurity in developed nations does not have a significant influence, at least in the national context. This finding further indicates that to be efficient, policy decisions should consider the perceptions of individual societies and local sustainability elements.

Although global perceptions of biotechnology are diverse, certain reflections and concerns about risks and benefits resulting from modern biotechnology are shared among stakeholders. Indeed, proponents of biotechnology are keen to ensure that biotechnologies become accepted by a sceptical and worried public;[55] members of the civil society and public want to ensure that their right for a healthy environment and welfare is assured. It then follows that policy makers must ensure that public concerns are considered in the decision-making process. Perceptions about biotechnology are still unsettled, and many questions remain unanswered either comprehensibly or publicly.

52 *Ibid.*

53 "In Europe, several scandals in the 80s and 90s (Anti-freeze agent in Austrian wine, HIV contaminated blood in France, mad cow disease in the UK, Dioxin in Belgian eggs and meat products, and contaminated Coke in Belgium and France) undermined trust in regulatory institutions significantly. This also explains why there is strong public opposition all over Europe, in spite of the different social, cultural and political backgrounds." Philipp Aerni, *Public Policy Responses to Biotechnology*, Chapter BG6.58.9.2 in *Our Fragile World: Challenges and Opportunities for Sustainable Development* (Paris: Encyclopaedia of Life Support Systems (ELOSS), UNESCO, 2001) 7.

54 Wouter Poortinga and Nick F Pidgeon, "Trust in Risk Regulation: Cause or Consequence of the Acceptability of GM Food?" (2005) 25 Risk Analysis 199 at 207.

55 Dominic Glover et al, *supra* note 33 at 24.

Although different parts of the world face similar challenges, no single approach underpins policy decisions for LMOs. Drawing out different perceptions of biotechnology in different countries emphasises the importance of country-driven processes, mainly when it comes to encourage awareness-raising activities, public education, and consultation. Increasing information about risks and benefits of biotechnology is essential for narrowing the gap between different public attitudes and to prevent equity issues, primarily among developing and developed nations.

Continuing advances in biotechnology will certainly impact agriculture, trade, food supply, human health, and the environment, and they will require the development and improvement of regulatory policies. Socioeconomic and environmental considerations must be addressed by policy decisions, because these directly affect public perceptions of biotechnology.

CONCLUSION

The foregoing provisions regarding public participation in international treaties on sustainable development underline the importance of public engagement in decision making for environmental matters. Principles of international law agreements also stand for each person's right to be informed and involved in such decision-making processes. In particular, the importance of public participation in policy decisions for LMOs is covered under the Cartagena Protocol, whereas public participation, awareness, and consultation are stressed as important components of regulatory biosafety frameworks. The Aarhus Convention also plays an important role in environmental democracy, especially in reinforcing public rights and participation in the decision-making processes for LMOs.

There is a significant challenge to countries that are Parties to the Cartagena Protocol. How can public participation be effectively implemented, as required in Article 23 of the Protocol? Despite provisions of international agreements, it is up to each country, as Parties or non-Parties to international conventions, to decide how they are going to implement public participation, especially in biosafety and biotechnology policy. Since public participation is driven by the people of that particular country, governments must determine the suitable approach and mechanisms to help them to promote and facilitate public awareness of, and education and participation in the biosafety debate, with consideration paid to local sustainability elements and capabilities.

Although public participation is country specific, policy makers must pay attention to the evolving global context within biotechnologies and approach the issue holistically. Local sustainability elements and the principle of sovereignty, as outlined in the Rio Declaration, shall be considered in the decision-making process for LMOs; however, consideration of the global interdependence affecting the debate is also fundamental.

A responsible regulatory policy, at national levels, regarding the safe use of biotechnology should be participatory, transparent, and accountable, based on scientific information, and should consider socioeconomic, cultural, and ethical issues, or at least establish mechanisms to address them. It is also important to look across other far-reaching principles, such as precaution, postcaution, sovereignty, and sustainable development. The public should be able to participate in a timely process and be confident that their opinions will be taken into account by policy makers, which will increase the level of trust afforded public institutions.

In the realm of biotechnology, it is important to define the role of trust in policy decisions. Having more trust in public institutions to regulate and control potential risks associated with modern biotechnology and ensure safety for the public and protection of the environment will most probably increase the acceptability of these risks. In a society in which the public sector is accountable for protecting the public from risks, trust in public institutions determines perceptions of risks and acceptability of new technologies. The public has great expectations about policy makers' decisions to weigh the short- and often uncertain long-term effects on human health and the environment.

How will informed decisions be made when there are still great scientific, social, ecological, and economical uncertainties surrounding biotechnology? Furthermore, how can these questions be presented to the public intelligibly and meaningfully? How can the public's informed involvement in the debate be increased? The challenges are plentiful; however, an open policy process, with greater interaction with the public and their values, interests, and concerns, is a step closer to the answers to these questions. Moreover, assessing experiences from other health and environmental matters, and lessons learned from the development and implementation of national biosafety frameworks, could be of great value in defining appropriate tools and mechanisms to foster public participation and genuine public engagement. Building accountability and legitimacy of policy decisions requires more effective public participation in decision making.

Effective participation also depends on the assumption that the public has been properly informed about content and the participation process itself. Increasing awareness, particularly among local communities, consumers, farmers, individual people, and policy makers, is crucial to a comprehensive integration of biotechnology issues with humanity and the environment. Community groups, nongovernmental organizations, faith groups, industry, and academia, and not only government, need to create space for awareness-raising activities. As mentioned before, government authorities are often expected to provide information for and conduct the process of public participation, yet other leading stakeholders should take responsibility and play a proactive role in this process, to inform decision makers and increase educated public opinion. Stakeholders must be persuaded to participate in the process, guiding the present and looking into future scenarios that might impact sustainable development.

Biosafety is a sustainable development issue, affecting the means to achieve a sustainable and equitable society as it permeates through the main aspects of sustainable development, namely, economics, social policy, and environmental conservation.

Credible public awareness, education, and participation enable a broader dialogue among all stakeholders, which facilitates rational decision making that benefits the greater collective interest. Stakeholders with cross-sectional perspectives and experiences, as well as the public with its values, interests, and concerns, can help to identify, clarify, and resolve many issues related to modern biotechnology. An effective policy process depends on the enabling of stakeholders and the public to frame the debate and accommodate diverse opinions, so that credibility and trust in public institutions legitimises policy decisions.

The future of biotechnology will be determined by public opinion and policy decisions. With goals of being effective and trustworthy, public opinion must be sufficiently informed and diligently elicited, and policy decisions must carefully consider public perception, present and future developments, risks and benefits, and look across precautionary and postcautionary principles. This authentic commitment to sustainable development and transparent reliable public governance requires the necessary inclusion of all aspects of public participation to influence biosafety decision making that is satisfying and effective for all, now and for the foreseeable future.

8

The Biosafety Clearing-House and Sustainable Development Law

Frederic Perron-Welch

INTRODUCTION

The Biosafety Clearing-House (BCH) is a mechanism set up by the Cartagena Protocol on Biosafety (the Protocol) as a part of the Convention on Biological Diversity (CBD) clearing-house mechanism (CHM) to facilitate the exchange of information on living modified organisms (LMOs) and assist the Parties to comply with their obligations under the Protocol.[1] Worldwide access to scientific, technical, environmental, legal, and capacity-building information on biosafety is provided online in all six UN languages.[2]

The BCH is established by Article 20 of the Protocol and maintained by the Secretariat of the CBD. Its operation and development are based on four key documents: the Modalities of Operation of the BCH adopted at COP-MOP 1;[3] the Multi-Year Programme of Work adopted at COP-MOP 2;[4] the outcomes of the meetings of the BCH Informal Advisory Committee (BCH-IAC);[5] and the Strategic plan for the Cartagena Protocol on Biosafety for 2011–2020 (2011–2020 Strategic Plan).[6] Parties are required to make available, through the BCH, information created by their regulatory processes, such as the advance informed agreement (AIA) procedure (described in Chapter 5) and any existing laws, regulations, or guidelines for the implementation of the Protocol.[7]

1 *Cartagena Protocol on Biosafety*, 29 January 2000 (entered into force 11 September 2003) at art 20(1).
2 The BCH homepage is: http://bch.cbd.int.
3 Cartagena Protocol, COP-MOP 1, Dec BS-I/3, Annex.
4 Cartagena Protocol, COP-MOP 2, Dec BS-II/2, Annex.
5 The BCH Informal Advisory Committee is constituted under Section 5 of the modalities of operation of the BCH.
6 Cartagena Protocol, COP-MOP 5, Dec BS-V/16, Annex.
7 *Cartagena Protocol on Biosafety*, supra note 1 at art 20(3).

The author would like to thank Andrew Bowers and Carolina Monardes for their assistance with this chapter.

From a sustainable development law perspective, the BCH plays a key role in implementing key principles underlying the Cartagena Protocol, such as the precautionary approach, public participation and access to information, good governance, capacity building, and common but differentiated responsibilities.[8] As explained in detail in Chapter 9, the BCH is expected to serve as a forum for information sharing, build capacity to assist implementation and decision making, provide evidence of compliance, and facilitate international trade. This is an ongoing task, and significant progress has been made toward these objectives over the past decade. One recent success is the development of online forums for decision makers for capacity building.

ORIGINS AND BASIS OF THE BIOSAFETY CLEARING-HOUSE

On the adoption of the Cartagena Protocol in 2000, the Executive Secretary of the CBD was requested to begin developing the BCH as an online tool.[9] The pilot phase ran until the first meeting of the Conference of the Parties serving as the meeting of the Parties to the Cartagena Protocol on Biosafety (COP-MOP 1) in February 2004. There, the Parties approved the transition of the BCH from the pilot phase to the fully operational phase and adopted its Modalities of Operation.[10]

The Modalities of Operation explain the role that the BCH plays in the provision and exchange of information supporting implementation of the Protocol. They state that the BCH has, at minimum, a role in providing access to information on[11]

(a) Existing national legislation, regulations, and guidelines for implementing the Protocol as well as information required by Parties for the AIA procedure[12]
(b) National laws, regulations, and guidelines applicable to the import of LMOs intended for direct use as food or feed, or for processing (LMOs-FFP)[13]
(c) Bilateral, multilateral, and regional agreements and arrangements[14]
(d) Contact details for competent national authorities, national focal points, and emergency contacts[15]
(e) Reports submitted by the Parties on the operation of the Protocol[16]

8 Refer to the Introduction for a deeper discussion of the principles of international sustainable development law.
9 International Institute for Sustainable Development (IISD) *Earth Negotiations Bulletin*, COP-MOP 1 of the Cartagena Protocol on Biosafety, vol. 9, no. 289, online: http://www.iisd.ca/download/pdf/enb09289e.pdf.
10 Dec BS-I/3, *supra* note 3.
11 *Ibid* at Part A.
12 Cartagena Protocol, *supra* note 1 at art 20(3)(a).
13 *Ibid* at art 11(5).
14 *Ibid* at arts 14(2) and 20(3)(b).
15 *Ibid* at arts 17(3)(e) and 19.
16 *Ibid* at art 20(3)(e).

(f) Decisions by a Party on regulating the transit of specific LMOs[17]
(g) Occurrence of unintentional transboundary movements that are likely to have significant adverse effects on biodiversity[18]
(h) Illegal transboundary movements of LMOs[19]
(i) Final decisions regarding the importation or release of LMOs[20]
(j) Information on the application of domestic regulations to specific imports of LMOs[21]
(k) Final decisions regarding the domestic use of LMOs that may be subject to transboundary movement for direct use as food or feed, or for processing[22]
(l) Final decisions regarding the import of LMOs-FFP that are taken under domestic regulatory frameworks or in accordance with Annex III of the Protocol[23]
(m) Declarations regarding the framework to be used for LMOs-FFP[24]
(n) Review and change of decisions regarding intentional transboundary movements of LMOs;[25] LMOs granted exemption status by each Party;[26]
(o) Cases in which intentional transboundary movement may take place at the same time as the movement is notified to the Party of import;[27]
(p) Summaries of risk assessments or environmental reviews of LMOs generated by regulatory processes and relevant information regarding products thereof[28]

Section 5 of the Modalities of Operation allows the Secretariat to seek assistance from an informal advisory committee (BCH-IAC), constituted and coordinated transparently by the Executive Secretary, focused on providing guidance on resolving technical issues associated with the ongoing development of the BCH. At its first meeting in 2004, the BCH-IAC adopted a set of operational principles that outlines terms of reference, make up, and organization. The terms of reference start with providing technical input and feedback on the development of the Internet site to ensure that it is accessible, user friendly, searchable, and understandable. The BCH-IAC also assists in identifying, reviewing, and designing common formats for reporting information, developing nodes, and facilitating cooperation with

17 *Ibid* at art 6(1).
18 *Ibid* at art 17(1).
19 *Ibid* at art 25(3).
20 *Ibid* at arts 10(3) and 20(3)(d).
21 *Ibid* at art 14 (4).
22 *Ibid* at art 11(1).
23 *Ibid* at arts 11(4), 11(6) and 20(3)(d).
24 *Ibid* at art 11(6).
25 *Ibid* at art 12(1).
26 *Ibid* at art 13(1).
27 *Ibid*.
28 *Ibid* at art 20(3)(c).

partner institutions through development and review of guidelines for information sharing.[29]

The Multi-Year Program of Work for the BCH (MYPOW) was adopted at COP-MOP 2, as the Annex to Decision BS-II/2, and contains five key programme elements. The first element pertains to the structure and function of the Central Portal and aims to improve ease of reporting and accessing information in the BCH. The second is related to information content and management and aims to increase the amount of information reported and ensure timeliness. The third is based on sharing information and experience with LMOs and aims to make a broader range of biosafety information accessible to users. The fourth addresses capacity building and non-Internet accessibility and aims to ensure that countries have the necessary capabilities to access the Internet-based Central Portal and can access information through the BCH in a timely manner. The fifth and final element aims at ensuring that the Programme of Work is achieving the goals of the BCH effectively.[30]

The 2011–2020 Strategic Plan, adopted at COP-MOP 5 in Nagoya, Japan, in Decision BS-V/16, is contained in the annex to the decision and is divided into five focal areas. The second and fourth areas, Capacity building and Information sharing, pertain to the BCH. The strategic objective relevant to the BCH for Capacity building is to ensure that all established stakeholders can easily access the BCH, in particular developing countries and countries having economies in transition. Outcomes include easily accessible tools through the BCH to facilitate implementation of the Protocol. The strategic objective for information sharing is to enhance the availability and exchange of relevant information by increasing the amount and quality of information submitted and retrieved, using the BCH as a tool for online discussions and conferences. The goal is to establish the BCH as a platform for assisting countries in implementation and to share information by other means of information exchange.[31]

THE BCH AND SUSTAINABLE DEVELOPMENT LAW PRINCIPLES

In 2002, the International Law Association's (ILA) Committee on the Legal Aspects of Sustainable Development released its New Delhi Declaration on Principles of International Law relating to Sustainable Development.[32] This statement notes that sustainable development has been amply recognized in various international legal

[29] Operational Procedures for the BCH-IAC, online: CBD http://bch.cbd.int/doc/operational procedures for the bch-iac.pdf.

[30] Cartagena Protocol, COP-MOP Dec BS-II/2, Annex.

[31] Cartagena Protocol, COP-MOP Dec BS-V/16, Annex.

[32] International Law Association, Resolution 3/2002, New Delhi Declaration of Principles of International Law Relating to Sustainable Development (ILA Declaration).

instruments, including the Cartagena Protocol. This declaration outlines seven principles that provide normative context for best policies. These principles, however, should be used in an integrated manner to determine sustainable legal solutions. This section explains the role of the BCH in the framework of the Cartagena Protocol in a holistic manner.

The New Delhi Declaration states that a precautionary approach is central to sustainable development because it requires that the absence of full scientific certainty shall not be used as a reason for postponing decisions if there is a risk of serious or irreversible harm to human health, natural resources, and ecosystems.[33] The declaration also states that decision-making processes should always endorse a precautionary approach to risk management and in particular should include the adoption of appropriate precautionary measures.[34] Furthermore, it states that precautionary measures should be based on up-to-date and independent scientific judgment and transparency.[35]

The Cartagena Protocol aims to protect biological diversity from the risks posed by LMOs resulting from modern biotechnology. Because the risks surrounding the introduction of LMOs in the environment are still undefined, the precautionary approach is reaffirmed in the preamble of the Protocol.[36] This approach enables Parties to the Protocol to employ precaution in practise through two different processes: AIA and LMO-FFP. Access to and exchange of information in this context is vital to putting the precautionary approach into practise. Because authorities are often asked to make decisions when confronted with scientific uncertainty, they can use the BCH to access up-to-date scientific information, if any, and exercise their decision-making capacities accordingly. The development work done in response to the four guiding documents of the BCH has addressed the availability of and access to up-to-date scientific information, as well as data related to the effective implementation of the Protocol. Public awareness, education, and participation are a principal elements for implementing the Cartagena Protocol and are facilitated by the BCH. The BCH promotes transparency and knowledge of the uses of LMOs, legislation on LMOs, and technical information on biodiversity, biotechnology, and biosafety.

The principle of good governance is an important sustainable development law theme expressed through the BCH. Transparency and public involvement in decision-making processes are essential to ensuring the public's confidence in State regulatory systems on biosafety. The information provided by Parties under the AIA and LMOs-FFP procedures promotes transparency in decision making and

33 *Ibid* at Principle 3.

34 *Ibid* at para 4.3.

35 *Ibid* at para 4.4.

36 Cartagena Protocol, *supra* note 1 at Preamble.

demonstrates compliance with international biosafety law.[37] Similarly, the BCH is a conduit to the global public that shows measures taken for the purposes of implementing the Protocol as well as demonstrating those Parties which have taken insufficient measures or are listed as being in non-compliance. Chapter 9 addresses how effective the BCH has been in performing this role.

[37] *Ibid* at arts 20(3)(a),(c) and (d). The Advance Informed Agreement procedure is described in Chapter 5.

9

Use of the Biosafety Clearing-House in Practise

Tomme Rosanne Young

This chapter briefly considers the roles of the BCH and its strengths and weaknesses as a tool for legislative analysis or an aid to national legislative development. This is an important topic for the implementation of the Cartagena Protocol, and others have argued (e.g., Introduction, Chapters 7 and 8) that access to information and good governance are key principles that have been operationalised in the Cartagena Protocol through provisions such as the BCH. As explained later, there are several reasons that reliance on the BCH alone might not yield a legally or practically accurate picture of national implementation of the Protocol. The following discussion is intended not as a criticism of the BCH, which has performed wonders in putting so much information online; rather, it is hoped that these comments can be of use as the BCH legislative database evolves.[1]

Through interviews with various delegates to the Protocol negotiations, it appears that Parties, as well as the staff of the Secretariat of the Convention on Biological Diversity (SCBD), expect the BCH to serve multiple objectives, including the following:

Compliance, that is, posting of documents that might be thought of as evidencing national compliance with Protocol requirements

Capacity building and assistance with implementation, that is, national legislators and implementing agencies can use posted instruments as examples and models for the development of national implementing legislation[2]

[1] Databases are, by definition, examples of "design for use." Consequently, this chapter indicates some of the challenges posed for designers with regard to the various uses of the BCH as a legal tool. The BCH staff must deal with two very difficult challenges, creating a tool and training for national officials who seek to upload data to the BCH, and also creating search tools for those using the database. Obviously, the former is of greater importance, and most national officials interviewed who have posted documents to the BCH have complimented the SCBD for the tool's excellence and sensitivity to national needs.

[2] This expectation underlies much (but not all) of UNEP's work. See UNEP-GEF, *Learning from Experience: The Global UNEP-GEF BCH Capacity Building Project* online: http://www.unep.org/

Facilitation of the LMO imports/exports, and the international movement of goods, that is, exporters/importers of LMOs and other LMO developers hope to use the BCH to help them make preliminary determination of the requirements of national legislation, before committing funds and manpower to development of LMOs and/or to getting permission to import, export and/or use or introduce them[3]

Information sharing, that is, Parties, analysts, and other members of the public can obtain information about the level of LMO use and the analyses of risks

Aid to decision making, that is, background information about each country's policies and positions on a particular LMO that may be useful to less developed countries, who hope to use information from the BCH to assist them in decisions about matters on which they might lack technical capacity.[4]

Tools of this type can achieve their objectives only when their practical design specifically promotes them. In some ways, the BCH has not yet been able to meet some of the above objectives but is well on the way to doing so. Three types of concerns should be addressed by the BCH in its continuing development: (1) the "metadata" describing the nature, source, and status of each instrument; (2) limitations on its usefulness as an analytical tool for academics and for Parties seeking to develop legislative documents and recommendations for their country's implementation of the Protocol; and (3) limitations on the legislative database's usefulness and value for persons proposing to export or import LMOs.[5]

biosafety/Documents/BCH/BCH_Learning_from_Experience_Feb08.pdf. See also Ruth Mackenzie, et al, *An Explanatory Guide to the Cartagena Protocol on Biosafety* (Gland, Switzerland and Cambridge, UK: IUCN, 2003) at paragraph 539, which identifies this as the "two main purposes of the BCH": "First, the BCH is designed to facilitate the exchange of information and experience concerning LMOs. The types of information to be exchanged are broadly described as scientific, technical, environmental and legal.... Second, the BCH is to assist Parties to implement the Protocol. Consistent with the Protocol's Preamble and other provisions, this ... function acknowledges the special needs of the following three groups: developing country Parties, in particular the least developed and small island developing States among them; Countries with economies in transition; [and] countries that are centres of origin and centres of genetic diversity."

3 Based on interviews at COP-MOP 4, LMO developers and commercial exporters/importers of LMO products, as well as many academic and nonacademic researchers working with LMOs, expect the BCH to provide a "one-stop shop" for legislation, policy, and other requirements relating to LMOs.

4 The author's review suggests that the final two objectives may be the most relevant. The Protocol assumes that many Parties lacking domestic technical capacity might hope to use another country's technical conclusions, but only if they hold similar policies about LMO risk. In Jarle Hårstad, et al., *Evaluation of GEF Support for Biosafety* (Washington, DC: GEF Evaluation Office, 2006), the evaluators noted that the single common factor preventing small and developing countries from forming regional approaches to Protocol implementation was their varying views (policy) on the level of risk posed by LMO introduction.

5 Many decision makers share a misunderstanding relating to the needs and costs of database development and maintenance. Discussion often focuses on creation rather than continued usefulness. (See Ruth Mackenzie et al, *supra* note 2 at paragraph 539: "The effective operation of the BCH will depend on the active participation of developed country parties, developing country parties and Parties with economies in transition. The availability of technological resources in developing country Parties and

"METADATA"

A challenge still to be addressed for the current BCH legislative database as a basis for legislative analysis relates to the "metadata"[6] for each of the posted instruments,[7] which should distinguish between in-force, binding instruments, and other documents that could be nonbinding or drafts from various sources or out-of-date legislation that has been amended or supplemented. In practice, it is not always possible to determine the legal status of particular posted documents in the law/regulation database. Random checking indicates that some of the posted documents are in the midst of an adoption process, whereas others have not been formally proposed to the relevant authorities at all.[8]

The BCH's metadata card lists the status of instruments but is not always clear on this point, listing most documents as *published*. The BCH uses this term to describe any document that has been printed or circulated; it does not convey the authority under which the document was posted on the BCH – that is, whether a particular instrument has been submitted by the country's NFP (and therefore presumably represents a government position regarding the instrument) or by some other body. Many of the documents posted (especially those labelled *draft* or *NBF*[9]) appear to have been provided by an intergovernmental body (commonly either UNEP or UNDP). The metadata do not indicate whether the document's posting has been reviewed and/or sanctioned by the national agencies who are to enact or adopt it, or how it relates to other (sometimes overlapping and sometimes inconsistent) legislation also posted in the BCH.

It would be useful if the metadata card focused on legally relevant status options (whether the legislation is in force, with terms including *enacted*, *adopted*, *amended*,

Parties with economies in transition is an important consideration in the design of the BCH. . . ."). The BCH Informal Advisory Committee (BCH-IAC) has been of inestimable and ongoing value in ongoing development.

6 As used in most discussions of legislative databases, the term *metadata* refers to information such as its legal status (e.g., whether adopted/enacted, repealed, proposed), official citation, source, and subsequent and related enactments. In some cases the metadata include keywords, references to specific sections, and a summary of the instrument.

7 Of the BCH's various databases, this chapter relies on the Laws and Regulations (http://bch.cbd.int/database/laws/) database and Country Decisions and Other Communications database (http://bch.cbd.int/database/decisions/).

8 For many countries, the process of adopting legislation begins with a formal vetting process and/or publication of the instrument, followed by hours, days, weeks, or years in which the legislation remains "under consideration." Some of the documents that are listed in the BCH as published are in the midst of such an adoption process, whereas others have been given this status by being publicly available, apparently.

9 The term *national biosafety framework* (NBF) was coined by the UNEP/GEF project to describe a single-source document created within the project, which was expected to include or at least describe the full range of national biosafety legislation and/or to propose new legislation. In fact, most NBFs proposed a single instrument that does not attempt to integrate with, repeal, or otherwise address the (overlapping and sometimes conflicting) legislation thus identified. See Jarle Hårstad, et al., *supra* note 4, at chapter 8.

published officially, *drafted officially*, and/or *recommended*). It would also be helpful if these status descriptives were updated regularly, providing an up-to-date picture of relevant legislative issues in each country. The most essential information is whether the document is in force, and if not, whether it has the direct official support of one or more authorized governmental units.[10]

Particularly with draft legislation, the intent underlying the posting of an instrument is critical to understanding and using it properly. This is true of the numerous BCH documents that do not appear to include any legislation or policy (even in draft form), and those that do not include any direct or indirect reference to LMOs. Within the BCH, the metadata supporting these multiple overlapping and inconsistent postings are unclear. This is particularly relevant with regard to multiple legislative provisions apparently addressing the same material,[11] as well as draft documents and nonadopted instruments at the national level. Postings from other sources (including former governments, international aid agencies, and other organisations) clearly have different relevance than those undertaken by current biosafety authorities. When that person or office formally posts draft legislation or some summary or descriptive document, it is fair to assume that the intent is to inform the world about how the relevant agencies intend to act pending adoption of the country's formal BCH legislation (law or regulation).[12] When posted by some other entity or project leader, the posting might not have this impact and could lead to confusion. This should not be taken as a negative comment on draft documents, however.[13] In countries in which more than one agency or ministry is vying for primary responsibility over LMOs, a draft posted by one of the contestant entities is of particular interest in conveying that entity's position and approach.

[10] One example is found in Cambodia, whose UNEP/GEF-sponsored draft National Biosafety Framework (herein dNBF) is listed only as "published" and described only as a "regulatory summary" and is not *named* as a "draft" in the BCH list. Although the document bears a Ministerial signature, it contains nothing that suggests that it has been formally adopted as national legislation. The dNBF in this case contains various documents that it proposes for national policy and/or national law, those documents are still defined as "draft" or "proposed." For one seeking information about Cambodia's biosafety requirements or researching their effectiveness, this document raises more questions than it answers, primarily due to lack of meaningful metadata. For purposes of this chapter, the author has assumed that all dNBF's and other NBFs referencing the UNEP/GEF Project have been posted, or their posting has been approved, by the country in whose name they were posted.

[11] Coverage is unclear, for example, between Indonesia, *Regulation of the Government of the Republic Indonesia Number 21 Year 2005 on Biosafety of Genetically Engineered Product* (BCH Record No. 11576) and another instrument posted by that country, *Joint Decree of the Minister of Agriculture, the Minister of Forestry And Estate Crops, the Minister of Health, and the State Minister of Food and Horticulture Concerning Biosafety and Food Safety of Genetically Engineered Agricultural Products* (BCH Record No. 9591).

[12] See, Bangladesh, *Biosafety Guidelines of Bangladesh* (Ministry of Environment and Forests, Government of the Peoples Republic of Bangladesh, 2005).

[13] Posting of such an instrument by some other person or agency will be considered as "literature," which may be consulted in legal analysis and decision making, but without the imprimatur of the relevant official.

A related problem arises when countries have made key decisions on the introduction of LMOs or individual permits in national legislation. This practice appears to have been common based on current postings in the BCH[14] but might have been a response to early concerns, before the BCH was fully operational, and this might decline in future. It also could reflect national practice regarding which decisions must be adopted in formal legislation and which might not.

It would also be valuable for the BCH instruments if the quality and completeness of the metadata for each posting were improved. This would not require the SCBD to formally validate the data,[15] if each record were more transparent about the source of both the instrument and the metadata provided with it. Even when the SCBD does not validate the data received, follow-up with those posting the instrument, when the metadata as submitted are incomplete or not entirely understandable, would still be helpful.

SEARCH MECHANISMS

The sheer volume of material in the BCH's legislative database suggests that it could be very useful to those studying and implementing biosafety policies and the Protocol. This value could be enhanced to the extent that the BCH evolves with the goal of improving accessibility for this type of use. As a tool for academic study or as input into national legislative development process, the BCH must have electronic marking and search mechanisms that function well, to enable both identification of a particular law or a single country's postings and examination of the breadth of legislative documents from many countries that address particular issues or Protocol elements. As yet, the BCH's macro-search mechanisms are imprecise and do not strongly support research by academics and/or by legislative draftsmen. At present, broad-scope search mechanisms in the BCH are somewhat imprecise, producing many irrelevant documents for every relevant document produced.[16]

The choice of search terms in the BCH is idiosyncratic from a legal perspective.[17] As a result, BCH users often find it necessary to read through the entire database, currently numbering more than 600 instruments. Posted documents range from

[14] Postings from Bolivia, (*Resolución Multiministerial No 1*. 2005. (BCH Record No. 12011), and *Resolución Administrativa* N° *044/2005*. 2005. (BCH Record No. 12010), and *Resolución Administrativa VRNMA No 135/05* (BCH Record No. 12012), BOLIVIA *Resolución Administrativa No 016/2005*. 2005. (BCH Record No. 12013)), for example.

[15] Ruth Mackenzie et al, *supra* note 2 at paragraphs 539 and 551. See paragraphs 547–552 for a discussion of the question of whether BCH data should be validated and by whom.

[16] Presumably, this glitch will be adjusted in the BCH development process.

[17] Documents are indexed by a set list of subject search terms, including specific exceptions from the Protocol's scope. Thus, it is possible to identify laws on pharmaceuticals and international transit, and a range of national laws that are not focused on biosafety and apply to all enterprises or activities in the country (e.g., national EIA legislation.) The use of the search function is further impaired by the fact that most documents, regardless of actual content, are keyworded only as "All functions related to the Cartagena Protocol," with only a portion of the records more exhaustively indexed.

ten to several hundred pages in length. Thus, a comprehensive analysis necessitates individual indexing – examination of several thousand pages of text, most of which appears irrelevant to biosafety and LMO issues and contains no reference to how LMO and biosafety issues might apply to it. This kind of searching would be daunting even to an academic, but for a national legislative draftsman, attempting to use the BCH for solutions to particularly challenging problems could be completely off-putting.

USE OF BCH DATA BY IMPORTERS, EXPORTERS, AND DEVELOPERS

Based on the rather unscientific process of interviewing people at the COP-MOP, it appears that there is great interest in the use and further development of the BCH as a tool for those who hope to develop, export, import, and/or market LMOs. This group includes LMO developers and commercial exporters/importers of LMO products, as well as many academic and nonacademic researchers working with LMOs (herein referred to as "the LMO sector"). This group has also been most outspoken regarding the deficiencies in the BCH.

Members of the LMO sector note that, with the exception of highly publicized products, such as Roundup Ready crops and LMOs designed for general drought resistance, LMO development to address food-security challenges is narrowly focused on locally relevant factors (e.g., crop varieties engineered to resist locally problematic pests, fungi, and diseases) for the particular area or region in which the LMOs are to be used.[18] As a consequence, most LMO developers think that it is important to be extremely sensitive to the national legislative and administrative situation regarding LMOs and their introduction in each individual country or locality in which they work. Before investing the time and expense of genetic engineering, the LMO sector must evaluate business matters and risks, a process that is highly dependent on legal certainty. The availability of relevant permissions and the administrative processes must be known or confidently predictable so that costs can be factored into the business decision to proceed with the development process, sometimes months or years before the variety is ready for contained or uncontained introduction.[19] Most of these searches will be limited to one or a few countries[20] but will be particularly concerned about conflicting and draft postings in

[18] Specific names withheld. This information was gathered in the course of 20 country visits undertaken in the global evaluation reported in Jarle Hårstad, et al., *supra* note 4.

[19] The relevance of issues of legal certainty in international environmental negotiations that may impact on commercial and noncommercial entities and contracts is discussed in detail in "Summary Analysis: Legal Certainty for Users of Genetic Resources under Existing Access and Benefit-sharing (ABS) Legislation and Policy," UN Doc. UNEP/CBD/WGABS/3/INF10. Reprinted in Tomme Young, ed. *Covering ABS: Addressing the Need for Sectoral, Geographical, Legal and International Integration in the ABS Regime: Papers and Studies of The ABS Project* (Bonn, Germany: IUCN-ELC, 2008).

[20] The BCH allows one to conduct searches that produce only documents posted by or for each particular country.

the name of these countries. Many have expressed doubt that the BCH documents do not identify the full range of policies, laws, regulations, administrative protocols, and practises that will be brought to bear on them, and indeed that the national legislation itself has not coordinated with such other legislation.

For BCH users from the LMO sector, improved metadata (as discussed previously) is essential for understanding whether and how particular posted documents (especially those labelled *draft* or which have not been formally adopted) apply. One commonly mentioned example is the relationship between LMO/biosafety and invasive species and/or alien species introduction.

Many members of the LMO sector appear to expect the BCH to provide at least a basic understanding of all legal matters relevant to LMO introduction, including basic points such as the type of legal system in the country, any processes for simplifying the regulatory process when multiple legal provisions overlap, and the nature of their rights and obligations following denial or acceptance of an application to introduce LMOs or to place them on the market. They note that LMO-related commerce will be best facilitated if this information is available in all UN languages (or at least in English, French, and Spanish).[21]

At present, the BCH understandably does not provide this type or level of information. Several countries have posted summaries of their national regulatory processes relevant to LMOs. In the future the COP-MOP or the BCH-IAC might determine that one purpose of the BCH should be to facilitate international commercial and academic development and introduction of LMOs (or perhaps such objectives shall be limited to LMOs for some purpose such as food security). In that case, it may be useful for the Secretariat, BCH-IAC, or international, regional, and bilateral aid programmes to promote the development of such summaries for all countries wanting to facilitate the importation and introduction of LMOs or some subcategory of LMOs.[22]

Arguably, it is not the obligation of the BCH to facilitate access of commercial and noncommercial exporters of LMOs. This task is often, however, undertaken by NFPs and other officials of the exporter country. Some means of coordinating their efforts, perhaps through more widespread use of the regulatory summary approach, might reduce the duplication that would result if each country undertook this task itself.

INITIAL CONTROVERSIES AND UNDERLYING OBJECTIVES

To understand fully the complex web of Protocol implementation, including both the provisions in the agreement and the implementing legislation adopted and/or

[21] All persons interviewed by the author on these matters were native speakers of or worked in one of these languages.

[22] As discussed later, the Protocol does not presently represent a consensus about LMO introduction, even with appropriate risk-management mechanisms in place.

drafted at the national level, it is useful to consider the reasons underlying the negotiators ultimate willingness to settle for such a complex, obscure, and limited instrument.

One of the most important background facts relating to the Protocol was that, even after the instrument was finalised, there remained a significant divergence of views among the negotiating countries regarding biosafety and LMOs, particularly about questions of risk and whether and how that risk should be addressed. Experience since adoption suggests that those divergences have not yet disappeared.[23] The Protocol provides an excellent example of the manner in which countries are able to arrive at some agreement, even in the face of unresolvable disagreements. In essence, it is not an international consensus in recognition of a risk or problem, so much as a reaffirmation of the Parties' mutual commitment to respect each country's formal position regarding the risks of LMOs and biosafety.

In the broader context of international policy, it is important to recognize the type of agreement the Protocol represents. Every international instrument is an agreement among sovereign nations; however, the nature of that agreement can vary widely. The four types most commonly recognised are as follows: (1) Conventions and Treaties that operate as promises not to take action or to undo actions previously taken. Peace Treaties and nonaggression pacts are classic examples of this kind of international commitment; (2) Instruments that operate as promises not to regulate or to keep regulation of a certain matter within agreed limits. The instruments within the General Agreements on Tariffs and Trade (GATT) treaty system, currently overseen by the World Trade Organization (WTO), are generally in this category, because countries have agreed not to adopt legislation, regulations, or other controls that have certain types of impacts on international movement and sale of goods and services; (3) Instruments the objective of which is primarily policy oriented, with countries agreeing to make all reasonable efforts to achieve certain agreed outcomes; and (4) some instruments include commitments of the countries to implement specific requirements.

The Convention on Biological Diversity (CBD) is generally thought to be the third type of instrument. Few of its provisions are expressed as mandatory commitments, and its primary organising force is the promotion of three objectives expressed as outcomes – the conservation of biological diversity, the sustainable use of its components, and the equitable sharing of the benefits arising from the utilisation of its genetic resources.[24]

The fourth category of agreement is a common common approach in addressing global environmental issues and conservation. Three well-known environmental

[23] See generally M. Bauer and G. Gaskell, *Biotechnology: The Making of a Global Controversy* (Cambridge, UK: Cambridge University Press, 2002).

[24] *Convention on Biological Diversity*, 5 June 1992, 1760 UNTS 79, 31 ILM 818 (entered into force 29 December 1993) at Art. 1.

instruments of this type are the Convention on International Trade in Endangered Species of Fauna and Flora (CITES),[25] the Montreal Protocol to the Vienna Convention for the Protection of the Ozone Layer on Substances That Deplete the Ozone Layer (Montréal Protocol),[26] and the Cartagena Protocol. It is no accident that two of these are protocols to global Conventions. In essence, like national legislative processes in which the underlying policy is (expressly or by implication) determined prior to adoption of functional regulatory instruments, it is normally necessary to adopt overall policy-style conventions (described in the third category previously set forth), before specifying particular mandatory actions in more detail. This is comparable to the national level process in some countries under which general and overarching principles and requirements are expressed in laws/statutes and details regarding the specifics of implementation in subsequently adopted regulations/subsidiary legislation. The fourth category is often divided into two groups: *action-coordinating instruments* (in which countries seek to eliminate disconnections and maximise the effect of national programmes that already exist), and *action-forcing instruments* (in which countries seek mutually to compel themselves and their peers to take some type of action).

Nearly all international instruments contain at least some aspects of all four types of instruments,[27] but the Protocol is generally considered to be the fourth type. Seen in this way, its most important unstated objective is to provide users and/or consumers with a basis for confidence that their government can provide and is providing appropriate controls on the safety of LMOs within the country.

The Republic of Korea's draft National Biosafety Framework (NBF), prepared under the UNEP/GEF Biosafety project, notes the underlying need for this confidence:

> Many benefits of GM crops are obvious or immediate and simple to quantify. Risks, on the other hand, involve events of very low probability, and are difficult to document and especially difficult to quantify. . . . This kind of labelling system is based on the premise that GM crops and processed foods are safe for consumption. It is therefore problematic that decision regarding the safety (or harmfulness) of GM processed foods be left to the consumer. The labelling system on GM agricultural products is a post-market control system that is still regulated by MAF. In contrast, the safety assessment system as a pre-market control system is regulated by KFDA. This dual GM products control system need to be revised in the long run and KFDA should be in charge of this. . . . [T]aking the problem of decision-making . . . on the

[25] *Convention on International Trade in Endangered Species of Fauna and Flora*, 3 March 1973, 993 U.N.T.S. 243, 12 I.L.M. 1085.

[26] *Montreal Protocol on Substances that Deplete the Ozone Layer*, 16 September 1987, 26 ILM 1550 (entered into force 1 January 1989).

[27] See Ian Brownlie, *Principles of Public International Law, 4th ed.* (Oxford, UK: Clarendon Press, 1990) at chapter 1 and J. Klabbers, Introduction to International Institutional Law (Cambridge, UK: Cambridge University Press, 2002) at chapter 2.

safety/harmfulness of a particular GM processed food together with . . . labelling system.[28]

Many such concerns were expressed within the Protocol negotiations. Commercial developers and marketers investing in LMOs were concerned that import country laws would damage their productivity and profitability. At the same time, those import countries were concerned about consumer fears of the introduction of unknown risks and uncertainties into their national environment and food production system. Unable to codify a single unified international view of the nature and extent of GMO risks, negotiators instead agreed on a flexible mechanism that each country could use. As long as individual people, agencies, NGOs, and consumers are confident about their government's ability and willingness to actively evaluate and/or manage risks, the decrease in fear benefits everyone: consumers, agriculturalists, markets, import companies, export companies, and GMO developers.

[28] Republic of Korea, Draft National Biosafety Framework (developed under the UNEP-GEF Biosafety Project). 2004. (BCH Record No. 43444), online: www.unep.org/biosafety/files/KRNBFrep.pdf at: pp. 21–23 and 57–58 further comments on this issue include the following: "Many benefits of GM crops are obvious or immediate and simple to quantify. Risks, on the other hand, involve events of very low probability, and are difficult to document and especially difficult to quantify. . . . Experiments to evaluate risks require long-term commitments; even then, science may be able to establish a probability of an untoward event, but it cannot prove a zero probability. Therefore, in order to decide whether a particular GM crop can be accepted, we need to compare benefits obtained from GM crops or agro-products with non-GM crops or agro-products. The decision must be made by society. But one should accept that even conventional agricultural practises pose risks to ecosystems. Risks from the use of biotechnology must logically be evaluated in that context (i.e., does it increase or reduce overall risk?) rather than in isolation. All technologies in use today have the potential for great benefit to humankind if applied properly or, if misapplied, the potential for harm. Regulation is necessary to prevent their misuse. So far, experience has indicated that there is a very low probability of risk to humans and the environment if appropriate policies are in place to govern its use."

10

Socioeconomics, Biosafety, and Sustainable Development

Frederic Perron-Welch

Biotechnology, particularly plant and animal biotechnology, is perceived differently by different cultures because of different assessments of its prospects and its ethics.[1] Many people, societies, and States express concern over the social and economic changes that will accompany the widescale use of biotechnology and its products. It is not a visible part of the international development agenda primarily because of the controversy surrounding the potential risks of living modified organisms (LMOs) to biodiversity and human health, as well as the potential relevance of LMOs for traditional farming sectors in developing countries.[2]

Chapter 16 of Agenda 21 on the environmentally sound management of biotechnology provides the following context:

> Like with most new technologies, research in biotechnology and the application of its findings could have significant positive and negative socio-economic as well as cultural impacts. These impacts should be carefully identified in the earliest phases of the development of biotechnology in order to enable appropriate management of the consequences of transferring biotechnology.[3]

[1] International Law Association (ILA), Res No 5/2010, "The Hague Recommendations on International Law on Biotechnology" at para 2 (Hague Recommendations), online: http://www.ila-hq.org/en/committees/index.cfm/cid/1016.

[2] Sakiko Fukuda-Parr, "Introduction: Genetically Modified Crops and National Development Priorities" at 4 in Sakiko Fukuda-Parr, ed, *The Gene Revolution: GM Crops and Unequal Development* (London, U.K.: Earthscan, 2007). Also see Chapter 26 of this book for a discussion of biotechnology and food aid.

[3] *Agenda 21: Programme of Action for Sustainable Development*, U.N. GAOR, 46th Sess., Agenda Item 21, UN Doc. A/Conf.151/26 (1992) at Chapter 16, para 38.

This chapter is the result of ongoing CISDL research on biosafety decision making and international sustainable development law and expands on the work of Martin Endicott, Christine Frison, and Kathryn Garforth with Marie-Claire Cordonier Segger, Jorge Cabrera, and Sylvestre Manga, *Innovations in Biosafety Law: A CISDL Working Paper* (Montreal: Centre for International Sustainable Development Law, 2005).

As a result, many countries that are adopting biotechnology insist on using a decision-making process that looks beyond the narrow range of impacts covered by the scientific risk assessment of a specific LMO. Deciding what is right or permissible in relation to biotechnology in many contexts requires bringing together the knowledge established by a scientific risk assessment of an LMO with socioeconomic information on that society and its culture. This is a reflection of the fact that when considering the use of new technologies, scientific, ethical, and social issues cannot be wholly separated from one another.[4] Socioeconomics, as a discipline, begins with the assumption that economics is not a self-contained system but is embedded in culture, government, and society. It does not assume that interests are necessarily complementary and harmonious, and recognises that societal sources of order are necessary for markets to function efficiently. It also recognizes that individual choices are shaped by values, emotions, social bonds, and moral judgments rather than presuming that people act rationally or pursue only their self-interest or pleasure.[5]

The Cartagena Protocol allows Parties to take socioeconomic considerations into account in biosafety decision making through Article 26, which reads:

1. The Parties, in reaching a decision on import under this Protocol or under its domestic measures implementing the Protocol, may take into account, consistent with their international obligations, socio-economic considerations arising from the impact of living modified organisms on the conservation and sustainable use of biological diversity, especially with regard to the value of biological diversity to indigenous and local communities.
2. The Parties are encouraged to cooperate on research and information exchange on any socio-economic impacts of living modified organisms, especially on indigenous and local communities.[6]

Biosafety risk assessment, described in greater detail in Chapter 4, aims to ensure that decisions about policies and activities pertaining to the development of biotechnology are taken with an understanding of the potential adverse impacts to the conservation and sustainable use of biological diversity, as well as risks to human health. In this light, socioeconomic information is needed to provide a more complete picture of the impact of biotechnology. An assessment of environmental, social, and economic impacts in biosafety decision making would be more consistent with the sustainable development principle of integration and interrelationship of social, economic, and environmental objectives.[7]

4 Nuffield Council on Bioethics (NCB), *Genetically modified crops: the ethical and social issues* (London, UK: NCB, 1999) at paras 1.2 and 1.5.

5 Society for the Advancement of Socio-Economics, "About SASE," online: http://www.sase.org.

6 *Cartagena Protocol on Biosafety to the Convention on Biological Diversity*, 29 January 2000, 2226 UNTS 208; 39 ILM 1027 (entered into force 11 September 2003) at art 26.

7 *ILA New Delhi Declaration of Principles of International Law Relating to Sustainable Development*, A/CONF.199/8, Annex at Principle 7 (New Delhi Declaration).

Application of the principle of public participation, discussed in greater detail in Chapter 7, in this context suggests that a country's policy about the use of biotechnology ought to be determined in consideration of the views and participation of a well-informed public. Such an approach is reflective of responsive, transparent, and accountable governance and is informed by the human right to hold and express opinions and to seek, receive, and impart ideas; the right to access appropriate, comprehensible, and timely information held by governments and commerce on economic and social policies about the sustainable use of natural resources and the protection of the environment; and the right to access to effective judicial or administrative procedures to challenge unsatisfactory decisions and seek compensation.[8]

The inclusion of socioeconomic considerations in the Cartagena Protocol allows for the creation of a decision-making process for biosafety that can be tailored to a broad range of States, societies, and cultures, so long as the measures taken are consistent with its other international obligations. It is also reflective of several sustainable development law principles, such as integration and interrelationship, common but differentiated responsibilities, the precautionary approach, public participation, and good governance.[9]

This chapter focuses on Article 26 of the Cartagena Protocol, fleshing out its meaning through its negotiating history, potential role in biosafety decision making, and the precepts of sustainable development law. The first section discusses the negotiations and the legal text of Article 26. The second section considers the role of socioeconomic considerations in biosafety regulatory regimes. The third section describes the nature of socioeconomic considerations in light of the principles of sustainable development law. The chapter concludes with a discussion of how the use of socioeconomic considerations in biosafety could be promoted as a tool to support sustainable development.

SOCIOECONOMIC IMPACTS AND BIOSAFETY

Negotiations on biosafety following the entry into force of the Convention on Biological Diversity (CBD) brought up many differing views on the socioeconomic impacts of LMOs and considerations relevant to biosafety decision making. At the first meeting on biosafety under the aegis of the CBD, the 1995 Madrid Group of Experts on Biosafety, no consensus could be reached between Parties on whether socioeconomic considerations should be a part of biosafety decision making.[10] With the receipt of the Madrid Group's report, the second Conference of the Parties to

[8] New Delhi Declaration, *supra* note 7 at Principle 5.

[9] See Introduction and Chapter 1.

[10] Secretariat of the Convention on Biological Diversity SCBD, *The Cartagena Protocol on Biosafety: A Record of the Negotations* (Montreal: SCBD, 2003) at 79 (A Record of the Negotiations). See also *Report of the Open-Ended Ad Hoc Group of Experts on Biosafety*, UN Doc UNEP/CBD/COP/2/7.

the CBD (COP 2) launched the Open-Ended Ad Hoc Working Group on Biosafety (BS-WG) to negotiate a Protocol, where socioeconomic considerations would be discussed further.

At the first meeting of the BS-WG (BS-WG 1), most developed countries argued that socioeconomics had little place in biosafety decision making and that including such considerations in the Protocol was not worthy of further study.[11] Many developing countries disagreed, because they believed that "in addition to economic impacts such as income distribution, the negative socio-economic impacts of LMOs could include erosion of agricultural and other biological diversity; risks to sustainable use of existing biodiversity; and the threats of transgenic animals and plants to the cultural and religious order of some countries."[12] The outcome of BS-WG 1 was a request that the Secretariat prepare a bibliography on both the positive and negative socioeconomic impacts of biotechnology for consideration at the following meeting.[13]

At BS-WG 2, the bibliography prepared by the Secretariat was considered alongside the written submissions of governments. The African Group proposed the most comprehensive provisions on socioeconomic factors, which were incorporated into the draft provisions on objectives, general obligations, notification, risk assessment and management, and liability and compensation. The African Group's draft article on socioeconomic considerations noted the time before impacts are felt and proposed a seven-year notification period prior to the export of LMOs, while also containing an extensive list of considerations for risk assessment. This list included anticipated changes in existing social and economic patterns; possible threats to biodiversity, traditional crops, or other products; impacts caused by the replacement of traditional crops, products, and indigenous technologies through modern biotechnology; anticipated social and economic costs resulting from loss of genetic diversity, employment, market opportunities, and the livelihoods of communities; disruptions to social and economic welfare; and possible effects contrary to the social, cultural, ethical, and religious values of communities.[14] Bolivia's submission noted that the sustainable use of biodiversity, particularly domesticated plants and animals, depends on the socioeconomic conditions of the people who have developed and conserved them for generations, and that the introduction of genetic engineering technologies and LMOs into countries with rich biological and genetic diversity can result not only in the depletion of that diversity but also threaten the economic situation of people who rely on biodiversity, which could lead to discontinuation of the traditional agricultural systems and a loss of genetic diversity.[15] On the other

[11] *Ibid.*

[12] *Report of the First Meeting of the Open-Ended Working Group on Biosafety*, UN Doc UNEP/CBD/BSWG/1/4 at para 88.

[13] *Ibid* at para 111.

[14] *Compilation of the Views of Governments on the Contents of the Future Protocol*, UN Doc UNEP/CBD/BSWG/2/2.

[15] *Ibid* at 70–1.

hand, Canada, the European Union, and Japan all weighed in against addressing socioeconomic considerations in the Protocol.[16]

Governments submitted draft text prior to BS-WG 3 and the submissions of the African Group, Cuba, India, Madagascar, Malaysia, and Sri Lanka all included reference to socioeconomic considerations.[17] It was agreed at that time that socioeconomic considerations would not be further researched, but rather included in the consolidated text of draft articles with the texts already submitted by governments set out as options.[18] A number of substantive options were drafted for consideration. The first reflected the submission of the African Group to BS-WG 2. A second option called for socioeconomic imperatives to be considered at all levels in the Protocol and that particular attention be paid to the displacement of particular agricultural resources, cultures, or livelihoods, and to the prevention and mitigation of possible adverse effects. A third option acknowledged the considerable variation in socioeconomic considerations among Parties and encouraged further research.[19]

Negotiations on socioeconomic considerations from this point forward were very difficult. The debate mainly centered on the need to include socioeconomic considerations in the text of the Protocol at all. Many developing countries thought the issue was central to the Protocol, while developed countries found the concept too vague and country specific to have its own provision.[20] The text and options were reduced over the course of BS-WG 4 and BS-WG 5 to one heavily bracketed compromise provision which referred to the prevention and mitigation of socioeconomic impacts, the assessment and management of risks with a long observation period, and encouraging further research on the topic. In addition, it called for Parties exporting LMOs-FFP to notify receiving Parties sufficiently in advance to allow appropriate measures to be taken, providing special assistance in cases involving developing countries.[21] Despite sympathy for the subject, countries could not agree about the need, place, and manner of handling the issue. Because of its bearing on the scope and other provisions for the Protocol, it was necessary for delegations to carefully consider their positions before BS-WG 6.[22]

The Chair's proposed text at BS-WG 6 significantly modified the draft option, providing only that Parties should take into account socioeconomic implications of adverse impacts of LMOs, also taking into account human health, especially for indigenous and local communities (ILC) as referred to in Article 8(j) of the CBD.

[16] *Ibid* at 71.

[17] See *Compilation of Government Submissions of Draft Text on Selected Items*, UN Doc UNEP/CBD/BSWG/3/3.

[18] *Report of the Third Meeting of the Open-Ended Ad Hoc Working Group on Biosafety*, UN Doc UNEP/CBD/BSWG/3/6 at para 27.

[19] *Ibid*, at 87–8.

[20] *Ibid*, at paras 29-38.

[21] *A Record of the Negotiations*, *supra* note 10 at 80-1.

[22] *Report of the Fifth Meeting of the Open-Ended Ad Hoc Working Group on Biosafety*, UN Doc UNEP/CBD/BSWG/5/3 at para 46.

It also encouraged Parties to cooperate on research and information exchange, including early warning to ILC that may be affected economically.[23]

The text was further revised and amended before being transmitted to the First extraordinary meeting of the Conference of the Parties (ExCOP). First, countries added a requirement that import decisions be consistent with international obligations. Second, socioeconomic considerations arising from the impact of LMOs was allowed, rather than only those linked to adverse impacts. Third, reference to human health was deleted. Fourth, the value of biological diversity to indigenous and local communities was added rather than a blanket reference to Article 8(j). Fifth, the general reference to research and information exchange on any socio-economic impacts of LMOs, especially on ILC, replaced earlier language on early warning and economic effects to those communities. Sixth, references to socioeconomic considerations were removed from Annex II to the Protocol.[24] At the ExCOP, a final addition was made during informal consultations, allowing countries to address socioeconomic considerations in domestic measures to implement the Protocol.[25]

As demonstrated by its history, Article 26 was the result of intense negotiation and compromise and is rather limited when compared to some of the legal text initially proposed by developing countries. Like the remainder of the Protocol, it represents a balance between the interests of industrialised and developing countries. Through this compromise, industrialised countries – and their biotechnology sectors – can be assured that scientific risk assessment remains the primary basis for biosafety decision making, while developing countries are also assured that socioeconomic considerations can be relied upon for some decision making processes. To bring clarity to the terms of Article 26, the particulars of biosafety decision-making and the relevance of socioeconomic considerations are discussed below.

BIOSAFETY REGIMES AND SOCIOECONOMIC CONSIDERATIONS

The main basis for biosafety decision making is scientific risk assessment.[26] The focus of scientific risk assessment is on the possible environmental and human health impacts of a new organism or its products, many of which can be measured by traditional scientific techniques. Opinions vary about the value and desirability of including socioeconomic considerations in decision-making processes and the introduction of broader socioeconomic considerations into biosafety analysis.[27]

[23] A *Record of the Negotiations*, *supra* note 10 citing Chair's draft text of the Protocol, UN Doc UNEP/CBD/BSWG/6/L.2 at Article 24.

[24] *Ibid*. Annex II of the Protocol is a list of information required concerning LMOs-FFP under Article 11 (Procedure for LMO-FFP).

[25] International Institute for Sustainable Development IISD, "Report of the resumed session of the Extraordinary Meeting of the Conference of the Parties (ExCOP) for the Adoption of the Protocol on Biosafety to the Convention on Biological Diversity" 9(137) Earth Negotiations Bulletin (ENB), at 9.

[26] See Chapter 4 for an in-depth discussion of the Cartagena Protocol's risk assessment procedures.

[27] José B. Falck-Zepeda, "Socio-economic Considerations, Article 26.1 of the Cartagena Protocol on Biosafety: What Are the Issues and What Is at Stake?" 12(1) AgBioForum 90.

Understanding the impacts of modern biotechnology and its products and processes in the context of sustainable development requires looking beyond the physical sciences. New crops and new foods have a trenchant impact on societies across the globe. Their socioeconomic effects are as yet undetermined but can be positive, negative, or neutral, and can be expressed by a variety of actors. Some of the concerns are relevant to numerous actors, whereas others are discrete, belonging to a particular group. Some of these concerns are common to both developed and developing countries, whereas others are relevant only to one or the other. Socioeconomic impacts are not unique to biotechnology, but taking them into consideration is of particular relevance given the rancorous debate over this technology. The International Law Association's Hague Recommendations on International Law on Biotechnology suggests accommodating a range of legislative responses while also considering different levels of social and economic development.[28]

The limited scope of Article 26 is important to recognise. First, the wording of Article 26(1) limits the use of socioeconomic considerations in biosafety decision making to those arising from the impact of LMOs on the conservation and sustainable use of biodiversity.[29] This can be interpreted to mean that where the introduction of LMOs under the Protocol affects biological diversity with potential or actual social and economic repercussions, Parties can justify taking into account such impacts when making decisions on LMO imports or in implementing domestic measures under the Protocol.[30] Second, the provision gives particular recognition to socioeconomic considerations arising from the value of biological diversity to indigenous and local communities.

Socioeconomic considerations beyond the Cartagena Protocol are those that concern products or activities falling outside the mandate of the Protocol, such as socioeconomic considerations related to organisms that do not fall within the definition of LMO, living organism, or modern biotechnology as used in the Protocol; considerations related to LMOs that are not subject to the activities covered by the Protocol (i.e., domestic or non-transboundary movement, transit, handling, and use); impacts of LMOs beyond the conservation and sustainable use of biodiversity; and LMOs that are pharmaceuticals for humans and are addressed by other international agreements or organizations. Countries can implement biosafety regulatory regimes that encompass socioeconomic concerns beyond those included in the Cartagena Protocol if their regulatory system complies with any other international obligations by which they may be bound. In the context of the AIA and LMOs-FFP procedures for import/export, socioeconomic considerations can be used by the importing Party as a basis for decision at that stage.[31]

[28] *Hague Recommendations*, *supra* note 1 at para 2.

[29] Ruth Mackenzie et al, *An Explanatory Guide to the Cartagena Protocol on Biosafety* (Gland, Switzerland and Cambridge, UK: IUCN, 2003) at para 628.

[30] *Ibid* at para 629.

[31] See Table, "Comparative Summary of the AIA and Article 11 procedures," in Chapter 16 of *UNEP Training Manual on International Environmental Law* (Nairobi, Kenya: UNEP, 2006) at 220–1.

A literature survey reveals a broad range of sometimes conflicting socioeconomic considerations relating to GMOs and LMOs. Careful research to clarify the socioeconomic issues related to biotechnology is essential but is not in itself sufficient to incorporate socioeconomic considerations into biosafety decisions. Practical steps are necessary for these considerations actually to be taken into account when decisions are made. These steps might include policies that mandate integration of socioeconomic considerations into decision-making processes; a clear definition of "socioeconomic considerations" and explicit criteria to determine when and where socioeconomic assessments are required; identification of the stages at which socioeconomic assessments should take place; efficient and cost-effective regulatory processes; and public participation mechanisms to ensure credible assessments and more widely accepted decisions.[32] The following methods have been proposed for taking socioeconomic considerations into account: procedures for assessing and addressing socioeconomic impacts in risk assessment and management, and subjecting decisions on import of LMOs to prior public consultation processes, especially with respect to communities directly affected by the import decision, for example the local community in which the LMO is destined for field trial or use, or that could be affected by any potential adverse impacts of the LMO on biodiversity.

In 2007, UNEP-GEF received funding from the U.K. Department for International Development to undertake a scoping exercise on socioeconomic considerations in biosafety decision making. The exercise included an information-gathering survey on countries' experience with socioeconomic considerations undertaken in cooperation with the SCBD.[33] At UNEP's request, the summary report was circulated as an information note for participants at the fifth Conference of the Parties to the Convention on Biological Diversity serving as the meeting of the Parties to the Cartagena Protocol (COP-MOP).[34] The responses are instructive of how socioeconomic considerations can or should be addressed as a part of the biosafety regulatory regime.

The survey first asked respondents their opinion on the importance of including socioeconomic considerations in decision making on LMOs in their country. Eighty-five percent indicated that socioeconomic considerations were either very important or somewhat important, with four-fifths selecting "very."[35] Of the countries having a decision-making system for LMOs in place, two-thirds of respondents stated that it was able to take socioeconomic considerations into account.[36]

32 L. Fransen et al, *Integrating Socio-Economic Considerations into Biosafety Decisions* (Washington, DC: World Resources Institute, 2005) at 3–4.

33 Paul J Thomassin, "Summary Report on the Survey on the Application of and Experience in the Use of Socio-Economic Considerations in Decision-Making on Living Modified Organisms", UN Doc UNEP/CBD/BS/COP-MOP/5/INF/10. The full survey is available online: http://bch.cbd.int/onlineconferences/socioeconomics.shtml.

34 *Ibid* at page 1.

35 *Ibid* at para 7.

36 *Ibid* at para 8.

In those countries where socioeconomic considerations were addressed in the decision-making system, most of respondents indicated that the regulatory framework addressed the concept in a general fashion, some indicated that socioeconomic considerations were treated on an ad-hoc or case-by-case basis without any explicit inclusion in the decision-making system, and the least number identified socioeconomic considerations as being covered through the assessment of particular elements during specific steps in the process.[37] The largest group of responses on how best to incorporate socioeconomic considerations into the decision-making process was in favour of a specific, identifiable mechanism; the next largest group supported a general framework for application within the risk assessment process; some respondents supported a participatory process that included a broad spectrum of stakeholders; a few respondents supported an ad-hoc or case-by-case approach; and a small number of respondents were against including socioeconomic considerations in biosafety decision making.[38]

The top five socioeconomic issues of importance in all countries with a biosafety system were identical: food security, health-related impacts, coexistence of LMOs and non-GM agriculture, impact on market access, and compliance with biosafety measures. More variation occurred with the importance of the next five issues, and depending on ranking, the following issues arose: conservation and sustainable use of biodiversity; economic impacts of changes in pest prevalence; macroeconomic impacts; farmers' rights; intellectual property rights (IPRs); impacts on consumer choice; economic impacts of changes in the use of pesticides and herbicides; and impacts to indigenous and local communities.[39]

Countries with a biosafety decision-making system had several different ways of considering socioeconomic considerations in the decision-making process. An assessment that happens during the risk assessment stage was identified most frequently, followed by an equal proportion of "prior to the risk assessment" and "after the risk assessment." During the appeal, review or renewal of the permit was identified the fewest number of times.[40] Lack of a mechanism for including socioeconomic considerations in decision making and lack of capacity were identified as important reasons why assessment was lacking from the decision-making system. Human resources and informational and financial challenges were identified as challenges to the inclusion of socioeconomic considerations in decision making. Capacity building and the development of guidelines are thus important mechanisms that could decrease the challenges to and the reasons for not including socioeconomic considerations in biosafety decision making.[41] Many countries have adequate capacity to perform socioeconomic assessments, but most answered "do not." It is also

37 *Ibid* at para 9.
38 *Ibid* at para 11.
39 *Ibid* at para 17.
40 *Ibid* at para 27.
41 *Ibid* at para 32.

clear, however, that once decision-making systems are put in place to account for socioeconomic considerations, skill levels increase, meaning that increasing the capacity for socioeconomic assessment could increase the number of countries that take socioeconomic considerations into account in their decision-making processes.[42]

Several conclusions can be drawn from this. First, socioeconomic considerations can be taken into account in the decision-making process if countries want to do so. Second, once countries start to incorporate them, they can do this across decisions, across multiple organisms, and across intended uses of the organism.[43] The current practice for including socioeconomic considerations in most countries seems to be for one entity to assess socioeconomic considerations and a different entity to evaluate the assessment. Three forms of government entity are most often involved in assessment and evaluation: the national competent authority, a multisectoral committee, and/or a government committee consisting of several departments.[44] Although there seems to be a strong consensus that socioeconomic considerations should be taken into account, it is less clear how and where socioeconomic information should be used. There was the strongest support for the consideration of socioeconomic information at the same time as scientific risk assessment information in the biosafety decision-making process, far less support for considering socioeconomic information after scientific risk assessment information, and a split between whether socioeconomic information should be evaluated separately from the scientific risk assessment or incorporated into the scientific risk assessment.[45]

A great degree of uncertainty exists over how socioeconomic information can be used in biosafety decision making and the weight that it should be assigned.[46] The capacity to undertake and evaluate socioeconomic assessments in biosafety decision making varies greatly. Countries that consider socioeconomic information in their decision-making systems score high in terms of having people with skills, professional training, or experience to undertake and evaluate that information. Most countries lack adequate capacity to perform socioeconomic assessments and might be reluctant to establish a decision-making system depending on socioeconomic information.[47] Countries are in need of capacity building on socioeconomic assessment capacity and are seeking a methodologic guide or toolkit on how to undertake socioeconomic assessments as a means to start the capacity-building process.[48] Elements that should be considered in biosafety decision making must be identified.[49]

42 *Ibid* at para 36.
43 *Ibid* at para 43.
44 *Ibid* at para 44.
45 *Ibid* at para 45.
46 *Ibid* at para 46.
47 *Ibid* at para 47.
48 *Ibid* at para 50.
49 *Ibid* at para 55.

A consensus is forming on the important socioeconomic issues and priority areas for capacity building. The socioeconomic issues that respondents identified as priorities for capacity building are similar to the socioeconomic issues being considered in existing decision-making systems. This will facilitate capacity building and provide guidance on the material to be included a methodologic toolkit.[50] Further research is needed to show how particular socioeconomic considerations can be used to address the priority areas of concern for capacity building and sustainable development.

BIOSAFETY, SOCIOECONOMICS, AND SUSTAINABLE DEVELOPMENT LAW

The social aspect of sustainable international biodiversity and biosafety law is the least developed aspect in national and international law.[51] This aspect overlaps other areas of sustainable development law, such as the right to food, access to medicines, the right of access to innovations, and corporate social responsibility (CSR). The following section elaborates on the socioeconomic aspect of the Cartagena Protocol in the context of emerging principles of sustainable development law.

Principle of Integration and Interrelationship

In accordance with the principle of integration,[52] the Protocol recognizes that economic and social development considerations can be pertinent to biosafety decision making in addition to the environmental considerations raised by scientific risk assessments. Specifically, it complies with the emerging customary norm of integration, in that the Cartagena Protocol requires States to "ensure that social and economic development decisions do not disregard environmental considerations, and not undertake environmental protection without taking into account relevant social and economic implications."[53] Because of its recognition of socioeconomic considerations as a fundamental basis for biosafety decision making, the Cartagena Protocol can be categorised as a highly integrative treaty.[54]

The Cartagena Protocol also reflects the interdependence of the needs of current and future generations. The socioeconomic impacts of modern biotechnology will be far ranging and affect States from this point on in history. This should be implemented at all levels of governance, because it is essential to the achievement of sustainable development.[55] The principle also calls for States to resolve

50 *Ibid* at para 48.

51 Jorge Cabrera Medaglia and Kathryn Garforth, "Sustainable International Biodiversity Law" at 349 in Marie-Claire Cordonier Segger and Ashfaq Khalfan, eds. *Sustainable Development Law: Principles, Practices and Prospects* (Oxford: Oxford University Press, 2004).

52 Cordonier Segger and Khalfan, eds., *Sustainable Development Law* at 106–7.

53 *Ibid* at 104.

54 *Ibid* at 109.

55 New Delhi Declaration, *supra* note 7, at para. 7.2.

apparent conflicts between competing imperatives through existing and appropriate new institutions,[56] including the WTO and the Codex Alimentarius, for example. Countries that include socioeconomic considerations alongside environmental risk assessments in their biosafety systems also will be creating sustainable development law. Some different ways to include the socioeconomic dimensions of biosafety in a domestic regulatory system offer differing degrees of integration.

Principle of Sustainable Use of Natural Resources

Under the principle of sustainable use of natural resources, States may manage natural resources pursuant to domestic environmental *and* developmental policies.[57] Integrating socioeconomic considerations into biosafety decision making plays an important role in determining whether the impacts of biotechnology on biodiversity are sustainable given the potential social and economic costs/benefits. This right is limited by the second component of the principle, the requirement that states must not cause irreparable damage to the territories of other states, or increasingly, the global environment.[58] States are under a related duty to manage natural resources in a rational, sustainable, and safe way to contribute to the development of their peoples, with particular regard for the rights of indigenous peoples, and to the conservation and sustainable use of natural resources and the protection of the environment, including ecosystems.[59] The specific language of Article 26 of the Cartagena Protocol maps directly onto this duty and is an excellent example of the principle in action. Complicating the principle's application is the requirement that States consider the needs of future generations in determining the rate of use of natural resources, and the related duty of all relevant actors (including States, industrial concerns, and civil society) to avoid wasteful use of natural resources and promote waste minimization policies.[60] Integrating socioeconomics into the biosafety decision-making process can account for the needs of future generations and help to implement the duty of all relevant actors to avoid waste and promote waste minimization. Ultimately, the integration of socioeconomics is of particular importance because centres of origin and centres of genetic diversity are of crucial importance to humankind, and the conservation and sustainable use of biologic diversity is a common concern of humankind.[61]

56 *Ibid* at para 7.3.
57 *Ibid* at para 1.1.
58 Cordonier Segger and Khalfan, eds., *supra* note 51 at 109.
59 New Delhi Declaration, *supra* note 7, at para. 1.2.
60 *Ibid* at para 1.2.
61 *Ibid* at para 1.3; *Convention on Biological Diversity*, 5 June 1992, 1760 UNTS 79, 31 ILM 818 (entered into force 29 December 1993) at Preamble; Cartagena Protocol, *supra* note 9, at Preamble.

Principle of Equity and Elimination of Poverty

The principle of equity and poverty elimination could significantly inform the use of socioeconomic considerations in biosafety decision making. This principle refers to both intergenerational equity (the right of future generations to enjoy a fair level of the common patrimony) and intragenerational equity (the right of all peoples within the current generation to fair access to the current generation's entitlement to the earth's natural resources).[62] The present generation has a right to use and enjoy the earth's resources but is obliged to take into account the long-term impact of their activities and to sustain the resource base and the global environment for the benefit of future generations of humankind. *Benefit* in this context should be understood in its broadest meaning, economic, environmental, social, and intrinsic benefits. Intergenerational equity requires the present generation not to introduce a technology that will irreparably harm the environment or the socioeconomic situation left for future generations. This equity also requires, however, that the present generation not deny future generations the possibility of benefitting from biotechnology and its socioeconomic gains. Intragenerational equity dictates that biotechnology should be employed to improve the ability of members of the current generation to access resources and reduce poverty and not exacerbate existing inequalities. Biosafety decision making must incorporate socioeconomic aspects for these assessments to be made.[63] The right to development must be implemented to meet developmental and environmental needs of present and future generations sustainably and equitably. This includes cooperating to eradicate poverty, in accordance with Chapter IX on International Economic and Social Co-operation of the Charter of the United Nations and the Rio Declaration on Environment and Development, as well as the duty to cooperate for global sustainable development and attaining equity in growth opportunities for developed and developing countries.[64] All States in a position to do so have a further responsibility, as recognized by the Charter of the United Nations and the Millennium Declaration of the United Nations, to assist States in achieving conditions of socioeconomic equity, and to ensure, at minimum, the eradication of poverty.[65]

Principle of Common but Differentiated Responsibilities

States and other relevant actors have common but differentiated responsibilities and are bound to cooperate in the achievement of global sustainable development and protection of the environment. The principle applies not only to environmental

[62] *Ibid* at para 2.1.
[63] *Ibid* at para 2.2.
[64] *Ibid* at para 2.3.
[65] *Ibid* at para 2.4.

protection but also to development of social goals.[66] International organisations, corporations (including in particular transnational corporations), NGOs, and the rest of civil society should cooperate with and contribute to this global partnership. Corporations have also responsibilities pursuant to the polluter-pays principle.[67] Differentiation of responsibilities, although based mostly on the contribution that a State has made to the emergence of environmental problems, must also consider the economic and developmental situation of the State.[68] Notably, the special needs and interests of developing countries and of countries with economies in transition, with particular regard to least developed countries (LDCs) and those affected adversely by environmental, social, and developmental considerations, should be recognised.[69] From a sustainable development law perspective, biosafety decision making that permits socioeconomic considerations may allow States to meet their common responsibility to ensure that there is adequate protection for the safe transfer, handling, and use of LMOs while providing a method for their economic and developmental situation. Ultimately, the principle suggests that the concept of common but differentiated responsibilities can be used to interpret and strengthen other corresponding principles of sustainable development law in situations of overlap or conflict between social, economic, and environmental regimes.[70]

The Precautionary Approach to Human Health, Natural Resources, and Ecosystems

Article 26 allows Parties to consider socioeconomic considerations when deciding on importing an LMO when the impact of the LMO on the conservation and sustainable use of biodiversity is concerned. In turn, both the AIA and LMO-FFP decision-making procedures allow precautionary decision making in light of insufficient scientific evidence on the extent of the potential adverse effects of an LMO on the conservation and sustainable use of biodiversity. Given the uncertainty over the socioeconomic impacts of LMOs and how these could affect biodiversity, the ambit of the allowable precautionary decision-making in Articles 10(6) and 11(8) seems that it include socioeconomic factors, at least to the extent that these can be considered "relevant scientific information and knowledge." Of course, States are also free to include socioeconomic considerations in determining whether to make a precautionary decision under their own decision-making processes outside the Protocol as long as this does not conflict with their other international obligations.

66 Cordonier Segger and Khalfan, eds., *supra* note 51 at 138.
67 New Delhi Declaration, *supra* note 7, at para 3.1.
68 *Ibid* at para 3.2.
69 *Ibid* at para 3.3.
70 Cordonier Segger and Khalfan, eds., *supra* note 51 at 143.

Public Participation and Access to Information and Justice

There are three dimensions to this principle. First, individual people should have the opportunity to participate in official decision-making processes and activities that affect and impact their lives and well-being directly.[71] Second, to participate effectively, the public must have access to adequate information concerning the issues, decision making, and policy making in which they are able to participate.[72] Finally, when people's rights have been infringed or when they have suffered harm, they should have access to administrative or judicial processes to challenge the measure and claim compensation.[73]

Numerous aspects of socioeconomic considerations in biosafety decision making relate to these three dimensions. The first dimension suggests that public participation should be a common component in all biosafety regulatory regimes, because these regimes relate to socioeconomic development. In other words, because biosafety decision making can directly affect people's lives and well-being, these same people should have an opportunity to participate in the decision-making process. Indeed, the Protocol requires public consultation in decision making as well as public participation on the safe transfer, handling, and use of LMOs in relation to the conservation and sustainable use of biologic diversity, also considering risks to human health. These are quite broad requirements that can be fulfilled in several ways.

The second dimension on public access to information is also reflected in Article 23 of the Protocol. This article requires that parties "[e]ndeavour to ensure that public awareness and education encompasses access to information on living modified organisms identified in accordance with the Protocol that may be imported." Much of the literature also recognizes that consumers want labelling of genetically modified foods or at least want to be able to find out what is in the food they consume. Consumers' reasons for wanting to know may well be based on socioeconomic factors (e.g., not wanting to consume something they feel runs counter to their philosophical values) rather than on environmental or health concerns. Biosafety regulatory regimes that are intended to support sustainable development need to consider how to fulfil the requirement of public access to information.

The final dimension is access to justice for people who think that their rights have been violated. Access to justice is necessary in cases of biosafety wrongdoing that harms socioeconomic rights. This could include the ability to be compensated for damages when there has been a socioeconomic biosafety injury (e.g., lost profits by organic farmers whose crops no longer command a premium because they contain introduced DNA).

71 *Ibid* at 158.
72 *Ibid* at 159.
73 *Ibid*.

Principle of Good Governance

The importance of the principle of good governance in the inclusion of socioeconomic considerations in biosafety decision making hinges primarily on its relationship with other principles of sustainable development law. The characteristics of good governance reinforce these other principles. Biosafety regulatory regimes that incorporate these principles result in systems that respond to the concerns of individual people and help to build consensus by acknowledging and not excluding the socioeconomic elements of biotechnology and biosafety. In turn, this encourages participation in the biosafety process rather than creating disillusionment and disenfranchisement. Ultimately, it can lead to national visions of biosafety that incorporate and reflect the ideas of all. Perhaps most important, building biosafety regulatory regimes around these principles of good governance helps to create regimes that integrate the other principles of international sustainable development law.

CONCLUSION

To varying degrees, many countries already consider socioeconomic factors within their biosafety regimes and recognize that socioeconomic considerations can be an important part of biosafety decision making. Although socioeconomic assessment of new and emerging technologies, including LMOs, is an invaluable tool supporting decision making, it can present a difficult hurdle if the assessment procedure is not clearly defined.[74] The main concern is to determine at which stage of the regulatory process inclusion is most useful, while maximising the functionality of the biosafety system.[75] The lack of a clear mechanism for the incorporation of socioeconomic information into the process and the limited institutional capacity for socioeconomic analysis in many countries are key barriers to the broader use of socioeconomic considerations in biosafety decision making.

Socioeconomic considerations in biosafety target values that states have already acknowledged as relevant and important. Numerous human rights treaties proclaim the rights of individual people to adequate food, work, and health. Nascent human rights to the environment are also emerging. These are just a few of the rights that could be affected by the socioeconomic impacts of the products and processes of biotechnology. States that are party to the different treaties in which these rights are enunciated are bound to uphold these rights. In the realm of trade, members of the WTO are bound by the terms of the Sanitary and Phytosanitary (SPS) Agreement, which lists relevant economic factors that members must consider in their risk assessments and determination of the appropriate level for sanitary or phytosanitary measures. These include the potential damage in terms of loss of production or sales,

74 Falck-Zepeda, *supra* note 27 at 91.
75 *Ibid.*

the costs of control or eradication, and the relative cost-effectiveness of alternative approaches.[76]

Given that the socioeconomic capacities, priorities, and processes of one region or area might not be identical to those of other regions, further analysis of biosafety decision making in different parts of the world is needed.[77] The elements of socioeconomic considerations relevant to biosafety decision making remain to be clearly defined, and an expert committee might be necessary to identify appropriate methods. Case studies and examples of different methods for assessing socioeconomic issues and their strengths and weaknesses should also be undertaken to guide practitioners.[78] The development of an institutional framework for the incorporation of socioeconomic considerations in biosafety decision making needs to be undertaken by first determining which institutions are most competent to address the issues. Methods should be identified and developed for use in the decision-making process, as well as methods for how to obtain stakeholder input during that process. Examples of decision-making frameworks and institutions that have been successful in incorporating socioeconomic considerations also should be identified and described.[79]

From a sustainable development law perspective, the consideration of socioeconomic information as a part of the biosafety decision-making process brings the field of biosafety closer to the goal of sustainable development through the integrated consideration of economic and social development, and environmental protection. Obtaining information and knowledge on socioeconomic issues further supports biosafety regulatory regimes that can contribute to sustainable development and demonstrate how to build synergies between economic, environmental, and social law.

[76] *Agreement on the Application of Sanitary and Phytosanitary Measures*, being part of Annex IA to the *Agreement Establishing the World Trade Organization*, 15 April 1994, 33 I.L.M. 1144 at Art. 5.3.

[77] Paul J. Thomassin, *supra* note 33, at para. 52.

[78] *Ibid* at para 53.

[79] *Ibid* at para 54.

11

The Compliance Mechanism of the Cartagena Protocol on Biosafety

Development, Adoption, Content, and First Years of Life

Veit Koester

The compliance mechanism of the Cartagena Protocol on Biosafety has been functioning for approximately 5 years. This period is probably too limited to draw any firm conclusions in respect of the future role of the compliance mechanism. On the other hand, the findings of the Compliance Committee have been considered by no less than three Conferences of the Parties serving as the Meetings of the Parties to the Protocol (2005, 2006 and 2008). Therefore, a preliminary stocktaking of the achievements of the compliance mechanism seems to be appropriate. This article attempts to provide such stocktaking, the conclusions being that while there are a number of positive indications, there are also some rather negative trends represented by the outcomes of the last Conference of the Parties serving as Meeting of the Parties (2008). In order to provide an informed background of the analysis, the article is introduced by a brief account of the negotiation history of the compliance mechanism and a presentation of its legal basis and main features.

INTRODUCTION

The first part of this article introduces the legal basis of the compliance mechanism (CM) of the Cartagena Protocol on Biosafety[1] (Cartagena Protocol). The second part of the article contains a summary of the negotiation history of the CM, and an analysis of the final result adopted by the first Conference of the Parties to the Convention on Biological Diversity[2] (CBD) serving as the Meeting of the Parties to the Protocol (COP MOP-1). The third part focuses on the first two reports of the Compliance Committee (CC) and the outcomes of the considerations of COP MOP-2 and COP MOP-3 of the reports. The fourth part presents the consolidated

[1] Cartagena Protocol on Biosafety (Montreal, 29 January 2000) (CPB).
[2] Convention on Biological Diversity (Rio de Janeiro, 5 June 1992) (CBD).

First published in (2009) 18(1) Review of European Community & International Environmental Law 77.
The views expressed herein are in the author's personal capacity.

report of the CC to COP MOP-4 based on the reports of the third and fourth meeting of the CC, and the outcomes of COP MOP-4 as related to the consolidated report. This part also includes some of the conclusions of the fifth meeting of the CC held after COP MOP-4. The final part contains some concluding remarks.

LEGAL BASIS OF THE COMPLIANCE MECHANISM

The legal basis of the CM of the Cartagena Protocol is Article 34, which provides that:

> The Conference of the Parties serving as the meeting of the Parties to this Protocol shall, at its first meeting, consider and approve cooperative procedures and institutional mechanisms to promote compliance with the provisions of this Protocol and to address cases of non-compliance. These procedures and mechanisms shall include provisions to offer advice or assistance, where appropriate. They shall be separate from, and without prejudice to, the dispute settlement procedures and mechanisms established by Article 27 of the Convention.

This article was not one of the most controversial articles during the negotiations of the Cartagena Protocol, since the main elements and options were already identified at the third meeting of the Open-Ended Ad Hoc Working Group on Biosafety (BSWG), which met six times between July 1996 and February 1999.[3] At BSWG-5 (1998) the article was almost fully developed, but was also bracketed to reflect a lack of consensus on whether the provision should be included in the Cartagena Protocol at all. At BSWG-6 (1999) the provision was slightly amended, the brackets were removed and the article provisionally adopted. The article remained unchanged until its final adoption by the resumed Extraordinary Conference of the Parties (ExCOP) on 29 January 2000.[4]

Article 34 is rather traditional in the sense that it is formulated as an enabling and mandatory clause, leaving the governing body to consider and approve procedures and mechanisms to address cases of non-compliance.[5] It also contains, however, elements which do not usually appear in comparable provisions of other treaties related to the environment and to sustainable development.[6]

3 FIELD, *The Cartagena Protocol on Biosafety: A Negotiating History* (Secretariat of the Convention on Biological Diversity, 2003) at 13.

4 See ibid., at 13 and 95 for a more detailed account of the negotiation history of Article 34 with references.

5 See Montreal Protocol on Substances that Deplete the Ozone Layer (Montreal, 19 September 1987) (the Montreal Protocol), Article 8; Rotterdam Convention on Prior Informed Consent Procedure for Certain Hazardous Chemicals and Pesticides in International Trade (Rotterdam, 10 September 1998) (the Rotterdam Convention), Article 17; Stockholm Convention on Persistent Organic Pollutants (Stockholm, 22 May 2001) (the Stockholm Convention), Article 17; and International Treaty on Plant Genetic Resources for Food and Agriculture (Rome, 3 November 2001) (ITPGR), Article 21.

6 Article 34 has more in common with ITPGR, Article 21 than the other provisions referred to in ibid., because the other provisions do not refer to the nature of the CMs or non-compliance response

According to Article 34, the CM was to be considered and approved at COP MOP-1.[7] COP MOP-1 completed its mandate by adopting Decision BS-I/7 on Procedures and Mechanisms on Compliance under the Cartagena Protocol on Biosafety (Section 4).

DEVELOPMENT OF THE COMPLIANCE MECHANISM

The Intergovernmental Committee for the Cartagena Protocol (ICCP)

By Decision EM-1/3 of the resumed session of the ExCOP, an Intergovernmental Committee for the Cartagena Protocol (ICCP) was established. COP-5 of CBD (2000) adopted a work plan for the ICCP. Among issues for consideration by the ICCP at its first and second meeting was compliance (Article 34).[8]

At ICCP-1 in 2000, there was consensus on the need for the establishment of an effective CM, that it should be of an advisory, facilitative, open, transparent, simple and non-confrontational nature, that the objective would be to promote and facilitate compliance, and that it should be triggered by parties.[9]

ICCP-1 decided that an open-ended meeting of experts should be held back-to-back with ICCP-2.[10] Accordingly, an expert group meeting was held immediately before ICCP-2 (September/October 2001) with a view to considering

measures. However, ITPGR, Article 21 does not address the inter-relationship between the CM and the dispute-settlement mechanism (DSM) as Article 34 of the CPB indeed does. The similarities between CPB, Article 34 and ITPGR, Article 21 indicate probably that ITPGR, Article 21 was influenced by Article 34 of CPB, which was adopted prior to ITPGR and is more related to the CBD and CPB than the other MEAs listed in ibid.

[7] This is in contrast to the Governing Body of the ITPGR, where ITPGR Article 21 provides that the Governing Body shall consider and approve the CM at its first meeting. The first meeting (in 2006), however, did not adopt a CM but only a resolution establishing a Compliance Committee (CC), which will commence its work following the approval of the CM. See Further Information on Experience regarding Cases of Repeated Non-Compliance under the Compliance Mechanisms of Other Multilateral Agreements (UNEP/CBD/COP-MOP/4/2 Add 1, 6 December 2007) (Note by the CC on Repeated Non-Compliance), para. 79. Article 17 of the Rotterdam Convention and Article 17 of the Stockholm Convention mandate the COP to approve the CM 'as soon as practicable'. However, none of these CMs has as yet been approved, although they have been negotiated at several COPs; see ibid., paras 64–66 and 82.

[8] See Decision EM-1/3, Adoption of the Cartagena Protocol and Interim Arrangements (UNEP/CBD/ExCOP/1/3, 2000), Annex, Section II, para. (a). Decision EM-1/3 and decisions of the first COP MOP (2004) are also included in the *Handbook of the Convention on Biological Diversity, 3rd edn* (CBD Secretariat, 2005) (*Handbook*). Decisions of the second COP MOP (2005) and of the third COP MOP (2006) are available in the booklets, respectively *Facing the Biosafety Challenge* (CBD Secretariat, 2005) and *Biosafety: Building further Consensus for Action* (CBD Secretariat, 2007).

[9] See *Plenary Report of ICCP 1* (UNEP/CBD/ICCP/1/9, 19 December 2000) (Report of ICCP 1), paras 44–55 and 104–111, as well as ICCP-1, Decision 4.5, Compliance, found in the Annex to the *Report of ICCP 1*.

[10] *Report of ICCP 1*, ibid., para. 49.

elements and options for a compliance regime based on a note by the Executive Secretary.[11]

The outcome of the expert group meeting was a draft recommendation, which was adopted by ICCP-2 as recommendation 2/11, however with the understanding that all elements of the draft would be subject to further discussion. The operative part of the recommendation dealt with the future process, while, more importantly, the outcome of substantive discussions was presented as 'Draft procedures and mechanism on compliance under Cartagena Protocol' (the Draft CM).[12]

At the time of ICCP-2, no further meetings of the ICCP had been scheduled because the outcomes of ICCP-1 and ICCP-2 as outlined in Recommendation 2/11 were to be submitted to COP MOP-1. However, since the Cartagena Protocol had not entered into force in time to hold a COP MOP in connection with COP-6 (2002) a third meeting of the ICCP (ICCP-3) was organized to be held immediately after that COP.

Only little progress was made at ICCP-3, and, not surprisingly, the most contentious issues remained unsolved. Among those were the inclusion of Principle 7 of the Rio Declaration in the CM (Draft CM, Section I, para. 3 on objective, nature and underlying principles); whether members of the CC should serve in their individual capacity (Section II, para. 3); the Party-to-Party trigger (Section V, para. 1 (a)); and some of the non-compliance response measures (issuing cautions, publication of cases of non-compliance and suspension of rights and privileges; Section VI, paras 2(b), (c) and (d)).[13]

First COP-MOP to the Cartagena Protocol (2004)

At COP-MOP 1 (Kuala Lumpur, 23–27 February 2004) many outstanding issues were resolved by trade-offs and/or compromise language.[14] The most contentious issues (see the preceding paragraph), however, were addressed by means of a package

[11] *Note of the Executive Secretary. Synthesis of Views Regarding Elements and Options for a Compliance Regime* (UNEP/CBD/ICCP/2/13, 17 July 2001) (Note ICCP 2/13).

[12] See *Report of the Open-Ended Meeting of Experts on a Compliance Regime* (UNEP/CBD/ICCP/2/13 Add 1, 28 September 2001), and *Report of ICCP 2* (UNEP/CBD/ICCP/2/15, 15 October 2001), para. 177, with Annex I containing Recommendation 2/11, which includes the Draft CM. The Draft CM includes a provision (para. 21) to the effect that the CC should develop its own rules of procedure to be submitted to the COP MOP for its consideration and approval (ibid., para. 21). This provision was never questioned. Accordingly, it is included in Decision BS-I/7 on *Establishment of Procedures and Mechanisms under the Cartagena Protocol on Biosafety with Annex on Procedures and Mechanisms* (UNEP/CBD/BS/COP-MOP/1/15, 27 February 2004), Section I, para. 7.

[13] *Report of ICCP 3* (UNEP/CBD/ICCP/3/10, 26 April 2002). Other unresolved issues included the question of a balance between importing and exporting countries (Section II, para. 2), triggering of the CM by the COP MOP (Section IV, para. 1), and consideration by the CC of information from non-governmental organizations (NGOs) and the Secretariat (Section V, para. 2).

[14] *Report of COP MOP-1* (UNEP/CBD/BS/COP-MOP/1/15, 27 February 2004), paras 116–126. The text is characterized as 'heavily bracketed' in S. Jungcurt et al., 'Summary of the First Meeting of the Conference of the Parties to the Convention on Biological Diversity serving as the Meeting of the

which included reference to Principle 7 of the Rio Declaration being replaced by language on paying particular attention to the needs of developing country parties, etc. (CM Section I, para. 3); provisions stating that members of the CC would serve in their personal capacity (Section II, para. 3); the party-to-party trigger restricted to a party which is affected or likely to be affected, with respect to another party (Section IV, para. 1(b)); the maintenance of issuing cautions and publication of cases of non-compliance, however, restricted to publishing such cases in the Biosafety Clearing-House (Section VI, paras 2 (b) and (c)); the replacement of the suspension of rights and privileges by a new paragraph leaving to COP MOP-3 the decision on measures to be taken in cases of repeated non-compliance (Section VI, para. 3).

COP MOP-1 adopted the outcome of the negotiations as Decision BS-1/7.[15]

Decision BS-I/7 on Establishment of Procedures and Mechanisms on Compliance

The annex to Decision BS-I/7 on the CM of Cartagena Protocol is divided into seven sections: objective, nature and underlying principles (I); institutional mechanisms (II); functions of the Committee (III); procedures (IV); information and consultation (V); measures to promote compliance and address cases of non-compliance (VI); and review of the procedures and mechanisms (VII).

Similarities and Differences of the Compliance Mechanism Compared to other Compliance Mechanisms

Many of the features of the CM are quite traditional, while others differ considerably from CM features of other MEAs.[16] The survey below of the CM focuses on some – but not all – similarities and differences:

Parties to the Cartagena Protocol on Biosafety: 23–27 February 2004', 9:289 *Earth Negotiations Bulletin* (1 March 2004), at 11, available at <http://www.iisd.ca/vol09/enb09289e.html>.

15 See *Report of COP MOP-1*, ibid., para. 128; and *Handbook*, n. 8 above, at 1404. See also Decision BS-I/7, n. 12 above.

16 Probably many differences should be seen from the perspective of the respective negotiation histories rather than, as commonly maintained, in the light of the nature of the respective MEAs. An analysis of this question might or might not substantiate this observation. CMs have for some years been among the items on the top of academic research agendas of international environmental law. The outcome of the most recent project, the Joint Research Project of University of Milan, University of Bologna and University of Parma under the direction of professors Tullio Treves, Attila Tanzi and Laura Pineschi, has been published as T. Treves et al. (eds), *Non-Compliance Procedures and Mechanisms, and the Effectiveness of International Environmental Agreements* (T.M.C. Asser Press, 2009). The publication includes analyses of existing and emerging CMs, including, inter alia, all CMs referred to in the present article, as well as analyses of a number of cross-cutting issues. The CM of the CPB is examined at 101 in C. Ragni 'Procedures and Mechanisms on Compliance under the 2000 Cartagena Protocol on Biosafety to the 1992 Convention on Biological Diversity'. The publication also includes at 277 V. Koester, 'The Compliance Mechanisms of the Aarhus Convention and the Cartagena Protocol on Biosafety: A Comparative Analysis of the Negotiation Histories and their Outcomes'.

Objective, Nature and Underlying Principles (Section I)

The common provisions in CMs on the objective and nature of the CM (that its objective is to promote compliance with the provisions of the Protocol, to address cases on non-compliance and provide advice or assistance, and that procedures shall be, *inter alia*, facilitative, non-adversarial and cooperative, and be guided by principles of transparency and predictability) are set out in Section I.[17]

Institutional Mechanisms (Section II)

Also the composition, size and election of the CC (15 members to be elected by the COP MOP on the basis of three members from each of the five regional groups of the UN) and most of the other provisions related to the members of the CC and the CC as such are quite common.[18] However, the requirement that members serve in a personal capacity[19] is not a commonly accepted requirement. The requirement was originally supported by a line in the budget of Cartagena Protocol to cover the costs of two annual meetings of the CC.[20] On rules of procedure (para. 7), see the section below entitled 'Draft rules of procedure presented by the CC'.[21]

[17] See Decision BS-I/7, n. 12 above, Section I, paras 1, 2 and 3.

[18] Ibid., Section II, paras 1, 2, 4, 5 and 6.

[19] Ibid., Section II, para. 3.

[20] For members to serve in their personal capacities is not a requirement under the CM of the Montreal Protocol (see MOP Decision IV/5 (UNEP/OzL.Pro.4/15, 25 November 1992), Annex IV, at 44, subsequently amended by Decision X/10 (UNEP/OzL.Pro.10/9, 3 December 1998), para. 5 (representative of Parties)), and the Convention on the Control of Transboundary Movements of Hazardous Wastes and their Disposal (Basel, 22 March 1989) (the Basel Convention) (see Basel Decision COP VI/12 (UNEP/CHW.6/40, 2002), Annex, para. 5 (members serve (only) objectively')). Members representing parties or not clearly serving in their individual capacity is also a feature of CMs of some regional conventions, e.g. the Implementation Committee of the Espoo Convention on Environmental Impact Assessment in a Transboundary Context (Espoo, 25 February 1991) (see MOP Decision II/4 (ECE/MP.EIA/4, 2001),Annex IV, at 72, revised by MOP Decision III/2 (ECE/MP.EIA/6, 2004), Annex II, at 49 (consolidated text)), and the Review Committee of the Alpine Convention (Salzburg, 7 November 1991) (Alpine Conference Decision II/4, reprinted in 33 Environmental Policy and Law (2003), at 179). Regarding the CC of the Aarhus Convention on Access to Information, Public Participation in Decision-Making and Access to Justice in Environmental Matters (Aarhus, 28 June 1998) (see MOP Decision I/7 (ECE/MP .PP/2/Add.8, 2 April 2004) as amended by MOP Decision II/5 (ECE/MP.PP/2005/2/Add.6, 13 June 2005), para. 12)), members are not only serving in their personal capacity, but furthermore cannot belong to any executive branch of government; see V Koester, 'The Convention on Access to Information' Public Participation in Decision-Making and Access to Justice in Environmental Matters (Aarhus Convention)', in G. Ulfstein et al. (eds), *Making Treaties Work: Human Rights, Environmental and Arms Control* (Cambridge University Press, 2007), 179, at 193. According to Section II, para. 5 of the CPB CM (see Decision BS-I/7, n. 12 above), the CC shall meet twice a year, unless it decides otherwise. The CC has held five meetings since its establishment in 2004. The fact that participation costs of members are funded by the budget of Cartagena Protocol and not by the respective governments secures at least the economic independency of members.

[21] On this provision, see also *supra* note 13.

Functions of the Committee (Section III)

The functions of the CC correspond to functions of other CCs, namely to consider individual cases of non-compliance, to review general issues of compliance (taking into account information provided in national reports), to make recommendations to the COP MOP, and to carry out any other functions assigned to it by the COP MOP.[22]

Procedures (Section IV)

The procedural safeguards contained in Section IV are quite usual, including the entitlement to reject consideration of submissions that are *de minimis* or ill-founded, and the handling of party-to-party submissions, i.e. making such submissions available to the parties concerned, providing an opportunity for those parties to comment (not later than 6 months after having received the submission), and entitlement of the parties concerned to participate in the deliberations of the CC; however, not in the elaboration of a recommendation.[23]

The party-to-party trigger is restricted to situations where a party is affected or likely to be affected.[24] This requirement is also included in some other CMs, but not all of them.[25] Probably, however, this limitation is not very important taking into account the general reluctance of parties to all MEAs to use the party-to-party trigger when this opportunity is available under the respective CMs.[26] Also, the Secretariat is not entitled to make submissions which, although such a right may present a political and tactical challenge to the Secretariat, might be considered as a weakness of the CM.[27] The CC, in undertaking all of its functions and activities, shall maintain the confidentiality of any information that is confidential under Article 21 of the Cartagena Protocol.[28]

[22] See Decision BS-I/7, *supra* note 13, Section III, para. 1(a)–(f).

[23] Similar to the CM of, e.g., the Montreal Protocol, n. 20 above, the CC of the Cartagena Protocol is only involved in a party-to-party submission after the period envisaged for comments by the party concerned has expired. The rationale of this rule, however, is quite obscure. The CC has interpreted Section VI, paras 1, 2 and 3 of Decision BS-I/7, n. 12 above, read together to mean that a submission of a party with respect to itself may be transmitted to members of the CC immediately. However, if the submission is not in English (i.e. in another official language of the UN) the Secretariat should transmit an English translation of the submission to the members of the CC within 90 days of receipt of the submission. See Report of the CC on the Work of its First Meeting (UNEP/CBD/BS/COP-MOP/2/2, 16 March 2006) (Report of First Meeting of Cartagena Protocol-CC), para. 11(d).

[24] See Decision BS-I/7, *supra* note 13, Section IV para. 1 (b).

[25] The party-to-party trigger is limited, e.g., in the Basel Convention CM, n. 20 above, para. 9 (b), but not limited, e.g., in the Montreal Protocol CM, n. 20 above, para. 1.

[26] V. Koester, 'Review of Compliance under The Aarhus Convention: A Rather Unique Compliance Mechanism', 2:1 Journal for European Environmental and Planning Law (2005), 31, at 41, n. 78.

[27] Even the Basel Convention CM, which is in many ways rather weak, offers its secretariat some possibilities, albeit limited, to make submissions; see Basel Decision COP VI/12, n. 20 above, para. 9(c)

[28] See Decision BS-I/7, *supra* note 13, Section V, para. 4. Rule 22 of the Rules of Procedure for Meetings of the CC (see Report of COP MOP-2 (UNEP/CBD/BS/COP-MOP/2/15, 6 June 2005), Annex I, at 29, Decision BS-11/1, Rules of Procedure for Meetings of the Compliance Committee (RoP)), provides

Information and Consultation (Section V)

Apart from sources of information that are specific to the Cartagena Protocol (i.e. the Biosafety Clearing House and the Biosafety Roster of Experts), the section on information and consultation does not differ much from comparable provisions of CMs of other MEAs. However, the non-explicit inclusion of the Secretariat in the list of sources of information is rather unique.

Measures to Promote Compliance and Address Cases of Non-Compliance (Section VI)

Section VI (para. 1) provides that the CC may take one or more of the following measures with a view to promoting compliance and addressing cases of non-compliance: (a) Provide advice or assistance to the Party concerned, as appropriate; (b) Make recommendations to the Conference of the Parties to the Convention serving as the meeting of the Parties to the Protocol regarding the provision of financial and technical assistance, technology transfer, training and other capacity-building measures; (c) Request or assist, as appropriate, the Party concerned to develop a compliance action plan regarding the achievement of compliance with the Protocol within a timeframe to be agreed upon between the Committee and the Party concerned; (d) Invite the Party concerned to submit progress reports to the Committee on the efforts it is making to comply with its obligations under the Protocol; and (e) Pursuant to paragraph 1 (c) and (d) above, report to the Conference of the Parties serving as the meeting of the Parties on efforts made by Parties in non-compliance to return to compliance and maintain this as an agenda item of the Committee until adequately resolved.

The power of the CC may be, at first glance, considered as being rather limited, but in fact it is not in comparison with the power of similar committees of other MEAs. Most notably the CC has the power to *request* the concerned party to develop a compliance action plan, although the timeframe for such a plan has to be agreed upon with the party concerned (para. (c)). The language of paragraph 2, furthermore, suggests that the CC may make recommendations to the COP MOP on measures being only available to the COP MOP.[29]

that in the event of a conflict between the RoP and any provision in the Cartagena Protocol or Decision BS-I/7, the provisions in the Cartagena Protocol or Decision BS-I/7 shall prevail. Due to the confidentiality requirements of Article 21 of the CPB, the CC at its first meeting decided that Rule 15 of the RoP on electronic means of communication would not be applicable for transmitting any confidential information. See Report of First Meeting of CPB-CC, n. 23 above.

[29] The CM of the Montreal Protocol only empowers the CC (Implementation Committee) to make recommendations to the governing body. So does the CM of the Espoo Convention. The CCs of the Alpine Convention and of the Basel Convention empower their committees to provide advice and support (on the CMs of these instruments, see n. 20 above). Also the power of the CC of the Aarhus Convention, n. 20 above, is, apart from its power to make recommendations on specific measures, which is closely related to the entitlement of members of the public to trigger the CM, rather limited. See V. Koester, n. 21 above, at 203 and 209.

The section on non-compliance response measures also provides for specific measures being available to the COP MOP, upon the recommendation of the CC, and taking into account, *inter alia*, such factors as the cause, type, degree and frequency of non-compliance. The following measures are available (para. 2): (a) Provide financial and technical assistance, technology transfer, training and other capacity-building measures; (b) Issue a caution to the concerned Party; (c) Request the Executive Secretary to publish cases of non-compliance in the Biosafety Clearing-House; (d) In cases of repeated non-compliance, take such measures as may be decided by the Conference of the Parties serving as the meeting of the Parties to the Protocol at its third meeting, and thereafter in accordance with Article 35 of the Protocol, within the framework of the review process provided for in Section VII below.

Measures in paragraphs (a)–(c) correspond to measures being available in comparable CMs, while paragraph (d) was elaborated in order to achieve a compromise (see the section above entitled 'First Conference of the Parties to the CBD Serving as the Meeting of the Parties to the Cartagena Protocol (2004)').[30] Consideration of the issue of non-compliance response measures in cases of repeated non-compliance was originally postponed to COP MOP-4, which was also due to review the effectiveness of the CM as provided for in Section VII of the CM (see sections below entitled 'Tasks Assigned to the CC by COP MOP-3' and 'Considerations and Conclusions of COP MOP-4').

First Report of the Compliance Committee and COP-MOP 2

Draft Rules of Procedure Presented by the Compliance Committee

The members of the CC were elected at COP MOP-1,[31] and the CC held its first meeting (CC-1) in March 2005. According to Decision BS-I/7, the CC shall develop and submit rules of procedure to the COP MOP for its consideration and approval.[32] Hence, the main item to be considered at the first meeting was rules of procedure

[30] Some of the measures correspond to measures included in the indicative list of measures annexed to the Montreal Protocol CM, Decision VI/5, and in the Basel Convention CM, para. 20 (see n. 20 above), as well as in the Aarhus Convention CM; see V. Koester, n. 21 above, at 210 on issuing cautions (and declarations of non-compliance). Reference is also made to the list of measures going beyond the measures listed in para. 2 of Section VII of the CM of CPB in *Note by the CC on Repeated Non-Compliance*, n. 7 above, Section III.

[31] See *Report of COP MOP-1*, n. 15 above, paras 128–129.

[32] See *Decision BS-I/7*, n. 12 above, Section II, para. 7, and *Report of First Meeting of Cartagena Protocol-CC*, n. 23 above, paras 8–13. Section II, para. 6 of Decision BS-I/7 provides that the CC shall submit its reports including recommendations to the COP MOP for consideration and appropriate action. Accordingly, meeting reports of the CC are serving both as a means of recording meetings for internal purposes of the CC and for reporting to the COP MOP. However, for COP MOP-4 the report of the CC was presented as a consolidated report encapsulating two meeting reports of the CC. See the section entitled 'Observations and recommendations of the CC' below.

for the meetings of the CC (RoP).[33] The CC developed such rules and adopted them by consensus (Draft RoP). Two rules which were not questioned by the COP MOP might be mentioned, while others that were specifically considered and/or amended by the COP MOP are discussed below (see the next section).

Rule 2 provides that the rules of procedure for meetings of the CBD COP, as applied, *mutatis mutandis*, to the COP MOP, shall apply, *mutatis mutandis*, to any meeting of the CC, except as otherwise provided in the RoP and in Decision BS-I/7.[34] The RoP contains also a provision to the effect that electronic means of communication may be used by members of the CC for the purpose of conducting informal consultations on issues under consideration, however, not for making decisions on matters of substance (Rule 15). This rule, which was not questioned by the COP MOP, was inspired by similar rules (but which are more far-reaching) under the Aarhus Convention CC.[35]

Apart from the Draft RoP, the report of the CC also contains some interpretative statements relating to Decision BS-I/7[36] or to the Draft RoP.[37] The report includes, furthermore, a request to the Executive Secretary to bring to the attention of the COP MOP a submission made involving a developing country party or a party with economy in transition, in which case funding for participation in the meeting of the CC should be provided by the Secretariat. However, this budgetary problem still remains unsolved.[38]

Rules of Procedure Approved by COP-MOP 2

At COP MOP-2 some concerns were raised relating to the Draft RoP and a group of Friends of the President was set up to address these concerns. The group considered a number of rules resulting, however, in only one substantive amendment and another rule being bracketed in its entirety.[39]

33 Another item was the development and adoption of a work plan for the CC, which, however, will not be discussed in the present article. Considerations of both items were based on notes by the Executive Secretary; see *Development of Rules of Procedure for Meetings of the Compliance Committee under the Cartagena Protocol* (UNEP/CBD/BS/CC/1/2, 14 January 2005) and Work Plan for the Compliance Committee under the Cartagena Protocol (UNEP/CBD/BS/CC1/3, Draft, 14 January 2005), respectively. The note dealing with the RoP included draft rules based, *inter alia*, on rules and practices of other compliance bodies.

34 See RoP, n. 28 above. The CC of the Montreal Protocol, n. 20 above, does not have its own rules of procedure implying that the rules of procedure of the governing body apply *mutatis mutandis*, while the approach of the Aarhus Convention CC is more in line with the above Rule 2. See V. Koester, n. 20 above, at 197 on *modus operandi* of the Aarhus Convention CC.

35 See V. Koester, ibid.

36 *Report of First Meeting of CPB-CC*, n. 23 above, para. 11 (a). Rule 15 is applicable in the framework of Rule 8, para. 3 concerning determination of relevance of information received in accordance with Decision BS-I/7, n. 12 above, Section V, para. 2.

37 See *Report of First Meeting of CPB-CC*, ibid., para. 11(a)–(e).

38 See ibid., para. 11(b) and the section entitled General issues of compliance below.

39 See *Report of COP MOP-2* (UNEP/CBD/BS/COP-MOP/2/15, 6 June 2005), paras 51–61. The other rules considered were Rules 6, 9, 11 and 13, while Rule 22 was only slightly amended.

Draft Rule 14, paragraph 1 provided that the CC should meet in closed session unless it decided otherwise, respecting, by virtue of Rule 13, the requirements of Decision BS-I/7 that parties concerned by submissions are entitled to participate in the deliberations of the Committee (see the section entitled 'Procedures (Section IV)' above).[40] Rule 14, para. 1 as amended by the COP MOP now reads:

> The Committee shall decide on whether it will meet in open or closed sessions. Such decisions, including reasoning, shall be reflected in the reports of the Committee.

Rule 14 as amended by the COP MOP is not very constructive from an operational perspective. The Rule was therefore addressed by the CC in its report to COP MOP-3 (see the section entitled 'Considerations and conclusions of COP MOP-3' below) and again in its report to COP MOP-4 (see the section entitled 'Procedural decisions' below). In addition, it is difficult to apply literally since almost all reasoning in favour of a closed session would simultaneously be reasoning against an open session and *vice versa*.

Rule 18, providing, *inter alia*, for decisions on matters of substance to be taken by a two-thirds majority of the members present and voting or by eight members,[41] whichever is the greater, if all efforts to reach consensus have been exhausted, was bracketed by the COP MOP in its entirety. The rationale of Rule 18 was (and still is) that since the rules of procedure of the COP of the CBD, and consequently of the COP MOP, apply *mutatis mutandis*, decision making on matters of substance would be subject to decisions taken by consensus by the CC,[42] unless a specific rule was included in the RoP. Furthermore, subsidiary bodies under the CBD (and accordingly under the Cartagena Protocol) are not required to take decisions by consensus, but may do this by a simple majority. The reasons for bracketing Rule 18 are not clear, since the powers of the CC are not significant (see the section entitled

[40] Based on the recollection of the author of the present article of the discussion of the CC on Rule 14, part of the rationale for not having a rule to the opposite effect (i.e. open sessions unless otherwise decided) was based on practical considerations (i.e. the size of the meeting room in the CBD Secretariat), as well as political considerations (i.e. only observers, parties and other governments as well as NGOs from the industrialized part of the world would be in a position to use the opportunity to be present at meetings of the CC). The last-mentioned concerns were also addressed in *Report of the Second Meeting of the CC* (UNEP/CBD/BS/COP-MOP/3/2, 8 February 2006) (*Second Meeting of Cartagena Protocol-CC*), para. 18.

[41] Due to the fact that 10 out of the 15 members (according to Rule 16) constitute a quorum, Rule 18 also provides that where consensus is not reached, the report shall reflect the views of all members.

[42] Rule 40, para. 1 of the rules of procedure for meetings of the COP (CBD Decision I/1, Annex, found in *Handbook*, n. 8 above, at 66) providing for decisions on matters of substance to be taken by a two-thirds majority, if all efforts to reach consensus have been exhausted, has from the outset been and still is bracketed, implying that such decisions must be taken by consensus. For instance, the Basel Convention CM, n. 20 above, Annex, para. 25 provides for substantive decision making by a two-thirds majority, and the Aarhus Convention CM entails a similar rule (three-fourths majority) by its application of the rules of procedure of the governing body *mutatis mutandis* (see V. Koester, n. 20 above, at 200).

'Measures to Promote Compliance and Address Cases of Non-Compliance (Section VI)' above). It is, of course, obvious that the CC is entitled to determine non-compliance, but such determination may be over-ruled by the COP MOP. Maybe some parties feared that acceptance of Rule 18 would be used as a precedent in future discussions on Rule 40 of the CBD RoP.

It might be argued, however, that taking decisions on matters of substance by qualified majority is not appropriate for determination of non-compliance, and may lead to rather absurd results. If, for example, only ten members of the CC are present at a meeting (representing a quorum),[43] and six of them are holding the view in a case triggered by a party-to-party submission that the party concerned is non-compliant, and the four other members represent the opposite view, the outcome would be a determination of no non-compliance. The report would, at the same time, state that six members were of the opposite opinion. Determination of non-compliance is, at least, of a quasi-judicial nature and should consequently be based on a decision taken by a simple majority.

In addition to the above amendments, the following statement was included in the report of COP MOP-2:[44]

> The Conference of the Parties serving as the meeting of the Parties to the Protocol invited the Compliance Committee to give further consideration to the issue of conflict of interest set out in rule 11 of the rules of procedure for meetings of the Compliance Committee.

Rule 11 is a rather simple, but at the same time also a very reasonable, rule providing that each member 'shall, with respect to any matter that is under consideration by the Committee, avoid direct or indirect conflicts of interest'. It further reads:

> Any interest that may constitute a conflict of interest shall be disclosed as soon as possible to the Secretariat which shall forthwith notify the members of the Committee. The concerned member shall not participate in the elaboration and adoption of a recommendation of the Committee in relation to such matter.

The nucleus of Rule 11 is an obligation of any member to disclose what he or she considers a direct or indirect conflict of interest. However, the philosophy of the COP MOP statement seems to be that it is possible to develop exhaustive and objective criteria for or definitions of the factors that would constitute a direct or indirect conflict of interest. Furthermore, the statement presupposes that it would be worthwhile undertaking such work. Both assumptions are, to put it mildly, extremely doubtful. Accordingly, the CC in its report to COP MOP-4 stated that the CC for the time being does not see any need for further consideration of Rule 11 (see the sections entitled 'Procedural decisions' and 'Other observations of the CC' below).

[43] See RoP, n. 28 above, Rule 16.

[44] See *Report of COP MOP-2*, n. 39 above, para. 59.

Second Report of the Compliance Committee and COP-MOP 3

The second meeting of the CC (CC-2) took place in February 2006.[45] Before the meeting, it was agreed by an electronic consultation process to hold the meeting in a closed session and information to that effect was made available on the Secretariat's website with other documents for the meeting.[46] One of the two main items of the meeting was review of general issues of compliance being one of the functions of the CC as specified in Decision BS-I/7,[47] and the other item being the RoP.

General Issues of Compliance

The basis for the CCs deliberations of general issues of compliance was interim national reports and information in the Biosafety Clearing House (BCH). In its analysis of the interim national reports, the CC emphasized four general issues relating to compliance that needed due attention in order to ensure a good start in the effective implementation of the requirements of Cartagena Protocol and the achievement of its objective: delay in putting in place the necessary regulatory and administrative measures to implement the Protocol, as required under Article 2 of the Cartagena Protocol; weakness in carrying out scientifically sound risk assessment (Article 15), and in designing and implementing appropriate risk management and monitoring measures (Article 16); limited or no steps taken by parties towards promoting and facilitating public awareness and participation concerning the safe transfer, handling and use of living modified organisms (LMOs), as well as facilitating public access to the BCH (Article 23); about two-thirds of parties to the Protocol did not fulfil their reporting obligation as required under Article 33 of the Protocol and Decision BS-I/9 of the first COP MOP.

The CC also identified some general issues that affected compliance with the requirements of making various types of information available to the BCH. These issues included: that several parties were still behind in fulfilling the requirements to make information available to the BCH, including time-bound requirements such as submitting information on national focal points and national competent authorities

45 See *Second Meeting of CPB-CC*, n. 40 above. Regarding CC-3, a provisional decision on the nature of the meeting was made at the preceding meeting and finalized by consultations conducted electronically in light of the provisional agenda of the upcoming meeting. The decision was then made available on the website, together with the meeting documents. At the following meetings of the CC it was decided to conduct the meetings in an open session; see the sections entitled 'Other observations of the CC' and 'Other outcomes of the considerations of COP MOP-4 of the Report of the CC' below.

46 *Note to Members of the CC* (UNEP/CBD/BS/CC/2/INF1, 6 January 2006). Procedures of the CC are transparent in the sense that documents for meetings as well as reports of meetings are made available on the website in the same manner as other official documents relating to the CBD and the CPB.

47 See Decision BS-I/7, n. 12 above, Section III, para. (d). General issues may, according to *Note by the Executive Secretary: Analysis of Information Contained in the Interim Reports* (UNEP/CBD/BS/CC/2/2, 29 November 2006), para. 5, be understood to mean issues relating to the problems of several parties to implement fully and effectively the Protocol that may require a concerned action or cooperation by all parties in order to overcome them and ensure compliance. However, the CC refrained from defining the notion of general issues of compliance.

no later than the date of entry into force of the Protocol for that party; and the lack of information on some LMOs that might have been approved for domestic use, including placing on the market, and that were subject to transboundary movement.

The concluding observations of the CC on general issues of compliance included, *inter alia*, the following concerns of the CC: a considerable number of parties were lacking adequate legal, administrative and technical frameworks to enable them to comply with the provisions of the Protocol; the relatively low number of transboundary movements of LMOs intended for direct use as food or feed, or for processing (LMOs-FFP) that had been reported to the BCH and the fact that not a single decision regarding the transboundary movement of LMOs intended for introduction into the environment had been reported; and a number of parties failed to submit interim national reports and several of those submitted did not contain sufficient information.[48]

Due to an ambiguity in the decision by COP MOP-1 on reporting, the CC also recommended to COP MOP-3 to clarify that all parties are under the obligation to monitor their implementation of the Cartagena Protocol and to report that through national reports in accordance with Article 33, regardless of when they became parties in a given reporting period.[49]

Based on its discussions, the CC submitted several recommendations to COP MOP-3.[50] The recommendations addressed a number of different subject matters, which influenced the way in which they were considered by the COP MOP (see the section entitled 'Considerations and Conclusions of COP MOP-3' below).

Rules of Procedure

The CC also considered the RoP as amended and approved by COP MOP-2, in particular Rules 14 and 18 (see the section entitled 'Rules of procedure approved by COP MOP-2' above). In its report to COP MOP-3, the CC outlined the practical difficulties involved in the application of Rule 14 and how the CC intended to overcome those difficulties, as well as its decision on who should be allowed to be present at open sessions. A specific recommendation was included in the list of recommendations regarding financial support to eligible, concerned parties to be represented in the deliberations of the CC.[51]

48 See *Second Meeting of CPB-CC*, n. 40 above, paras 31–38.

49 Ibid., para. 40(l). The problem arose from the deadline prescribed for submitting national reports, which could lead to the conclusion that States becoming parties after the expiry of the deadline were not under an obligation to submit national reports. In connection to reporting, it is interesting to note that parties did not react to the fact that the CC in its report also addressed the 'obligation of parties' under the relevant decision to submit their reports within a certain deadline (see ibid., para. 32), although the CC, according to Decision BS-I/7, n. 12 above, Section III, para. (d), should review (only) compliance with 'obligations under the Protocol'.

50 Ibid., Annex (Draft Decision).

51 Ibid., paras 13–19. As a general rule, parties that express their desire to the Secretariat to be present at open sessions of the CC may attend meetings leaving open the possibility of observers also being

The CC recommended to the COP MOP to remove the square brackets around Rule 18 (see the section entitled 'Rules of procedure approved by COP MOP-2'),[52] arguing that to its knowledge, all CMs under other MEAs provide for a majority or qualified majority voting procedure. The COP MOP did not accept this recommendation. However, according to COP MOP-3 Decision BS-III/1,[53] Rule 18 was due to be addressed at COP MOP-4 as part of its review of the effectiveness of the CM (see the section entitled 'Measures to promote compliance and address cases of non-compliance' above and the section entitled 'Other outcomes of the considerations of COP MOP-4 of the report of the CC' below).

Considerations and Conclusions of COP-MOP 3

The recommendations of the CC were considered by COP MOP-3 under various agenda items. Consequently, the recommendations, which were accepted, were included, directly or indirectly, in decisions relating to the same subject matters that were addressed by individual recommendations of the CC. Almost all recommendations were accepted, either directly or indirectly.[54] A few recommendations were not explicitly accepted or rejected and/or not considered, including, not surprisingly, two recommendations dealing with budgetary matters; one of them dealing with the lack of funds to enable eligible, concerned parties to participate in meetings of the CC (see the section entitled 'Draft rules of procedure presented by the CC' above).[55] With this great number of recommendations adopted, overall the outcome from COP MOP-3 may be considered a success seen from the perspective of the CC.[56]

represented. Those who wish to attend are required to register in advance. The nature of a meeting of the CC is announced on the website.

52 Ibid., paras 22–25.

53 Decision BS-III/1 on Compliance (Decision BS-III/1), para. 1. In *Report of COP MOP-3* (UNEP/CBD/COP-MOP/3/15, 8 May 2006), Annex, The Annex to the *Report of COP MOP-3* includes all COP MOP-3 decisions, including, *inter alia*, those referred to in n. 54 below.

54 A list of recommendations is presented in the Annex to *Second Meeting of CPB-CC*, n. 40 above. Recommendations 2, 3, 5 are reflected in Decision BS-III/1, paras 1, 3 and 4; and Recommendation 11 is reflected in Decision BS-III/2. The operation and activities of the Biosafety Clearing House, para. 1, which in para. 7 also includes the recommendation in *Second Meeting of CPB-CC*, para. 41; Recommendation 6 is in Decision BS-III/3 on capacity building, para. 6; Recommendation 7 is in Decision BS-III/4 on roster of experts, para. 7; Recommendation 8 is Decision BS-III/5 on matters relating to the financial mechanism, para. 6; and Recommendations 12, 13 and the recommendation on reporting in para. 40 of *Second Meeting of CPB-CC* is in Decision BS-III/14 on monitoring and reporting, paras 2 and 4. Recommendations 9 and 10 were rejected. Decisions of COP MOP-3 are included in *Report of COP MOP-3*, ibid., and in the booklet *Biosafety: Building further consensus for action* (CBD Secretariat, 2007).

55 See ibid. on recommendations 9 and 10.

56 'News, events and resources at the intersection of trade and biodiversity', 6:6 *Bridges Trade BioRes* (3 April 2006), available at <http://www.ictsd.org/biores/index.com>, which contains the following review of the work of the CC: 'The work of the Compliance Committee embodied the most important, comprehensive and detailed analytical recommendations provided to the MOP . . . [T]he Committee

Report of the Compliance Committee to COP-MOP 4 and Outcome

Tasks Assigned to the Compliance Committee by COP-MOP 3

By Decision BS-III/1,[57] parties at COP MOP-3 decided to postpone the review of the effectiveness of the CM, according to Section VII of Decision BS-I/7 (see the section entitled 'Measures to promote compliance and address cases of non-compliance' above) scheduled for COP MOP-3, to COP MOP-4,[58] in order to include this review within the framework of the overall evaluation of the effectiveness of the Cartagena Protocol in accordance with Article 35 scheduled for COP MOP-4.[59] Decision BS-III/1 included a request from the COP MOP to the CC to compile further information on experiences of other MEAs regarding cases of repeated non-compliance for consideration at COP MOP-4. Consequently, this request should be seen in the context of both the review of the effectiveness of CM and the overall evaluation of Cartagena Protocol. Furthermore, COP MOP-3 also requested the CC to prepare a report on general issues of compliance by parties with their obligations under the Cartagena Protocol, in accordance with Decision BS-I/7, Section III.

Observations and Recommendations of the Compliance Committee

To carry out the tasks assigned to it by COP MOP-3, the CC held two meetings in 2007.[60] Following prior consultations, the third meeting was conducted in an open session. One observer, who was also allowed to make some interventions, attended the meeting. As decided at the third meeting also the fourth meeting was conducted in an open session (no observers were present).

raised a wide range of concerns, including matters of procedure . . . and Protocol administration . . . The Committee's most important concerns, however, focused on the relative sluggishness of national development needed for full and formal implementation of the Protocol, with particular attention to national legal frameworks, administrative capacity, national reporting and the Biosafety Clearing House. These points were identified by both developed and developing country Parties to be the most critical issues facing the Protocol, and to bear an integral relationship to the problem of capacity development needs . . .'. On *Second Meeting of Cartagena Protocol -CC*, n. 45 above, see also L. Boisson de Chazournes and U.P. Thomas, 'WTO Law, Science and Risk Communication', *Special Edition 2006 of Journal of Trade and Environment* (EcoLomics International in cooperation with Geneva University Law Faculty, 2007), at 71, available at <http://www.ecolomics-international.org/>.

57 See Decision BS-III/1, n. 54 above.

58 COP MOP-4 took place in Bonn, 12–16 May 2008.

59 Modalities for the review are outlined in Decision BS-III/15, Assessment and Review, found in *Report of COP MOP-3*, n. 53 above.

60 The third meeting of the CC was held in March 2007 (Kuala Lumpur) and the forth meeting in November 2007 (Montreal). Meeting reports are available on the website as: *Report of the Third Meeting of Cartagena Protocol -CC* (UNEP/CBD/CC/3/3, 7 March 2007) (*Third Meeting of Cartagena Protocol -CC*) (on open sessions, see paras 7 and 41 and the section entitled 'Rules of procedure approved by COP MOP-2' above) and *Report of the Fourth Meeting of CPB-CC* (UNEP/CBD/BS/CC/4/3, 23 November 2007) (*Fourth Meeting of CPB-CC*). *Report to COP MOP-4*(UNEP/CBD/BS/COP-MOP/4/2, 27 November 2007) presents the outcomes of the two meetings.

Measures in Cases of Repeated Non-Compliance

At its third and fourth meetings, the CC reviewed and compiled information on experience regarding cases of repeated non-compliance under other MEAs. The CC also included observations as well as an indicative list of non-compliance response measures adopted under existing CMs or suggested under emerging CMs of MEAs reviewed by it with the purpose of facilitating consideration of the issue by COP MOP-4. Observations include, *inter alia*, that: the indicative list of non-compliance response measures is only illustrative; stringent measures that exist in most CMs have rarely been invoked or applied; stringent measures in CMs may provide a strong incentive for parties to comply; and it may not be advisable to oblige the COP MOP to apply measures in the order in which they appear in the CM, even if they are listed in an ascending order.[61]

General Issues of Compliance

At its fourth meeting, the CC also discussed general issues of compliance based, *inter alia*, on a note by the Executive Secretary[62] and reviewed in this connection information contained in the first regular national reports.[63]

The outcome of the deliberations includes a number of recommendations, representing a balanced approach. The number of recommendations was reduced compared to the number of recommendations to COP MOP-3. Recommendations dealing with issues raised in previous recommendations were, generally speaking, not repeated, even if the situation had not improved, in order to focus on what the CC considered to be most essential. This also applies to recommendations rejected by COP MOP-3, with the exception of the recommendation to remove the square brackets around Rule 18 of RoP (see the section entitled 'Rules of Procedure' above).

Recommendations included those addressing, *inter alia*, the following issues:[64] non-compliance with reporting obligations under Article 33 and COP MOP decisions regarding the deadline for submitting the reports (only approximately 35% of parties complied with the deadline); illegal transboundary movements of LMOs (reports on incidents of illegal movements suggested a lack of capacity to detect and identify the presence of LMOs); lack of submissions from parties to the CC with respect to themselves; funding for covering the participation costs of an eligible party

[61] See *Note by the CC on Repeated Non-Compliance*, n. 7 above. The CC did not provide any reasoning for the advice not to apply measures in a specific order. The advice is, however, probably linked to Decision BS-I/7, n. 12 above, Section VI, para. 1 providing for, *inter alia*, not only the frequency, but also the cause, type and degree of non-compliance to be taken into consideration when applying non-compliance response measures.

[62] Preliminary Suggestions and Information for Consideration (UNEP/CBD/BS/CC/4/Add 1, 14 November 2007).

[63] See *Monitoring and Reporting under the Protocol* (UNEP/CBD/BS/COP-MOP/4/13, 14 November 2007); and *Summary of Responses Submitted in the First National Reports* (UNEP/CBD/BS/COP-MOP/4/INF 11, 1 November 2007).

[64] See *Report to COP MOP-4*, n. 60 above, Annex.

in the consideration of a submission regarding its compliance (funds are still not available to cover such costs); and replacement of members of the CC during an inter-sessional period in an expeditious manner.[65]

Other Observations of the CC

The CC Report to COP MOP-4 included a number of observations not related to the subjects referred to in the previous two sections, i.e. observations related, *inter alia*, to:[66] facilitating access to resources from the Global Environment Facility (GEF) for activities regarding the implementation of the Cartagena Protocol, including preparation of national reports; compilation of experience on the rates of national reports under other MEAs; continued existence of significant gaps regarding the obligation to put in place at the national level the necessary legal, administrative and other measures to implement the Cartagena Protocol, being a serious issue of non-compliance; and, the unsatisfactory level of promotion of public awareness and participation.

Procedural Decisions

Finally, the CC decided that there was no need, at that moment, to do anything as regards Rule 11 of the RoP (see the section entitled 'Rules of procedure approved by COP MOP-2' above) on conflicts of interest taking into account, *inter alia*, that it would almost be impossible to foresee the different circumstances under which conflicts of interest could possibly arise.[67] The CC also decided, based on experience from conducting its third and fourth meetings in open session, to conduct, as a general practice, its upcoming meetings in open sessions unless specific circumstances require otherwise. The CC was of the view that such a decision is consistent with Rule 14.[68]

Considerations and Conclusions of COP-MOP 4

In accordance with the practice established at COP MOP-3 (see the section entitled 'Considerations and conclusion of COP MOP-3' above) recommendations and observations of the CC were considered by COP MOP-4 (Bonn, May 2008) under various agenda items and were consequently mirrored in three COP MOP-4

65 So far, Rule 10, para. 2 of the RoP, n. 28 above, has not worked well. The rule provides that the Bureau of the COP MOP shall, in consultation with the appropriate regional group, appoint a replacement in the event of a vacancy. The CC has had some vacancies, which were not filled in. This might have caused a problem of quorum. CC-1 to CC-5 were attended by 11, 12, 12, 11 and 13 members, respectively.

66 See *Report to COP MOP-4*, n. 60 above, paras 12, 14, 16 and 17.

67 The CC noted that, on the one hand, normal principles of conflict of interest apply, and, on the other hand, that being a citizen of the State whose compliance was to be discussed would not in itself be considered as a conflict of interest. See *Third Meeting of CPB-CC*, n. 60 above, para. 23.

68 Ibid., paras 24–26.

decisions.[69] The outcomes of COP MOP-4[70] were discussed by the CC at its fifth meeting (November 2008) concluding that: [o]ver all, members felt that their recommendations were received favourably and appropriately integrated into various decisions of the COP-MOP 4.[71] To some extent, however, outcomes and non-outcomes of COP MOP-4 in connection with recommendations and observations of the CC demonstrated some rather negative trends. The following comments focus on these trends.

Measures in Cases of Repeated Non-Compliance

In its decision on the report of the CC, COP MOP-4 takes note of the information compiled by the CC on the experiences of other MEAs in addressing cases of repeated non-compliance (see the section entitled 'Measures in cases of repeated non-compliance' above). The conclusion, however, is very weak, namely:

> to defer consideration or, as appropriate, adoption of measures on repeated cases of non-compliance until such time as experience may justify the need for developing and adopting such measures.[72]

Quite apart from the fact that any amendment to the CM would need consensus, the decision likely means that the COP MOP is never going to agree on and adopt stringent measures. A justification based on experience implies that the COP MOP by definition is prevented from dealing with this issue *in abstracto*, which, of course, would add considerably to the difficulties. Furthermore, the scope of application of such measures would raise difficult legal issues, for example to what extent measures might be applied to a pending case. In other words: when consideration of stringent measures is deferred until stringent measures are needed, the momentum for adopting such measures has passed.

This decision of COP MOP-4 may be seen as a rather typical decision of an intergovernmental political body in the sense that if it were possible to identify a

[69] See, respectively, the following COP MOP-4 decisions (also available in the booklet *Biosafety: Taking Further Steps towards Effective Implementation of the Protocol* (CBD Secretariat, 2008): Decision BS-IV/1, Report of the Compliance Committee; Decision BS-IV/9, Handling, transport, packaging and identification of living modified organisms, Article 18(2)(a); and BS-IV/14, Monitoring and reporting under the Protocol, found in *Report of COP-MOP 4* (UNEP/CBD/BS/COP-MOP/4/18, 25 June 2008), Annex.

[70] The discussions were based on *Review of the Outcomes of the Fourth Meeting of the Conference of the Parties Serving as the Meeting of the Parties to the Cartagena Protocol on Biosafety as they relate to the Work of the Compliance Committee (from the Chair with assistance of the Secretariat)* (UNEP/CBD/BS/CC/5/INF2, 22 October 2008).

[71] *Report of the CC on the Work of its Fifth Meeting* (UNEP/CBD/BS/CC/5/4, 21 November 2008) (*Report of Fifth Meeting of CPB-CC*), para. 11.

[72] See Decision BS-IV/1, n. 69 above, para. 3. The Preamble refers to, *inter alia*, the absence of any submissions relating to compliance to the CC, but COP MOP-4, at the same time, completely ignored the observation of the CC that stringent measures may provide a strong incentive for parties to comply (see the section entitled 'Measure in cases of repeated non-compliance' above).

common way of escaping from difficult negotiations and decisions, an agreement to utilize this opportunity would normally be reached.

General Issues of Compliance

The discussion of the COP MOP as well as the outcome of the discussion in respect of the recommendation of the CC relating to non-compliance with reporting obligations[73] is noteworthy in several respects.

First, Brazil and Japan stated at the COP MOP Plenary dealing with the report of the CC that since the role of the CC was facilitative in nature it was not within the powers of the CC to make the statement that failure to submit a national report constituted non-compliance with Article 33 of the CBD. The statements ignored completely that the CC in its report to COP MOP-3 identified lack of submission of national reports (interim reports) as a general issue of compliance entailing a decision by COP MOP-3 addressing this issue,[74] thus accepting both the nature of the problem and the entitlement of the CC to deal with the problem as a general issue of compliance.

Second, while the statements of Brazil and Japan were not supported by other parties, neither were they contradicted, which was rather strange. Accordingly, the COP MOP decision dealing with monitoring and reporting, and, according to the Preamble of the decision, taking into account the recommendation of the CC, carefully avoids the notion of non-compliance in connection with non-fulfilment of reporting obligations.[75]

73 See the section entitled 'General issues of compliance' above. See also *Report to COP MOP-4*, n. 60 above, Annex, para. 1. It reads: 'Remind each Party of its obligations to submit national reports in accordance with Article 33 of the Protocol, emphasizing that failure to do so constitutes non-compliance, and urge Parties to respect relevant decisions on reporting including provisions on timeframes for the submission of national reports'. The statements of Brazil and Japan are not included in *Report of COP MOP-4*, n. 69 above, but are to some extent reflected in A. Appleton *et al.*, 'COP/MOP 4 Highlights: Monday, 12 May 2008', 9:437 *Earth Negotiations Bulletin* (13 May2008), at 2, available at <http://www.iisd.ca/vol09/enb09437e.html>, which reads: 'Many [delegations] expressed concern about the low number of national reports. Japan and Brazil opposed stating that failure to report constitutes non-compliance'. The statements of Brazil and Japan as summarized above are based on the author's notes of the Plenary after his presentation of the report of the CC to the COP MOP. It almost goes without saying that Brazil and Japan at the time of the COP MOP were in a state of non-compliance with respect to their reporting obligations. Both parties, however, submitted national reports within 3 months after the COP MOP.

74 See the section entitled 'General issues of compliance' above and *infra* note 76.

75 See Decision BS-IV/14, *supra* note 70 above. Paragraph 3 of the decision reminds parties that not submitting a national report within the deadline prescribed in decisions on reporting does not absolve them from fulfilling reporting obligations, which is in line with Decision BS-III/14, n. 54 above, para. 4, based on Recommendation 12 of the CC to COP MOP-3 (see *Second Meeting of CPB-CC*, n. 40 above), Annex, para. 13. As a passing remark, it might be added that the Chair of the CC was not offered any chance to react to the comments of parties to the report of the CC which is, of course, very criticizable. This caused the CC to include in its report of its meeting after COP MOP-4 a suggestion to the effect that the Chair of the CC should be given the opportunity during the sessions of the COP MOP that consider recommendations from the CC to provide information to parties regarding the

Third, stating that a body mandated to review general issues of compliance and being confronted with the fact that only approximately 35% of the parties had submitted national reports (see section entitled 'Rules of procedure' above) is not entitled to observe in general terms that non-fulfilment of reporting obligations constitutes non-compliance is obviously absurd. The statement of the CC is a pure and simple factual statement,[76] the truth of which is unquestionable, and a *conditio sine qua non* for dealing with general issues of compliance is, of course, that the issue at hand constitutes or includes elements of non-compliance. General issues of compliance are by their very nature not formally linked to individual parties, since they are not based on submissions from parties to the CC. Hence, a facilitative role cannot and should not be exercised *vis-à-vis* individual parties. Furthermore, one of the recommendations of the CC to COP MOP-4 actually did address non-compliance with reporting obligations in a facilitative (and general) manner, and this recommendation was accepted by the COP MOP.[77]

At its fifth meeting in November 2008 (after COP MOP-4), the CC:

> reiterated its strong belief that failure to fulfil . . . [reporting obligations] still constitutes a state of non-compliance . . . [and] that reaching such a general conclusion was not only consistent with the functions of the Committee specified in section III, paragraph (d) to the Annex to decision BS-I/7, but was also within the scope of decision BS-III/14 of COP-MOP on monitoring and reporting.[78]

The outcomes of COP MOP-4 included considerations of two other recommendations of the CC that were also rather disappointing.

One recommendation of the CC concerned an authorization of the Executive Secretary to use any balance funds being available in the budget allocated for two meetings of the CC in any given year for covering the participation costs of an eligible party concerned in consideration of any submission regarding its compliance before the Committee (see the section entitled 'General issues of compliance' above). This recommendation, however, resulted in a decision by

nature and implications of any compliance-related recommendation under consideration (see *Report of Fifth Meeting of CPB-CC*, n. 71 above, para. 13).

[76] See V. Koester, 'Some Insider Reflections on the Compliance Mechanism under the Cartagena Protocol on Biosafety', 3:5 *Cartagena Protocol News* (2008), 5, available at <http://www.cbd.int/doc/newsletters>.

[77] See the section entitled 'General issues of compliance' above. See also *Report to COP MOP-4*, n. 60 above, Annex, para. 2, and Decision BS-IV/14, n. 69 above, para. 6, urging the GEF to make financial resources available with a view to enabling eligible parties to prepare their national reports (see the section entitled 'Other observations of the CC'). This recommendation was later included as para. 4(b) in Section C of the Annex to COP-9 Decision IX/31. See Decision IX/31, Financial resources and the financial mechanism and guidance to the financial mechanism, found in *Report of COP-9* (UNEP/CBD/COP/9/29, 9 October 2008), Annex.

[78] See *Report of Fifth Meeting of CPB-CC*, n. 71 above, para. 12. On general issues of compliance, see n. 47 above, and on Decision BS-III/14, see n. 54 above.

the COP MOP which actually reduces the possibilities of the CC completing its functions.[79]

Another recommendation of the CC addressed the lack of submissions to the CC by parties with respect to themselves.[80] It recommended that the COP MOP request parties to submit views on how to make better use of the CM and that the CC make observations and suggestions on the basis of these views.[81] While accepting the essence of this recommendation, the COP MOP, however, apparently did not wish to entrust the CC to analyse parties' views, but limited itself to request the Executive Secretary to compile the views and make them available to COP MOP-5,[82] which obviously is not a very constructive or forward-looking decision. Nevertheless, at the fifth meeting, the CC included in its work plan for 2010 an item on consideration of observations of parties on how the supportive role of the CC might be improved.[83]

Other Outcomes of the Considerations of COP MOP-4 of the Report of the CC

COP MOP-4 did not address all of the observations of the CC,[84] and RoP 18 (see the sections entitled 'Rules of procedure approved by COP MOP-2' and 'Rules of

79 See Decision BS-IV/1, n. 69 above, para. 5, which encourages the CC to meet less than twice a year, notwithstanding the two annual meetings of the CC provided for (and referred to in the section entitled 'Institutional mechanisms' above) in Decision BS-I/7, n. 12 above, Section II, para. 5. Also note that the budget adopted for the biennium 2009–2010 only provides for one annual meeting of the CC. Even though the decision might be considered also in the light of budgetary constraints in general, it is regrettable that the concerns of the CC were not addressed through a proper dialogue with parties concerned due to lack of financial resources, which resulted in a further reduction of the possibilities of the CC to handle submissions.

80 See Decision BS-I/7, n. 12 above, Section IV, para. 1(a) and see the section entitled 'General issues of compliance' above. See also n. 23 above.

81 See *Report to COP MOP-4*, n. 60 above, Annex, para. 5.

82 See Decision BS-IV/1, n. 69 above, para. 6.

83 See *Report of Fifth Meeting of CPB-CC*, n. 71 above. Self-triggering of the CM in accordance with the provisions of the CM (see Decision BS-I/7, n. 12 above, Section IV, para. 1(a)) might entail, at least from a theoretical point of view, some risks. The non-compliance response measures outlined in the CM (see the section entitled 'Measures to promote compliance and address cases of non-compliance' above) are probably not limited to party-to-party submissions. Hence, self-triggering of the CM could, at the very least, lead to the issuance of a caution. Whether a submission might be withdrawn, and whether a withdrawal of a submission prevents the CC from further considerations of the submission, is questionable. The Aarhus Convention CC, n. 20 above, has decided that in the event that a communicant withdraws its communication (i.e. submission), the CC would have the option either to proceed with consideration of the communication or not to do so, and that the choice would depend upon the extent to which it considered that the compliance issues raised were important (see V. Koester, 'The Compliance Committee of the Aarhus Convention: An Overview of Procedures and Jurisprudence', 37:2–3 *Environmental Policy and Law (March* 2007), 83, at 88). At its fifth meeting, the CC decided that in the event of allegations received from non-party sources concerning the state of compliance of a party, which the CC under the rules of the CM would lack the mandate to consider, the CC 'may invite the party concerned to indicate, if the party so wishes, to the Committee to consider the information received with a view to providing advice and assistance to that Party, as appropriate'. See *Report of Fifth Meeting of CPB-CC*, n. 71 above, para. 24. Probably an acceptance of such an invitation would not be considered as a submission to the CC.

84 COP MOP-4 ignored, *inter alia*, the observation of the CC on continued failure of many parties to implement properly the CPB (see the section entitled 'Other observations of the CC' above).

procedure' above) remained bracketed.[85] The fact, however, that the COP MOP did not address the decisions of the CC regarding the nature of its meetings,[86] and Rule 11 of the RoP on conflicts of interest,[87] indicates that the COP MOP has implicitly accepted these decisions.

CONCLUSION

The CM of the Cartagena Protocol was adopted in 2004. Consequently, the CC has only been functioning for a limited period of time and has only held five meetings, the first in March 2005 and the fifth in November 2008. On the other hand, the work of the CC has been considered by no less than three COP MOPs, and the effectiveness of any CM depends, ultimately, upon the reactions of the governing body to the recommendations and observations of the CC. A number of recommendations of the CC have been accepted by the COP MOP, and, furthermore, the COP MOP has not interfered very much with the working methods of the CC.[88] In contrast, however, COP MOP-4 demonstrated to some extent a general lack of interest in an effective and workable CM and an opposition to identifying non-compliance as non-compliance. So, the experience so far embodies both positive and negative signs.

The CC has, however, as yet not received any submission from a party with respect to itself or any party-to-party submission. The real test may depend on the outcomes of the first submissions, in particular the first party-to-party submissions and the reaction of the COP MOP to recommendations of the CC in such cases. In this respect, it is noteworthy that the CM suffers from some inherent weaknesses that may ultimately put its credibility at risk, in particular that no financial resources are available to the CC or the COP MOP itself for implementing recommendations to provide assistance to non-compliant parties. Furthermore, the possibilities for the CC to consider submissions, if any, in depth, were considerably reduced by COP MOP-4. The test of the success of the CM may also depend upon whether a party

[85] See Decision BS-IV/1, n. 69 above, para. 4.

[86] See, for instance, Rule 14, para. 1 of the RoP, n. 28 above, on open or closed sessions (the sections entitled 'Rules of procedure approved by COP MOP-2', 'Rules of procedure' and 'Procedural decisions' above). The fifth meeting of the CC was, in principle, conducted in open session. One observer organization attended the meeting. However, part of the agenda item 'other matters' was discussed in closed session (see *Report of Fifth Meeting of CPB-CC*, n. 71 above, para. 24).

[87] See the sections entitled 'Rules of procedure approved by COP MOP-2' and 'Procedural decisions' above.

[88] As a strictly personal observation, I may add that the CC itself has been working extremely well implementing fully the provisions of the CM that members serve in their personal capacity. So, the fact that the composition of the CC is based on nominations of three members from each of the five regional groups of the UN (see the section entitled 'Similarities and differences of the CM compared to other CMs' above) has had no impact on the decisions of the CC, but has, of course, enriched the discussions of the CC with the regional experience of individual members.

having serious non-compliance problems implicating real economic interests *vis-à-vis* another party would use the World Trade Organization's Dispute Settlement Mechanism instead of the CM.[89] Hence, overall, it is probably too early to attempt to draw any firm conclusions.

[89] In theory, non-compliance with provisions of the CPB could at the same time constitute a violation of World Trade Organization (WTO) provisions, although from different perspectives. However, in spite of the fact that the WTO Dispute Settlement Mechanism (DSM) is obligatory for WTO parties, the WTO has recommended that an effort be made to resolve such conflicts within the framework of the MEA. See V. Koester, 'The Cartagena Protocol: A New Hotspot in the Trade–Environment Conflict', 31:2 *Environmental Policy and Law* (2001), 82, at 87. Whether the fact that a dispute is *lis pendens* before a DSM prevents the same dispute from being considered by a CM and *visa versa* is a different issue. This is addressed briefly in V. Koester, n. 20 above, at 213 (with further references) in respect of the Aarhus Convention CM and the DSM of the Convention, and in V. Koester, n. 26 above, at 34, referring to the commonly held view that compliance procedures, by their very nature, should not be prevented from considering matters that are, at the same time, being considered by DSMs, at least, not as long as the matter is not *sub judice*. However, the DSM of CPB is not very relevant in this context because it is not compulsory beyond seeking solution by negotiation (see CBD, n. 2 above, Article 27(1) and CPB, n. 1 above, Article 32). For an in-depth analysis of the 'without prejudice' clause in Article 34 of the Cartagena Protocol and similar clauses of other MEAs, and the relationship between DSMs and CMs in general, see the analysis of T. Treves, 'The Settlement of Disputes and Non-Compliance Mechanisms', in T. Treves *et al.*, n. 16 above, at 499.

12

Biosafety, Liability, and Sustainable Development

Frederic Perron-Welch and Olivier Rukundo

The internalization of environmental costs into economic decision making is a basic principle of sustainable development. The potential of modern biotechnology to contribute to sustainable development thus cannot be met in its entirety without the establishment of clear legal and financial guarantees to handle harm to biodiversity and the natural environment. This requires actors engaged in the field of biotechnology to be held accountable for damage caused to biodiversity by living modified organisms (LMOs) through a system that apportions liability and stipulates corresponding redress obligations. Such accountability is easily given effect domestically through the normal precepts of the law of civil liability, but ensuring accountability when actors are located in different jurisdictions can be more challenging. Although virtually every country has a civil liability regime, the increasing transboundary flow of LMOs predicates a need to have international norms regarding non-State liability because of the limited availability of remedies for transboundary damage within the limits of national civil liability frameworks and international fora.

Liability and redress for environmental harm was one of the issues that remained unaddressed in the negotiations to the Cartagena Protocol, yet negotiating Parties recognized that it was a significant issue and mandated a four-year process to elucidate clear and functional rules on liability and redress.[1] Seven years after the Cartagena Protocol came into force, the legal lacuna in liability and redress for environmental damage from the transboundary movement of LMOs was addressed in October, 2010 by the adoption of the *Nagoya – Kuala Lumpur Supplementary Protocol on Liability and Redress to the Cartagena Protocol on Biosafety* (Supplementary Protocol).[2] The main focus of the Supplementary Protocol is to lay out the grounds

[1] *Cartagena Protocol on Biosafety to the Convention on Biological Diversity*, 29 January 2000, 2226 UNTS 208, 39 ILM 1027 (entered into force 11 September 2003) at art 27.

[2] *Nagoya – Kuala Lumpur Supplementary Protocol on Liability and Redress to the Cartagena Protocol on Biosafety*, 15 October 2010, UN Doc UNEP/CBD/BS/COP-MOP/5/17, Annex online: CBD <http://bch.cbd.int/protocol/NKL_text.shtml [*Nagoya – Kuala Lumpur Supplementary Protocol*].

The authors would like to thank Jennifer S. Bond, LL.B. (Dalhousie) for her assistance in writing this chapter.

and underlying considerations that should be addressed by Parties in drafting domestic liability regimes.

INTERNATIONAL LAW ON LIABILITY AND REDRESS FOR ENVIRONMENTAL HARM

States have sovereign rights over their natural resources. This right is circumscribed by the corresponding obligation not to cause irreparable damage to the territories of other States. Increasingly, this limitation is being extended to broad swaths of the global environment.[3] The Convention on Biological Diversity (CBD) is one such limitation, asserting that the conservation of biological diversity is a common concern of humankind.[4]

In international law, the term *liability* is often linked to the obligation to provide compensation for damage caused by activities that carry a risk to persons, property, or the environment. To that end, international law has accepted some overarching principles of good neighbourliness for state liability, mainly through reference to the maxim *sic utere tuo ut alienum non laedas*.[5] This maxim literally means "use your own property not to injure others" and has been recognized in the Trail Smelter[6] arbitration decision on transboundary air pollution and other seminal decisions. It is now widely accepted that a State must pay reparations for injury caused outside its jurisdiction by industrial activities.

The principles of good neighbourliness have been further developed by the International Court of Justice (ICJ) in several cases. These include the Corfu Channel[7] case, in which the Court held that every state has the obligation not to knowingly allow its territory to be used for acts contrary to the rights of other States;[8] the Nuclear Tests[9] case, in which France was barred from conducting further atmospheric nuclear tests over Australian territory; and the Advisory Opinion Legality of the Threat or Use of Nuclear Weapons, in which the Court affirmed that "the general obligation of states to ensure that activities within their jurisdiction and control respect the environment of other States or of areas beyond national control is now a part of the corpus of international law relating to the environment."[10]

3 Marie-Claire Cordonier Segger & Ashfaq Khalfan, eds, *Sustainable Development Law: Principles, Practices and Prospects* (Oxford, UK: Oxford University Press, 2004) at 109.

4 *Convention on Biological Diversity*, 5 June 1992, 1760 UNTS 79, 31 ILM 818 (entered into force 29 December 1993) at Preamble.

5 Patricia Birnie, Alan Boyle, and Catherine Redgwell, *International Law and the Environment*, 3rd ed. (Oxford, UK: Oxford University Press, 2009) at 137.

6 *Trail Smelter Arbitration (United States v. Canada)*, 3 R Int'l Arb Awards 1911 (1938), reprinted in 33 AJIL 182 (1939), 3 R Int'l Arb Awards 1938 (1941), reprinted in 35 AJIL 684 (1941).

7 *Corfu Channel Case (UK/Albania)*, [1949] ICJ Rep 4.

8 *Ibid* at 22.

9 *Nuclear Tests Case (Australia/France)*, [1974] ICJ Rep 253.

10 *Advisory Opinion on the Legality of the Threat or Use of Nuclear Weapons*, [1996] ICJ Rep 226 at para 29.

The Gabcikovo-Nagymaros[11] case similarly noted that "in the field of environmental protection, where damage can often be irreversible, vigilance and prevention are required on account of the often irreversible character of damage to the environment and of the limitations inherent in the very mechanism of reparation of this type of damage."[12] The 2010 ICJ judgment Pulp Mills on the River Uruguay *(Argentina v. Uruguay)*[13] explains that the principle of prevention, as a customary rule of international environmental law, has its origins in the due diligence required of a State in its territory, and requires States to avoid having activities in areas under jurisdiction that cause significant damage to the environment of another State.[14]

Despite this line of decisions on the principles of good neighbourliness and the prevention of environmental damage, the one area of judicial activity that the ICJ has shied away from is that of liability for damage to the environment.[15] This leaves a lacuna in ICJ jurisprudence on the remedies available for incidents of transboundary environmental harm, what constitutes a precautionary approach, and the "polluter-pays" principle.

The customary principles of good neighbourliness, prevention, precaution, and polluter-pays are, however, also listed in numerous United Nations declarations. The 1992 Rio Declaration on Environment and Development (Rio Declaration),[16] for example, affirms that, with the sovereign rights of States over natural resources, States have the responsibility to ensure that activities within their jurisdiction or control do not cause damage to the environment of other States or of areas beyond the limits of national jurisdiction.[17] It also affirms that States will immediately notify other States of any natural disasters or other emergencies that are likely to produce sudden harmful effects on their environment.[18] Further, the Rio Declaration affirms that States must give prior and timely notification and relevant information to potentially affected States relating to activities that might have a significant adverse transboundary environmental effect, and to consult with them at an early stage and in good faith.[19] Finally, States pledged to cooperate in discouraging or preventing the relocation and transfer of any activities and substances that cause severe environmental degradation, or are harmful to human health, to other States.[20]

The Rio Declaration principles of a precautionary approach to the environment and the polluter-pays principle are also pertinent to the issue of liability and redress

11 *Case Concerning the Gabčíkovo-Nagymaros Project (Hungary/Slovakia)*, [1997] ICJ Rep 7.

12 *Ibid* at para 140.

13 *Pulp Mills on the River Uruguay (Argentina v Uruguay)*, Judgment of 20 April 2010, online: International Court of Justice, http://www.icj-cij.org/docket/files/135/15877.pdf.

14 *Ibid* at para 101.

15 Birnie, Boyle, & Redgwell, *supra* note 4, at 140.

16 *Rio Declaration on Environment and Development*, UN Doc A/CONF.151/26 (Vol I), 31 ILM 874 (1992).

17 *Ibid* at Principle 2.

18 *Ibid* at Principle 18.

19 *Ibid* at Principle 19.

20 *Ibid* at Principle 14.

for transboundary environmental harm.[21] The Declaration explains that, to protect the environment, a precautionary approach should be applied by States according to their capabilities. Where there are threats of serious or irreversible damage, States may not use the lack of full scientific certainty to postpone cost-effective measures to prevent environmental degradation.[22] National authorities should promote the internalisation of environmental costs and the use of economic instruments, taking into account the polluter-pays principle, with due regard to the public interest and without distorting international trade and investment.[23]

In the Rio Declaration, States also committed to developing national law on liability and redress for the victims of pollution and other environmental damage, and to cooperate to develop further international law on liability and redress for environmental harm caused by activities within their jurisdiction or control to areas beyond their jurisdiction.[24] This commitment was a consideration in the minds of those drafting the Nagoya – Kuala Lumpur Supplementary Protocol.[25]

The Johannesburg Declaration and Plan of Implementation of the World Summit on Sustainable Development (WSSD) provide further support for the principles of the Rio Declaration. The 2002 New Delhi Declaration on Principles of International Law on Sustainable Development, appended to the outcomes of the WSSD, asserts that sustainable development requires a precautionary approach to the environment, including accountability (and, when appropriate, State responsibility) for harm caused to the environment.[26] The Declaration further supports the notion that the principle of common but differentiated responsibilities in global sustainable development and protection of the environment requires corporations (in particular transnational corporations) to cooperate in and contribute to global sustainable development and the protection of the environment, also recognizing that they have responsibilities pursuant to the polluter-pays principle.[27] Several treaties have also adopted the principles of good neighbourliness, prevention, precaution, and polluter-pays when assigning liability and obtaining redress for environmental damage in relation to particular activities, but many of them are not in force.[28] As a practical result, imposing liability for environmental harm is still heavily weighted on an ex post facto assessment of damage rather than front-end rules on preventive measures relating to events that might cause transboundary environmental harm.

[21] Birnie, Boyle, & Redgwell, *supra* note 4, at 53.

[22] Rio Declaration, *supra* note 15, at Principle 15.

[23] *Ibid* at Principle 16.

[24] *Ibid* at Principle 13.

[25] Nagoya – Kuala Lumpur Supplementary Protocol, *supra* note 1, at Preamble.

[26] *ILA New Delhi Declaration of Principles of International Law Relating to Sustainable Development*, UN Doc A/CONF.199/8, Annex at para 4.2(a) (New Delhi Declaration).

[27] *Ibid* at para. 3.1.

[28] See Annex IV to this book for a detailed list of treaties on liability for environmental harm.

DEVELOPMENT OF THE LIABILITY AND REDRESS SUPPLEMENTARY PROTOCOL

Compensation for damage to biodiversity from LMOs was a recurrent point in the negotiation of the Cartagena Protocol on Biosafety. Although consensus could not be reached on the inclusion of a liability and redress regime to address environmental damage resulting from the transboundary movement of LMOs, the issue was recognised as being important. This recognition came first because of the outcomes such a regime can promote. First, this regime can promote normative compliance, the implementation of the precautionary approach, and trigger prevention and/or reparation in case of a breach or conduct that causes harm. Second, it can serve a reparative function by shifting the costs of environmental damage from society at large to the person or persons responsible for the activity causing the damage, thus implementing the polluter-pays principle. Last, holding the author of environmental harm responsible for its redress will have a deterrent effect and lead to investment in preventive measures.[29] Negotiators agreed on Article 27, an enabling provision requiring further negotiations on liability and redress through the Conference of the Parties serving as the meeting of the Parties to the Protocol (COP-MOP). More specifically, Article 27 provides a legal framework for the elaboration of international rules and procedures in the field of liability and redress for damage resulting from transboundary movements of LMOs, within four years after the Protocol's entry into force.

After the adoption of the Cartagena Protocol but prior to its entry into force, the COP to the CBD established the Intergovernmental Committee for the Cartagena Protocol (ICCP) as an interim governing body.[30] The issue of liability and redress was not addressed within this process until the third meeting, which adopted a recommendation on the issue. Again, liability and redress proved to be the most contentious issues of discussion.[31] Accepting the recommendations of the ICCP, COP-MOP 1 established the open-ended Ad Hoc Working Group of Legal and Technical Experts on Liability and Redress (Working Group) to elaborate on the options for international rules and procedures on liability and redress.[32] The Working Group's terms of reference were fairly broad in scope. They requested an analysis

[29] *Liability and Redress for Damage Resulting from the Transboundary Movements of Living Modified Organisms: Review of Existing Relevant Instruments and Identification of Elements: Note by the Executive Secretary*, UN Doc UNEP/CBD/ICCP/2/3 at para 6; René Lefeber, *Transboundary Environmental Interference and the Origin of State Liability* (Cambridge, MA: Kluwer Law International, 1996) at 313.

[30] *Adoption of the Cartagena Protocol and interim arrangements*, UN Doc UNEP/CBD/ExCOP/1/3, Annex.

[31] IISD Reporting Services, *A Brief Introduction to the Cartagena Protocol on Biosafety, Including the Negotiation Process, ICCP, COP/MOP and the Working Group on Liability and Redress*, online: IISD, http://www.iisd.ca/process/biodiv_wildlife-biosafetyintro.html.

[32] *Establishment of an Open-Ended Ad Hoc Working Group of legal and technical experts on liability and redress in the context of the Protocol*, COP-MOP Dec BS-I/8.

of general issues relating to damage scenarios to identify situations for which international rules and procedures might be needed, and how to apply rules and procedures to those situations. The Working Group was also asked to elaborate the elements of such rules and procedures, which could include

1. Definition and nature of damage, including scope of damage resulting from transboundary movement of LMOs
2. Valuation of damage to biodiversity and to human health
3. Threshold of damage
4. Causation
5. Channelling of liability
6. Roles of Parties of import and export
7. Standard of liability
8. Mechanisms of financial security
9. Standing/right to bring claims

Given the complexity of the issue, the Working Group failed to reach consensus within its four-year mandate despite meeting five times between 2005 and 2008. The negotiations were characterised by deep divisions between Parties who wanted a legally binding agreement and those who did not, making progress very slow. Several issues proved particularly difficult in the negotiations, and these were emblematic of the deep gulf that existed among the Parties. These included issues such as whether "products thereof" (products derived from LMOs) should be included in the scope of the Supplementary Protocol, time limitations, damage to human health, the concept of imminent threat of damage, response measures, the definition of operator, exemptions and mitigation, and financial security. At its final meeting in Cartagena, Colombia, the Working Group agreed to some core elements of international rules and procedures in the field of liability and redress for damage resulting from transboundary movements of LMOs. These core elements included a primary compensation scheme based on an administrative approach; the negotiation of a supplementary protocol to the Cartagena Protocol on Biosafety; guidelines on civil liability, including an enabling clause on private international law; a supplementary compensation scheme for damage to the conservation and sustainable use of biodiversity; and an institutional arrangement for complementary capacity-building measures.[33]

A group of the Friends of the Co-Chairs on Liability and Redress in the Context of the Cartagena Protocol on Biosafety (Friends of the Co-Chairs) was established at COP-MOP 4 to further negotiate international rules and procedures in the field of liability and redress for damage resulting from transboundary movements of living

[33] *Report of the Open-Ended Ad Hoc Working Group of Legal and Technical Experts on Liability and Redress in the Context of the Cartagena Protocol on Biosafety on the Work of Its Fifth Meeting*, UN Doc UNEP/CBD/BS/WG-L&R/5/3 at para 78.

modified organisms.[34] The Friends of the Co-Chairs met four times to negotiate the text of an agreement on rules and procedures on liability and redress. At their first meeting in Montreal, the negotiating Parties agreed to work toward a legally binding instrument on administrative measures pertaining to liability and redress in the form of a supplementary protocol.[35] The draft text was further negotiated by the Friends of the Co-Chairs in Kuala Lumpur, Malaysia, from 8 to 12 February 2010 and 15 to 19 June 2010.[36] The Friends of the Co-Chairs met one more time just prior to COP-MOP 5 from October 6 to 11, 2010 in Nagoya, Japan, to finalize the Supplementary Protocol.[37]

At that last meeting, two proposals in the text on liability and redress remained. The first part laid out a legally binding supplementary protocol on liability and redress for damage resulting from transboundary movements of LMOs, now the Nagoya – Kuala Lumpur Supplementary Protocol. The second proposal contained draft guidelines on civil liability and redress for damage resulting from the transboundary movements of LMO, which addressed core issues in the implementation of domestic laws on civil liability.[38] In the end, although the group consolidated the draft guidelines, they were not considered a part of an "administrative approach" and thus were not included in the final decision. This decision, however, does not preclude future steps in the development of binding rules on civil liability for damage from LMOs.[39]

NAGOYA – KUALA LUMPUR SUPPLEMENTARY PROTOCOL

At the end of the long negotiation process, the Supplementary Protocol was adopted, which reflects the diversity of interests among Parties and takes into account the concerns of Parties, civil society and the private sector.[40] The aim of the Supplementary Protocol is to contribute to the conservation and sustainable use of biodiversity,

[34] *Liability and redress under the Cartagena Protocol on Biosafety*, UN Doc UNEP/CBD/BS/COP-MOP/IV/12.

[35] *Report of the Group of the Friends of the Co-Chairs on Liability and Redress in the Context of the Cartagena Protocol on Biosafety on the Work of its First Meeting*, UN Doc UNEP/CBD/BS/GF-L&R/1/4.

[36] *Report of the Group of the Friends of the Co-Chairs on Liability and Redress in the Context of the Cartagena Protocol on Biosafety on the Work of its Second Meeting*, UN Doc UNEP/CBD/BS/GF-L&R/2/3; and *Report of the Group of the Friends of the Co-Chairs on Liability and Redress in the Context of the Cartagena Protocol on Biosafety on the Work of its Third Meeting*, UN Doc UNEP/CBD/BS/GF-L&R/3/4.

[37] *Report of the Group of the Friends of the Co-Chairs on Liability and Redress in the Context of the Cartagena Protocol on Biosafety on the Work of its Fourth Meeting*, UN Doc UNEP/CBD/BSGFLR/4/3.

[38] *Consolidated draft guidelines on civil liability and redress*, Annex II, Appendix I to UN Doc UNEP/CBD/BS/GF-L&R/4/2.

[39] *Report of the Fifth Meeting of the Conference of the Parties to the Convention on Biological Diversity Serving as the Meeting of the Parties to the Cartagena Protocol on Biosafety*, UN Doc UNEP/CBD/BS/COP-MOP/5/17 at para 133.

[40] Statement of the Brazilian Representative, *ibid* at para 142.

taking into account risks to human health, by providing international rules and procedures in the field of liability and redress related to LMOs.[41] That objective accords with the objective of its overarching framework, the CBD: to promote the conservation and sustainable use of biodiversity.[42] It also complements the objective of its parent instrument, the Cartagena Protocol, by providing protection for the safe transfer, handling, and use of LMOs that could have adverse effects on the conservation and sustainable use of biodiversity.[43]

Relationship with the CBD, the Cartagena Protocol, and International Law

Before putting forward the terms of the Supplementary Protocol, the Parties to the Cartagena Protocol took into account Principle 13 of the Rio Declaration on liability and compensation;[44] reaffirmed the precautionary approach contained in Principle 15 of the Rio Declaration; recognized the need to provide for appropriate response measures when there is damage or sufficient likelihood of damage envisioned by the Cartagena Protocol; and recalled Article 27 of the Cartagena Protocol, which was the basis for negotiations.[45]

The definitions in the CBD and Cartagena Protocol are incorporated by reference into Article 2 of the Supplementary Protocol.[46] Except where otherwise noted the provisions the CBD and Cartagena Protocol also apply to the Supplementary Protocol.[47] It is supplementary to the Cartagena Protocol and neither modifies nor amends its terms,[48] nor does it affect the rights and obligations of its Parties under the CBD and the Cartagena Protocol.[49] Last, it does not affect the rights and obligations of a Party under international law,[50] particularly those pertaining to the responsibility of States for internationally wrongful acts.[51]

The COP-MOP to the Cartagena Protocol will serve as the meeting of the Parties to the Supplementary Protocol and will keep the implementation of the Supplementary Protocol under regular review. In addition, the COP-MOP

41 *Nagoya – Kuala Lumpur Supplementary Protocol*, *supra* note 2 at art 1.

42 *Convention on Biological Diversity*, *supra* note 3 at art 1.

43 *Cartagena Protocol*, *supra* note 1 at art 1.

44 *Rio Declaration*, Principle 13 reads: "States shall develop national law regarding liability and compensation for the victims of pollution and other environmental damage. States shall also cooperate in an expeditious and more determined manner to develop further international law regarding liability and compensation for adverse effects of environmental damage caused by activities within their jurisdiction or control to areas beyond their jurisdiction."

45 *Nagoya – Kuala Lumpur Supplementary Protocol*, *supra* note 2 at Preamble.

46 *Ibid* at art 2(1).

47 *Ibid* at art 16(3).

48 *Ibid* at art 16(1).

49 *Ibid* at art 16(2).

50 *Ibid* at art 16(4).

51 *Ibid* at art 11.

will, within its mandate, make the decisions necessary to promote the Supplementary Protocol's effective implementation. It will also perform the function assigned to it by the Supplementary Protocol and, *mutatis mutandis*, the functions assigned by paragraphs 4(a) and (f) of Article 29 of the Cartagena Protocol.[52] The Secretariat to the CBD also serves as the secretariat to the Supplementary Protocol.[53]

The right of Parties to, in their domestic law, provide for financial security for damage must be exercised in a manner consistent with their rights and obligations under international law, taking into account the three concluding sentences of the preamble to the Cartagena Protocol, which recognise that trade and environment agreements should be mutually supportive to achieve sustainable development; emphasize that the Protocol does not imply a change in the rights and obligations of a Party under any existing agreements and that this statement does not subordinate the Protocol to other international agreements.[54]

Scope of the Supplementary Protocol

The scope of the Nagoya – Kuala Lumpur Supplementary Protocol covers damage resulting from LMOs intended for food, feed, or processing (LMOs-FFP), contained use, or intentional introduction into the environment, which engage in a transboundary movement.[55] Damage from any authorized use of the LMOs is included,[56] as is damage from unintentional and illegal transboundary movements.[57] The jurisdictional scope is limited to damage that occurs in areas within the limits of the national jurisdiction of Parties resulting from a transboundary movement of LMOs, while the temporal scope is limited to those movements that began after the Supplementary Protocol's entry into force for the Party into whose jurisdiction the movement was made.[58] Criteria set out in domestic laws may used to address such damage.[59] Legislation implementing the Supplementary Protocol shall also apply to damage resulting from movements of LMOs from non-Parties.[60] In the negotiations, Parties had different views of the application of Article 27 of the Cartagena Protocol to

[52] *Ibid* at art 14. Article 29(4)(a) and (f) of the Cartagena Protocol gives the power to "make recommendations on any matters necessary for implementation" and "exercise such other functions as may be required for implementation."

[53] *Ibid* at art 15.

[54] *Ibid* at art 10(2).

[55] *Ibid* at art 3(1).

[56] *Ibid* at art 3(2).

[57] *Ibid* at art 3(3). These forms of transboundary movement are referred to in Articles 17 and 25 of the Cartagena Protocol, respectively, and are discussed in earlier chapters.

[58] *Ibid* at arts 3(4) and 3(5).

[59] *Ibid* at art 3(6).

[60] *Ibid* at art 3(7).

products of LMO origin. One such view was that Parties may apply the Supplementary Protocol to damage caused by processed materials, provided that a causal link is established between the damage and the LMO in question.[61]

Harm

Harm is the basis on which liability in tort is founded and for which compensation is required; therefore, establishing the nature of the harm covered by the Supplementary Protocol is of primary concern. Damage is defined as

> an adverse effect on the conservation and sustainable use of biological diversity, taking also into account risks to human health, that:
>
> (i) Is measurable or otherwise observable taking into account, wherever available, scientifically established baselines recognized by a competent authority that takes into account any other human induced variation and natural variation; and
> (ii) Is significant as sent out . . . below.[62]

Whether the harm is significant is established by considerations such as whether long-term or permanent change occurred that cannot be redressed through natural recovery within a "reasonable" period, the extent of adverse impacts to biodiversity, the reduction in the ability of components of biodiversity to provide goods and services, and the extent of adverse effects on human health in the context of the Cartagena Protocol.[63]

Causation

In the framework of tort law, cause must be determined and fault assigned. Under the Supplementary Protocol, a causal link between the harm and the LMO in question shall be established in accordance with domestic law.[64] Response obligations and financial liability attach to the operators in control of the LMO, which could be "any person in direct or indirect control of the [LMO] which could, as appropriate and as determined by domestic law, include, *inter alia*, the permit holder, person who placed the [LMO] on the market, developer, producer, notifier, exporter, importer, carrier or supplier."[65]

[61] *Report of the Group of the Friends of the Co-Chairs on Liability and Redress in the Context of the Cartagena Protocol on Biosafety on the Work of its Fourth Meeting*, *supra* note 37 at para 12.
[62] *Nagoya – Kuala Lumpur Supplementary Protocol*, *supra* note 2 at art 2(2)(b).
[63] *Ibid* at art 2(3).
[64] *Ibid* at art 6.
[65] *Ibid* at art 2(2)(c).

Response Measures

Because of the environmental nature of the harm, response measures are an integral aspect of the Supplementary Protocol. Response measures are defined as reasonable actions to

(i) Prevent, minimize, contain, mitigate, or otherwise avoid damage, as appropriate
(ii) Restore biological diversity through actions to be undertaken in the following order of preference:
   a. Restoration of biological diversity to the condition that existed before the damage occurred, or its nearest equivalent; and where the competent authority determines this is not possible;
   b. Restoration by, *inter alia*, replacing the loss of biological diversity with other components of biological diversity for the same, or for another type of use either at the same or, as appropriate, at an alternative location.[66]

In the event of damage, Parties shall require the appropriate operator(s), subject to any requirements of the competent authority (CA), to immediately inform the CA, evaluate the damage, and take appropriate response measures.[67] The CA in turn has a responsibility to identify the operator that has caused the damage, do its own evaluation of the damage, and then determine which response measures should be taken by the operator.[68] Where relevant information indicates that there is a sufficient likelihood that damage will result if timely response measures are not taken, the operator shall be required to take appropriate response measures to avoid such damage.[69] The CA may implement appropriate measures, particularly when the operator has failed to do so,[70] and has the right to recover from the operator the costs and expenses related to the evaluation of damage and implementation of appropriate remedial measures.[71] Parties also have the choice to provide, in domestic law, for situations when the operator may not be required to bear these costs and expenses.[72]

Decisions of the CA requiring the operator to take response measures should be reasoned, and the operator should be notified. Domestic law shall provide for remedies, including the opportunity for administrative or judicial review of these decisions, which the CA should communicate to the operator. Recourse to such remedies, however, shall not impede the taking of response measures in appropriate

66 *Ibid* at art 2(2)(d).
67 *Ibid* at art 5(1).
68 *Ibid* at art 5(2).
69 *Ibid* at art 5(3).
70 *Ibid* at art 5(4).
71 *Ibid* at art 5(5).
72 *Ibid*.

circumstances, except where otherwise provided by domestic law.[73] In implementing Article 5, on response measures, Parties may assess whether specific response measures are already addressed by their domestic law on civil liability, and ensure that all measures are implemented in accordance with domestic law.[74]

Exemptions from Liability

Provisions for exemption from liability are largely left to Parties to implement in their own domestic laws. Such exemptions may include acts of God or force majeure, and acts of war or civil unrest,[75] but also any other exemptions or mitigations as deemed fit.[76] Claims for redress may be prevented by the use of limitation periods, and Parties may provide in their domestic law for relative and/or absolute time limits, including for actions related to response measures, and for the commencement of the period to which a time limit applies.[77] There is also the option to provide for financial limits for the recovery of costs and expenses related to response measures.[78] Finally, the Supplementary Protocol does not limit or restrict any right of recourse or indemnity that an operator might have against any other person.[79]

Financial Security

Provisions for financial security have not yet been fully addressed by the Supplementary Protocol. Although Parties have the right to provide for financial security in their domestic law,[80] this right must be exercised in a manner consistent with their international obligations. The first COP-MOP to the Cartagena Protocol after the entry into force of the Supplementary Protocol shall request the Secretariat of the CBD to undertake a comprehensive study that shall address, *inter alia*, the modalities of financial security mechanisms; their environmental, economic, and social impacts, in particular on developing countries; and identify the appropriate entities to provide financial security.[81]

Implementation of the Supplementary Protocol and Relationship with Civil Liability

The commitments made under the Supplementary Protocol require Parties to provide for rules and procedures in their domestic law that address damage caused by

73 *Ibid* at art 5(6).
74 *Ibid* at art 5(7) and 5(8).
75 *Ibid* at art 6(1).
76 *Ibid* at art 6(2).
77 *Ibid* at art 7.
78 *Ibid* at art 8.
79 *Ibid* at art 9.
80 *Ibid* at art 10(1).
81 *Ibid* at art 10(3).

LMOs. In implementing this obligation, Parties must provide for response measures in accordance with the rest of the Supplementary Protocol. They may apply existing domestic law, including general rules and procedures on civil liability, apply or develop civil liability rules and procedures specifically for this purpose, or apply or develop a combination of both.[82] The same goes for the provision of adequate rules and procedures on civil liability in domestic law for material or personal damage from LMOs,[83] but when developing such rules, Parties shall be mindful to address, as appropriate, the elements of damage, the standard of liability (including strict or fault-based liability), channelling of liability, and the right to bring claims.[84]

COMPLEMENTARY LIABILITY AND REDRESS RULES

In negotiating the Nagoya – Kuala Lumpur Protocol, negotiators looked to other liability and redress regimes in the field of international environmental law for guidance in framing the mechanisms under the Cartagena Protocol.[85] Two developing regimes that were considered by the negotiators were the United Nations Environment Programme Draft Guidelines for the Development of Domestic Legislation on Liability, Response Action and Compensation for Damage Caused by Activities Dangerous to the Environment,[86] and the European Union Environmental Liability Directive (ELD).[87]

Other complementary international instruments that address the issue of liability and redress have been developed by the International Law Commission. This includes Draft Articles on the Prevention of Transboundary Harm from Hazardous Activities[88] and the Draft Principles on the Allocation of Loss in the Case of Transboundary Harm Arising out of Hazardous Activities.[89] The Draft Articles concern themselves with activities not prohibited by international law that involve a risk of causing significant transboundary harm through their physical consequences,

[82] *Ibid* at art 12(1).

[83] *Ibid* at art 12(2).

[84] *Ibid* at art 12(3).

[85] *Recent Developments in International Law Relating to Liability and Redress, Including the Status of International Environment-Related Third Party Liability Instruments*, UN Doc UNEP/CBD/BS/GF-L&R/3/INF/1.

[86] United Nations Environment Programme Governing Council, *Draft guidelines for the development of domestic legislation on liability, response action and compensation for damage caused by activities dangerous to the environment*, UN Doc UNEP/GCSS.XI/5, Annex B.

[87] *Study on the Implementation Effectiveness of the Environmental Liability Directive (ELD) and Related Financial Security Issues*, online: European Commission, http://ec.europa.eu/environment/enveco/others/pdf/implementation_efficiency.pdf.

[88] International Law Commission, *Prevention of Transboundary Harm from Hazardous Activities*, UNGAOR, 56th Session, Suppl No 10, UN Doc A/56/10 (2001), online: United Nations, http://untreaty.un.org/ilc/texts/9_7.htm.

[89] International Law Commission, *Draft principles on the allocation of loss in the case of transboundary harm arising out of hazardous activities*, ILCOR, 58th Sess, UN Doc A/61/10 (2006), online: United Nations, http://untreaty.un.org/ilc/texts/9_10.htm.

and place an onus on States of origin to take all appropriate measures to prevent or minimize the risk of such harms. Preventive measures should be decided by consultations between states and based on an equitable balance of interests.[90]

The Draft Principles on the Allocation of Loss in the Case of Transboundary Harm Arising out of Hazardous Activities are intended to contribute to the process of developing international law in the field of transboundary harm and loss caused by hazardous activities by providing appropriate guidance to States regarding those activities not covered by specific agreements.[91] Its main purposes are to ensure the prompt and adequate compensation to victims of transboundary damage and mitigate any environmental damage that might be caused by such activities.[92] The Draft Principles stress the imposition of strict liability on the operator, and the establishment of industry-wide compensation funds when appropriate.[93] In addition, they state that the mitigation of damage caused by transboundary activities should be borne by both the State of origin and any affected States.[94]

Additionally, the major multinational firms involved in the plant biotechnology sector, including BASF, Bayer CropScience, Dow AgroSciences, DuPont, Monsanto, and Syngenta, have entered into a private sector liability and redress scheme called The Compact.[95] The Compact, a voluntary instrument implemented in October 2010 that complements the Supplementary Protocol, allows any state to submit a claim against a member of the agreement if the release of an LMO by that member caused damage to biological diversity in that state. If settlement cannot be reached, the matter is referred to the Permanent Court of Arbitration and the Compact. The Compact only covers damage to biological diversity and not "traditional" tort damages, such as personal injury or property damage, and does not adopt a standard of strict civil liability.[96]

CONCLUSION

The principles of sustainable development law explain why liability and redress rules are important in implementing biosafety policies and standards. The adoption of the Nagoya – Kuala Lumpur Protocol at COP-MOP 5 could be a significant event in both international environmental and sustainable development law, but the history of liability and redress treaties suggests caution. First, financial guarantees are not environmental guarantees and do not prevent damage to biological diversity. Many

[90] *Prevention of Transboundary Harm from Hazardous Activities*, *supra* note 88 at arts 1, 9, and 10.
[91] *Ibid* at Principle 1.
[92] *Ibid* at Principle 3.
[93] *Ibid* at Principle 4.
[94] *Ibid* at Principle 5.
[95] Croplife International (CI), "The Compact," online CI, http://www.croplife.org/the_compact.
[96] CI, "Liability and Redress," online CI, http://www.croplife.org/public/liability_and_redress.

treaties pertaining to liability and redress for environmental harm have never come into force, and most of the remaining treaties have a small number of signatories, limiting their effectiveness.[97]

Implementation of the Supplementary Protocol will have to be consistent with the requirements of other international agreements, particularly those on trade, to avoid creating costly and burdensome barriers. Implementation should thus be guided by the tenet of sustainable development law that international agreements on trade and on the environment should be mutually supportive, with a view to achieving sustainable development.

The Nagoya – Kuala Lumpur Protocol was open for signature at the United Nations Headquarters in New York from 7 March 2011 to 6 March 2012 and will enter into force 90 days after being ratified by at least forty Parties to the Cartagena Protocol on Biosafety.[98] The COP-MOP to the Cartagena Protocol will undertake a review of its effectiveness five years after its entry into force and every five years thereafter as long as information requiring a review has been made available. The first review shall include a review on the effectiveness of the articles on financial security and implementation/relation to civil liability.[99] The implementation of the Supplementary Protocol into domestic frameworks will be a daunting task. Because response and restoration measures are placed in the hands of domestic administrative agencies, Parties will be required to take broad legislative measures regarding new or existing agencies and provide new funds. Careful monitoring and research will be needed to determine the effectiveness of this new international treaty for sustainable development.

97 See Annex IV to this book for a list of liability and redress treaties.

98 See Biosafety Clearing-House for procedures for deposit of the instrument of ratification, acceptance, approval or accession with the Depositary, online: CBD, http://bch.cbd.int/protocol/ratification.shtml.

99 *Nagoya – Kuala Lumpur Supplementary Protocol*, *supra* note at art 13.

PART III

Domestic Implementation of Biosafety Regulatory Aspects

13

Legislative Options for National Implementation

Tomme Rosanne Young

The first question to be considered in regard to legislative options for national implementation of the sustainable development objectives of the Cartagena Protocol[1] is the nature of legislative tasks required by the Protocol. The Parties identified all legislative and administrative tasks under the Protocol in 2002, in a list adopted by the third Intergovernmental Committee for the Cartagena Protocol on Biosafety (ICCP),[2] reproduced in Annex II as Tables I to IV.[3]

This list of tasks from the Protocol provides the apparent basis for many later Cartagena Protocol and UNEP papers discussing the Parties' implementation of the Protocol.[4] As further discussed in this chapter, these lists include many tasks that do not require legislative action, and tasks that might not necessarily be mandatory in all cases or in all countries. In addition, some countries have concluded that some or all of these tasks can be addressed without legislative action.

1 *Cartagena Protocol on Biosafety to the Convention on Biological Diversity*, 29 January 2000, 2226 UNTS 208, 39 ILM 1027 (entered into force 11 September 2003), online: http://bch.cbd.int/protocol/text/ [*Cartagena Protocol*].

2 Intergovernmental committees (ICs) of this type conduct the organizational work of an international instrument that has been adopted but has not yet entered into force. An IC operates in essentially the same way as a Conference of Parties (COP) and its purposes are essentially the same. Yet, it is created in a slightly different form because it is not legally possible to create a COP until after the instrument has formally entered into force (see art 29(6) of the *Cartagena Protocol, ibid*). There were three ICCP meetings under the Protocol prior to entry into force.

3 The tables in Annex II are reproduced from ICCP 3, Recommendation III/5, *Capacity-building (Article 22 and Article 28, paragraph 3)*, Annex III, UN Doc UNEP/CBD/ICCP/3/10, online: http://www.cbd.int/recommendation/iccp/?id=7513.

4 For example, see UNEP-GEF, *Toolkit for Development of National Biosafety Frameworks*, online: http://www.unep.org/biosafety/Toolkit.aspx [UNEP-GEF Toolkit], especially Phase III, Part I: "Formulation of the Regulatory Regime" at pages 57–8 (stating without citation that the list was endorsed by COP-MOP 4). See also R Mackenzie, et al, *An Explanatory Guide to the Cartagena Protocol on Biosafety* (Bonn, Germany: IUCN-ELC, 2003) and S Johnston, et al, *Internationally Funded Training in Biotechnology and Biosafety – Is It Bridging the Biotech Divide?* (Yokohama, Japan: UNU-IAS, 2008).

OUTCOME ORIENTATION: IS LEGISLATION REQUIRED?

In most of its provisions relating to the tasks listed in Tables I to IV, the text of the Cartagena Protocol does not specifically focus on formal legislative action (adoption of any policy, law, or regulation). Instead Parties agree to bring about specific *outcomes* – to take specific action, as in one example, "the Party of import shall acknowledge receipt of the notification, in writing, to the notifier within ninety days of its receipt."[5] This provision calls on the State to take action (acknowledge receipt) but does not specify legislative action (i.e., it does not require any country to adopt laws, regulations, or guidelines governing the acknowledgement, unless needed under that country's law to authorise the relevant official to send an acknowledgement). A country whose agencies actually send acknowledgement has met its obligation, regardless of whether it has adopted legislation requiring this action.

Implementation Approaches

Some governments are not obligated to adopt implementing measures, presuming themselves fully bound by virtue of their ratification, accession, or acceptance to the Cartagena Protocol. This so-called monist approach embodies a general presumption that treaties enter directly into, and bind, the country in question.[6] In the event that such a country receives a notification, or discovers an unpermitted introduction or cross-border movement of LMO, for example, these officials expect to simply apply the Protocol by its terms, without the need for preexisting national legislative measures. The fact that international instruments are generally written in a way that gives governments some flexibility regarding how they implement the required provisions is generally assumed to give that flexibility to the national implementing agency, without legislation.

By contrast, other countries adopt specific implementing legislation for every convention or treaty they join.[7] Holders of this view are most often countries that apply the "strict rule of law" as their primary legislative approach. Companies and

[5] *Cartagena Protocol*, *supra* note 1 at art. 9(1). The current list of Parties is located online: http://www.cbd.int/biosafety/parties/list.shtml. The BCH legislative and decisional databases are also located online: http://bch.cbd.int/database/.

[6] For one example, see ST Shemirani, "Review of the Iranian Legislation Relating to Alien Invasive Species" (Report to UNEP's Caspian Environment Programme, 2006), which states that "Treaties, once approved by the Parliament, have the force of law and, hence are enforceable within the Iranian territory. Article 9 of the Iranian Civil Code stipulates that provisions of treaties between the Iranian government and other governments in accordance with the Constitution shall have the effect of law." A similar approach has been taken by Armenia on the Cartagena Protocol. According to Armenia's draft National Biosafety Framework (BCH Record ID 43421) it has opted to "follow" the May 2003 Declaration of the Fifth UNECE Ministerial Conference "Environment for Europe," which committed Parties to enhance cooperation in environmental protection, in lieu of adopting its own national policy on GMOs.

[7] See generally A Aust, *Modern Treaty Law and Practice* (Cambridge, UK: Cambridge University Press, 2000) at 144–61.

individual people expect to be able to determine, on the basis of unambiguous legislation, exactly what the law requires and exactly which powers and benefits arise from compliance.

Some countries' legal traditions and resulting legislative documents are difficult to interpret this point. For example, a clause in the Vietnamese biosafety law states that "Where treaties which the Socialist Republic of Vietnam has signed or acceded to contain provisions different from this Regulation, the provisions of such treaties shall apply."[8] It is probably not advisable for external actors to attempt to predict how to apply this provision using strict-rule-of-law principles. It is probably necessary to contact Vietnamese officials directly and obtain a full understanding of how the country interprets its Protocol-based responsibilities.

Effect of Nonlegislative Implementation

The legal acceptability or relevance of the two implementation approaches described earlier, however, is not the most important impact of those options. Of much greater concern is the impact of this legislative decision on commercial and biosafety issues affecting that country and its people. The goal of "legal certainty" for would-be importers, exporters, and introducers of genetic resources has a significant impact on the extent to which those actors will come to the country.

For many, the Cartagena Protocol is primarily seen as a control on commercial and potentially precommercial (research and development) activities. As such, it relies on a central mechanism that is used in virtually every country, when attempting to place environmental or social restrictions on commercial actors, without preventing them from operating. That mechanism is the regulatory permit. Thus, for example, air or water pollution is not controlled by complete prohibition of all pollutant discharges or emissions, but by requiring dischargers/emitters to get permits setting limits on polluting behaviour. The extent to which a particular commercial activity is inhibited by such a permit framework depends on (1) the ability of the actor to control emissions, (2) the level of emissions allowed under the permit legislation, and (3) the effectiveness of government oversight/enforcement.

National biosafety law operates in the same way. An enterprise, agriculturalist, or researcher must be able to control certain impacts of his or her activity to the level of safety (nonimpact) required under national biosafety law. The level of safety/risk set by the government determines how many researchers, agriculturalists, and other enterprises will be able to obtain the needed permission. As noted previously, Parties to the Cartagena Protocol were unable to agree on a single view regarding the minimal level of safety to be applied in permitting introduction of GMOs. Instead, they agreed to respect each other's individual sovereign determinations of the level

[8] Vietnam, *Master Scheme of Strengthening Capacities on Biosafety Management over Genetically Modified Organisms and Products and Goods Originating from Genetically Modified Organisms until 2010 in Order to Implement Cartagena Protocol on Biosafety*, BCH Record ID 41471.

of protection applicable in each country. National legislation (i.e., formally adopted or published policy, law, and/or regulations) is the primary tool by which a country's sovereign decisions on this topic are communicated to researchers, agriculturalists, and other enterprises. The BCH is designed to aid in this communication process. A country that has not clearly formulated and publicised its national approach to GMOs and biosafety might not offer the certainty needed by foreign companies proposing export or other activities within that country.

Normally, a commercial entity or one investing in research and development related to commercial and precommercial GMO activities in another country must have a sufficient level of "legal certainty"[9] that he or they will be able to gain and comply with all requirements at a reasonable cost. In strict-rule-of-law countries, biosafety law serves as, in effect, a legal guarantee that if these entities comply with specific controls and other requirements, they will be permitted to engage in LMO-related activities. Like most environmental permit statutes, biosafety legislation often represents a political contest between a country's commercial/agricultural/economic interests and its environmental/social interests. The monist approach can be very uncertain from this perspective – the official making the decision might have so much latitude that his or her decisions may not be commercially predictable for those who need to plan their business decisions and related costs. A country that does not have detailed and specific implementing legislation in place might discourage commercial and other entities from investing time and effort to develop and market LMOs in that country. On the other hand, there could be some importers, exporters, and introducers who interpret the country's failure to adopt legislation to mean that the country imposes no control on the movement or introduction of LMOs.

In the BCH Legislative Database

As noted previously in chapter 9, although 161 countries have ratified the Protocol, only 32 Parties and 5 nonparties have reported any decisions. Most of these decisions are reported by an even smaller list of countries, with 14 countries posting 1092 decisions,[10] and the other 25 countries posting a combined total of 117 decisions and communications.

This level of traffic, coupled with the factors and uncertainties noted before, understandably could motivate national governments (especially in developing countries

9 Discussed in detail in another biodiversity context in T Young, "Summary Analysis: Legal Certainty for Users of Genetic Resources under Existing Access and Benefit-sharing (ABS) Legislation and Policy," UN Doc UNEP/CBD/WG-ABS/3/INF/10. Reprinted in: T Young, *Covering ABS: Addressing the Need for Sectoral, Geographical, Legal and International Integration in the ABS Regime: Papers and Studies of The ABS Project* (Bonn, Germany: IUCN-ELC, 2008).

10 China (23), Australia (29), Germany (30), Brazil (40), Colombia (46), Costa Rica (52), New Zealand (68), Mexico (69), European Union (96), Philippines (96), Canada (113), United States of America (119), Republic of Korea (136) and Japan (175).

with low levels of involvement in LMO development and industries[11]) to economize on the costs and time involved in formally creating biosafety legislation and institutions by delaying the development, adoption, and implementation of formal Cartagena Protocol measures and institutions until practical factors indicate that the country's level of LMO activities and traffic is sufficient to justify these actions. Presumably, the ninety-plus countries whose only postings in the BCH include only draft documents and general legislation on environment, phytosanitary and veterinary issues, penalties, and marketing have chosen this approach.

WHICH CARTAGENA PROTOCOL TASKS ARE MANDATORY?

Normally, it is relatively simple to determine whether a clause is mandatory by closely examining the text of that clause. Mandatory positive provisions are normally expressed using mandatory terms, such as "The Parties shall . . . ," "The Parties must . . . ," or "the Parties agree to. . . . " Mandatory limitations normally use similar language, adding the word *not* (e.g., "The Parties shall not . . . "). Nonmandatory provisions are sometimes identified by hortatory terminology, including "The Parties should . . . ," or "The Parties may . . . " and at other times can be identified by the fact that the provision includes one or more options that the Parties can choose for achieving a particular result.

After identifying the tasks in the Protocol (Tables I to IV), the next step in national implementation would be to allocate the tasks among the following four categories: *mandates and duties* (provisions that specifically require legislative or administrative development by the country); *limitations* on action (provisions that impose limitations on national legislative options); *options and recommendations* (provisions that suggest or recommend a variety of nonmandatory actions); or *permissions* (provisions that make it clear that the Parties may take particular types of action). Armed with these determinations, the person or entity drafting a country's implementing legislation need focus only on (1) complying with the mandates and duties, and (2) considering the relevance and value for his/her or their country of those tasks listed in each of the other three categories.

Within the Cartagena Protocol, however, this second step is not a simple matter of parsing the specific text. For most of the tasks identified in Tables I to IV, other provisions within the Protocol operate to complicate the "mandatory or not?"

[11] As noted in GEF, *Evaluation of GEF's Support to the Cartagena Protocol on Biosafety* (2006), online: http://www.thegef.org/gef/node/2140 at para 4.3, the most relevant and useful way to consider Cartagena Protocol compliance relates to each country's level of direct involvement in LMO technology (development and testing), evaluated at the national level. Many of the developing countries reviewed as "focus countries" for this chapter were found to have a high level of involvement in LMO technological activities.

determination.[12] The following discussion examines the varying levels of mandate within the Protocol.

Overarching Mandates Underlying the Cartagena Protocol

The strongest level of mandate in the Cartagena Protocol is reserved for three types of provisions: those that concern the Protocol's own administration (mostly listed found in Table I of Annex II); clauses expressing overarching general commitments; and those provisions that impose direct obligations from one State to the others. Although the Parties' obligations with regard to international administration of the Protocol are not unimportant, they are relatively straightforward and well understood. The following discussion turns to the Cartagena Protocol's overarching mandates, which are expressed in two forms – as positive mandates and as "antinegative" mandates.

Positive Mandates: Biosafety

Two of the Cartagena Protocol's substantive provisions are phrased as unequivocal obligations of all Protocol Parties. As such, they appear to be unaffected by the flexibility mechanisms described here. These provisions specifically require each Party to "take necessary and appropriate legal, administrative and other measures to implement its obligations under this Protocol,"[13] and, collectively, to "ensure that the development, handling, transport, use, transfer and release of any living modified organisms are undertaken in a manner that prevents or reduces the risks to biological diversity, taking also into account risks to human health."[14]

Expressed so broadly, these clauses have been cited as the justification for the position that all of the tasks listed in Tables I to IV are essentially mandatory and must be adopted in fulfilment of the Protocol's stated objective and binding on the Parties[15] to "contribute to ensuring an adequate level of protection in the field of the safe transfer, handling and use of living modified organisms resulting from modern biotechnology that may have adverse effects on the conservation and sustainable use of biological diversity, taking also into account risks to human health."[16]

It is notable, however, that both of these substantive provisions are extremely general, which makes it difficult to use either as a basis for evaluating or challenging national Cartagena Protocol implementation. The Protocol does not set any

[12] Tables I through IV do not distinguish between provisions that are apparently mandatory and those that are not. In all likelihood, the omission of this distinction could have been intentional, to avoid raising complex and controversial questions that would delay adoption.

[13] *Cartagena Protocol*, *supra* note 1 at art 2(1).

[14] *Ibid* at art 2(2).

[15] Article 26 of the *Vienna Convention on the Law of Treaties*, 23 May 1969, 1155 UNTS 331 (entered into force on 27 January 1980) establishes that every treaty in force is binding upon the parties to it and must be performed by them in good faith.

[16] *Cartagena Protocol*, *supra* note 1 at art 1; as discussed in Mackenzie et al, *supra* note 4 at paras 159–74.

standards for determining which actions are "necessary and appropriate," which level of risk is posed by LMOs, or which measures will be necessary to control or reduce those risks. Lacking internationally agreed-on standards on these points, each Party's obligation is necessarily determined under its own internal standards, policies, and other judgments. Strictly speaking, a State must normally be considered to be in compliance, as long as a Party can state that its national agencies have acted in accordance with domestic mandates and positions regarding risk and safety. In other words, a State that has not formally adopted legislation specifically addressing LMO risk assessment and/or risk management would not necessarily be found noncompliant with the Protocol.

Presumably, a good gauge of each Party's view regarding LMO risk would be the extent to which it has adopted legislative instruments mandated or recommended by the Cartagena Protocol, and/or the extent to which they have applied those instruments since the Protocol went into effect for that country. Only thirty Parties and two nonparties have reported any decisions relating to implementation in the BCH. Of these, twenty-four Parties have reported operative decisions (direct approval or rejection of any application to for introduction, assessment, and transit of LMOs). Nearly all countries reporting any decisions have also posted at least one legislative record.[17]

Based on initial review of the volume and content of decisions, legislative documents, and drafts submitted to the BCH, it appears that some countries presume that LMOs present a high level of risk that must be assessed, managed, and controlled, whereas others appear to have imposed relatively few biosafety- and risk-related measures and consequently have made little or no change to their pre-Protocol legislative framework and procedures. This suggests that strong differences could remain among countries regarding biosafety risk. Thus, the main value of the convention as it currently stands lies in the Parties' commitments to recognise each other's sovereign decisions on the risks of LMOs and their management. As to the latter, the Protocol's mandate appears to be unequivocal.

[17] Countries' LMO decisions (i.e., those directly relevant to LMO introduction, assessment, and/or transit) are posted under six headings. The raw figures may be deceptive, since at the time of writing, a total of 39 countries posted all of the "decisions and communications" included in the BCH (1209 in total): Japan has posted 175; Republic of Korea, 137; United States of America, 119; European Union, 96; Philippines; 96; Mexico, 69; New Zealand, 68; Brazil, 40; Australia, 29; Germany, 26; Nicaragua, 15; Netherlands, 13; South Africa, 13; Sweden, 12; Argentina, 11; Switzerland, 8; Austria, 7; Finland, 5; Romania, 3; Belgium; 2; El Salvador, 2; Bulgaria, 2; Indonesia, 2; Norway, 2; Barbados, 1; Cyprus, 1; Czech Republic, 1; Denmark, 1; Dominican Republic, 1; Libyan Arab Jamahiriya, 1; Niger, 1; Saint Lucia, 1; Spain, 1; and Syrian Arab Republic, 1. Eight of these countries (Barbados, Cyprus, Dominican Republic, Libyan Arab Jamahiriya, Niger, Saint Lucia, Spain, Syrian Arab Republic) have posted only communications regarding organisational matters – that is, decisions about future decision making (i.e., the decision to use Cartagena Protocol options under Article 11, and/or notifying their ratification of the Protocol). None have formally posted any notifications or decisions indicating that they have approved the import or introduction, of any LMO, or LMOs-FFP for any type of use, introduction or transit.

"Antinegative" Mandates: Precaution under the Cartagena Protocol

One of the most often-discussed aspects at the time of the Cartagena Protocol negotiations was the Protocol's incorporation of "precaution" (dubbed "the precautionary approach") into the operative text of the instrument. Precaution, as discussed in the Introduction and Chapter 1, is an important principle of many international treaties on sustainable development, including the Cartagena Protocol. Included in the Protocol in four places,[18] the expression of this principle is not so much a positive mandate as an "antinegative" one. Specifically, it is generally expressed as follows:

> Lack of scientific certainty due to insufficient relevant scientific information and knowledge regarding the extent of the potential adverse effects of a living modified organism on the conservation and sustainable use of biological diversity in the Party of import, taking also into account risks to human health, shall not prevent that Party from taking a decision, as appropriate, with regard to the import of the living modified organism in question as referred to in paragraph 3 above, in order to avoid or minimize such potential adverse effects.[19]

In effect, the clause merely states that uncertainty about the risk "shall not prevent" the biosafety agency from "taking a decision . . . in order to avoid or minimise" such risks. Much discussion in the wake of the Protocol's adoption has focused on the extent to which this provision mandates that all Parties adopt some type of preemptive regulatory or mitigation system for LMOs introduced from foreign countries.

Regulatory Flexibility Provisions in the Cartagena Protocol

The greatest divergence of views concerning Cartagena Protocol legislative implementation relates to the interpretation of the text to determine which provisions are mandatory and which are not. The Cartagena Protocol contains provisions that some commentators might argue give Parties a high degree of flexibility for purposes of implementation. Some of these "flexibility provisions" enable and allow Parties to impose stricter and more inclusive requirements than those stated in the Protocol, whereas others allow them to opt for more lenient treatment of biosafety matters.

18 *Cartagena Protocol, supra* note 1 at arts 1, 10.6, 11.8, and Annex II, art 8(f).

19 *Ibid* at art 10(6). The other references are comparable, Article 11(8), is identical to the quoted text up to the words "with regard to," which are followed in Art 11(8) with the phrase "the import of that living modified organism intended for direct use as food or feed, or for processing, in order to avoid or minimize such potential adverse effects." Article 1 merely incorporates the precautionary approach by reference. In Annex III, more streamlined language is expressed more permissively: "Where there is uncertainty regarding the level of risk, it may be addressed by requesting further information on the specific issues of concern or by implementing appropriate risk management strategies and/or monitoring the living modified organism in the receiving environment."

Stricter Legislation and Broader Coverage

Some commentators have cited several key provisions as broadly enabling countries to legislate in areas that are specifically excluded from the coverage of the Protocol and to impose stricter controls than are stated in the Protocol. Other clauses appear to allow countries to adopt a more lenient approach in their domestic regulatory frameworks.

The primary provision that enables Parties to impose broader coverage and stricter controls is Article 2.4, which specifically provides that

> [n]othing in this Protocol shall be interpreted as restricting the right of a Party to take action that is more protective of the conservation and sustainable use of biological diversity than that called for in this Protocol, provided that such action is consistent with the objective and the provisions of this Protocol and is in accordance with that Party's other obligations under international law.[20]

This broad authority has been interpreted both to legislate in areas that are specifically excluded from the coverage of the Protocol, and to impose stricter controls than are stated in the Protocol.[21] This provision distinguishes the Cartagena Protocol from most international instruments that address risks arising from commercial and non-commercial movement of materials and products, because most such instruments recently have focused on limits of such controls.[22]

As discussed later, BCH review indicates that the limiting language of Article 2.4, which purports to restrict its scope to "conservation and sustainable use of biological diversity," might not have limited countries in their legislative development. Looking only at new legislation and drafts directly addressing biosafety, virtually all such documents contain at least some measures that are significantly broader in coverage. Many of these new legislative instruments and drafts have taken advantage of Article 2.4, to impose the full AIA process to all introductions of any LMOs (including LMOs-FFP).

In taking these actions, legislators and draftsmen have supported this extended coverage with reference to Article 2.2's general statement (quoted previously) that

20 *Ibid* at art 2(4).

21 Mackenzie et al, *supra* note 4 at paras 184–5. The authors of that treatise note that the phrases "consistent with the objective and the provisions of this Protocol" and "in accordance with that Party's other obligations under international law" were intended to limit the scope of each country's authority under the Protocol.

22 See *International Plant Protection Convention (New Revised Text)* (1997 IPPC) as discussed in T Young et al, *Independent Evaluation of FAO Corporate Strategy B-1 (International instruments concerning food, agriculture, fisheries and forestry, and the production, safe use and fair exchange of agricultural, fishery and forestry goods)* (Rome: FAO-PBEE, 2008). The 1997 IPPC's provisions are generally interpreted as a tool for integrating national food/agriculture importation rules with the relevant WTO principles. As such, it is discussed as, in effect, an upper limit (or, more properly, a "safe harbor") on the plant protection issues that can be addressed in national legislation affecting trade.

all Cartagena Protocol Parties are obligated to address and oversee "the development, handling, transport, use, transfer and release of LMOs to ensure that they are undertaken in a manner that prevents or reduces the risks to biological diversity, taking also into account risks to human health." They note that this general mandate can be met only by countries whose law extends well beyond the basic framework of Articles 7 to 18. As such, they appear to consider the Cartagena Protocol's specific tasks to be only a small part of their overall national biosafety framework. In this, they take advantage of Article 12.4, in which the Protocol specifically notes each Parties option to extend the coverage of their biosafety legislation to all introductions (including later introductions, following the first international introduction).

Another set of flexibility mechanisms in the Protocol relates to scope and coverage of national biosafety legislation. Many Cartagena Protocol provisions attempt to focus the Protocol's scope; however, these narrowing provisions are counterbalanced by other statements that appear to recognise broader authority, including several beyond Articles 2.4 and 2.2, described earlier, which specifically allow Parties to consider a broader range of activities than those specified in Articles 7 to 18, including the following specific extensions of scope: subsequent introductions of an LMO (after its first international introduction), controls on the handling, use, and release of LMOs on the domestic level (regardless of origin), and implementation of specific regulatory frameworks for LMOs that are excluded from the protocol (e.g., pharmaceuticals).

These extensions are clearly permissible, under specific flexibility clauses in the Protocol. For example, Article 5 specifically excludes pharmaceuticals from the Protocol's coverage but is expressly adopted "without prejudice to any right of a Party to subject all living modified organisms to risk assessment prior to the making of decisions on import."

Similar examples of the right hand taking and the left hand giving back are found in Article 6.1's exclusion of LMOs in transit, which was adopted "without prejudice to any right of a Party to subject all living modified organisms to risk assessment prior to the making of decisions on import." A third such provision is Article 6.2, which excludes LMOs for contained use from coverage under the AIA and Article 11 processes of the Protocol. Again, this clause was adopted without prejudice to the Parties' right to subject those LMOs to risk assessments and import standards, which many have done, as further discussed later in this chapter.

The expansiveness of the Cartagena Protocol flexibility provisions enabling stricter measures is an important characteristic of the Protocol and reflects the concerns that drew the Parties to the negotiating table, despite significant divergence in their objectives. Some countries' participation was driven by the GMO-related fears and concerns of the general public. Other countries, not experiencing these pressures, expressed concern that GMO development activities, designed to improve conditions, could face trade restrictions and other obstacles that might have repercussions throughout the realm of international trade.[23] Flexibility mechanisms were a practical result of the fact that these pressures were neither resolved by nor integrated into the Protocol. It was strongly in the interest of both sides for the Protocol to find a way to assuage these fears, to create a mechanism that citizens could count on to protect them against unknown risks, and to ensure that each of these mechanisms would be accepted and adopted throughout the LMO subsector. The widespread use of these flexibility options in documents posted by the BCH could be indicative of the effectiveness of this approach.

More Lenient Approaches

The Protocol also includes several provisions that allow countries to impose fewer restrictions or act more leniently toward LMOs within their borders. These provisions also arose from the fact that the Protocol did not adopt any position on risks posed by LMOs. Thus, each Party's commitment regarding its domestic activities, by definition, turns on its own analysis of the level of risk potentially involved in LMOs and its determination of which level of risk is "acceptable," even when the level selected is very low and necessitates lenient treatment of LMOs.[24] To date, there has been less analysis of these types of flexibility. A review of the documents posted in the BCH, however, suggests that many of the countries that have posted formally adopted laws, regulations, and practices have chosen to take advantage of these mechanisms.

The overarching mandates for Cartagena Protocol implementation is stated in terms of each party's determination of several matters. The Protocol provides no agreed standard for risk nor for determining which measures are "necessary and appropriate." Thus, each country must begin its implementation by determining these two standards, building its own implementation based on that determination. In practice, this provision can enable either stricter or more lenient measures, because each country has the full spectrum of options available, from determining that LMOs are extremely risky and imposing the strictest possible limitation ("no LMOs") on one hand, to identifying formally virtually the entire class of LMOs as acceptable for introduction and imposing no domestic risk management measures (discussed later).

In addition to these provisions, the Protocol includes other exceptions and simplifications to allow more lenient implementation in certain situations. These include a power of each Party or the COP collectively to notify the Parties that it will not apply the strict rules of the Protocol to certain cases. Other potential options,

[23] Mackenzie et al, *supra* note 4, has an excellent summary of global trade issues most relevant to LMOs at paras 67–70 and 844–914 that is highly recommended for those who have not worked closely in that field.

[24] R Cooney, ed, *Biodiversity and the Precautionary Principle: Risk and Uncertainty in Conservation and Sustainable Use* (London: Earthscan, 2005).

simplifications, and exceptions arise from the Protocol's provisions regarding each country's "domestic regulatory framework" and special procedures for streamlined acceptance of LMOs-FFP in developing countries. These measures provide excellent examples of the flexibility that the Protocol negotiators agreed to, and of the reasons that flexibility was thought to be necessary.

The Protocol includes other directly usable mechanisms for lenient decision making. The most direct of these is found in Article 13, under which any Party-of-import may "specify in advance to the Biosafety Clearing-House: (a) Cases in which intentional transboundary movement to it may take place at the same time as the movement is notified to the Party of import; and (b) Imports of living modified organisms to it to be exempted from the advance informed agreement procedure."[25] To date, two countries (i.e., Colombia and the Republic of South Africa) have posted a total of twelve decisions using this authority.[26] The Protocol goes beyond that, specifically allowing that Party to stipulate that such notifications "may apply to subsequent similar movements to the same Party."[27] This phrase may operate to allow any party to decide not to regulate or control LMOs in any way. As such, it can conflict with the Parties' general duties to inform other Parties when LMOs might have intentionally or inadvertently crossed from their territory into that of another country – an event that a country might be unaware of if that country imposes no controls on LMOs.

Another flexibility measure is found in the Parties' collective power as the COP-MOP to predefine one of more categories of permissible introductions. Essentially as an optional exemption to the entire AIA process (apart from the Article 7 notification), the Protocol gives all Parties acting together the option to adopt a pre-agreed list of LMOs that are automatically approved. In essence, this provision allows any party to define any size class of LMOs, which can be considered to be permitted either with a relatively limited AIA process[28] or none at all.[29]

There are strict prerequisites to the use of this option – that is, that "adequate measures are applied to ensure the safe intentional transboundary movement of living modified organisms in accordance with the objective of this Protocol." Here also, however, the fact that the Protocol does not contain any agreed-on language providing a standard of "adequacy" suggests that each country's own determination normally governs. As such, it seems likely that a country that is capable of identifying

[25] *Cartagena Protocol*, *supra* note 1 at art 13(1).

[26] Colombia: BCH Record IDs 8151 and 8155; South Africa: BCH Record IDs 5666, 5679, 5712, 5715; 8164, 8167, 8170, 40513, 40514, and 40516.

[27] *Cartagena Protocol*, *supra* note 1 at art 13(1).

[28] Article 13(1)(a) allows a Party to identify "Cases in which intentional transboundary movement to it may take place at the same time as the movement is notified to the Party of import." Subsequent provisions in Article 13 note that this determination, once made, "may apply to subsequent similar movements to the same Party."

[29] Article 13(1)(b) allows a Party to identify imports of living modified organisms to it to be exempted from the advance informed agreement procedure.

and controlling movement of and overseeing LMOs enforcing compliance with relevant laws can be considered as applying adequate measures for their safety and control. In essence, this proviso operates as a commitment by parties using this simplification process that, although they do not intend to control LMOs (or some group of them) within their own borders, they will prevent activities in their country from having an impact on other countries or areas beyond national jurisdiction.

Domestic Regulatory Framework

Another important flexibility mechanism that has operated as a justification for the adoption of more lenient measures is the Protocol's reference to the Parties' "domestic regulatory frameworks" (or similar wording).[30] The Protocol's provisions regarding the domestic regulatory framework are relatively sketchy and not expressed as an implementation option. On their face, however, these provisions appear to maximise each country's options. The most basic statements on this point are found in Article 14.4, which states, "Any Party may determine that its domestic regulations shall apply with respect to specific imports to it and shall notify the Biosafety Clearing-House of its decision."

Article 9 (notifications) provides a further gloss on this power, in two statements. First, it instructs that the domestic regulatory framework "shall be consistent with this Protocol."[31] Second, it provides that a Party's acknowledgement of a notification under Article 8 shall inform the notifier, *inter alia*, "[w]hether to proceed according to the domestic regulatory framework of the Party of import or according to the procedure specified in Article 10." This clause appears to give countries the option of choosing either the Protocol's AIA process or the country's domestic regulatory framework, and even of making this decision on a case-by-case basis.[32]

The Protocol makes numerous other references to the *domestic regulatory framework*. For example, it specifically states that the domestic regulatory framework may indeed be different from the framework specified in the Protocol, including that it may require additional information prior to any decision.[33] Regarding the Article 11 procedure, the Cartagena Protocol makes specific provision for LMOs-FFP introductions when a country has no domestic regulatory framework,[34] provisions that

30 Given that the phrases "domestic regulatory framework," "national regulatory measures" and "domestic regulations" are used without definition, many commentators have generally assumed them to have the same general meaning. See generally UNEP-GEF Toolkit, *supra* note 4; and Mackenzie et al, *supra* note 4 at paras 13(vii), 86, 94, 298–304, and Boxes 8 and 24. The authors of Mackenzie et al, noticed that the UNEP-GEF project coined the term "national biosafety framework" or "NBF" rather than choose one of the terms used here. Presumably this terminology also avoided the need to identify any distinction between countries relying on their domestic regulatory frameworks, and those developing new frameworks whose scope and coverage differs from that set forth in the Protocol.

31 *Cartagena Protocol*, *supra* note 1 at art 9(3).

32 *Ibid* at art 9(2)(c).

33 *Ibid* at art 10(2)(c).

34 *Ibid* at art 11(6).

are apparently intended to enable importation of FFPs, without the delay that would be caused if the country first needed to adopt legislation for implementation. The Cartagena Protocol specifically notes and authorizes countries, in their "national domestic measures," to address social and economic considerations, in addition to the issues addressed in the Cartagena Protocol.[35]

A further tool in the interpretation of the concept of "domestic regulatory frameworks" can be found in Article 14, which specifically authorizes parties to enter into bilateral, regional, or multilateral agreements or arrangements regarding the intentional transboundary movement of LMOs, so long as such instruments are "consistent with the objective of the Protocol" and do not result in a lower level of protection than that provided in the Protocol.[36] (Given the Protocol's ambiguity about levels of risk and flexibility opportunities, the limiting language might be relatively difficult to apply with any legal or judicial certainty).

The "domestic regulatory framework" concept as discussed in the Protocol has various interpretations. Some commentators presume that it creates an exception to the provisions calling on Parties to adopt and apply the AIA process. Others have taken the view that terms such as "domestic regulatory framework," "national regulatory measures," and "domestic regulations" are simply generic references to each Party's national biosafety legislation, including a country's national version of the Article 7–18 procedures after the adoption of the Protocol.[37] Although widespread, this latter view appears to conflict with the text of the Protocol, which appears to recognize the Article 7–18 framework and the "domestic regulatory framework" as options, calling on the Parties to select one.

All international agreements (unless they specifically require the parties to adopt specific and harmonised legislation) are implemented by national regulatory measures, whether newly adopted or in existence (possibly adjusted to meet the Parties' commitments under the international instrument). Under the Cartagena Protocol, the entire regulatory system envisioned in the Protocol expresses this practice by identifying three options: (1) use of the country's existing "domestic regulatory framework," (possibly amended as needed to be sure that it is Protocol consistent); (2) adoption of new legislation directly implementing the Cartagena Protocol (and possibly addressing other biosafety issues as well); or (3) application of the Cartagena Protocol's LMOs-FFP framework directly (without legislation). Any of these

35 *Ibid* at art 26(1).

36 *Ibid* at art 14. Parties must inform the BCH of any such arrangements. The specific AIA provisions of the Protocol will not apply to intentional transboundary movements of LMOs between Parties to those agreements or arrangements.

37 Mackenzie et al, *supra* note 4 at paras 298–304; see also UNEP-GEF Toolkit, *supra* note 4, especially Phase III, Part I "Formulation of the Regulatory Regime," at pages 56–7 and 66–75. Although the Toolkit does not use the term *domestic regulatory framework*, it clarifies the view of the concept put forward in Mackenzie et al.

three can, if properly applied, constitute full compliance with the country's Cartagena Protocol obligations. Apart from the need to make the proper disclosure under Article 9, there appears to be no need to distinguish between the domestic regulatory framework and the nationally adopted Article 7–18 framework.

Based on interviews and contemporaneous literature, the concept of "domestic regulatory framework," as variously expressed, was thought at the time to refer to preexisting national biosafety legislation, amended as necessary for Protocol consistency. As such, it is sometimes assumed to refer primarily to developed country legislation. Among the developing countries, the legislation of which is the focus of this chapter, however, several appear to have had preexisting domestic regulatory frameworks – particularly those having the longest histories of LMO involvement. In particular, South Africa, Argentina, Republic of Korea, the Philippines, and Bangladesh began their involvement in LMO development, introduction, and use long before the Protocol negotiations.[38] Developing countries with the highest level of involvement in LMO activities have more complete and comprehensive regulations, having adopted regulations addressing many different types of introductions, as well as individualized regulations for permitting and regulating each.[39]

Most commentators discussing national implementation have assumed that adoption of the Article 7–18 framework in national legislation is the most prevalent (and perhaps the easiest) approach.[40] Without the ability to survey all Parties, it is impossible to determine whether in fact this is the case. Although many countries have submitted legislation in force to the BCH, and many draft instruments in the database specifically follow the basic Article 7–13 framework, few appear to have formally implemented the administrative processes formally.

38 Republic of South Africa, *Genetically Modified Organisms Act (No. 15 of 1997)*, BCH Record ID 255; Republic of South Africa, *Genetically Modified Organisms Act, 1997, (Act No. 15 of 1997) Regulations*, BCH Record ID 269; Argentina, *draft National Biosafety Framework, 2004*, BCH Record ID 43463 (and other postings on the components of the country's domestic regulatory framework); Bangladesh *Draft National Biosafety Framework, 2006*, BCH Record ID 43436 (detailing the country's history with regard to LMOs, including national policy adopted in 1992 and revised thereafter) and Bangladesh *Biosafety Guidelines of Bangladesh, 2005*, BCH Record ID 43880.

39 See Argentina, *Tratamiento diferenciado para el otorgamiento de la autorización de comercialización para aquellas acumulaciones provenientes del cruzamiento de parentales que contienen eventos que en forma individual ya tienen autorización para la comercialización en la República Argentina. Resolución 60/2007*, BCH Record ID 40386; Argentina, *Liberación al Medio de Organismos Vegetales Genéticamente Modificados (OVGM) / Resolución n° 39*, BCH Record ID 5649; Argentina, *Solicitud para la concesión de permisos para Experimentación y/o Liberación al Medio de Microorganismos Genéticamente Modificados y/o sus productos para aplicaciones en animales. / Resolución N° 656*, BCH Record ID 11812; Argentina, *Experimentación y/o Liberación al Medio de Organismos Animales Genéticamente Modificados. / Resolución N° 57*, BCH Record ID 11911. Compare with Bahamas, *Draft National Biosafety Framework (Developed under the UNEP-GEF Biosafety Project)*, BCH Record ID 43464 at para 4.2.1 (pp 20–7).

40 See Johnston et al, *supra* note 4; and UNEP/GEF Toolkit, *supra* note 4.

Article 11 Alternative Procedure

In considering this clause, we should recall that many countries in the negotiations knew that they possessed little or no domestic technical capacity relative to LMOs and LMO risks. Thus, one of the most important objectives served by the adoption of optional simplifications and exceptions is to assist countries that might have difficulty in complying with all aspects of the Protocol. Despite its perceived urgency, the Protocol might be thought of as a "technology forcing" instrument, that is, one that expects and requires countries to adopt technology and institutions, and to undertake compliance and enforcement measures that are beyond their current capacity. In essence, the Protocol is built on assumptions regarding the capacity of each country to identify LMOs; to control their import and export; to analyse their risks; and to impose, oversee, and enforce compliance with specific obligations on those who import or introduce them. The Protocol does not specifically release countries from their respective duties (pending acquisition of capacity) but does recognize that countries could be called on to consider international introductions long before they have developed sufficient capacity to enable sound legislative action for LMO introduction. A sizable percentage of Protocol Parties, even those with a high level of involvement in LMO research/development and use, admit to being unable to comply fully with the Protocol's provisions.

For developing countries and countries with economies in transition, the Protocol offers options allowing them to take action before they have adopted a complete regulatory framework to address LMOs within their borders and those from outside. First, such a country may "declare, through the Biosafety Clearing-House, that its decision prior to the first import of a living modified organism intended for direct use as food or feed, or for processing," will be taken through a simplified process, referred to in Cartagena Protocol circles as the Article 11 (LMOs-FFP) process. This provision is specifically directed at developing country Parties (or Parties with economies in transition), which have not adopted or enacted domestic regulatory framework to date. Such countries may condense their Article 11 process to the following: "A risk assessment undertaken in accordance with Annex III; and . . . a decision made within a predictable timeframe, not exceeding two hundred and seventy days."[41] Second, these simplifications may be taken without such declaration (by default). Regarding this type of approval, the Protocol specifically states that "failure by a Party to communicate its decision according to paragraph 6 above, shall not imply its consent or refusal to the import of a living modified organism intended for direct use as food or feed, or for processing, unless otherwise specified by the Party."[42]

This stands in dramatic contrast to other Protocol provisions, which take pains to state that "[a] failure by the Party of import to communicate its decision within

[41] *Cartagena Protocol, supra* note 1 at art 11(6).

[42] *Ibid* at art 11(7).

two hundred and seventy days of the date of receipt of the notification shall not imply its consent to an intentional transboundary movement."[43] According to the BCH, it appears that two countries have reported a total of twelve decisions using this simplification process,[44] whereas the BCH identifies two others that have posted notifications under the heading "Declaration that, in the absence of a domestic regulatory framework, decisions on LMOs-FFPs will be taken according to Article 11.6."[45] Neither of these latter countries has since reported any decisions regarding introduction or other activities involving either LMOs or LMOs-FFP.[46]

NONMANDATORY PROVISIONS

Finally, it is necessary to consider collectively the provisions within the Protocol that are optional or nonmandatory. The Protocol's most prominent nonmandatory provisions generally fall into four categories: *simplifications or exceptions*: provisions that identify some lesser performance that can substitute for a mandatory provision, as long as the Party meets some specific prerequisites; *options:* provisions allowing the parties to select, among more than one clause, their preferred means of satisfying a particular requirement; *interpretative provisions:* clauses included primarily to clarify particular points, especially to formally recognise that a particular (optional or voluntary) action is not precluded by the basic framework created by the Protocol (e.g., "The Parties may . . . "); and *hortatory provisions:* clauses that recommend or encourage particular actions or performance, but do not require it (e.g., "The Parties should . . . ").

The "flexibility mechanisms" described previously represent the first two categories (i.e., simplifications and options). In addition to these, the Protocol includes interpretative and hortatory provisions to indicate or clarify what is permitted or prohibited under the Protocol in particular situations. Several of these provisions constitute very strong statements recognizing or reiterating the rights of Parties.[47] In some cases, they provide information that a general right exists or that a common activity or option is permitted. Thus, for example, Article 12.4, although generally a "scope" provision, actually focuses on underscoring or clarifying the meaning of Article 2.4, by providing that "The Party of import may, at its discretion, require a

43 *Ibid* at arts 9(2)(4) and 10(5).

44 Columbia (2 decisions) and South Africa (10 decisions).

45 Barbados, BCH Record ID 37647; and Saint Lucia, BCH Record ID 48418; and see two postings of pre-Protocol decisions: Dominican Republic, "Decisión para la introducción de pez cebra genéticamente modificado," BCH Record No 43828; and Indonesia, BCH Record ID 9612.

46 Of the countries named in the previous note, only Indonesia has reported any other (post-Protocol) decision. See Indonesia, "Risk Assessment of Bt-Cotton (pre-Protocol decision; limited environmental release 2001, 2002 and 2003)," BCH Record ID 44147.

47 One of the strongest of these is the provisions is Article 12(1), which allows the importing country to review and possibly change a final decision on LMO imports on the basis of new scientific information that it receives.

risk assessment for subsequent imports."[48] As discussed in Part IV, the BCH includes many biosafety laws and drafts that have taken advantage of this option.

Most interpretative provisions in the Protocol relate to activities that may be taken to streamline the administrative process or otherwise to clarify particular steps within it. For example, the Parties are assumed to be permitted to fold their LMO-FFA process into their AIA process, applying the more detailed AIA requirements to all.[49] This approach is based on the apparent assumption that the AIA process is stricter than Article 11.[50] Other interpretative and/or hortatory provisions encourage parties, "inter alia, to take into account, as appropriate, available expertise, instruments and work undertaken in international forums with competence in the area of risks to human health;"[51] request "additional information from the authority identified in paragraph (b) of Annex II;"[52] and "indicate its needs for financial and technical assistance and capacity-building with respect to living modified organisms intended for direct use as food or feed, or for processing."[53] The intent of these provisions is clarification, because it is doubtful that any country would fail to take these actions, whether specifically authorized to do so or not.

Another important clarifying provision is Article 14.1, which speaks of the unalienable right of all sovereign nations, that is, to "enter into bilateral, regional and multilateral agreements and arrangements regarding intentional transboundary movements of living modified organisms." Some of the Protocol's hortatory provisions represent issues on which there was a lack of consensus in the negotiations, despite a general desire for the Protocol to discuss the issue. Many of these recommendations are not trackable, because there is currently no definitive information on what the parties "take into account" in their decision making.

One final nonmandatory provision should be noted. Article 17.3 recommends the content that should be included in intergovernmental notifications regarding "unintentional transboundary movement of a living modified organism that is likely to have significant adverse effects on the conservation and sustainable use of biological diversity." This recommendation is an excellent example of the combination of mandatory and nonmandatory items in a single clause. The notification itself is mandatory, but

[48] This provision essentially authorizes countries to extend the AIA process to subsequent introductions of a species that has been subjected to the AIA process on its first international introduction.

[49] See the UNEP-GEF Toolkit, *supra* note 4, Phase III. This assumption generally relies on the statement in Article 11(4) that "A Party may take a decision on the import of [LMOs-FFP] under its domestic regulatory framework that is consistent with the objective of this Protocol," coupled with permission in Article 2(4) to impose stricter or more extensive measures, appears to enable and encourage this choice.

[50] This assumption tacitly appears in Table IV. Normally, advocates of broad acceptance of the "strict rule of law" assume that the presence of a written rule that is binding on the relevant governmental decision maker prevents him or her from imposing unduly harsh restrictions (as well as from acting in a way that is unduly favourable to his or her friends).

[51] *Cartagena Protocol*, *supra* note 1 art 2(5).

[52] *Ibid* at art 11(3).

[53] *Ibid* at art 11(9).

this particular discussion of the contents of that notification is clearly hortatory (*should* instead of *shall*). This phrasing reflects the fact that the content of intergovernmental communications is normally a matter of highest-level diplomacy – that is, a very important part of national sovereignty.[54]

54 The search mechanism of the BCH has variously either reported no such communications or pulled up approximately ninety communications and decisions posted under other headings (most posted as "other decisions on the transport of LMOs" or "other decisions on contained use of LMOs"); a sampling through the documents provided indicated that none of them reported either intentional (illegal) or unintentional transboundary movement of LMOs.

14

National Biosafety Regulatory Systems in Central and Eastern Europe

David Duthie and Liina Eek

Analysis of the national biosafety regulatory systems of countries in central and eastern Europe (CEE) provides a useful insight into the challenges that exist when implementing such systems in "transitional" countries, in light of the sustainable development objective and principles of the Cartagena Protocol on Biosafety.[1] The diverse range of external circumstances affecting implementation include the breakup of the Soviet Union, the fragmentation of the former Yugoslavia, the strong gravitational pull of the European Union (EU), and a complex and changing mosaic of intensive and extensive agricultural landscapes and policies. An additional layer of complexity is added by the adoption of the Cartagena Protocol in January 2000, its entry into force in September 2003, and opportunities for funding by the Global Environment Facility (GEF) for biosafety capacity building.

The precise composition of the CEE varies in different fora. For example, the CEE as defined by the UNEP-GEF Biosafety Project excluded several countries of central Asia, but included Malta and Turkey (as aspiring EU countries) for a total of nineteen countries. Three CEE countries (Bulgaria, Hungary, and Poland) developed their biosafety strategies as a part of the UNEP Pilot Biosafety Project from 1997–9. This chapter reviews the experience of countries from a broader CEE region, which includes an analysis of twenty-four countries (See Annex V) over the period from 2000–8.

Information has been gathered from National Biosafety Frameworks (NBF), National Biodiversity Strategy and Action Plans (NBSAPs),[2] and the global Biosafety Clearing-House (BCH).[3] As of April 2008, all twenty-four CEE countries have

[1] *Cartagena Protocol on Biosafety to the Convention on Biological Diversity*, 29 January 2000, 2226 UNTS 208, 39 ILM 1027 (entered into force 11 September 2003), online: http://bch.cbd.int/protocol/text/ [*Cartagena Protocol*].

[2] NBSAPs for various countries can be found on the CBD Secretariat website, online: http://www.cbd.int/nbsap/search.

[3] The Biosafety Clearing-House (BCH) is an information-exchange mechanism that provides open access to key information about LMOs, online: http://bch.cbd.int.

ratified the Cartagena Protocol. Of the nineteen CEE countries that participated in the UNEP NBFs Development Project, seventeen had finalized their NBFs by April 2008. Only Azerbaijan and Ukraine are still in the NBF review process (see Annex V, Table 4).

The UNEP-GEF Biosafety Project began in June 2001. CEE countries' biosafety policies were at varying stages when this project began, creating a diverse range of challenges for this project. Some CEE countries that were scheduled to become EU members in 2004 had already prioritized harmonisation with existing and pending EU regulatory systems.[4] Until 2007, Bulgaria and Romania were aspirants to EU membership but did not meet the Copenhagen criteria for entry. A third group of countries are scheduled as "candidate" countries and are receiving preaccession assistance.[5] A fourth group, "potential candidate" countries, are also eligible for preaccession assistance but have a longer negotiation procedure for membership.[6] The remaining CEE countries have not yet initiated EU membership negotiations and are not expected to achieve EU membership in the near future.[7]

Countries approached the drafting of NBFs differently, depending on the existence of a national system or a complete lack thereof. Acknowledging country-specific circumstances, needs, and priorities, each NBF consists of five common elements:

1. A policy on biosafety, either a stand-alone policy or one that is a part of other relevant national policies;
2. An operational regulatory regime for biosafety consistent with both the CPB and national laws;

4 These countries were the Czech Republic, Estonia, Hungary, Latvia, Lithuania, Poland, Slovakia, and Slovenia.

5 These "candidate" countries were Croatia, Former Yugoslav Republic of Macedonia, and Turkey. Croatia and Turkey started membership talks in October 2005; Croatia is now undergoing the accession process and Turkey remains a candidate for accession. The Former Yugoslav Republic of Macedonia officially gained candidate status in December 2005 but has so far failed to receive a start date for accession talks.

6 These "potential candidates" were Albania, Bosnia and Herzegovina, Kosovo, Moldova, Montenegro, and Serbia.

7 The remaining countries are Azerbaijan, Armenia, Belarus, Georgia, Russian Federation, and Ukraine. None of the European countries that currently fall under the European Neighbourhood Policy (ENP) – Azerbaijan, Armenia, Belarus, Georgia, Moldova, and Ukraine – are being considered as a potential candidate state. Moldova's position vis-à-vis the EU is not as clear-cut as it seems, however, because Moldova also participates in other integrative processes in southeast Europe (most notably the Stability Pact for South-East Europe), which include both EU accession candidates and potential accession candidates. Armenia, Georgia, Moldova, and Ukraine have stated their interest in joining the EU, but the EU's response has been lukewarm. Although Belarus and Russia are also seen as eligible to join, the lack of democratic structures in Belarus works against this, whereas the EU would like to bind Russia more strongly to its own policies and goals through partnership and cooperation agreements.

3. Systems for handling the release of genetically modified organisms (GMO); (administrative, risk assessment, and decision making);
4. A system for enforcement and post-release monitoring; and
5. Functional and transparent systems for public awareness and participation in decision making.

This framework is used for the comparative analysis in the next sections.

1. *Biosafety Policy*

Sound biosafety policy is the first necessary component when establishing an NBF. Some CEE countries have chosen to develop a stand-alone biosafety policy, whereas others have integrated it into other sectoral policies (e.g., agricultural, biotechnology, or food safety). Biosafety was a comparatively new area prior to the adoption of the CPB in January 2000, and most countries had not developed a specific biosafety policy before participating in the NBF project. Bulgaria and Poland were the exceptions because they had participated in the UNEP Pilot Biosafety Project from 1997–9.

In some CEE countries, the first elements of a biosafety policy appeared in their NBSAPs, which were developed in response to Article 6a of the CBD. In general, NBSAPs focus on classical *in situ* and *ex situ* conservation of biologic diversity. For example, if genetic resources or biotechnology are identified as important issues (e.g., in the NBSAPs of Albania, Croatia, Lithuania, and Romania), then there is little, if any, focus on GMO and biosafety. In several cases, biosafety and GMO-related activities are referred to only vaguely; for example, the NBSAPs of the Czech Republic, Estonia, and the Republic of Moldova each refer to genetic resources and biologic security but concentrate on the erosion of genetic resources rather than on biosafety in the sense of the Cartagena Protocol. Some NBSAPs do provide general strategic direction with respect to biosafety, as in the case of Slovakia – "strengthen the control over use of GMOs" – and Armenia – "defining the need to further develop legislation and build capacity."[8] The only CEE countries to use NBSAPs as a real biosafety policy document with clear guidance, strategic goals, and activities to achieve these goals were Georgia and Turkey.

With the adoption of the CPB and the pressing need to harmonise legislation with the EU, CEE countries were among the first to put national biosafety policies in place. A description of the development of biosafety policy in CEE, beginning with the countries which have already joined the EU, follows.

[8] Slovakia, *National Biodiversity Strategy of Slovakia*, 1997, online: http://www.cbd.int/doc/world/sk/sk-nbsap-01-p1-en.pdf; Armenia, *Biodiversity Strategy and Action Plan for the Republic of Armenia*, 1999, online: http://www.cbd.int/doc/world/am/am-nbsap-01-en.pdf.

EU Countries

At the Lisbon Summit in 2000, the European Community set a strategic goal for becoming the most competitive and knowledge-based economy in the world. Recognizing the role of modern biotechnology in achieving this objective, the European Commission launched a Strategy for Life Sciences and Biotechnology in 2002.[9] This initiative established an eight-year strategic vision to increase Europe's competitiveness in life sciences and biotechnology, while addressing the ethical and societal concerns of its citizens. By acceding to the EU, CEE countries are signing on to this long-term strategy.

The Commission of the Life Sciences and Biotechnology Strategy identified five main priorities for the governance of modern biotechnology:

1. Societal scrutiny and dialogue should accompany and guide the development of life sciences and biotechnology;
2. There must be responsible development of life sciences and biotechnology in harmony with ethical values and societal goals;
3. Informed choice should facilitate demand-driven applications;
4. Science-based regulatory oversight should enhance public confidence; and
5. There must be respect for basic regulatory principles and legal obligations to safeguard the Community's single market and international obligations.

The EU now has a comprehensive regulatory system for biosafety, which regulates or provides guidance for most GMO uses (i.e., GMO field trials, marketing, contained use, transboundary movements). Some of those regulations are binding across the EU, whereas others are used only as guidance documents by member states in the development of their own detailed rules that incorporate the general principles adopted by the EU. The EU has legislated over GMO since the early 1990s – the latest iteration entered into force in April 2004. In 2006, the European Commission adoped the EU policy on biotechnology,[10] which may be better defined as a policy on GMO, for it does not cover "white" (industrial) or "red" (medical) biotechnology issues. It can be considered a biosafety policy, *sensu* Cartagena Protocol.

Stavros Dimas, the European Commissioner for the Environment, states in the foreword to the policy that the EU GMO legislation "sets out clear, transparent and stringent rules on genetically modified food, feed and crops. It is possibly the strictest GMO legislation in the world."

9 European Commission (EC), "Life sciences and biotechnology – A strategy for Europe," COM (2002) 27, online: ec.europa.eu/biotechnology/pdf/com2002-27_en.pdf.

10 EC, Directorate General Environment, "EU policy on biotechnology," online: http://europa.eu.int/comm/environment/env-informa.

The main components of the legislation are:

- Regulation (EC) No 1829/2003 on genetically modified food and feed regulates the placing of GMOs on the market for food and feed use, as well as food and feed containing, consisting of or produced from GMOs. If one of the uses of a GMO concerns food or feed, the applicant may file a single application for the GMO and all its uses (cultivation, processing into industrial products, and feed use) under this regulation.
- Regulation (EC) No 1946/2003 on transboundary movements of GMOs regulates unintentional transboundary movements of GMOs and the exportation of GMOs to third countries.
- Directive 90/219/EEC, as amended by Directive 98/81/EC, regulates the contained use of genetically modified microorganisms (GMMs). This Directive regulates research and industrial work activities involving GMMs (e.g., genetically modified viruses or bacteria) under conditions of containment, i.e., in a closed environment (e.g., a laboratory) in which contact with the environment and population is avoided.
- Regulation (EC) No 1830/2003 and amending Directive 2001/18/EC addresses the traceability and labelling of genetically modified organisms and the traceability of food and feed products produced from genetically modified organisms. Regulation 1829/2003 identifies specific labelling requirements for GM food and feed.

Many implementing measures to support operation of this framework have also been adopted by Member States over the past few years. These include guidelines for risk assessment, monitoring, formats for submission of notifications, sampling and detection, and coexistence measures.

OVERVIEW OF BIOSAFETY POLICIES OF CEE COUNTRIES THAT ARE EU MEMBERS

Slovenia is the only country among the new EU Member States that has a clear, stand-alone biosafety policy. In most CEE countries, biosafety is part of the biotechnology policy. Biotechnology is identified as a priority in some countries (e.g. Estonia, Lithuania, and Slovenia), although there is still a conflict between the promotion of biotechnology and the use of GMOs.

Lithuania does not yet have a unified opinion regarding the use of GMOs and genetically modified plants (GMPs). Although various governmental programs have identified national priorities for the expanding biotechnology sector, the government continues to promote and encourage ecological and organic farming.

The EU is an open market economy, and the development and safe use of modern biotechnology should be promoted; however, defining national priorities for the use of modern biotechnology products in agriculture and the food/feed industry is necessary. The current state of affairs in the countryside of many CEE countries would favour the development of organic farming (e.g. there are many small/medium-sized farms and a high rate of employment in the agricultural sector in the Baltic States). Therefore, national priorities for the use of modern biotechnology products in agriculture and the food/feed industry must be defined, because there is a need to promote the coexistence of organic, conventional, and GM farming.

Romania was one of the first countries in eastern Europe to establish and implement its national biosafety framework. In 1999, Romania enacted its "Law on Obtaining, Testing, Use and Commercialization of GMOs and of the Products Resulting Thereof" and adopted a modified law in 2001. Unfortunately, Romania has had difficulty including biosafety in its biotechnology policy. Romania now plans to develop a specific strategy for modern biotechnology that includes biosafety; however, there are few funds available for research in the field of modern biotechnology. The 2005–8 Romanian Government Programme briefly describes the development policy and strategy at that time. Chapter 18 focuses on the environmental protection policy, although one of the main concerns with Romania's environmental policy is its failure to incorporate biosafety. This omission highlights the importance of implementing the provisions of the Cartagena Protocol.

Some CEE countries that are EU Members have biosafety policies that are part of many broader policies and strategies. In the Czech Republic, for example, biosafety is covered, in part, by the State Environmental Policy, the Strategy of Sustainable Development, the State Nature Conservation and Landscape Protection Programme, and the Strategy of Food Safety. In Estonia, biosafety appears in the Environmental Strategy (2007), the Biotechnology Strategy (under preparation for the last 10 years), and the Coexistence Policy. In Latvia, biosafety is part of wider policies in the field of environmental protection, biodiversity conservation, and food safety. In Lithuania, biosafety is part of the national biodiversity conservation strategy and action plan, National Program for Environmental Protection, National Food and Nutrition Strategy and Action Plan for 2003–2010, and the National Program for Agricultural Development Strategy for High-Tech Science and Modern Technology.

NON-EU COUNTRIES (INCLUDING CANDIDATE AND POTENTIAL CANDIDATE COUNTRIES)

As one would expect, biosafety policy is more heterogeneous outside the EU, where there are no generally agreed-on and accepted framework-shaping policies. Even countries that plan to join the EU have taken a different approach to biosafety policy and have not necessarily modelled it after that of the EU. To appreciate the differences between the countries, each national policy is analysed separately.

It can be said that as in the EU, only a few non-EU countries have clearly defined biosafety policies (Turkey and Serbia), and of those remaining CEE non-EU countries, biosafety usually forms part of a larger biotechnology and/or agricultural policy.

Turkey has two policies that address biosafety. One is a national biosafety policy that prioritizes the protection and sustainable use of biological diversity and environment and also protects human, animal, and plant health. The other is a science and technology policy. This second policy, which supports biotechnology and includes a provision to protect biosafety, is determined by the High Council of Science and Technology under the chairmanship of the Prime Minister. Turkey's eighth Five-year Development Plan of Turkey (2000) foresees the establishment of legislative, institutional, and practical arrangements as integral for implementing a regulatory system on biotechnology. The plan aims to minimise potential risks associated with the use of modern biotechnology.

Serbia's biosafety policy is based on measures that prevent and reduce the possible adverse effects that GMOs might have on human health and the environment. As in many other countries, modern biotechnology is an area of strategic importance for Serbia's economic development. Modern biotechnology in its agricultural production is important. In 2005, Parliament adopted the Strategy of Development of Serbian Agriculture. The strategy recognizes the importance of modern biotechnology and biosafety in contemporary agriculture and in this regard proposes to "develop strategic, legal, administrative and technical instruments to provide adequate level of protection and control in the area of work with GMOs." The importance and benefit of GM plants in the development and use of agriculture is also emphasized in the strategy.

As is true of some CEE EU Member States, a group of countries has integrated biosafety into other national policies. In Croatia, biosafety policy is integrated into the policies of biodiversity conservation, biotechnology, science and technology, food production, food safety, environment protection, and sustainable development. These various policies are the responsibility of several ministries and this creates major contradictions because of a lack of mutual coordination. There are no clearly defined priorities and targets for biosafety. In the National Environmental Strategy (adopted in 2002), the preservation of existing biological diversity is the priority, whereas GMOs are secondary. The National Strategy for Forestry (2003) explicitly forbids the use of GMOs in forestry.

Azerbaijan has currently included biosafety in its environmental and health policy but is considering the important economic role of agriculture there. The government will likely attempt to incorporate biosafety policy into its agricultural policy.

In Armenia, biosafety is included in the national environmental policy, specifically its policy on conservation and sustainable use of biological diversity. Armenia is currently undertaking measures to join the European Economic Partnership and is therefore harmonising its policies and legislation with those of the EU.

Belarus' biosafety policy is one of the most interesting of all the CEE countries, because the country is not bound by any EU agreements and has chosen to use biotechnology to improve its economy. Although GMOs can be perceived as a potential environmental threat, Belarus refers to the UNEP International Technical Guidelines for Safety in Biotechnology that state that the risks to human health and the environment resulting from genetic engineering activities do not differ from those arising from traditional breeding technologies. The policy recognises that these risks can be identified and assessed in the early phases of the breeding process or even at the planning stage of experiments and defines how to avoid or minimise possible adverse effects of genetic engineering activities. This has resulted in a biosafety policy that does not refer to or incorporate precautionary approaches, which is unlike most other countries in the region.

The biosafety policy in Belarus is part of the overall policy in the field of health care and environmental protection. The Republic of Belarus has identified two main biosafety objectives. One is to create an enabling environment for deriving maximal benefit from the achievements of modern biotechnology and to foster development of genetic engineering as a research priority. The other objective is to ensure that human and environmental health is protected when genetic engineering activities are performed, new biotechnologies are implemented, and genetically engineered products are consumed (Belarus Strategy of Sustainable Development, 2003).

The Republic of Belarus is very aware of the important role of biotechnology in the sustainable development of its economy. Biotechnology and genetic engineering are on the list of priority research lines and technologies. There have been three main directions of modern biotechnology formed in Belarus: (1) the creation of new effective strains of microorganisms for synthesis of biologically active compounds and for their use in industry, agriculture, and environmental protection; (2) the breeding of genetically engineered varieties of agricultural and ornamental plants; and (3) the application of genetic engineering biotechnologies in medicine for diagnostics and treatment of diseases and the creation of fundamentally new medical drugs.

In 2001, the Parliament of the Republic of Moldova adopted the Concept of Environmental Policy. The implementation of this policy will become the basis for all local and national environmental activities in the Republic of Moldova. Considering the importance of agriculture and agribusiness for Moldova's national economy and in view of the objective to enter the international market of agricultural products, Moldova has approved the National Concept for Natural Farming, Production and Distribution of Environmentally Clean and Non-GMO Food (August 21, 2000). The document clearly declares that the use of genetic engineering is considered inadmissible for the purposes of agricultural production. This statement does not conform in full with the EU GMO policy, although Moldova does realise that a precondition for entrance into the European market is the introduction of new agricultural technologies. The Parliament of the Republic of Moldova has approved several strategic priorities for the period from 2004 to 2010, which include governmental programs for research and development in the field of agriculture, including:

a. Advanced biotechnology with intended application in agriculture, food industry and medicine;
b. Development of modern technologies based on the use of agricultural biodiversity gene pools in the field of phytotechnology and animal breeding; and
c. New technologies for producing food products and ensuring their safety.

Additionally, according to the Strategy of Social-Economic Development of the Republic of Moldova (Government Decision No. 1415 of 19 December 2001), the future development of the Moldovan economy will be "innovational, based on the use of scientific research findings and advanced technologies."

Countries without a Biosafety Policy

The last group of countries to be discussed are those that do not have a biosafety policy or have not yet incorporated biosafety into other policies or strategic and/or political documents. Albania might be one of the countries planning to draft a stand-alone biosafety policy, although it has not yet done so. Albania is currently developing a National Strategy for Socio-Economic Development that will incorporate a strategy for agriculture and food and will probably include a biosafety policy.

The Former Yugoslav Republic of Macedonia and the European Community signed an Association and Stabilization Agreement in 2001, obliging the Former Yugoslav Republic of Macedonia to harmonise environmental (and other) legislation with that of the EU and to adopt a programme that aligns the Former Yugoslav Republic of Macedonia's national legislation with that of the EU, including its biosafety policy. This has yet to occur. The national biosafety "vision" is in its initial phase, forming only a segment of other strategies, such as environment protection, biodiversity conservation, and food safety. The integration of a biosafety system into other (nonenvironmental) sectors such as health, agriculture, food and feed, research and development (R&D), and industry is of great importance to the Former Yugoslav Republic of Macedonia.

Despite having a strong biotechnology sector, Ukraine does not have a complete policy or political agreement on biosafety. Ukraine's national law "On State Biosafety System of Creation, Trials, Transportation and Use of GMOs" was adopted in June 2007, after being modified and postponed for several years. The law was to take into account the requirements and regulations of EU GMO policy, but it still differs greatly from current EU legislation. Ukraine might have established this legal framework to avoid accusations from the WTO of having purposely implemented technical trade barriers for biotechnology products. There are, however, several other legal acts establishing trade barriers and restricting the use of GMOs for human consumption in different regions of Ukraine.

Georgia was one of the very last countries in CEE to ratify the Cartagena Protocol, in 2010, and it does not yet have an official biosafety policy.

Summary

The adoption of high-level national strategies and policies on modern biotechnology without clearly separating out issues on GMOs and GM crops has left countries struggling with policy gaps and conflicts across a wide range of sectors. In many CEE countries, the regulatory regime ran ahead of the policy framework because of the need for harmonisation with EU systems. This necessity to meet existing standards resulted in a lack of wide consultation and debate with the public and even within government, which has caused much tension between economic sectors in some countries. Various countries have discovered, for example, that their agricultural sector has several contradictory policies, such as the promotion of both GM crops and organic agriculture, whereas others have actual or de facto national-level bans on GM crops that conflict with EU legislation.

Coexistence rules provide a potential way around this problem. Several EU countries have put in place coexistence measures (e.g., the Czech Republic and Slovenia, and such measures are under preparation in Hungary, Latvia, Lithuania, and Estonia). EU member states have the right to set their own rules for coexistence. Normally, these rules set certain parameters for growing both conventional and organic crops (e.g., isolation distances, different growing times, GMO field registers), to prevent the unintentional presence of GM materials from rising above certain thresholds, for labelling purposes. The labelling system is quite complex: 0.9 percent of adventitious presence is allowed for GMOs marketed within the EU; 0.5 percent is allowed for GMOs that have not yet been approved but have undergone risk assessment (i.e., approval by the European Food Safety Authority); and zero percent is the threshold for unapproved GMOs that have not undergone risk assessment.

In most CEE countries, the safe use of modern biotechnology (and hence biosafety) is seen as a priority area for sustainable development, but this has not usually been translated into a priority for mainstream governmental funding. There is a further conflict between prioritizing biotechnology for its potential economic benefits and the desire to keep GMOs out of agriculture. The entire Caucasus region wants to declare itself GMO free to promote organic farming and to protect their population and environment from potential GMO-related threats. Other countries ban the use of GMOs in certain areas only. For example, Croatia forbids its use in forestry, Turkey wants to ban GMO use in baby food, and the Ukraine has adopted several laws forbidding the use of GMOs for human consumption. As members or potential members of the WTO (see Annex V, Table 3), these regulations raise possible conflict with WTO rules in the fields of veterinary protection (under the OIE, Office International des Épizooties), phytosanitary systems (International Convention on Plant Protection), and food safety (Codex Alimentarius).

REGULATORY SYSTEMS

A biosafety regulatory system should develop from biosafety policy. Countries in the CEE region have taken different approaches while working out their regulatory systems for biosafety. In most countries, some legislation was already in place when the Cartagena Protocol was adopted, permitting simple amendments to existing acts. The remaining countries, having no biosafety legislation, have had to draft completely new biosafety statutes. Most of the countries have tried to avoid creating new institutions and have instead attempted to fit biosafety into existing administrative and surveillance systems. Most countries have assessed the most efficient method for creating their biosafety system as that which does not require major amendments to existing legislation nor requires the establishment of new institutions. The initial approach was to divide tasks and responsibilities between currently available bodies, granting them (if needed) proper authority. This task was very difficult, however, and many countries changed their competent authorities because the original system did not work. For example, in Latvia the Cartagena Protocol focal point (FP) was initially the Latvian Food Centre, but then the Food Center was reorganised, and the FP became the Ministry of Agriculture. There have been interesting reasons for changing FPs – in some cases, ministries saw biosafety as an additional funding possibility and increase in power, whereas elsewhere it was seen as an additional burden of work and political pressure, unappealing to any ministry.

National Focal Points and Competent National Authorities

In CEE, the CPB national focal point (NFP; *sensu* CPB Art 19.1) is usually the Ministry of the Environment, although sometimes it is the Ministry of Agriculture (see Annex V, Table 2). There have been some interesting cases, however, in which the Ministry of Culture (Croatia) or the Ministry of Foreign Trade and Economic Relations (in Bosnia and Herzegovina) have been nominated as NFPs.

The location of the CPB NFP does not appear to correlate with any particular biosafety policy position. Czech Republic and Slovakia are the only CEE countries currently growing GM crops, whereas Belarus, Estonia, and Romania have relatively pro-GM policies, and Albania, Croatia, Georgia, Hungary, and Moldova have policies that are effectively de facto moratoria. In some countries, such as Belarus, there appears to be a spreading of responsibility, with the Ministry of Foreign Affairs, the Ministry of Natural Resources and Environmental Protection, and the Academy of Sciences being nominated at different levels, reflecting the need to address biosafety issues at the international, national, and technical levels.

In some countries, all CBP functions are the responsibility of one authority (e.g., Lithuania, Poland), but in most countries, biosafety functions and responsibilities are divided among different competent national authorities (CPB Art 19.1), as is appropriate, given the cross-sectoral nature of biosafety. In some countries, the existence of a single competent authority (CA) is the result of the country's use of interim measures when transitioning to a more distributed designation of responsibilities (e.g., Serbia, Armenia). The designation of responsibilities has been hindered by the rapidly changing political landscape in the region, resulting in the fragmentation of countries and a high level of political "turnover." For example, in Serbia and Montenegro, the Federal Ministry of Agriculture was responsible for contained use, deliberate release, and the placing of GMOs and products derived from GMOs that are not made for human consumption on the market (i.e., seed, planting material, and feed), whereas the Federal Ministry for Labour, Health and Social Care was responsible for GMOs and products derived from GMOs used as food. In 2006, Serbia and Montenegro separated, and the federal ministries were reorganised. During this process, neither the Republic of Serbia's new Ministry of Health nor the Ministry of Agriculture took authority over GM food. It is expected that the new Government will resolve this omission by assigning the Ministry of Agriculture as the authority.

In many countries in the region, institutions not recognized as CAs have biosafety responsibilities. For example, in Romania, in addition to the Ministry of Environment, several other authorities have been promulgating the GMO Act, including the Ministries of Agriculture, Forestry, and Rural Development; National Sanitary-Veterinary and Food Safety Authority; and the National Authority for Consumer Protection. In Slovakia, responsibility for human health lies with the Ministry of Health, which shares the responsibility for food safety with the Ministry of Agriculture. The regulation of seed, feed, and other agricultural application of GMOs is the sole responsibility of the Ministry of Agriculture.

In Croatia, the situation is quite unusual. The Ministry of Health and Social Welfare issues marketing authorizations for pharmaceuticals, cosmetics, and food. The Ministry of Agriculture, Forestry, and Water Management issues authorizations for GMO feeds and GMO use in agriculture, forestry, and fisheries. The Ministry of Culture is the competent authority responsible for the placing of GMOs or GMO derivatives on the market, with no input from the Ministry of Environment.

Many countries have identified their competent authorities in their NBFs but have not yet nominated them through the BCH (e.g., Belarus, Croatia, Georgia, Malta, Montenegro, Serbia, and Turkey). It is not clear if this is simply an accidental oversight or an intentional omission of biosafety issues by countries.

Legislation

A regulatory regime for GMOs and biosafety consists of binding international treaties and relevant national legislation. In the EU and countries aspiring to join the EU, there is an additional "layer" – the regional EU legislation. The provisions of international treaties, EU directives, and other EU legislative acts that are not directly applicable must be incorporated into national legislation. EU regulations are directly applicable in EU countries, and relevant national legislation might not regulate the same aspects of GMOs and biosafety issues. Hence, the regulatory system in the EU countries consists of three "layers" – international agreements (e.g. Cartagena Protocol, WTO, International Plant Protection Convention (IPPC), and International Union for the Protection of New Varieties of Plants; UPOV), EU regulations that are directly or indirectly applicable, and national legislation. Despite the prevalence of international and regional biosafety legislation, the regulation of biosafety and (the safe use of) modern technology varies greatly among CEE countries. Table 1 in Annex V lists the major changes in regulatory framework that have occurred in the region, using Protocol ratification as a benchmark.

It is clear from this that countries that have developed stand-alone biosafety legislation have still had to make changes to existing legislation. Harmonisation with international legislation and the new language of genetic modification are the two major reasons for the required changes. Alternatively, attempts to incorporate biosafety into broader environmental protection legislation have also proved difficult, because politicians react to public opposition to GM technology at the constituency level.

For example, Albania drafted a general nature protection law that contained provisions for GMOs. This law has been before Parliament for more than two years but has yet to be adopted because of the GMO provisions. Owing to the strong political opposition to the law, the GMO provisions were removed from the draft legislation and developed into a new biosafety law. Although now completed, the draft law has been under revision since October 2007 and has yet to be sent to parliament for approval.

Biosafety is a cross-cutting area that requires the regulation of food, feed, and processing, as well as contained use and environmental releases. The regulation of these activities requires amendments to other regulations, such as seed, food, and feed regulations. Most countries have identified in their NBFs the need to amend legislation if only to add a reference to GMOs – this could include acts that regulate general nature protection, food, feed, fertilizers, organic farming, forestry, wine production, and laboratory conditions.

Countries have many options when developing and implementing a national biosafety regulation. Choices include the creation and introduction of a new biosafety (GMO) act that addresses all uses of GMOs (usually excluding pharmaceuticals) or the amendment of existing legislation. They also must decide whether they would regulate the transboundary movements of GMOs, including their export, import, and transit. All CEE countries have gone beyond the minimal requirements identified in the Cartagena Protocol and regulate contained use and field releases that do not include transboundary movements. In most CEE countries, all GMO uses are regulated under the same act, even if various ministries have been involved in the decision-making process or the implementation of law and surveillance. For example, in Lithuania, management and administration of activities involving the use of GMOs is delegated to the Ministry of Environment, which is the National Competent Authority that works in conjunction with other institutions, including the Ministry of Agriculture, the Ministry of Health, and the State Food and Veterinary Service. The roles and responsibilities of these ministries are defined in the Framework Law on GMOs.

Although most countries have introduced all-encompassing GMO legislation, some countries still have multiple sets of laws regulating GMOs and their uses. For example, in Estonia, uses of GMOs in the environment and in contained areas are divided between Ministry of Environment and Ministry of Social Affairs (equivalent to the Ministry of Health). The Ministry of Agriculture is responsible for overseeing the regulation of GM seeds, GM fertilizers, and GM food and feed.

Most CEE countries do not regulate the transit of GMOs. Some countries state that they also regulate transit in their laws, but in reality they do not have explicit provisions defining the transit process, because most legislation is vague and has common provisions for the transit, export, and import of GMOs (e.g., Albania, Azerbaijan, Moldova, and Ukraine). Turkey, however, regulates transit explicitly in Article 31 of its draft law, which states that: "Transits of GMOs and products thereof shall be subject to conditions specified in the written consent of the competent authority that would be given within 60 days pursuit [*sic*] to the submission of consents of importer and exporter countries and shall be subject to provisions of the law number 4458 on Customs. Shipments in transit not having such a consent are considered as smuggling."

Albania's draft law states that transit of GMOs may take place only in transport safeguarded against undesirable leakage of GMOs into the environment and against loss or theft. Azerbaijan has made provisions in its draft law for special export and transit permits, but this regulatory system has not yet been clearly defined.

Administrative System

In the EU, the administrative and decision-making system has two layers – member state level and EU level. Decisions regarding environmental releases, the placement of GMOs on the market, and risk assessments are made at national level, but the final approval decision for the entire EU is made at the EU level. For novel food and feed, the European Food Safety Agency (EFSA) undertakes overall risk assessment for the EU, although competent authorities in member states have an opportunity to provide their opinions. This, however, does not apply to GMOs in contained use, for which decisions are made solely at the national level.

As a general rule, an administrative system for getting permits for use of GMOs consists of two distinct and separate entities – the competent authority, which is responsible for receiving applications and making final decisions, and the advisory bodies to the competent authority. The advisory body could be in the form of a committee, commission, council, or expert panel from which people are chosen according to the speciality of the particular application. The advisory body in most CEE countries is an expert commission, but in many cases they include members from competent authorities, surveillance bodies, and NGOs. This could lead to a situation in which political advice is given at two different and potentially conflicting levels – at the expert level (opinion) and then at the decision level, where the expert opinion could be influenced by political considerations.

In most CEE countries, applications are submitted to the CA and then forwarded to an advisory body or bodies, which are then responsible for analysing risks, evaluating the risk assessment submitted by the applicant, and providing an opinion on whether to grant a permit, request more information, or refuse the application. Based on the advisory body's opinion, the CA makes a final decision. Sometimes, the decision is made at a higher level than the Ministry (the CA) – for example, by an intersectoral council or commission.

There are numerous administrative systems used by CEE countries. It is difficult to give a general scheme, because there are as many different schemes as there are countries, and no scheme is identical in any two countries. Every country differs – from an additional coordinating political advisory body to different schemes for expert advice.

Albania proposed a simple system in its draft law. It consists of only one entry point – the National Biosafety Committee – to which all applications, regardless of their scope, are submitted. This is a new public entity, independent from any other authority except the Council of Ministers. The National Biosafety Committee makes decisions on the basis of advice from the Scientific Commission, which is composed of scientists from different fields – microbiology, genetics, medicine, biochemistry and molecular biology, pharmacy, agriculture, veterinary science, biotechnology, and safety at work. A possible weakness of the system is that it lacks a formal consultation process with other ministries, although three competent authorities are registered in the BCH – the Ministries of Environment, Agriculture, and Health.

Georgia has also proposed a "user-friendly" system in its draft GMO Act. It appoints one central administrative body, the Ministry of Environmental Protection and Natural Resources (MEPNR), to consider all applications. This central body would receive the application and then forward it to the appropriate competent government authority. This body could also keep a central GMO registry and database for the BCH. The MEPNR forwards the application to the Ministry of Labour, Health and Social Protection (MLHSP) and/or to the Ministry of Agriculture (MoA) for their opinions.

The applications are also forwarded to the appropriate advisory body, the Scientific Commission for GMOs, which will provide expert opinion on safety. Different experts are used for different GMO applications. The Scientific Commission for GMOs is responsible for conducting the risk assessments and for providing conclusions and recommendations. Once it has made its recommendation, and after other ministries have given their opinion, the application goes to Intersectoral Advisory Council. This is an interministerial body that provides a final opinion based on the recommendations from the advisory body and other ministries. The MEPNR then has the right to make a political decision based on the various opinions submitted. As a result, the MEPNR issues one permit that covers all areas, thereby eliminating additional applications or requests. For seed registration, the MoA issues the permit, but the MEPNR approves it.

Moldova has also chosen a centralised approach. The National Biosafety Committee (NBC) was established in 2003. This committee is responsible for receiving applications, conducting public hearings, and issuing permits for all uses of GMOs, including deliberate release of GMOs into the environment; the placing on the market of GMOs and products made thereof; and imports and exports of GMOs and/or products made thereof. The NBC consists of scientists and members of national authorities.

The NBC has a technical committee that will receive applications, check whether the application is compliant with requirements, and then forward the compliant applications to the NBC. The NBC then chooses one of two options when deciding how to proceed. Under the simplified procedure, the NBC requests an examination and opinion of the document from the technical biosafety committees in the relevant ministries (opinion papers from the national authorities on agriculture, food industry, health and consumer protection), the scientific councils in the relevant research institutions, and from NGOs. Selection of the reviewers is determined on a case by case basis and depends on the type of application.

The second option is the detailed examination procedure, which not only includes all the procedures identified in the simplified option but also requires a formal risk assessment. In this case, an opinion is requested from the National Testing and Risk Assessment Center in addition to the assessments performed by the biosafety technical committees of the relevant Ministries, the scientific councils of research institutions, and the NGOs accredited as experts. The NBC submits its opinion to the Ministry of Ecology and Natural Resources, and this opinion is formalised by an order issued by the Ministry. Based on this order, the National Biosafety Committee, acting on behalf of the ministry, will issue an authorization.

In Estonia, the Advisory Committee for Genetic Modification is the central advisory committee responsible for conducting risk assessments of GMOs and products containing GMOs or consisting of GMOs. The advisory committee was established under the MoE and advises the three ministries that have been identified as competent authorities: the Ministry of the Environment, which is responsible for the issuance of permits for the deliberate release and marketing of GMOs or products containing GMOs or consisting of GMOs; the Ministry of Agriculture, which is responsible for conducting tests on animals; and the Ministry of Social Affairs, which is responsible for issuing permits for the contained use of genetically modified microorganisms (GMMs). Prior to joining the EU, Estonia had established the Novel Food Committee, which conducted risk assessments of GM food and feed. This committee was the responsibility of the Ministry of Agriculture, but since Estonia joined the EU, many of this committee's responsibilities now lie with the EFSA, which conducts most of the risk assessments.

A relatively simple scheme has also been used in other EU countries. In Lithuania, the Ministry of Environment receives applications and is then advised by GMO Regulatory Management (Steering) Committee and a GMO Experts Committee. The GMO Steering Committee is a political advisory body for the development and enforcement of the national biosafety regulatory system. The Steering Committee consists of members appointed by relevant state authorities (e.g., Ministry of Agriculture or Ministry of Health), subordinated organizations, the national biotechnology industry (R&D), NGOs, universities, and scientific institutes. The GMO Experts Committee is also a consultative advisory body to the Ministry, which focuses on environmental risk assessments.

One of the most cumbersome biosafety systems is that proposed by the GMO Act of Croatia. This is a new system that was not in place before. The act identifies three national competent authorities: the Ministry of Science and Technology, responsible for GMOs in contained use; the Ministry of Health and Social Affairs, responsible for the deliberate release and marketing of GMOs and food or pharmaceuticals for humans; and the Ministry of Agriculture, Forestry and Water Management, responsible for feed or plant propagation material used in agriculture, forestry, or veterinary medicine. Applications are submitted to the relevant CAs.

Croatia has is a two-step system of advisory bodies: the Council of Genetically Modified Organisms and the subcommittees of GMOs. The government nominates the Council for GMOs, which is a general advisory body that first receives the applications. The council then nominates, according to proposals from the CAs, the Scientific Committee for Contained Use of Genetically Modified Organisms and the Scientific Committee for the Release of Genetically Modified Organisms into the Environment. The purpose of these committees is to monitor status and developments in the field of GMO handling and to provide technical assistance to biosafety competent authorities.

These specialized advisory bodies share their opinion with the council, which then organizes public hearings and consults relevant ministries when appropriate to provide the minister with a well-researched proposal. In addition to these two committees, the Croatian Food Agency also appoints a Scientific Committee and Scientific Panels, which are expert bodies that generate agency-relevant scientific opinions. Permits for the placing of cosmetics, pharmacy, and human health care products containing GMOs on the market are issued by the Minister of Health. All other uses of GMOs are governed by a system that requires two levels of consent. Permits for GMOs used in agriculture, veterinary medicine, forestry, and fisheries are granted by the Minister of Agriculture with the consent of the Ministry of Culture and Ministry of Environmental Protection, Physical Planning & Construction (MEPPPC). The advisory body in these cases is the Committee for Release of GMOs into the Environment. Permits for placing on the market GM foodstuffs and products that are used in food processing industry or are a product thereof are granted by the Minister of Health, with the consent of the Ministry of Agriculture. The advisory body in this case is the Croatian Food Agency. Permits for placing GMOs or products containing GMOs on the market that are not listed here are granted by the Minister of Culture and Minister of Environment.

Belarus has a different approach from many other countries, reflecting its more "permissive" overall biosafety policy (see discussion earlier).[11] The classification of GMOs (more precisely, genetically modified microorganisms, GMMs) differs from most other European countries. According to EU legislation (Council Directive 90/219/EEC on the contained use of genetically modified microorganisms; Council Directive 98/81/EC amending Directive 90/219/EEC), GMMs used in contained conditions are divided into four risk classes (1–4), 4 being the riskiest. In Belarus, GMMs are divided into pathogenic GMOs, defined as GMMs 2–4 risk class, and nonpathogenic GMOs – defined as GMM 1 risk class. Nonpathogenic GMMs do not need a permit, although as in the EU system, Belarus' laboratories and institutions that use GMOs do require a permit.

Belarus has three CAs: the Ministry of Natural Resources and Environment Protection (responsible for nonpathogenic GMOs field trials, any first-time releases, and import), the Ministry of Public Health (responsible for pathogenic GMOs), and the Ministry of Agriculture and Foodstuffs (responsible for state trials and registration of GM varieties).

GMOs intended explicitly for food and feed do not need a special permit provided that labelling and packaging requirements are met. In other words, GM food is classified as conventional food and therefore requires only a conventional food permit from the Ministry of Health.

Other applications are forwarded to the Biosafety Co-ordination Centre under the Institute of Genetics and Cytology (Academy of Science). This centre appoints experts to conduct risk assessments. Experts are chosen from a national roster of experts adopted by the government. Every ministry has a set of experts (expert committee), who are individually appointed for each case. Experts forward their conclusions to the centre for approval, and the assessments are then sent to the ministry for a final decision.

Surveillance System

Surveillance and monitoring is perhaps the most difficult section to implement of the Cartagena Protocol. The field of biosafety is a new area, with unique obligations such as monitoring, relatively new even to scientists. It is a concept that uses sampling

[11] Azerbaijan intends to use the same approach as has been taken in Belarus, although its law is still in an early form.

methods to generate answers that might not be reliable in court cases (i.e., because the error in sampling results is too large). Understandably, post-release monitoring and surveillance has caused difficulties for national legislators. Most CEE countries have chosen to use existing national inspection and enforcement mechanisms rather than create new institutions (e.g., environmental inspectorates, quarantine, phytosanitary measures, seed management, animal health, and customs). Although in many cases the system appears successful on paper, where it is regulated by law and secondary legislation, unfortunately it might not be functional, having not yet been used or tested. Insufficient training and financial constraints are at fault, because a single quantitative analysis can cost hundreds of euros.

Serbia has an unusual and (seemingly) effective system for border control in place. It consists of a mandatory sampling of seed, mercantile corn, soybean, and feed products that are transported over the state border. The samples are sent to authorized laboratories, where they are analyzed for the presence of modified genes. Every shipment of maize and soybean, including feed product, is analysed. The product's importer is responsible for the financial costs of the analysis. If unauthorized modification is detected, the shipment is either returned to the sender or destroyed.

Public Participation

Protection of the public's right to access biosafety information and participate in the decision-making process is, as discussed in the Introduction and Chapter 7 of this book, an important principle of sustainable development law[12] and has been a priority in many CEE countries for the past several years. Europe's concern with the protection of public rights to access and transparency has led to the adoption of the Convention on Access to Information, Public Participation in Decision-Making, and Access to Justice in Environmental Matters (Aarhus Convention). All countries in CEE except Turkey are parties to the Aarhus Convention (see Annex V, Table 3). In May 2005, a special amendment to the Convention was adopted for GMOs, although few countries have become parties to the amendment.[13]

Additionally, all EU countries have transposed Directive 2003/4/EC of the European Parliament and of the Council on Public Access to Environmental Information. This directive ensures access to environmental information. The objectives of the Directive are (1) to guarantee the right of access to environmental information held by or for public authorities and to set out the basic terms and conditions of, and practical arrangements for, its exercise; and (2) to ensure that environmental information is progressively made available and disseminated to the public to achieve the

[12] See Introduction and Chapter 7 of this book.

[13] *Amendment to the Convention on Access to Information, Public Participation in Decision-Making and Access to Justice in Environmental Matters*, Second Meeting of the Parties to the Aarhus Convention, 2005, Decision II/1, UN Doc ECE/MP.PP/2005/2/Add.2.

widest possible systematic availability and dissemination of environmental information. Computer telecommunication and/or electronic technology, when available, shall therefore be promoted.

In EU countries, the public has thirty days from the day of CNA's announcement to comment on an application before a decision is made. Most CEE countries have included this timeframe in their legislation, although some have been reluctant to (see later discussion). Countries have approached the protection of public access to information and the right to participate in the decision-making process in various ways. Some CEE countries have found it difficult to accept the public's right to participate, as evidenced by this omission from early drafts of biosafety laws. In one case, it was proposed that the public should have the right to comment only when the CA, through its advisory body, found there to be a possible risk, but not in cases when it was considered to be safe. All current versions of CEE laws and draft laws include a section that allows some form of public participation, but these sections vary in three crucial ways: (1) the level or stage at which the participation occurs, for example, whether the public is invited to comment on applications, draft decisions, or both; (2) how the participation occurs, for example, through the Internet, public hearings, newspapers, and other venues; and (3) how the information is treated. For example, is a decision made without considering the public, is there is an obligation to respond to public comments, or is there an obligation to summarise public opinions and include them with short answers in the decision-making document? These issues have been addressed differently in various countries.

For example, in the Czech Republic, applications are published regularly on the MoE website and at the official desk of the Ministry of the Environment. The public has thirty days to make comments, which are then discussed at a public hearing. The results from the public hearings are considered by the ministry, which issues the final decision. The Ministry of Environment then publishes the final decision and the lists of authorized GMOs on its website. A similar system is used by all EU countries and many others, including Estonia, Latvia, Lithuania, Former Yugoslav Republic of Macedonia, Malta, Moldova, Romania, Slovakia, and Slovenia.

Some countries outside the EU have proposed promising systems in draft laws, but unfortunately no information about their success is available. In Albania, the Biosafety Council must publish their decision in the Official Gazette and in some of the most widely circulated environmental publications. Similarly, the scientific committee decisions are published by the National Biosafety Committee in some of the most widely circulated environmental publications.

Before a final decision is determined, the National Biosafety Committee must ensure that all authorized persons have had the right to be heard on the subject of the proceedings. The National Biosafety Committee can make decisions only on issues that have been commented on by the public. Applications involving environmental releases and marketing must be subject to public consultation. The

National Biosafety Committee shall summarize the results of the public input, with particular emphasis on the participation and conclusions of the consultation.

The Albania draft law does not require public consultation prior to a decision, although public authorities are encouraged to involve the public more actively by exploring other mechanisms and measures such as consensus conferences, round-table discussions, stakeholders' dialogues, and citizens' juries on issues relating to, for example, the risk assessment and management of GMOs.

Belarus is another country with a biosafety system that is not modelled on the EU system. Under national legislation, the public has sixty days to comment instead of the thirty days required by EU legislation. The National Co-ordination Biosafety Centre posts applications on the information website of the National Co-ordination Biosafety Centre. Experts responsible for the review must consider the comments and proposals submitted. In cases in which it is impossible to consider particular comments and proposals, the experts are obliged to provide written reasons to the centre. The comments, proposals, and results of the expert committee should be reflected in the final expert opinion submitted to the National Co-ordination Biosafety Centre.

In Georgia, applications shall be made available to the public through a specially designated website and by publication in both the *Official Gazette of Georgia* and in at least two nationwide newspapers. Public announcements shall include the name and address of the CA contact person who is responsible for providing the public with the required information. The public must submit their opinions, observations, and concerns within ninety days of the announcement. Additionally, the CA can, if it so chooses, organise consultations with public representatives.

In summary, all CEE countries have some provisions in their biosafety laws that include public participation, with differences in the amount and timing of the involvement. In general, applications and decisions for environmental releases and placement of GMOs on the market are made public, although the approaches vary for contained use. In most cases, applications for GMOs used in contained conditions are not made public, although some countries do publicly share information about GMOs from higher risk classes (third and fourth risk classes). This type of system is used in Croatia and Moldova, where third- or fourth-risk class GMOs in contained use are open to public comment.

CONCLUSION

Although the CEE region is perhaps the most homogenous biosafety region in the world, there is still considerable heterogeneity, especially in the design of national biosafety frameworks and in the implementation of the CP obligations. All CEE countries are in the process of creating or implementing biosafety frameworks, whether still composing the first drafts of the national legislation or introducing comprehensive and functioning systems. These states have publicly announced that biosafety is a national priority, although owing to either a lack of public awareness and understanding of the issues or opposition within government, it is not a high priority in state budgets. Most countries are aware of the possible risks associated with the use of GMOs but realize the potential economic, ecologic, and intellectual benefits that can also come from the safe use of modern biotechnology, and hence they are attempting to balance the opposing issues.

15

Implementing the Cartagena Protocol in West Africa

National and Regional Activities

Gregory Jaffe and Papa Meissa Dieng

Over the past five years, the eight West African Economic Monetary Union (WAEMU) countries (i.e., Benin, Togo, Mali, Burkina Faso, Niger, Senegal, Guinea Bissau, and Côte D'Ivoire) and other West African institutions have worked vigorously to develop national systems that regulate the safe introduction of genetically modified organisms (GMOs)[1] in accordance with the sustainable development objective, principles, and provisions of the Cartagena Protocol. Effort at the national levels has been funded by the UNEP-GEF Biosafety Project and other donors. This has resulted in National Biosafety Frameworks (NBFs) in the eight WAEMU countries, one biosafety law (Burkina Faso), five draft biosafety bills, and numerous other legal documents (e.g., decrees, guidelines, and regulations) that support the proposed national biosafety regulatory systems. Work has also commenced at the Institute of the Sahel (INSAH), which is part of the Permanent Inter-State committee for the Fight Against Drought in the Sahel (CILSS). Under a mandate from the Economic Community of West African States (ECOWAS) to regionally coordinate biosafety regulation, INSAH was able to secure CILSS approval of a regional framework for the Cartagena Protocol Convention in March 2006.

This chapter summarizes the key legal aspects of the proposed National Biosafety Regulatory Systems of the WAEMU countries and analyzes the regional regulation of biosafety in West Africa in light of the sustainable development objective enshrined in the Cartagena Protocol, and in the national development plans of the countries themselves. Similarities and differences between the various nationals systems are identified, and issues concerning the implementation of the Cartagena Protocol

[1] In this chapter, the term *living modified organisms* or LMOs is used when discussing documents that refer to that term, such as the *(Cartagena Protocol) on Biosafety to the Convention on Biological Diversity*, 29 January 2000, 2226 UNTS 208, 39 ILM 1027 (entered into force 11 September 2003), online: http://bch.cbd.int/protocol/text/ [*Cartagena Protocol*]. The term *genetically modified organism* or GMO is used when discussing national or regional documents (e.g., draft laws, conventions, or NBFs) that use those terms.

requirements are discussed. A legal analysis of a centralized regional regulatory system that would fall under the umbrella of the WAEMU Treaty is presented. Finally, established regional regulatory systems, such as the WAEMU regional regulatory system that monitors veterinary drugs and the INSAH regional regulatory system that monitors chemical pesticides, are analysed and considered as models in the design of a regional biosafety regulatory system.

WAEMU BIOSAFETY REGULATORY SYSTEMS

Each WAEMU country has begun to develop a national biosafety regulatory system. With funds from the UNEP-GEF Biosafety Project, WAEMU countries have analysed their existing legislation to determine how best to fulfil their obligations under the Protocol. Although significant progress has been made in many of the countries, only Burkina Faso has passed a biosafety law and has put in place the institutional structures capable of supporting a functional biosafety regulatory system. The table in Annex VI titled "Current Status of Biosafety Regulation in the Eight WAEMU Countries" summarizes each country's level of implementation of Protocol obligations at the national level and development of national biosafety regulatory systems.

Although only Burkina Faso has an operating biosafety regulatory system, the efforts of the remaining WAEMU countries to develop their own national systems has generated enough information to allow the analysis of proposed compliance with the Cartagena Protocol and the regulation of biosafety generally. This discussion summarizes and analyses the proposed national biosafety regulatory systems of the eight WAEMU countries. The primary sources of information used for the analysis are the National Biosafety Framework documents developed through the UNEP-GEF Biosafety Project; biosafety laws; draft documents of biosafety bills; decrees; regulations currently being developed; and interviews with key biosafety stakeholders. Information from the analysis is summarized in Tables 2 and 3 titled: "Summary of Key Components of the Biosafety Regulatory Systems in WAEMU Countries."

Benin

According to interviews with stakeholders, Benin adopted a five-year moratorium on GMOs in 2002. In 2004, ministers from various departments adopted a decree creating the National Biosafety Committee (NBC; Decree 2004–293 of May 20, 2004).[2] The mandate of the National Biosafety Committee is to define Benin's biosafety policy and provide recommendations for the regulation of activities related to GMOs.

[2] Republic of Benin, *National Biosafety Framework and Action Plan for Benin* (Cotonou, Benin: Ministry of the Environment, Housing and Urban Planning, 2005) at 13, online: http://www.unep.org/biosafety/files/BJNBFrepEN.pdf.

The National Biosafety Committee, charged with biosafety policy development, also frequently advises the Ministry of the Environment, Housing, and Urban Planning.

With assistance from the NBC, Benin produced a National Biosafety Framework in 2005 and a draft biosafety bill. Under the draft bill, the Ministry of the Environment, Housing, and Urban Planning is identified as the national competent authority for biosafety and has the legal authority to make decisions regarding requests for the use of GMOs in the country. The draft biosafety bill requires government authorization for all GMOs and/or products derived from GMOs used in confined-use experiments, disseminated into the environment, imported, exported, or released as commercial products.

The draft bill also requires completion of a risk assessment prior to authorization. The purpose of the risk assessment is to determine "the potential unfavourable effects of GMO and GMO products on conservation and sustainable use of biodiversity in the milieu including risks on human health, environment, and mainly consequences on the socio-economic environment." The draft bill also states that decisions on applications should be made "to avoid unfavourable effects on the conservation and sustainable use of biodiversity and human health." Finally, the draft bill does not require that the competent authority share with the public any portion of a GMO application; instead, it provides the Ministry of Environment, Housing, and Urban Planning with the discretion to consult the public prior to rendering a decision.

Burkina Faso

In 2006, the President of Burkina Faso signed Law No. 005–2006, the first biosafety law in West Africa.[3] This law establishes the National Biosafety Agency (NBA) and identifies the regulation of biosafety in Burkina Faso as its mandate. The NBA is supported by several different governmental and nongovernmental advisory committees.[4]

Under the Burkina Faso Biosafety Law, almost every activity involving a GMO or a product of a GMO requires notification to and authorization from the NBA prior to the commencement of the activity. In most if not all cases, authorization is granted only after a risk assessment is conducted, risk management measures are explored; a study of ethical, social, and economic impacts is completed; and the public has been given an opportunity to comment on the submission. The risk assessment considers all potential risk issues, including environmental, biodiversity, and human and animal health. Proposed activity is also reviewed by the National Scientific Committee on Biosafety, a technical advisory committee, and any relevant Scientific

[3] Regina Birner et al, "Biofortified Foods and Crops in West Africa: Mali and Burkina Faso" (2007) 10(3) AgBioForum 192 at 194.

[4] Gouvernement du Burkina Faso, *Cadre National Pour la Prévention Des Risques Biotechnologiques au Burkina Faso* (Ouagadougou, Burkina Faso: Comité National de Biosécurité, 2005), online: http://www.unep.org/biosafety/files/BFNBFrep.pdf.

Committee on Biosafety that is housed in the various government ministries. The committees share the results of their consultation with the NBA, which then uses the information to decide whether to grant or deny authorization. The standards guiding the NBA's decisions are defined in Law No. 005–2006.[5] The law requires the NBA to establish that the GMO or a derivative product can (1) benefit the country without causing any damaging risk to human or animal health, biodiversity, or the environment; (2) contribute to sustainable development; (3) cause no harm to socioeconomic development; and (4) comply with ethical rules.

To implement Law No. 005–2006, the Burkina Faso NBA has drafted eight decrees and three regulations. One draft decree addresses public participation and requires that a survey be conducted where the release is proposed to occur. A chief surveyor collects comments from the public through means that can include written submissions, public hearings, and personal interviews. The chief surveyor then submits a report identifying the public's position to the NBA in advance of its decision on a GMO application. The decree, however, does not specify the scope of the surveyor's activities (e.g., whether the inquiry only explores relevant risk issues for the particular crop in question, or whether it addresses opinions or concerns that the public might have regarding GMOs in general) or how the report will be used by the NBA. Although an extensive process has been established that allows public participation in a GMO application, Law No. 005–2006 does not clearly define the consultation process and fails to identify how the results of the report will be incorporated in the decision making process.

Côte D'Ivoire

Côte D'Ivoire completed its NBF in July 2005.[6] The purpose of the framework is to develop a law that establishes a national biosafety regulatory system that includes all activities involving GMOs and their products. It also states that "any use of [a] GMO . . . for whatever reason . . . is subject to an advance informed agreement." The biosafety regulatory system outlines mechanisms to safeguard the environment, biodiversity, and human health and to ensure that modern biotechnology and ethical values are compatible. The proposed system's requirements are more comprehensive and arguably more stringent than those of the Cartagena Protocol.[7]

The NBF creates a biosafety regulatory system that establishes, within the Ministry of Environment, a National Biosafety Commission (CNBIOS) responsible for the management of biosafety issues. The CNBIOS consists of an interministerial

5 International Monetary Fund (IMF), *Burkina Faso: Poverty Reduction Strategy Paper – Annual Progress Report* (IMF Country Report No. 08/212) at 40, online: http://www.imf.org/external/pubs/ft/scr/2008/cr08212.pdf.

6 Republic of Côte d'Ivoire, *Biosafety Framework* (Abidjan, Côte d'Ivoire: Ministry of State/Ministry of the Environment, 2005), online: http://www.unep.org/biosafety/files/CINBFrepEN.pdf.

7 *Ibid.*

monitoring committee, a permanent secretariat, and specialized scientific commissions. Authorization of a GMO is made by the relevant technical ministry (e.g., agriculture, health, animal product, or industry), on discussions with the Ministry of Environment, who would have previously received advice from the relevant branch of the CNBIOS. The results of consultations with the public are also considered by the relevant technical ministry during the decision-making process. The NBF does not identify the ministry responsible for ensuring compliance with and enforcement of decisions.

Guinea-Bissau

According to an interview with a Guinea-Bissau government official, the Ministry of Environment is the primary government office responsible for biosafety and will make decisions on applications in consultation with the Ministers of Agriculture and Health. A National Biosafety Committee with broad government and nongovernmental representation also advises on biosafety policy matters but is not a technical committee and therefore does not conduct risk assessments. The biosafety regulatory system established in Guinea-Bissau is not expected to differ significantly from the systems currently being implemented in other WAEMU countries.[8]

Mali

Mali completed its NBF in April 2005.[9] This NBF includes a draft biosafety bill that is based in part on the African Model Law on Biosafety, a document developed by the African Union in 2001 to help countries to comply with the Cartagena Protocol.[10] As defined in the biosafety bill and the NBF, the Ministry of Environment is the authority responsible for the biosafety regulation and must therefore review and authorize any activity involving a GMO or a product of a GMO. GMO activities specified in the draft bill include the importation, transportation, contained use, and release and placement of GMO products in the market. The notification and authorization requirements for contained use and confined field trials are identical to those that apply to product release in the market.

A National Biosafety and Biotechnology Committee, made up of representatives from different government ministries as well as nongovernment personnel, provides

[8] Republic of Guinea-Bissau, *National Biotechnology and Biosafety Framework of Guinea-Bissau* (Bissau, Guinea-Bissau: Ministry of Natural Resources and Environment, 2008), online: http://www.unep.org/biosafety/Documents/NBFs/GB_NBF_En.pdf.

[9] Regina Birner et al, *supra note* 3; Republic of Mali, *National Biosafety Framework Project* (Bamako, Mali: Ministry of Environment and Sanitation, 2005), online: http://www.unep.org/biosafety/files/MLNBFrepEN.pdf.

[10] African Union, *Revised African Model Law on Biosafety* (Draft) (Addis Ababa, Ethiopia: African Union, 2008), online: http://www.africa-union.org/root/au/auc/departments/hrst/biosafety/AU_Biosafety_2b.htm.

advice and recommendations to the Ministry of Environment on both policy issues and applications for GMO activities. According to the draft bill, the public is consulted during the decision-making process, and their opinions and concerns are considered by the Ministry of Environment. To authorise an activity, the Minister must find that the GMO or the product of such GMO has no significant risk to human health, biologic diversity, or the environment (Article 25) and that the GMO (1) benefits the country without risk; (2) participates in sustainable development; (3) does not harm the socioeconomic environment; and (4) responds to ethical values and community concerns, and does not threaten the community knowledge and technologies (Article 27). The provision to "benefit[s] the country without risk" requires that the minister determine that the GMO is not only safe but beneficial. Once a GMO is approved, the Ministry of Environment has the legal authority to ensure compliance with, and enforcement of, the regulations.

To encourage development of GMOs prior to the enactment of the national biosafety law, the Ministry of Agriculture has drafted a decree that regulates research on genetically modified plants. This decree was developed as an interim measure to allow GM testing while the biosafety bill becomes law. Strangely, this decree establishes a dramatically different legal structure for authorization of GM activities than that outlined in the biosafety bill. Under the decree, contained testing and field trials using genetically modified plants can be done only with written authorization of the Interministerial Committee. The Interministerial Committee, established by the decree, is chaired by the Minister of Agriculture. Members of the committee include representatives from eight other ministries. Decisions are implemented by the Subcommittee on Genetic Engineering, which consists of scientific specialists who also implement risk assessments and provide technical advice to the Interministerial Committee. The decree, which defines the Ministry of Agriculture as the responsible department, therefore establishes an interim structure that is completely different than the permanent structure outlined in the proposed biosafety bill.

Niger

Niger completed its NBF in January 2005 and has since drafted a biosafety bill.[11] Under the draft bill, the Ministry of Water and Environment is identified as the National Competent Authority (NCA) for biosafety and has the legal authority to make decisions on applications in collaboration with other specified administrations. The draft biosafety bill requires government authorization for all GMOs and/or products derived from GMOs involved in confined-use experiments, laboratory activities, field trial disseminations, voluntary releases into the environment,

[11] Republique du Niger, *Cadre National de Biosecurité* (Niamey, Niger: Ministre de L'Hydraulique, de L'Environnement et de la Lutte Contre la Desertification, 2005), online: http://www.unep.org/biosafety/files/NENBFrep.pdf.

importation, exportation, or releases as commercial products. The National Technical Biosafety Prevention Authority is a scientific committee made up of government and nongovernment experts that provide the Ministry of Water and Environment with advice regarding the authorization of applications. The draft bill states that all decisions require a risk assessment of the potential impacts on "human, animal and vegetal health, biodiversity, and environment." The biosafety bill does not identify standards or guidelines to be used by the National Technical Biosafety Prevention Authority during the authorization process. It also fails to identify which government entity is responsible for compliance with and enforcement of authorizations, although it does envision the implementation of a separate decree to address this issue.

Senegal

Senegal, like Niger, completed its National Biosafety Framework in 2005 and has since drafted a biosafety bill.[12] The biosafety bill identifies the National Biosafety Authority (Senegal NBA) as the legal authority responsible for the regulation of biosafety and authorization of activities involving GMOs. The bill, however, does not identify whether Senegal NBA is a new or existing agency or office within an existing ministry, or whether it is a newly created independent government agency.

Under the draft biosafety bill and the NBF, GMOs and/or products derived from GMOs involved in contained-use experiments, dissemination into the environment, importation, exportation, or released as commercial products must be authorised.

The bill establishes the National Biosafety Committee (NBC) and identifies, as part of its mandate, the provision to supply the Senegal NBA with technical advice on biosafety issues. The NBC assesses, in cooperation with the National Scientific Advisory Committee, individual applications prior to the authorization of GMO activities.

The bill requires that a risk assessment be conducted and that the evaluation shall include analysis of "the direct and indirect risks to human and animal health, the environment, biological diversity and the socio-economic conditions and ethical values of the country." The bill does not clearly articulate a safety standard to be used in the analysis of applications but does state that the Senegal NBA can approve only a GMO or a derivative product if it will "benefit the country without causing any damaging risks to human and animal health as well as the biodiversity and environment," and if it will contribute to sustainable development. The bill requires the Senegal NBA to make all GMO applications available to the public for consultation prior to an application's approval.

[12] Republique du Senegal, *Cadre National de Biosecurité* (Dakar Hann, Senegal: Ministre de L'Environnement et de la Protection de la Nature, 2005), online: http://www.unep.org/biosafety/files/SNNBFrep.pdf.

Togo

Togo completed its National Biosafety Framework at the end of 2004 and completed a draft biosafety bill in April 2006.[13] Under the draft bill, the Ministry of Environment and Forest Resources is identified as the NCA for biosafety and has the legal authority to approve or reject applications. The draft biosafety bill requires government authorization from the Ministry of Environment and Forest Resources for all GMOs and/or products derived from GMOs involved in confined-use experiments, dissemination into the environment, importation, exportation, released as commercial products, or released voluntarily or involuntarily into the environment. The Ministry of Environment and Forest Resources receives biosafety advice from the Scientific and Technical Committee, a technical and scientific body composed of experts from public and private institutions. In addition to the Scientific and Technical Committee, each relevant Ministry has an internal biosafety committee that can be consulted depending on the nature of the GMO. The Ministry of Environment and Forest Resources also has the authority to establish public committees, although their role in the authorization process is not defined.

The draft bill requires a risk assessment for each application to determine "if the GMO and/or derived products are beneficial without causing harm; will contribute to sustainable development; will not cause socio-economic harm; will not threaten traditional knowledge and values; and will comply with ethical rules." The draft bill states that decisions on applications should be made "to avoid unfavourable effects for human and/or animal health, biodiversity, socio-economic and cultural values." Finally, the draft bill establishes two separate risk evaluation procedures, one for pharmaceutical GMOs and one for nonpharmaceutical GMOs.

ISSUES RAISED DURING LEGAL ANALYSIS OF DRAFT LAWS AND NBFs

Comparisons between the proposed national biosafety systems of the WAEMU countries and the obligations outlined in the Cartagena Protocol have raised several important legal issues, which are outlined and discussed in the next section.

Scope of the Biosafety Regulatory System

A regulatory system must define its scope adequately and accurately. Which products are included? Which actions require oversight? Which risks should be evaluated? The answers to these questions, which can be found in the legal documents establishing the system, determine the comprehensiveness of the system. Although the Cartagena Protocol does identify certain minimal requirements, many WAEMU

13 Republic of Togo, *National Biosafety Framework* (Lomé, Togo: Ministry of the Environment and Forest Resources, 2004), online: http://www.unep.org/biosafety/files/TGNBFrepEN.pdf.

countries have proposed broader, more stringent biosafety systems than the international standard.

Which Products Should Be Regulated?

The Cartagena Protocol applies to all LMOs, which are defined as "living organisms that have a novel combination of genetic material obtained by either *in vitro* nucleic acid techniques or cell fusion." The Protocol does not apply to products created using traditional breeding methods, such as cross-breeding or "conventional" breeding methods (i.e., irradiation and chemical mutagenesis). It is also not applicable to products of LMOs that are not living or capable of replication, such as finished food products or clothing derived from LMOs.

Under the laws, draft laws, and NBFs of the WAEMU countries, many national biosafety regulatory systems regulate a broader range of products than defined in the Cartagena Protocol. In the Burkina Faso biosafety law, a "genetically modified or transgenic organism" is defined as "any biological entity or organism which genetic material has been modified in a way other than natural multiplication or recombination." A plain reading of this definition suggests it could include crops developed using chemical mutagenesis, irradiation, or other techniques that scientists use to change the DNA of organisms. Therefore, all organisms of which the DNA has been modified in an unnatural manner, regardless of the involvement or addition of foreign DNA, could require authorization for any activity, including contained use or confined field trials. This is a much broader scope than most biosafety regulatory systems around the world, which usually limit the scope to the novel introduction of DNA through in vitro recombinant techniques. The biosafety regulatory systems of some countries, such as Canada, include novel products based on their phenotypes, but even those systems apply to a smaller range of products than is included in the Burkina Faso biosafety law. Therefore, one issue for the design of the national biosafety regulatory systems is whether that system limits itself to LMOs as defined in the Cartagena Protocol or include other "conventional" processes that produce organisms with altered DNA.

An issue that arises during the development of a national biosafety regulatory system is whether the new system incorporates the definition of LMOs as currently defined in the international arena or whether it broadens the scope to include products derived from those organisms, whether they are living or capable of replication. The plain reading of the legal documents from WAEMU countries (e.g., the Burkina Faso Biosafety Law), could be interpreted to require notification and authorization of the importation or commercial sale of paper, wooden furniture and wooden picture frames derived from GM trees, or the maize oil, starch, and syrup derived from GM corn. If each product derived from a GMO must be approved using the same procedures, risk assessments, and information requirements as are currently required, the system could be inundated with thousands of applications. In most, if not all, of those cases, there could be little or no potential risk to the environment or biodiversity, because the product cannot replicate or transfer its DNA to a living organism.

Most countries with an established biosafety regulatory system require approval only for a particular GE organism at each stage of development. Premarket safety approval for each individual product derived from GMOs is not required. Instead, the food safety and environmental risk assessments for the particular GE organism includes an environmental or food safety analysis of the different products that could be produced or derived from the GE organisms. Under this system, only the GMO itself is subjected to the regulatory process, although analysis of all potential products derived from that specific GMO is included in the assessment.

Which Activities Should Be Regulated?

The Cartagena Protocol targets the transboundary movement of LMOs by focusing on two particular types of activities: (1) the intentional release of the LMO into the environment (e.g., the commercial release of seeds, plants, or animals) and (2) the use of LMOs for food, feed, or processing (LMOs-FFP, e.g., commodity grain for human or animal consumption). To ensure that the previously mentioned activities are implemented safely, the Protocol places a legal burden on both the importing and exporting countries. The scope of the Cartagena Protocol also includes LMOs that transit a country or are used in a contained facility, although these activities are exempt from most safety obligations. The Cartagena Protocol therefore sets forth a proportionate risk-based regulatory system in which activities with a greater relative risk to biodiversity (e.g., intentional releases of commercial LMOs) require detailed review and approval procedures, whereas relatively low-risk activities (e.g., the contained use of an LMO in a laboratory) require significantly less regulation.

The national biosafety regulatory systems reviewed for this report generally regulate imported and domestically produced GMOs and their products. At minimum, most of the regulated activities include (1) contained use (research laboratories); (2) limited releases into the environment (confined field trials); and (3) the placement of products on the market (the commercial release or the importation of a commodity grain shipment). The regulatory treatment by WAEMU countries of these activities is more stringent than the Cartagena Protocol. For many of the WAEMU countries, the regulatory requirements for contained-use experiments, confined field trials, and commercial releases are the same, with each activity requiring advanced informed consent that involves detailed applications, risk assessments, and risk management plans. Although similar stringent treatment can be found in biosafety regulatory systems established by other developing countries, they differ significantly from the biosafety regulatory systems in developed countries. The systems in developed countries require detailed and comprehensive procedures for commercial releases but contain less stringent risk assessment and decision processes for confined field trials. Regulation is minimal, and preapproval is usually not required for research done in contained laboratories.

What Issues Should the System Address?

The definition of biosafety varies greatly from country to country. The Cartagena Protocol is primarily an environmental agreement with a mandate to avert adverse harm to biodiversity and the environment. The Protocol also addresses human health risks and socioeconomic considerations if they result from environmental and biodiversity concerns. The issues targeted by the national biosafety regulatory systems determine the comprehensiveness of the regulation and whether this covers the full range of possible risks from GMOs.

All WAEMU countries state that their biosafety system includes an assessment of the potential risks to biodiversity and the environment, which is the minimum required by the Cartagena Protocol. Some of the countries' NBFs and draft or final laws also tackle other risk areas, including human and animal health, socioeconomic considerations, and compatibility with ethical values. Addressing those additional areas can be consistent with the Cartagena Protocol and allows countries to establish systems that are more comprehensive or more protective.

One substantive area addressed by most national biosafety regulatory systems is food safety. To date, many GMOs are food crops intended for human and animal consumption, and the public demands that government ensure the safety of these products. By including a food safety regulatory process or review as a component of the biosafety regulatory system (although not necessarily the responsibility of the same government agency), GE organisms that are ready for market can be assessed for food-safety and environmental risk assessment at the same time. Food safety must also be considered when GE organisms are imported solely for food and feed purposes, such as grain shipments or food aid.

Most of the WAEMU countries' NBFs and/or biosafety laws address issues related to human health, including food safety. Although there are significant provisions detailing how environmental and biodiversity risk issues are to be tackled, the biosafety laws fail to outline the same provisions for food safety concerns (e.g., they do not identify who will conduct and/or review the food safety risk assessment). Similarly, interviews with members of the national regulatory bodies and key stakeholders found that the current systems do not consider food safety issues thoroughly. They think that additional discussions are needed when considering whether to include GMO food safety in biosafety regulation. If it is decided that food safety issues should be included, the regulatory process must integrate this.

There are two reasons why the food safety regulation of GMOs could be ideal for centralized regional decision making. First, products are involved at the commercialization stage, such as commerce and cross-border trade. Second, as opposed to environmental issues, which vary according to geographic region and agricultural practices, food safety risks are generally uniform throughout the region. Therefore, if WAEMU countries include food safety in their biosafety regulatory systems, they might want to consider regional regulation as a more efficient method of safety assessment.

In addition to food safety, the public is also concerned about the potential socioeconomic impact of the commercial release of GMOs. The proposed biosafety regulatory systems in Benin, Burkina Faso, Senegal, and Togo all state that they will consider these potential impacts in the decision-making process. Unfortunately, none of the proposed systems identify which socioeconomic issues are to be considered, how they will be analysed, or how they will factor into the decision-making process. Without these details, it is impossible to determine whether these proposals comply with the Cartagena Protocol or other international obligations.

Article 26 of the Protocol allows for the use of socioeconomic considerations presented by LMOs in decision-making but places conditions on that analysis. First, it limits the socioeconomic considerations to those effects that arise "from the impact of living modified organisms on the conservation and sustainable use of biological diversity, especially with regard to the value of biological diversity to indigenous and local communities."[14] The language of Article 26 restricts the socioeconomic issues of LMOs to only those that are a direct result of impacts on biologic diversity, but some stakeholders think that the impacts of LMOs are much broader than that. Although the broader socioeconomic considerations might be valid societal concerns relevant to GE organisms, they are not likely to be included in a biosafety regulatory system consistent with the Protocol.

Second, Article 2 of the Protocol states that inclusion of socioeconomic considerations must be done in a manner that is consistent with other international obligations.[15] In general, World Trade Organization (WTO) rules state that decision standards shall rely on scientific risk assessments, thereby greatly limiting a country's ability to make a decision based on issues that do not directly relate to the safety of the product. WTO rules sometimes permit consideration of nonsafety concerns, but those exceptions are narrowly defined and primarily consider only cost-benefit analyses. To ensure consistency with the Cartagena Protocol and other international obligations, countries that include socioeconomic considerations in their national biosafety regulatory systems should limit them to those as defined under Article 26.

Compliance with ethical values or rules is another criterion included in at least two of the proposed national biosafety regulatory systems (Burkina Faso and Côte D'Ivoire). Neither the Burkina Faso biosafety law nor the Côte D'Ivoire NBF provide details on or guidance outlining the analysis, procedure, or weight that ethical values will be given in the decision-making process. Such a requirement may or may not be consistent with the Cartagena Protocol or other international obligations, for the same reasons identified in the analysis of socioeconomic considerations. Thus, although some WAEMU countries are concerned about the ethical implications of GMOs, attempts to protect these rights within a national biosafety regulatory system actually could make them inconsistent with international law.

[14] *Cartagena Protocol*, *supra* note 1 at art 26(1).

[15] *Ibid* at art 2.

Labelling and liability and redress are two final concerns included in many of the biosafety regulatory systems of WAEMU countries. In most cases, the labelling provisions are enacted to provide consumers with information and choice, and they do not address safety concerns. The liability and redress provisions in national biosafety regulatory systems vary widely among the participating countries. Both issues, however, are very controversial, as was revealed in the negotiations for the Nagoya – Kuala Lumpur Supplementary Protocol on Liability and Redress.

The Decision-Making Process

An important part of the biosafety regulatory system is the process that is followed when authorizing the use of a GMO. The roles and responsibilities for the agency or committee legally responsible for making decisions, as well as other institutions, must be clearly identified. Clear guidelines outlining the authorization standard and the extent of public participation are also instrumental in ensuring fair and due process.

Each biosafety regulatory system proposed by the various WAEMU countries outlines a decision-making process. Most of the regional systems identify the agency responsible for final decisions (in most cases either the Ministry of Environment or the National Biosafety Agency) and the requirement that the GMO needs authorization before any activity is begun. Almost every country establishes advisory committees with government and nongovernment representatives that provide policy and/or technical guidance to the agency or committee. Some issues about the decision-making process do arise during the review and analysis of the proposed systems, and those issues are identified in the next section.

What Is the Decision Standard?

A biosafety regulatory system should establish a standard that is used to guide decisions in the approval process. Because no activity is risk free, and the concept of safety varies greatly, the standard should identify the level of protection necessary for authorization. To ensure consistency, the regulatory system should define the factors to be considered and should identify the minimum level of risk acceptable. The standard should specify whether a cost-benefit analysis of the possible introduction of a GMO is to be considered. In a functional system, all interested parties can expect the legal standard to be applied uniformly and fairly. Unfortunately, the Cartagena Protocol does not specify a safety standard to be used in the assessment of individual applications, resulting in the establishment of various national standards. The reason for this is because the concepts of safety and risk vary greatly from country to country.

The countries participating in this study have begun to discuss which factors should be considered important in the decision-making process. Both the Burkina Faso law and the Mali draft law specify that a GMO must (1) benefit without causing harm; (2) contribute to sustainable development; (3) not cause socioeconomic harm; and (4) comply with ethical rules. Although the factors are identified, the legal documents do not define key terms, such as *sustainable development, socio-economic considerations*, or *ethical rules*, thereby weakening the effect of the standard. The lack of clear, detailed definitions hinders the decision-making process and limits the ability to predict the likelihood of authorization of a particular GMO. Some countries, such as Côte D'Ivoire and Senegal, do not identify the criteria to be used in the approval process of an application, making this an area that needs further consideration.

Public Participation in the Decision Process

Article 23 of the Cartagena Protocol states "[t[he Parties shall, in accordance with their respective laws and regulations, consult the public in the decision-making process regarding living modified organisms."[16] The Protocol, however, provides little guidance on the public consultation process or the inclusion of public concerns in the decision-making process. Public participation can include the submission of written comments, participation in a hearing, or public interviews conducted by government officials. In most countries around the world, public input does inform the decision-making process, but the regulatory agency making decisions is not obliged to consider the information. Public participation can help to garner consumer and public trust in both the regulatory system and its decisions.

There are varying amounts of public participation in WAEMU country biosafety regulatory systems. Some incorporate public participation in the decision process through the inclusion of nongovernment representatives on their advisory committees. Others notify the public of an application and invite comments. Burkina Faso has drafted a separate decree that specifies how it will incorporate the public's opinions in the decision process. Under the decree, a chief surveyor will collect written and oral comments from citizens living in the area of a proposed GMO release, draft a report analysing the comments, and share it with the National Biosafety Agency in advance of the decision. Thus, there is a range of mechanisms used by WAEMU countries that allow the public to participate in biosafety decision making

Harmonisation of National Biosafety Regulatory Systems

Tables 2 and 3 in the Annex and the previous discussions identify the key legal components of the different national biosafety regulatory systems. The similarities among the national biosafety regulatory systems are likely to facilitate harmonization of the systems. The country that might be most affected by a regional harmonization effort is Burkina Faso, because it would likely need to amend existing legislation to conform with the other members of WAEMU.

[16] *Ibid* at art 23.

The substantive sections of the national biosafety regulatory systems that would not require significant changes to achieve harmonization are (1) the identification of applicable products (GMOs and derived products); (2) the listed activities (i.e., contained use, confined field trials); and (3) the decision-making structures (the incorporation of advisory committees and technical expert advice in the decision-making process). The various draft laws treat these legal areas very similarly, if not identically. Substantive areas requiring some effort to harmonise include (1) the risk considerations (human health, food safety, socioeconomic); (2) the decision-making standard; (3) the incorporation of proportionality; (4) transparency and public participation; and (5) confined field trials. An agreement identifying both the path and language to be used would be necessary to achieve harmonisation among national legislations and regulations.

REGIONAL ANALYSIS

In addition to national level activity, regional bodies in West Africa are also contemplating the introduction of a regional regulatory biosafety system. This section explores the possible legal implications of establishing such a system, including any inconsistencies with the Cartagena Protocol, and also considers the regional biosafety work completed by CILSS through INSAH and analyses the existing regional regulatory systems of other substantive areas to determine whether they could function as a model for a WAEMU regional biosafety regulatory system.

A Potential WAEMU Regional Regulatory System and the Cartagena Protocol Obligations of Individual WAEMU Countries

Each WAEMU country is a Party to the Cartagena Protocol and must implement its own obligations to promote sustainable use of biotechnology, in particular LMOs. Analysis of the Protocol confirms that a regional biosafety regulatory system established by a regional body such as WAEMU can satisfy each country's Protocol obligations. Article 14 of the Protocol states that Parties may enter into "bilateral, regional, and multilateral agreements regarding intentional transboundary movements of living modified organisms." A regional agreement satisfies the obligations of a particular country under the Protocol if the following conditions are met: (1) the agreement is "consistent with the objectives of this Protocol"; (2) the agreement does not result in a lower level of protection than the Protocol; and (3) the Parties to the Protocol are informed about the agreement through the Biosafety Clearing-House (Article 14).[17]

The Protocol does not specify how a regional agreement can ensure that it is "consistent with the objectives under the Protocol." Based on a plain reading of the

[17] *Ibid* at art 14.

Protocol's objective, scope, and obligations, however, several conditions are likely to need to be met to ensure that the regional agreement is "consistent" with the Protocol. First, any regional biosafety regulatory system should apply to all LMOs, such as animals, plants, seeds, and microorganisms. Second, it ought to cover the two major types of transboundary movements: the commercial release of LMOs and the use of LMOs-FFP. Third, it should establish procedures that authorize such movements once an adequate level of environmental and biodiversity protection is confirmed. Finally, it should contain a decision-making process consistent with the Protocol, involving scientific risk assessments and risk management measures. The establishment of a comprehensive regulatory system that incorporates the Protocol obligations would best ensure individual national compliance.

A WAEMU-led regional biosafety regulatory system could address some or all of the major obligations under the Protocol. For example, a regional system could regulate LMOs-FFP – or address only confined field trials, and refrain from addressing the commercial release of LMO products into the environment, because this is not likely to occur in many WAEMU countries for several years. In such a situation, each individual country must introduce measures that satisfy these remaining Protocol obligations. Each WAEMU country would need to continue enacting biosafety laws and establish national competent authorities, thereby significantly decreasing the value of a regional biosafety mechanism. Although a regional agreement might address some national obligations under the Protocol, each Party to the agreement must ensure that it is meeting all of the obligations outlined under the Protocol.

The Legal Authority for WAEMU to Establish a Regional Biosafety Regulatory System

WAEMU is the major regional body analyzed in this chapter. Does it have the legal authority to establish a regional biosafety regulatory system? The representatives' interviewed for this study are of the opinion that a binding regional biosafety regulatory system can be established and enforced under the WAEMU Treaty. An analysis of this authority follows.

Legal Authority under WAEMU

Although the WAEMU Treaty is primarily an economic instrument, it contains two sections that may permit the establishment of a regional biosafety system. First, Article 4(d) states that one of its objectives is to coordinate national sectoral policies in several fields, including the environment and agriculture. Protocol No. 2 of the WAEMU Treaty states that the goal of environmental coordination is to improve the environment within the Union (Article 9). The Conference of the Heads of State and Government has the authority to pass acts that become addenda to the treaty to improve the environment. The council also has the legal authority to make regulations that achieve that goal (Article 11). Protocol No. 2 specifies that one of the objectives under the environmental sectoral policy is to protect natural resources and *biodiversity* (Article 10; emphasis added). Thus, to the extent that a regional biosafety regulatory system would improve the environment in the Union and/or protect biodiversity, the establishment of acts by the Conference and/or regulations by WAEMU Council or Commission would be authorized by the treaty.

The second potential authority permitting the creation of a regional biosafety regulatory system can be found in the treaty's objectives. Article 4(c) of the treaty identifies, as one of the main objectives, the need "to create between Member States a common market based on freedom of movement of people, goods, and services." GMOs, such as engineered corn or cotton seeds, are clearly agricultural goods whose producers could benefit economically from access to the common markets if these goods were determined safe for human health and the environment. It is possible that WAEMU could regulate GMOs to promote a common market.

Compliance and Enforcement of WAEMU Acts and/or Regulations

Compliance with and enforcement of biosafety acts and/or regulations in each member country is a major concern that is addressed in several WAEMU Treaty articles. First, Article 43 states that regulations established by the WAEMU Council are obligatory in all their elements and are directly applicable in all Member States. Second, Article 6 states that acts adopted by the union bodies intended to ensure compliance with objectives of the present treaty and that are in accordance with the established rules and procedures shall be applied in each Member State notwithstanding any contrary, former, or subsequent national law. This means that any act adopted by WAEMU takes precedence over national laws on the same subject. Thus, if WAEMU established a biosafety regional regulatory system, it would take precedence over any former or subsequent biosafety law in any of the WAEMU countries. If a country refused to abide by the WAEMU biosafety regulatory system, the Court of Justice of the union and the commission could be used to solve the dispute.

Biosafety Topics to Be Included in a WAEMU Regional Biosafety Regulatory System

As identified earlier, WAEMU could establish a binding regional biosafety regulatory system. The easiest way to accomplish this would be for the WAEMU Council or Commission to establish a biosafety regulation under Article 42 of the treaty and Article 11 of Protocol No. 2. The most legally defensible scope probably would involve the regulation of the commercial release of GMOs and their potential impact on biodiversity and the environment. This would necessitate a regulatory system that included (1) a common market for the movement of agricultural goods and (2) protection of the environment. Such a system could also include the regulation of confined field trials, although a regulatory system composed of only a field trial regulation would likely not fall within the legal mandate of the WAEMU treaty, because confined trials do not involve commercial goods nor do they pose considerable risks to the environment or biodiversity.

The WAEMU regional biosafety regulatory system could also address food safety concerns, because one of the objectives of the agricultural sectoral policy is to achieve food safety within the Union (Article 14). To date, however, WAEMU has not established many regional environmental laws or regulations.

A Model for Consideration

If WAEMU is to establish a regional biosafety regulatory system, the recent WAEMU regional regulation for veterinary drug approvals might serve as a model. Approximately five years ago, WAEMU began to harmonize the approval process for veterinary drugs throughout the union. Prior to 2006, if a drug manufacturer wanted to market a veterinary drug in the eight WAEMU countries, the manufacturer had to obtain eight authorizations from the eight different countries. The WAEMU regulation adopted in March 2006 established a regulatory system that allows a veterinary drug to be marketed in all eight countries once approved by the WAEMU Commission. A permanent Secretariat to handle applications has been established, as well as a Regional Committee of Veterinary Drugs. The committee members are experts from the scientific community who review applications and advise the commission on the safety of the product. The WAEMU regulation provides each country with a one-year transition period. The regulation does not address precommercial testing of veterinary drugs nor does it contain a mechanism to enforce the commission's decisions.

A regional biosafety regulatory system could establish an expert committee to review and approve applications, as the veterinary drug approval process has. Responsibilities of the expert committee could include the review of commercial products, whereas pre-commercial activities (e.g., contained research experiments and confined field trials) and postmarket enforcement would be overseen by individual country regulatory processes. A significant shortfall with the regional veterinary drug approval regulation is that there is little, if any, public participation or transparency in the decision-making process. The committee conducts its review behind closed doors, and the public does not have the opportunity to review or comment on the application. No public input is provided to the committee or the commission. Although this system, with no public participation, may work well for veterinary drug approvals, the public expects both transparency and the opportunity to participate in the regulation of biosafety. Therefore, if the veterinary drug regulation is used as a model for the development of a regional biosafety regulatory system, mechanisms for transparency and public participation in the decision-making process might be added.

CILSS/INSAH REGIONAL BIOSAFETY INITIATIVE

Background and Current Status

On behalf of CILSS, INSAH has been working for several years on a regional convention to establish a common biosafety regulatory system for CILSS countries as well as a regional coordination framework to carry out that convention. In January 2006, INSAH brought together representatives from the CILSS countries to review and revise three documents: (1) the Framework Convention Introducing a Common Biosafety Regulation for the Prevention of Biotechnological Risks in the CILSS Countries; (2) the Framework Convention Instituting Common Regulations for Conventional and Transgenic Seeds in the CILSS Area; and (3) the Structure and Operation of the Regional Consultative Committee for Seeds and Biosafety. That meeting resulted in near-final documents that were presented to and adopted by the CILSS Council of Ministers in March 2006.[18] Thus, the output from the INSAH biosafety effort to date has been the consensus adoption by the CILSS Council of Ministers of three documents that would, on ratification, regionally regulate biosafety and seeds for CILSS member countries.

According to officials at INSAH, the next steps for the CILSS Biosafety Convention are the following: (1) the CILSS Ministers for each country will be presented with the final Convention for their signature; (2) each country will need to ratify the Convention and pass national legislation consistent with the Convention; and then (3) the Convention will come into force after ratification by five of the nine CILSS countries. It is anticipated that signatures from the nine countries could be obtained relatively soon, but ratification will take longer. Once the Convention is ratified by five countries, a regional regulatory system will be established to assess projects involving GMOs.

While CILSS and INSAH begin the long and slow process of bringing the Biosafety Convention into force, INSAH is working to implement the Regional Consultative Committee (RCC). The RCC will include in its membership a nonvoting member from WAEMU as well as other relevant regional bodies. It is proposed that because the RCC will not have legal authority to make binding decisions during the time it is operational until the Convention comes into force, it can act as an expert regional body and voluntarily help member states with technical biosafety issues.

Summary and Legal Analysis of the Biosafety Convention

The objective of the INSAH/CILSS Biosafety Convention is to establish common biosafety regulations in the CILSS region and set forth a comprehensive model to

[18] Sahel and West African Club, *Agricultural Biotechnology and the Transformation of West African Agriculture: Synthesis of the Regional Consultation with West African Actors* (Paris, France: Sahel and West African Club Secretariat, 2006).

harmonise the national biosafety regulatory systems. The CILSS Biosafety Convention requires advance authorization from the national competent authority for any activity involving a GMO, including use in contained laboratories, confined field trials, commercial releases into the marketplace, and all imports and exports. The convention also addresses GMOs and products derived from GMOs, although it does limit derived products to those that are eaten as food by humans or as feed by animals. For authorization, the Convention requires that the risk assessment analyses the risks to human and animal health (including food safety), the risks to the environment, and when releasing a commercial product into the environment, an in-depth study of the potential ethical and socioeconomic consequences. It also states that national competent authorities must engage the public in the decision-making process and that any inspections and enforcement are the responsibility of the national competent authority. The Convention has detailed provisions on risk assessment, risk management, labelling, compliance, and enforcement. If implemented as written, the Convention would meet the objectives and requirements of the Cartagena Protocol.

A second INSAH/CILSS document, "Structure and Operation of the Regional Consultative Committee for the Regulation and Control of Seeds and GMOs," establishes an RCC to implement the Convention. In addition to implementing the Convention, the RCC's powers and responsibilities include promoting and facilitating the circulation of seeds, defining the requirements for registering products, and "controlling the circulation of GMO products" with the CILSS countries. The RCC is composed of national seed committee members, members of national agencies of biosafety and scientific experts, and has an office and secretariat at INSAH. The RCC will meet several times a year to conduct its duties, which include reviewing files that require authorization for GMO activities in CILSS countries.

As a model for regional biosafety regulation, the most interesting feature of the INSAH/CILSS Biosafety Convention is how it balances the roles of the regional institution – the RCC – and the national competent authorities. First, the Convention states that each country must designate an authority to implement and enforce the convention. That national authority will make decisions authorising the use of GMOs in laboratories, greenhouses, and field trials after giving the RCC thirty days' notice to submit comments on the proposed decision. If a country does not have a regulatory framework in place, then the decisions for laboratory, greenhouse, and field trials are made by the RCC. Decisions for "deliberate releases or marketing at the country level" (assumed to mean commercial releases) are also decided by the national competent authority, but applications for marketing a product in "other CILSS countries" (assumed to mean more than one country) are determined by the RCC. In addition, any provision specifying prior authorisation for an activity using a GMO states that authorisation must come from the national competent authority, not the RCC. Similarly, the INSAH/CILSS document establishing the RCC defines the implementation of the Convention and collaboration with the national biosafety authority as the RCC's responsibility, although it is not responsible for the final decisions regarding applications. In fact, that same document states that the mandate of national agencies includes "giving authorizations at the national level." Therefore, the regional system proposed at INSAH/CILSS requires the establishment of national biosafety authorities that are empowered to make decisions on GMO activities and are guided by the procedures and technical expertise of the RCC.

In summary, the current draft of the INSAH/CILSS Biosafety Convention sets forth a regional regulatory system in which (1) each country establishes its own national biosafety regulatory system, using the procedures, definitions, and responsibilities for their national competent authority set out in the convention; (2) the national authorities make most of the decisions regarding authorisation of activities using GMOs; (3) the INSAH/CILSS RCC reviews and advises on proposed national decisions on particular GMOs and provides general technical and policy support to the national competent authorities; and (4) the RCC makes some authorisation decisions for countries without a regulatory framework in place or when products will be marketed throughout the region. Thus, for most situations, the proposed regional system is decentralized and nonbinding, placing the legal authority for authorisations on each member country. The benefit of an INSAH/CILSS Biosafety Convention is that it provides harmonisation, technical support, and regional oversight of the procedures used in the authorisation process. The RCC has only limited powers to introduce binding decisions.

The CILSS Common Regulation for the Registration of Pesticides

The work at INSAH on the Biosafety Convention builds on the experience and success INSAH has had with the 1999 Common Regulation for the Registration of Pesticides. Although the INSAH pesticide regulation process is significantly different from the proposed INSAH Biosafety Convention, a review of that Convention and its procedures can help to identify the potential strengths and weaknesses of a centralised regional biosafety regulatory decision process that could be implemented by either INSAH/CILSS or WAEMU.

The CILSS Ministers signed the first common pesticide regulation in 1992, which at that time was ratified by Gambia only. The document was then revised in 1999, resubmitted to the Council, and ratified by eight of the nine countries (all except Guinea Bissau). Under the Common Pesticide Regulation, the approval and registration of pesticides for the CILSS countries is performed by the Sahelian Pesticide Committee (CSP), which consists of two voting members from each country that has ratified the Regulation and other experts in relevant scientific areas. Applicants complete a dossier of information required for registering a pesticide, which is reviewed by the CSP at one of its biannual meetings. The CSP is then authorised by the regulation to determine whether to register the pesticide for five years, provisionally register it for three years, obtain additional information, or deny the registration. Their consensus decision is based on a safety standard elaborated under Article 11 of the Common Pesticide Regulation. Their decision is then communicated to the member countries and the applicant. If the applicant is successful, the pesticide is added to the list of registered products and can then be sold for its intended purpose in any of the nine countries.

The most significant benefit of the Common Pesticide Regulation is the establishment of a centralised regional process for the registration of pesticides. Instead of an applicant submitting individual applications to multiple national regulatory bodies and receiving multiple registration decisions (some which might have reached different conclusions), one application is submitted to a committee of experts, whose decision is applicable to all participating CILSS countries. Such a system decreases transaction costs for the applicants and national governments, applies scarce technical resources in the region in a targeted manner, provides applicants with access to a larger market for their products, and allows for the analysis of environmental and health and safety concerns that are not limited by country boundaries. A regional Biosafety Convention that is responsible for approving the commercial release of genetically engineered organisms could see benefits similar to those arising from the Common Pesticide Regulation.

Although the Common Pesticide Regulation does provide for a regional approval process, it does not cover all aspects of pesticide regulation at the regional level comprehensively. First, as with the CILSS Biosafety Convention, it is not self-executing. Adoption and signature by the CILSS Council of Ministers do not make the regulation legally binding. Countries must ratify the Common Pesticide Regulation and pass or modify national legislation to comply with its terms. To date, one CILSS country has not ratified the Common Pesticide Regulation, which means the decisions of the CSP are not binding in that country. Similarly, until the CILSS Biosafety Convention is ratified and incorporated into national legislation, any decisions by the RCC have no legal effect. For pesticide regulation in the CILSS countries, only one country, Senegal, had a fully operational pesticide regulatory system at the time of the Common Pesticide Regulation, which means that for most countries, ratification and adoption of national legislation were relatively easy. For biosafety, the extensive work already conducted by individual CILSS countries on their own biosafety regulatory systems through the UNEP-GEF Biosafety Project might make ratification and further modification of national legislation much more difficult.

Second, the Common Pesticide Regulation has an opt-out provision, in which a country can refuse to follow the CSP's registration decision. The Common Pesticide Regulation provides each member state with the right not to allow a pesticide that is registered by the CSP from being sold in their country if they can provide arguments that (1) no areas of use for the pesticide exist in their country; (2) it is impossible to satisfy the conditions related to its use; (3) the ecologic conditions in the country are substantially different from those used in the evaluation of environmental risks by the CSP; or (4) the use of the pesticide contradicts national policies in agriculture, environment, or public health (Article 8). Given the broad nature of the possible reasons for opting out of a CSP decision, almost any country could opt out of any decision it did not like for any reason, scientific, political, or otherwise. In fact, this provision has been invoked a few times, and the CSP has either affirmatively accepted the member state's justification or quietly refused to challenge the member state's justification. If a similar provision were included in a regional biosafety regulatory system, member states could use it to reduce significantly both the advantages and importance of the centralized regional approval process.

A third feature of the Common Pesticide Regulation to consider is that it does not cover pre- or post-registration activities. Regulation of trials or experiments conducted for the purpose of research and development is left to the competent authority in each member state (Article 21). The monitoring of registered pesticides is also delegated to each state as well as the enforcement of registration conditions and the potential negative impact on the environment (Article 24 and 25). The Biosafety Convention being considered by INSAH encourages an even greater sharing of authority between the regional and national biosafety authorities, which could reduce the potential value of a regional approach. For example, the cost savings generated from a regional regulatory system will be less if each country must establish a competent biosafety authority to address field trial applications and to monitor and enforce regional decisions.

A final possible limitation of the Common Pesticide Regulation to consider is the lack of public participation or transparency in its decision-making process. The applications received by INSAH are not made public, the CSP meets in private, and there is no opportunity for the public to comment on pesticide applications. Although the lack of such procedures might be appropriate for the registration of pesticides, the Cartagena Protocol and the NBFs of the CILSS countries all call for significant degrees of transparency and public participation in the regulation of GMOs. It might be difficult for INSAH to provide an adequate level of transparency and public participation across the region.

In conclusion, if West Africa were to establish a regional biosafety regulatory system modelled after the INSAH Pesticides Convention, such a system would centralise the decision making for commercial GMO products but leave pre- and post-development activities to national regulators. This type of a regional system would not eliminate the need for national biosafety authorities, and those agencies would still regulate confined field trials, which is probably the most important biosafety issue in the region for the next five to ten years. As the INSAH pesticide system's decision-making process is neither transparent nor open to the public, the proposed regional biosafety regulatory system would need to incorporate inventive mechanisms and procedures to address transparency and public participation concerns.

CONCLUSION

In summary, several conclusions can be drawn from the WAEMU countries' biosafety activities to date and their possible future national and regional level activities. There are many more similarities than differences between the proposed national biosafety regulatory systems of the different WAEMU countries, which can make harmonisation possible. Although the proposed national biosafety regulatory systems for the WAEMU countries comply with many of the Cartagena Protocol requirements, they have broader scopes and more stringent provisions than the obligations established in the protocol.

It can be observed that when countries embark on establishing a national biosafety regulatory system, consideration must be given to the types of products to be included, activities to be covered, issues of concern, the inclusion of public participation, and standards to be applied. The Cartagena Protocol allows countries to enter into a regional agreement to satisfy their obligations under the protocol as long as that agreement is "consistent with the objectives of the Protocol." The WAEMU Treaty gives WAEMU adequate legal authority to establish a binding regional biosafety regulatory system. If WAEMU establishes a regional biosafety act or regulation, it will be binding on member countries and will replace any existing laws that are inconsistent with its terms. In contrast, the INSAH/CILSS biosafety convention is only binding in each country after it is ratified and national legislation consistent with the Biosafety Convention is introduced.

The regional regulatory system to be established by the INSAH/CILSS Framework Biosafety Convention is a decentralised decision-making system that is primarily only responsible for coordinating technical and procedural aspects of biosafety regulation while entrusting the national biosafety authorities with most of the legal authority to make decisions. This system has been approved by the CILSS countries' Ministers of Agriculture.

Two legal models that demonstrate a coordinated and centralized process for approving products to be marketed throughout the WAEMU region are the WAEMU veterinary drug regulation and the INSAH/CILSS chemical pesticide registration process.

None of the three regional regulation models – the CILSS/INSAH Framework Biosafety Convention, the WAEMU veterinary drug regulation, and the CILSS/INSAH chemical pesticide regulation – contains a level of transparency and public participation in their decision-making process that would be considered acceptable for a biosafety regional regulation and would fully satisfy a firm commitment to the sustainable development law principle of public participation backed by access to information and justice.

16

Comparative Analysis of the National Biosafety Regulatory Systems in East Africa

Gregory Jaffe

Over the past several years, three countries in East Africa – Kenya, Tanzania, and Uganda – have made it a priority to establish a working biosafety regulatory regime for genetically engineered (GE) and genetically modified organisms (GMOs). These governments established committees responsible for drafting laws, regulations, guidelines, and other documents and authorised government agencies to implement the biosafety regulatory regime. To develop and implement a biosafety regulatory system, those committees and government agencies have, among other things, analyzed their current laws and regulations applicable to agricultural products and food and their obligations under international conventions and treaties; conducted meetings to hear from interested stakeholders and the public; and reviewed biosafety systems already in place in other countries (e.g., the United States and the European Union) as well as model laws developed by the Organization of African Unity (the African Model Law) and experts.

This chapter analyses the current and proposed biosafety regulatory systems in Kenya, Tanzania, and Uganda. The first section briefly sets forth the current status of biosafety regulation in each country. In light of the sustainable development focus of this book, particular emphasis is placed on the way that countries have operationalised important principles of sustainable development law such as public participation and good governance, which, as argued in the Introduction and Chapter 1, are reflected and indeed operationalised in the Cartagena Protocol on Biosafety (Cartagena Protocol). The second section discusses some issues that arise when the biosafety regulatory systems in the three study countries are analysed using the components from Section 1 and the international obligations set forth in Section 2. Finally, the last section discusses possible ways forward for East African countries to efficiently and effectively implement coordinated biosafety regulatory systems that protect the environment and human health while complying with international obligations.

BIOSAFETY REGULATORY SYSTEMS IN KENYA, TANZANIA, AND UGANDA

All three East African countries discussed in this chapter have worked for years to establish portions of their national biosafety regulatory systems. The following is a summary of the status of the biosafety regulatory systems in each country.

Kenya

Until mid-2010, in Kenya, the National Council for Science and Technology (NCST) was the government agency responsible for overseeing the implementation of the biosafety regulatory system. It issued Regulations and Guidelines for Biosafety in Biotechnology in 1998 under the Science and Technology Act of 1980, despite that act having no regulatory authorities and no means to enforce compliance with the regulations.[1] The NCST also established the National Biosafety Committee (NBC) to develop the country's biosafety policy and review applications for GMOs.[2] The membership of the NBC included representatives from relevant government ministries as well as representatives from civil society and the national universities.[3]

Under the former interim Kenyan biosafety regulatory system, applications to import or release GMOs (including applications for confined field trials) were submitted to the relevant Institutional Biosafety Committee (IBC), where they were reviewed and assessed for compliance to the guidelines before submission to the NBC.[4] Those applications were then forwarded to the NBC, where they were reviewed by the NBC and/or a technical subcommittee of the NBC. A recommendation was made by the NBC, and the NCST Secretary decided whether to approve the application.[5] Kenya approved many confined trials using those procedures, but not any commercial products.[6]

In July 2002, Kenya began working on several legal documents to turn its interim biosafety regulatory system into a permanent and comprehensive system. Those documents included revised regulations, a biosafety law, and a national biotechnology and biosafety policy. In 2007, the Kenya Draft Biosafety Bill was published by the government.[7] The Kenyan Biosafety Act was approved by the Kenyan Parliament

1 P. Traynor & H.K. Macharia, *Analysis of the Biosafety System for Biotechnology in Kenya: Application of a Conceptual Framework, ISNAR Country Report 6* (The Hague, The Netherlands: International Service for National Agricultural Research, 2003).

2 M. Harsh, "Formal and Informal Governance of Agricultural Biotechnology in Kenya: Participation and Accountability in Controversy Surrounding the Draft Biosafety Bill" (2005) 17 J. Int. Dev. 661, 677.

3 P. Traynor & H.K. Macharia, *supra* note 1; H.K. Macharia, "East Africa PBS Roundtable" (Presentation at PBS Roundtable in Entebbe, Uganda, 18 April 2005).

4 Republic of Kenya, *National Guidelines for the Release of Genetically Modified Organisms (GMOs) into the Environment* (Nairobi, Kenya: National Council for Science and Technology, 1998).

5 P. Traynor & H.K. Macharia, *supra* note 1.

6 H.K. Macharia, *supra* note 3.

7 Republic of Kenya, *The Biosafety Bill 2007* (Nairobi, Kenya: National Council for Science and Technology).

in 2008 and signed by the Prime Minister in 2009. That legislation calls for the establishment of a National Biosafety Authority (NBA) to regulate activities with GMOs and ensure safety to both humans and the environment. The NBA would be managed by a board composed of government officials from relevant agencies as well as outside experts and stakeholder representatives. The NBA and its board were officially constituted in the first half of 2010.

The Kenyan Biosafety Act of 2009 provides legal authority to the NBA to comprehensively regulate GMOs.[8] The act discusses contained-use activities, releases in the environment, as well as imports, exports, and transit of GMOs. It sets up regulatory procedures for approvals and discusses the use of risk assessment in the process. It also provides for several different regulations to implement the act. Kenyan government officials have been working on those regulations since 2009, and some of them were adopted in 2011, including the regulation for the contained-use activities and the regulation for introductions into the environment.

Tanzania

The structure of Tanzania's biosafety regulatory system is described in its National Biosafety Framework (NBF) issued in March 2005.[9] In that system, the National Biosafety Focal Point (NBFP) is the Ministry of the Environment. They are subsequently tasked with review and approval of applications, as well as implementation of biosafety-related issues and programs.[10] The NBFP gets advice on technical and policy issues from the NBC, composed of government and nongovernmental organizations, as well as the Competent Authorities (CA), which are other agencies with areas of relevant expertise within the government. The NBF also discusses the IBC, which performs biosafety functions within any institution conducting genetic engineering.

Tanzania also has established a specific interim biosafety regulatory process for permitting small-scale confined field trials of plant and plant products. Under the authority of the Plant Protection Act, the Ministry of Agriculture and Food Security promulgated Schedule 18.[11] That document puts in place a review and approval process for all small-scale confined field trials involving GE plants. It requires the completion of a detailed application, which is reviewed by the Agricultural Biosafety Advisory Committee (ABSAC), a technical advisory committee that is a CA of

8 Republic of Kenya, *The Biosafety Act 2009* (Nairobi, Kenya: National Council for Science and Technology).

9 United Republic of Tanzania, *The National Biosafety Framework for Tanzania, 2005* (Division of Environment, Vice President's Office: Dar es Salaam, Tanzania, 2005).

10 *Ibid.*

11 United Republic of Tanzania, *Schedule 18: Requirements for the Conduct of Confined Field Trials of Genetically Engineered Plants in Tanzania* (Ministry of Agriculture and Food Security, Dar es Salaam, Tanzania: United Republic of Tanzania).

the Ministry of Agriculture and Food Security (MAFS). The application is also reviewed by the National Biotechnology Advisory Committee (NBAC), which is a national committee on biotechnology issues based under the Ministry of Science, Technology, and Higher Education (MSTHE). With the advice of the ABSAC and the NBAC, the minister then decides whether to allow the field trial and issue a permit.[12] Schedule 18 also gives the Tropical Pesticides Research Institute (TPRI), which is a Plant Biosafety Office consisting of biosafety inspectors, the ability to require risk management measures to ensure that the field trial does not affect the environment or human health.

Finally, Tanzania has established its permanent biosafety regulatory system using authority established in the Tanzanian Environmental Management Act of 2004 (EMA), which was signed by the President of the Republic on July 1, 2005.[13] That law provides the legal authority for the Ministry of Environment to regulate GE organisms in the broader legislation that covers many environmental issues. In 2009, the government published its biosafety regulations under that Act, titled The Environmental Management (Biosafety) Regulations, 2009.[14] Those regulations comprehensively set forth the rules and procedures for ensuring the biosafety of GMOs in Tanzania. There are provisions on the responsibilities of the Ministry of Environment, the National Biosafety Committee, the use of experts, and Institutional Biosafety Committees. There is extensive discussion about the approval process for GMOs for different activities (e.g., contained use; releases in the environment; placing on the market; food, feed, and processing), and the use of risk assessment and risk management. Finally, the regulations even include a discussion on issues such as labelling, socioeconomic assessments, and liability and redress. At least on paper, therefore, Tanzania has the most comprehensive biosafety regulatory system of the three countries.

Uganda

The interim biosafety regulatory system in Uganda is coordinated by the Uganda National Council for Science and Technology (UNCST). In 1995, that office established the National Biotechnology Committee (NBC) made up of representatives from other government agencies and civil society.[15] The NBC is the "national administrative arm on matters concerning biosafety."[16] The main function of the

[12] *Ibid.*

[13] United Republic of Tanzania, *The Environmental Management Act, 2004* (Parliament of the United Republic of Tanzania, Dar es Salaam, Tanzania: United Republic of Tanzania, 2004).

[14] United Republic of Tanzania, *Government Notice* 265 published on July 24, 2009.

[15] D. Wafula & N. Clark, "Science and Governance of Modern Biotechnology in Sub-Saharan Africa-the Case of Uganda" (2005) 17 J. Int. Dev. 679, 694; P. Nampala, C. Mugoya and T. Sengooba, "Biosafety Regulatory System in Uganda" (Presentation at PBS Roundtable in Entebbe, Uganda, 2005).

[16] Republic of Uganda, *Draft National Biotechnology and Biosafety policy 2004* (Kampala, Uganda: Uganda National Council for Science and Technology, 2004).

NBC is technical advice on biosafety issues, including the assessment of individual applications for activities with GE organisms.[17]

The UNCST and the NBC have been responsible for writing the National Biotechnology and Biosafety Policy (Republic of Uganda, 2006), the Draft Biosafety Bill, Guidelines on Biosafety in Biotechnology for Uganda,[18] and several draft manuals addressing specific issues surrounding biosafety regulation, such as confidential business information. Those documents set forth the current and proposed biosafety regulatory framework for Uganda. Under the most recent public draft of the Uganda National Biotechnology and Biosafety Bill of 2008,[19] the UNCST will be the competent authority to implement biosafety regulation. The UNCST will be advised on specific GMO applications and on policy matters by the NBC, which has been transformed into an advisory committee of no more than eleven people, with specific areas of expertise from a group of listed government and nongovernmental stakeholders.

COMPARATIVE ANALYSIS OF EAST AFRICAN BIOSAFETY REGIMES AND THE AFRICAN MODEL LAW

Kenya, Tanzania, and Uganda are each moving forward to establish functional and effective biosafety regulatory systems that allow them to benefit from safe applications of GE. Their current and/or proposed systems have reached a stage of development where they can be analysed to determine how well those systems will achieve a functional biosafety system that ensures an adequate level of protection to humans and the environment and how well they incorporate the characteristics and components that were identified earlier in this book. Some of the major issues that arise from that analysis are set forth below and summarized in Annex VI, Table 1.

Issues Surrounding the Biosafety Systems of Kenya, Tanzania, and Uganda

Although each country has made tremendous progress in establishing a functional and protective biosafety regulatory system, those systems continue to evolve and mature. By comparing those systems with the key characteristics and components found in preexisting biosafety regulatory systems, issues are identified regarding those evolving systems that can help them to achieve biosafety systems that are functional, comprehensive, transparent, participatory, and efficient. Those issues are discussed in the next section.

[17] *Ibid.*

[18] Republic of Uganda, *Guidelines on Biosafety in Biotechnology for Uganda 2002* (Kampala, Uganda: Uganda National Council for Science and Technology, 2002).

[19] Republic of Uganda, *Draft National Biotechnology and Biosafety Bill 2008* (Kampala, Uganda: Uganda National Council for Science and Technology, 2008).

Comprehensiveness – The Need to Address Potential Food Safety Risks

One goal of the regulatory systems in all three countries is to address all potential risks from any GE organism comprehensively. For all three countries, the current or proposed regimes have definitions that are broad enough to capture all different GE organisms, whether plants, animals, or microorganisms. In addition, those systems do not distinguish GE organisms based on the products they produce or the intended purpose, and so they capture GE organisms that are engineered for food or feed use, engineered for industrial purposes, and engineered to produce a pharmaceutical. Each of the biosafety regulatory systems also captures the range of activities that might occur with a GE organism. The Kenyan Biosafety Act of 2009 and the Ugandan Draft Biosafety Bill of 2008 cover contained use, introduction into the environment, and the placing of GE organisms on the market.[20] The Tanzanian Environmental Protection Act and the promulgated regulations address all activities involving GMOs, such as contained research, field trials, and releases into the environment. In addition, Schedule 18 and the Plant Protection Act address small-scale confined field trials,[21] although it is unclear how the plant protection regulation will operate under the next regulatory system set up by the promulgated regulations. Each country's system has the ability to address the full range of potential environmental issues that might arise from any type of activity involving a GE organism.

Although all three countries' biosafety regulatory systems address environmental issues, they have not yet set forth in detail how they will assess any potential food safety risks possibly arising from some GE organisms. Although there is a small likelihood that eating a particular GE organism could have a harmful health effect, biosafety regulatory systems should analyse all potential risks to humans, including food safety issues. By including a food safety regulatory process for GE organisms as a component of the biosafety regulatory system (although not necessarily in the same statute or with the same government agency), GE organisms that are ready to be commercialized can have food safety issues addressed simultaneously with environmental issues. Food safety also is an important issue when GE organisms are imported solely for food and feed purposes, such as grain shipments for food aid.[22]

All three East African countries state that potential risks to human health are to be addressed through a comprehensive biosafety regulatory system. In Kenya's Biosafety Act of 2009, *biosafety* is defined in part as "the avoidance of risk to human health."[23] Similarly, in the Uganda National Biotechnology and Biosafety Policy, and the Draft

[20] *Supra* note 8; *ibid.*

[21] *Supra* note 13; *supra* note 15.

[22] K Von Grebmer & S Were Omamo, "Lessons and Recommendations" in S. Were Omamo and K. von Grebmer (eds), *Biotechnology, Agriculture, and Food Security in Southern Africa* (IFPRI/ FANRPAN: Washington, DC/Harare, Zimbabwe, 2005).

[23] *Supra* note 8.

Biosafety Bill of 2008, biosafety is defined to include risks posed to human health.[24] The Tanzanian NBF identifies allergenicity and toxicity as two food safety concerns with GE organisms and the Environmental Protection Act (EPA) defines biosafety to include "risks to human health."[25] In none of the legal or policy documents for any of the three countries, however, is there any discussion about which agency is to be responsible for ensuring the food safety of GMOs, the procedures to be used to conduct any food safety assessment, or which type of risk assessment will be required. This is in sharp contrast to the extensive discussion about how each system addresses environmental risks. For example, the Tanzanian regulations do discuss ensuring that the GMO is safe for humans and animals and do require the application to include food safety data, but generally the focus is on environmental risks, and there is no discussion about the involvement of particular ministries on food safety issues.

One reason for the absence of any discussion on how the regulatory system will address food safety could be that biosafety activity in East Africa has been focused on complying with obligations set forth in the Cartagena Protocol. Although the protocol mentions that it can also apply to "risks to human health," it is primarily an environmental agreement, and its provisions do not mention anything about assessing or addressing food safety issues surrounding GE organisms.[26]

To ensure that a biosafety regulatory system is in place that can address the full range of potential risks that can arise from a GE organism, the three East African countries need to identify how food safety issues will be addressed. In their biosafety policy, those countries could identify the existing food laws and agencies that focus on the analysis of food safety concerns for GE organisms and summarise the procedures to be used for any review and approval process. Alternatively, those countries could empower the same agency conducting the environmental risk assessment under their biosafety law to also conduct food safety assessments, providing a one-stop regulatory process that handles all safety issues, as seems to be proposed in the Tanzanian regulations. It should be noted that whichever path is chosen, food safety assessments need not be conducted for GE organisms at all stages of development. Food safety assessments are usually conducted when a GE organisms is released either in large-scale unconfined field trials or prior to release as a commercial product.

Adequate Legal Authority – Establishing New Laws

The three countries in East Africa each decided that existing laws were inadequate for establishing biosafety regulatory systems. Therefore, each country has proposed and/or adopted new legislation to regulate GMOs adequately, although that

[24] *Supra* note 19; Republic of Uganda, *National Biotechnology and Biosafety Policy* 2006 (Kampala, Uganda: Uganda National Council for Science and Technology, 2006).

[25] *Supra* note. 13; *supra* note 9.

[26] Gregory Jaffe, "Implementing the Cartagena Cartagena Protocol through National Biosafety Regulatory Systems: An Analysis of Key Unresolved Issues" (2005) 5 Journal of Public Affairs 299 and 311.

legislation differs in each country. An analysis of those bills and/or laws provides insight into the possible options available to ground biosafety systems legally, and the advantages and disadvantages of those options.

Until late 2005, Uganda had decided not to develop a specific biosafety law but instead proposed to promulgate biosafety regulations under the legal authority contained in the UNCST Statute No. 1 of 1990.[27] A review of that statute, however, finds that the UNCST statute does not contain typical regulatory functions and powers (e.g., the authority to protect the environment, the power to issue permits and conduct inspections,) and therefore does not provide adequate legal authority for the regulatory system being proposed. Although there might be other existing Ugandan statutes that provide authority for the regulations, the proposed regulations are not legally supported by the Statute.

If enacted, the Ugandan Draft National Biotechnology and Biosafety Bill of 2008 would provide adequate legal authority to regulate the safety of GMOs. The bill specifically identifies the UNCST as the competent authority over the regulation of GMOs and requires their approval for all activities with GMOs. It requires detailed risk assessments for each application, as well as a benefits assessment; all are needed to ensure that there is adequate protection of human health and the environment from GMOs. It is sufficiently detailed legislation and would put in place a strong legal and regulatory system.[28] It also is very similar to the Kenyan Biosafety Act of 2009, which is discussed later.

Kenya initially made the same decision as Uganda, drafting Regulations and Guidelines for Biosafety in Biotechnology, which were issued in 1998 under the Science and Technology Act of 1980.[29] The Science and Technology Act does not confer regulatory authority to the NCST, however, and therefore it was determined that the regulations carried no legal weight and were not enforceable.[30] Thus, the NBC and other government representatives decided that current legal authorities would not legally sustain a biosafety regulatory system and that a specific biosafety law needed passage by Parliament.

The Kenyan Biosafety Act of 2009, signed and published by the government in February 2009, provides adequate legal authority to regulate the biosafety of GE organisms.[31] The bill specifically sets up a NBA, identifies the functions of that authority, and sets forth the legal requirement to obtain government approval before conducting activities with GE organisms. It discusses the information required of applicants, the risk assessment process, and the role of different regulatory agencies in ensuring compliance. Finally, the Act identifies as one of its objectives "to ensure

27 *Supra* note 15; Republic of Uganda, *Statute No. 1990* (Kampala, Uganda: Uganda National Council for Science and Technology, 1990).

28 *Supra* note 19.

29 *Supra* note 1.

30 *Ibid.*

31 *Supra* note 8.

an adequate level of protection" where GE organisms could have an adverse effect on human health and the environment. This new law, when fully operational, therefore provides adequate legal authority for the government to assess and address risks posed by GE organisms.

In Tanzania, a review was conducted of existing legislation that might address biosafety and concluded that no single legislative instrument adequately addressed biosafety concerns.[32] In addition, it was determined that while existing legislation might provide for some interim measures for a biosafety regulatory framework, there was a need for new biosafety-specific legislation.[33] In contrast to Kenya or Uganda, however, Tanzania decided not to draft and enact an independent biosafety bill but to include provisions for regulating GE organisms in the much broader Environmental Management Act of 2004 (EMA). By taking this path, Tanzania was able to establish the legal authority for its biosafety regulatory system faster than its two neighbours, both of which began addressing biosafety regulatory concerns much earlier than Tanzania.

The EMA addresses numerous environmental issues, one of which is GE organisms.[34] As a regulatory statute, EMA provides the government with the authority to obtain information, grant approvals, inspect, and enforce. The provisions that directly address GE organisms also provide legal authority to ensure that those organisms are reviewed and approved before they are released into the environment. Under the general authority to conserve and protect biologic diversity, EMA gives the Ministry of the Environment authority to set forth regulations that manage the biosafety risks associated with LMOs, which are defined to include risks to human health and the environment.[35] EMA also defines *risk assessment* to include "the evaluation of the direct and indirect risks to human and animal health, the environment, biological diversity and to the socio-economic conditions and ethical values of the country or its populace which may be posed by the import, contained use, deliberate release or placing of the market of GMOs or its products."[36] Finally, EMA requires that "major developments in biotechnology including the introduction and testing of genetically modified organisms" require the developer to conduct a detailed "environmental impact analysis."[37] Thus, EMA illustrates that only a few statutory provisions in a broader statute might be all that is needed to establish the legal authority for a biosafety regulatory system. With those legal provisions, the details of the system can be provided in promulgated regulations.

[32] *Supra* note 9.
[33] *Ibid.*
[34] *Supra* note 11.
[35] *Ibid.*
[36] *Ibid.*
[37] *Ibid.*

The Standard for Making an Approval Decision: Establishing an Adequate Level of Protection

A key component of a biosafety regulatory system is a clearly articulated safety standard that is used to judge applications for approval by the government. The Kenyan and Ugandan biosafety regulatory systems do not provide such a standard. The Tanzanian system provides some criteria or benchmarks for an approval decision, although those criteria are potentially restrictive and not well defined.

In Uganda, the Draft Biosafety Bill states that an approval is required before any activity involving a GE organism but it does not provide the criteria which will be used to determine whether to grant the approval.[38] Similarly, the Kenyan Biosafety Act of 2009 sets forth what the authority shall take into account in reaching a final decision – the information submitted by the applicant, the risk assessment, the relevant comments from the public – but does not provide the legal benchmark or safety standard that must be met for an approval.[39]

The Tanzanian NBF does set forth some specific criteria that must be met before an approval is granted. The NBF states that no approval should be given unless the GMO will (1) benefit the country without causing any risk/significant risk to human health and animal health, biologic diversity, and the environment; (2) contribute to sustainable development; (3) not have adverse socioeconomic impacts; and (4) accord with the ethical values and concerns of communities and does not undermine community knowledge and technologies."[40] The Tanzanian biosafety regulations state that an "approval shall be given where there is firm and sufficient evidence that the GMOs or their products pose no risk to human and animal health, the environment and biological diversity."[41] They also state that in evaluating the risk assessment report, which forms the basis of any approval decision, the decision maker should assess whether the GMO or products thereof will "benefit the country; and contribute to, and not undermine, sustainable development." Thus, the regulations restate in a slightly different manner both the requirements for firm evidence of safety and also assessing whether the GMO will benefit the country and its sustainable development.

How to apply those criteria of benefitting the country and its sustainable development is not discussed anywhere in the NBF, the promulgated regulations, or in any other documents that establish Tanzania's biosafety regulatory system. Without a more thorough explanation, it is thus hard to know how an application for the release of a GE organism can meet those criteria. For example, *sustainable development* should be defined if different stakeholders are to be able to understand how GE organisms will be judged to fit within sustainable development. Depending on how they are interpreted, the criteria described in the NBF

38 *Supra* note 19.
39 *Supra* note 8.
40 *Supra* note 9.
41 *Ibid* at s 32.

are potentially restrictive and could prevent many GMOs that are safe for humans and the environment from being approved either because they do not "benefit the country," they have some socioeconomic impacts, or they neither contribute to nor hurt sustainable development.

Whereas the Tanzanian NBF and the promulgated regulations includes the statement set forth in the previous paragraphs, the confined field trial permitting process established under Schedule 18 make no mention of those criteria or that standard. Nowhere in Schedule 18 is there any discussion about any safety standard that needs to be met or any criteria that must be satisfied when the agency is determining whether a request for a permit will be approved or rejected.[42]

One interesting part of all three of the proposed biosafety regulatory systems is that they each affirmatively state that a GMO's benefit should be considered in the approval process. The Tanzanian NBF states that implementation of the biosafety framework should recognize both the "potential benefits and risks of modern biotechnology."[43] The Kenyan Biosafety Act of 2009 states in several provisions that an application may contain information "that the applicant or the Authority may consider necessary for the assessment of the potential risk or benefit" of the GMO activity proposed.[44] That language suggests that the assessment addresses not just risks but benefits. The Ugandan Draft Biosafety Bill has language that is almost identical to the Kenyan law on this issue.[45] Thus, the East African biosafety regulatory systems all propose explicit consideration of the benefits of a GMO, not something that some biosafety regulatory systems around the world consider.

Addressing Socioeconomic Considerations in the Decision-Making Process

All three countries state that their biosafety regulatory systems need to address socioeconomic considerations that might arise from individual GE organisms. The Tanzanian NBF states that "the social, economic, and ethical considerations shall be taken into account in Biosafety decisions" and that it is one of the responsibilities of the NBC to facilitate the undertaking of "socio-economic impact assessment."[46] This is reinforced in the Tanzania biosafety regulations, which discuss the details of the socioeconomic impact assessment and analysis in several places. The regulations define *socioeconomic impact* as "the direct or indirect effects to the economy, social, or cultural practices, livelihoods, indigenous knowledge systems, or indigenous technologies as a result of the import, release, contained use, handing, or placing on the market of GMOs or their products." The regulations further state in section 26 that prior to the release of a GMO, the CA shall conduct "a thorough study of (a) their ethical and social-economic impact on the local population

[42] *Supra* note 11.
[43] *Supra* note 9.
[44] *Supra* note 8.
[45] *Supra* note 19.
[46] *Supra* note 9.

concerned; (b) the transitional market and export earnings; (c) health; (d) production systems; (e) ethical, moral, and social considerations; (f) the actual economic value of traditional species likely to be affected by the introduction of the GMOs." The Third Schedule on Risk Assessment Parameters also gives details and guidance on the scope of the assessment of socioeconomic considerations.

The Kenyan Biosafety Act of 2009 states that "in reaching a final decision, the Authority shall take into account . . . (e) socio-economic consideration arising from the impact of the GMO on the environment" for decisions involving the introduction of a GMO into the environment.[47] The Ugandan Biosafety Bill mentions that for decisions on applications, the CA shall take into account "socio-economic and cultural considerations."[48] The bill also defines "socio-economic impact" as "a direct or indirect effect on the economy or on social or cultural conditions or on the livelihood or indigenous knowledge systems or technologies of a community or communities and on the economy of the country."

Other than the references mentioned in the previous paragraphs, however, no country provides practical details about how socioeconomic considerations will be analyzed, and how they will be factored into the decision-making process. What is the meaning of "socioeconomic considerations?" Which information will be used to analyze any socioeconomic impacts, and what is the baseline for the analysis? Who will perform the analysis, and at which stage in the development process will this occur? For the biosafety regulatory system to be fair, predictable, and transparent, the details surrounding the inclusion of socioeconomic considerations in the decision-making process should be spelled out in more detail than is currently available in the documents from Kenya, Tanzania, and Uganda. Without sufficient details, the system could be perceived as unfair to the applicants and the public, both of whom will not know how specific applications are to be judged in this area.

Without more details on the interaction between socioeconomic considerations and the decision-making process, it is unclear whether the three biosafety systems might violate the Cartagena Protocol or World Trade Organization (WTO) obligations. Article 26 of the protocol allows consideration of socioeconomic considerations of LMOs but places conditions on that analysis. First, it limits the socioeconomic considerations to those effects that arise "from the impact of living modified organisms on the conservation and sustainable use of biological diversity, especially with regard to the value of biological diversity to indigenous and local communities."[49] Thus, the plain language of Article 26 does not allow all socioeconomic considerations of LMOs to be considered, but only those that directly arise from impacts on biologic diversity.[50] Some stakeholders believe, however, that the socioeconomic

47 *Supra* note 8.

48 *Supra* note 19.

49 Secretariat of the Convention on Biological Diversity (SCBD), *Cartagena Protocol on Biosafety to the Convention on Biological Diversity: text and annexes* (Montreal, Canada: SCBD, 2000).

50 R Mackenzie et al, *An Explanatory Guide to the Cartagena Protocol on Biosafety*, (Gland, Switzerland and Cambridge, UK: FIELD and IUCN, 2003).

impacts of LMOs are much broader and might include concerns such as "impacts on farmers' incomes and welfare, cultural practices, community well-being, traditional crops and varieties, domestic science and technology, rural employment, trade and competition, the role of transnational corporations, indigenous peoples, food security, ethics and religion, consumer benefits, and ideas about agriculture, technology, and society."[51] Although those broader socioeconomic considerations might be valid societal concerns relevant to GE organisms, they might not properly be part of a biosafety regulatory system consistent with the protocol.

Second, the Protocol states that inclusion of socioeconomic considerations must be done in a manner consistent with other international obligations.[52] WTO rules generally emphasise procedures for decision making that primarily rely on scientific risk assessments and appear limit the ability to make decisions based on nonsafety concerns. WTO rules sometimes do allow the narrow use of nonsafety concerns. For example, the SPS Agreement does set forth a risk assessment procedure that includes both scientific and socioeconomic considerations.[53] Some of the relevant economic factors under the SPS Agreement include "the potential damage in terms of loss of production or sales in the event of entry, establishment or spread of a pest or disease; the costs of control or eradication in the territory of the importing Member; and the relative cost-effectiveness of alternative approaches to limiting risks."[54] That exception, however, may be interpreted narrowly, allowing cost-benefit analysis to play a role in certain decisions.

Therefore, to try to comply with WTO rules and the plain language of Article 26, countries that include socioeconomic considerations in their domestic regulatory systems might tailor what they will consider to only what Article 26 allows (e.g., socioeconomic concerns directly linked to impacts on biodiversity). That might greatly narrow the socioeconomic issues that the three countries can address in their biosafety systems, however. Broader socioeconomic considerations, however, might be addressed through other means, such as voluntary processes implemented by research institutions and companies or other laws and regulations.[55]

Any assessment of socioeconomic considerations that is to be conducted should occur when a GE organism is ready to be commercially released. This type of limitation is included in the Kenyan Biosafety Act and is a well-thought-out provision.

51 A La Vina & L Fransen, *Integrating Socio-Economic Considerations into Biosafety Decisions: The Challenge for Asia* (World Resources Institute: Washington, DC, 2004).

52 F. Kalibwani et al, *Governing Biotechnology in Africa: Toward Consensus on Key Issues in Biosafety* NEPAD, (Johannesburg, South Africa, Harare, Zimbabwe, and Washington, DC, 2004); F Kalibwani et al, "African Policy Dialogues on Biotechnology-Southern Africa" (Harare, Zimbabwe, September 20–21, 2004).

53 The World Trade Organization, *General Agreement on Tariffs and Trade (GATT)* (1994), online: http://www.wto.org/english/tratop_e/gatt_e/gatt_e.htm; The World Trade Organization, *Technical Barriers to Trade Agreement* (1995), online: http://www.wto.org/english/tratop_e/tbt_e/tbt_e.htm.

54 *Ibid.*

55 L Fransen et al, *Integrating Socio-Economic Considerations into Biosafety Decisions* (Washington, DC: World Resources Institute, 2005).

When a GE organism is used in a contained laboratory or in a confined field trial, the GE organism is not expected to persist in the environment, so there are minimal effects of that organism on the population of the country and few significant socioeconomic considerations to analyze. If a country is concerned about conducting GE organism research as a whole, those issues should be addressed in a policy context. If socioeconomic considerations are to be analysed for individual GE applications, the proper time is when the producer of the organism seeks approval for commercial release.

Proportionate Risk-Based Review

Proportional risk-based reviews of GE organisms allow the regulator to streamline the approval of safer applications while spending more time and resources on applications that pose greater relative risk. The current and proposed regulatory systems in the three countries differ tremendously on how extensively they have incorporated that component into their biosafety regulatory systems.

The Kenyan Biosafety Act of 2009 clearly differentiates aspects of its review and approval procedures based on the relative risk of either the GE organism or the specific activity to be conducted with the GE organism. The bill specifically defines *contained use* and sets forth separate procedures for approval of contained-use applications.[56] Those applications need only to provide the limited information required by the Third Schedule (paragraph 18), and there is no need for both notice to the public and an opportunity for public comment. Those applications still require approval from the competent authority, however, and they must conduct a risk assessment (although it is assumed to be a limited one). The Biosafety Act of 2009 also distinguishes between applications for introduction into the environment (paragraph 19) and placing of the GMO on the market (paragraph 21).[57] When one reads those provisions and the schedules to which they refer, however, there is no distinction in procedures or information required for those two different situations, although they have different levels of potential risk. Thus, although the Act establishes different procedures for field trials and commercial releases, the reality is that the procedures are virtually identical. It is possible that the implementing regulations proposed to be released in 2010 might clarify how different procedures apply to different activities with GMOs. The Biosafety Act of 2009 does allow, however, for a risk assessment to be omitted if "sufficient experience or information exists to conclude that the GMO or contained use activity concerned do not pose a significant risk," but it is unclear if this provision is only limited to contained-use activities.[58]

The Ugandan Draft Biosafety Bill is similar to the Kenyan Biosafety Act of 2009 on some aspects of the issue of proportionality. The Ugandan Bill establishes separate

56 *Supra* note 8.

57 Based on the definitions in the Bill, it is assumed that introductions into the environment are confined and unconfined field trials while "placing on the market" is the commercial sale of a GMO.

58 *Supra* note 8.

procedures and separate data requirements for the different activities with GMOs (i.e., contained use; introduction into the environment; GMOs for direct use in food, feed, and processing) using almost the same language from the Kenyan Act. The Ugandan Biosafety Bill establishes several exemptions for GMO activities from the authorization procedures and numerous situations in which simplified approval procedures can apply (paragraph 21). For example, simplified procedures can be used for a GMO that has been authorized by the CA of another country having a comparable ecosystem.[59]

The best example of a proportional risk-based review process in the region is the Tanzanian approval process for confined field trials under Schedule 18. That document sets forth a procedure to allow for the review and approval of small-scale confined field trials, which, because of their size and confined nature, pose relatively little risk.[60] Under that procedure, there is no need for a thorough risk assessment but only the completion of an application with limited information requirements. No food safety assessment is required, nor are socioeconomic issues considered or analysed. Thus, the procedures in Schedule 18 creditably balance the regulatory process with the potential risk of the activity.

Although Schedule 18 puts in place a proportional risk-based review process, the Tanzanian NBF and Biosafety Regulations require that all GMOs and their products, regardless of their intended use, must have detailed risk assessments and prior authorizations. The NBF specifically states that imports of GMOs for contained use and for food, feed, or processing require application of the advance informed agreement (AIA) procedure.[61] WTO rules require that imports are treated the same as domestic products, and so if imported GMOs for contained use need a full risk assessment as set forth in the protocol, then the same would be required for the contained use of a domestically engineered organism. The EMA biosafety regulations treat all GMO applications – contained use, confined field trials, and releases into the environment – with many of the same requirements in some parts of the regulations and with differences based on use in other parts. Thus, the language in the Tanzanian NBF and EMA biosafety regulations might contradict what Schedule 18 and the regulation of confined field trials under the Plant Protection Act are trying to accomplish. Any inconsistencies, it is hoped, will be addressed as the biosafety regulations are applied in practice along with the Plant Protection Act requirements.

Public Participation

As Chapters 1 and 2 of this book note, the commitment to facilitate public participation, supported by access to information and access to justice, is an important principle of many international treaties on sustainable development, including the Cartagena Protocol.[62] Public participation in a biosafety regulatory system usually

[59] *Supra* note 19.

[60] *Supra* note 11.

[61] *Supra* note 9.

[62] See the Introduction and Kathryn Garforth et al.

involves two separate components: (1) the opportunity to provide comments and opinions on the laws, regulations, and policies before they are adopted; and (2) the opportunity to provide comments before an application for a GE organism is approved by the regulatory agency. Public participation is different from public awareness, in which the government educates and informs the public about biosafety, biotechnology, and the regulatory process.

All three countries in this study have involved the public in the process of drafting their biosafety regulatory system (e.g., bills and regulations) and their biosafety policy. Kenya has conducted several public stakeholders meetings, beginning in 2001, to address the country's biosafety policy and biosafety law,[63] although some commentators have criticised Kenya for not making their process more inclusive and participatory.[64] Similar workshops and stakeholder meetings have been conducted in Uganda and Tanzania.[65]

All three countries also provide for some level of public participation in the review and approval process for individual GE organisms. In Tanzania, EMA biosafety regulations specifically provide for the regulatory body to provide all applications to the public and allow for a three-month period for the public to provide comments. There is also a provision that requires the regulator to provide for public consultations before any decision is made, although there is no definition of what is meant by a "public consultation." Nowhere in Schedule 18 is there any discussion about public participation, however, or the opportunity for the public to comment on an application before the regulatory decision is made.[66]

Kenya and Uganda also provide for public comment on applications. In Kenya, the Biosafety Act of 2009 provides for public comment in two provisions. First, it states that, for applications to "introduce GMOs into the environment," a person had thirty days from the publication of notice of the application to make representations to the NBA. Second, a separate provision addressing public awareness and participation states that any person had thirty days to "submit written comments on a proposed decision for any application for placing a genetically modified organism on the market." Thus, under the Kenyan Biosafety Act of 2009, the public is provided the opportunity to comment on applications for introduction into the environment and for placing a GMO on the market, but there are no similar provisions for contained use or import applications.[67]

63 *Supra* note 1; Kenya Council for Science and Technology, *Implementation of the National Biosafety Framework for Kenya: Proceedings of the UNEP-GEP Workshop, 14th-18th April 2003* (UNEP-GEF: Nairobi, Kenya, 2003).

64 M. Harsh, "Formal and Informal Governance of Agricultural Biotechnology in Kenya: Participation and Accountability in Controversy Surrounding the Draft Biosafety Bill" (2005) 17 J Int. Dev. 661 and 677.

65 *Supra* note 15; Consent Stakeholder Consultative and Awareness Initiative, *First Biotechnology and Biosafety Stakeholders Dialogue Report* (Report of 18 June 2003) (Consumer Education Trust: Kampala, Uganda, 2003).

66 *Supra* note 11.

67 *Supra* note 8.

In Uganda, the Biotechnology and Biosafety Policy is silent about public participation in the decision-making process for GMO applications, but it does provide for an organized public awareness campaign to ensure that the public is knowledgeable about GMOs and biosafety.[68] The Uganda Draft Biosafety Bill of 2008 has the same provision as the Kenyan Act, which allows for public participation in decisions for introduction into the environment. The Draft Biosafety Bill also specifies that the public be given sixty days to comment on any measure that is proposed to implement the act.[69] Thus, the three countries provide a range of public participation options for the decisions made on individual GMOs. The plain reading of the Tanzanian Regulations suggests that there is public comment for all applications whereas the Kenyan Biosafety Act of 2009 provides for public comment for applications for introduction into the environment or placing on the market of a GMO. In contrast, Uganda merely provides for public comment for introductions into the environment and for legislative implementation proposals.

Although all three countries provide the opportunity for the public to comment on applications, the value of that opportunity depends on how those comments are used in the decision-making process. In Kenya, the Biosafety Act of 2009 states that the CA "shall take into account . . . (d) any relevant representations submitted by members of the public."[70] The addition of the word *relevant* is important because it rightfully allows the authority to take into accounts comments specific to the issues it is deciding (e.g., the scientific risks of a GE organism) and ignore irrelevant comments, such as general opinions that GE organisms are bad or harmful. In Tanzania, however, EMA biosafety regulations state that all comments made shall be taken into account by the decision maker in making its decision. There is no qualification that the comments be relevant.

Transparency

Public participation in the biosafety regulatory system cannot occur unless the regulatory system is transparent. The legal documents from all three countries establish provisions ensuring that their systems are transparent. All three countries state that they will provide to the public the applications for GE organisms, as well as the final decisions concerning all applications.[71] In addition, the Tanzanian biosafety regulations specifically state that they will make available to the public the risk assessment of the GE organism.[72]

All three biosafety regulatory systems also balance the rights of the public to information with the rights of the developer or applicant to protect confidential

68 *National Biotechnology and Biosafety Policy 2006*, *supra* note 24.

69 *Supra* note 19.

70 *Supra* note 8.

71 *Supra* note 8; *supra* note 9; *supra* note 19.

72 *Supra* note 9; Republic of Uganda, *Draft National Biosafety Regulations* (Uganda National Council for Science and Technology: Kampala, Uganda, 2004).

business information. Each country has detailed provisions in its documents to protect confidential business information.[73] In Uganda and Tanzania, however, the confidentiality provisions ensure that some basic information (e.g., a description of the GE organism and a summary of the risk assessment) cannot be claimed confidential. Those provisions are consistent with the Cartagena Protocol, which promotes transparency but allows for protection of some confidential information in Article 21.

Scope of the Biosafety Regulatory System – Inclusion of Products of GMOs

In most countries with established biosafety regulatory systems, those systems require approval only for each particular GE organism at each stage of development and do not require premarket safety approval for each individual product derived from those GMOs. (The term *product* generally means foods and nonreproducing materials such as edible oils, industrial materials, and pharmaceuticals.) Instead, the food safety and environmental risks assessments for the particular GE organism include an analysis of the different products that might be produced from the GE organisms and whether those products might pose any environmental or food safety risks.

Only the Tanzanian biosafety regulatory system explicitly includes products of GMOs in its regime. The scope section of their NBF states that the biosafety regulatory systems includes "products of GMOs" and the approval processes that it describes require advance informed consent for a GMO, a product of a GMO, or any material derived by processing a GMO.[74] The definitions in EMA also mention that risk assessments are to be performed on both GMOs and their products. Similarly, the EMA biosafety regulations requires approvals and all other major legal requirements for not just GMOs but also "products thereof." If the Tanzanian biosafety system is supposed to approve each product of a GMO using the same procedures, risk assessments, and information requirements, as would be done for a GMO, the system could receive thousands of applications from persons and companies who do not know very much about the GMO. For example, a GE tree could be used to make dozens of different products, including paper, wooden furniture, and wooden picture frames. Similarly, GE maize could be used to make maize oil, maize starch, and maize syrup and could be found in Coke, Fanta, Pepsi, Corn Flakes, corn meal, and hundreds of other food products. Is it anticipated that under the Tanzanian biosafety regulatory system, soda or paper would require an approval if they were made from a GMO? The person who makes the paper or soda would know about the wood he or she used or where he or she got the corn syrup, but those persons would not have information about the genetic construct of the GMO or information about the engineered protein. From both a practical and a scientific

[73] *Supra* note 8; *supra* note 9; *supra* note 19.

[74] *Supra* note 9.

viewpoint, it might therefore not make sense to approve products that are derived from the GMO.

Interim Regulatory Systems

The long-term goal is to have a working biosafety regulatory system that is comprehensive, covering potential risks for potential organisms at different stages of development. For many African countries, including the three in this study, currently there is limited research and product development using GE,[75] and thus there might not be an immediate need for a comprehensive regulatory system at this time. In those situations, interim regulatory systems that address immediate needs – such as confined field trials for GE plants – can play an important role in allowing biotechnology to advance while more permanent regulatory systems are established.

Kenya, Tanzania, and Uganda have each established interim biosafety regulatory processes to address any immediate biosafety needs. Kenya's NCST issued regulations and guidelines for activities with GE organisms and approved several confined trials.[76] It established the NBC both to review applications for GE organisms and to draft the government's biosafety policies (the Biosafety Policy and Draft Biosafety Bill). Similarly, Uganda established its NBC in 1995, along with functional national biosafety guidelines.[77] The NBC has reviewed several applications for GE organisms, and recently approved an application by the National Banana Research Program for a special license to grow genetically modified organisms for research purposes.[78]

Tanzania's interim biosafety regulatory system includes an approval process for small-scale confined field trials using the legal authority granted by the Plant Protection Act. Those requirements, which are identified in Schedule 18, state that "no person can conduct a confined field trial of any GE plant without authorization from the Minister under this Schedule."[79]

There are two issues that arise with the interim measures. First, there is a question about whether legal authority supports them. For example, although the Kenyan regulations and guidelines have been used to approve confined trials, there is general consensus that those procedures are not legally supported by the Kenyan Science and Technology Act.[80] By contrast, there is probably adequate legal authority in the Tanzanian Plant Protection Act for Schedule 18, because it is limited to small-scale confined field trials for GE plants. The Tanzanian NBF, however, found the Plant Protection Act inadequate to address biosafety, "particularly with regard to risk assessment and management, (. . .) issues of environmental impact assessment,

75 I. Sithole-Niang, J. Cohen, and P. Zambrano, "Putting GM technologies to work: public research pipelines in selected African countries" (2004) 3 African Journal of Biotechnology 564,571.

76 *Supra* note 4.

77 P. Nampala, C. Mugoya, and T. Sengooba, *supra* note 15.

78 See Nature, "Uganda prepares to plant transgenic bananas," 1 October 2010, online: http://www.nature.com/news/2010/101001/full/news.2010.509.html.

79 *Supra* note 11.

80 *Supra* note 1.

(. . .)."[81] Thus, that law would probably not support more complete regulation of GE plants as they move toward commercialization.

The second issue that arises with interim measures is what happens to them when new legislation is put in place. For example, Tanzania recently promulgated EMA biosafety regulations, which provide for regulation of the release of GE organisms into the environment. Will those regulations now regulate small-scale confined field trials instead of Schedule 18? It would be inefficient if two reviews were required. Similarly, the Kenyan NCST and the NBC have gained experience regulating GE organism experiments under their interim processes. The Kenyan Biosafety Act calls for the NBA managed by a board from several government agencies.[82] Will that authority make use of the existing expertise from the NCST and the NBC? Will the documents that make up the current Kenyan biosafety regulatory system become regulations and guidelines promulgated under the Biosafety Act? While the biosafety regulatory systems in the three countries transition from interim procedures to new legal authorities, it is important that the knowledge, experience, and work already completed are transferred and built on.

Efficiency: The Value of Institutional Biosafety Committee

Biosafety regulation has the potential to be costly and time-consuming, both for the researcher who is trying to produce a useful GE organism and for the government that is trying to ensure adequate protection of humans and the environment. It might therefore be beneficial if regulatory procedures were as streamlined as possible, while not compromising safety. The biosafety regulatory systems in all three countries have multiple steps with many different people and committees conducting different parts of the process. One part of the regulatory process in all three countries that should be analysed more closely for its added value is the Institutional Biosafety Committees (IBC), particularly for private institutions.

All three East African countries require every institution that conducts activities with GE organisms, both public and private, to establish an IBC. An IBC is usually made of scientists and other people from that institution, who provide oversight to biosafety aspects of any GE that is occurring at the institution. The Kenyan Guidelines for release into the environment state that the "IBC will ensure that experiments relating to GM and releases by the institution conform to the provisions of the Guidelines."[83] The IBC is also required to undertake a risk assessment, monitor release sites, and provide its findings to the NBC.[84] Similarly, in Uganda, the IBCs' functions include reviewing and clearing prospective applications before they go to the NBC, reporting regularly to the NBC, training, and oversight. The Ugandan Draft Biosafety Bill of 2008 specifically states that one of the functions

[81] *Supra* note 9.
[82] *Supra* note 8.
[83] *Supra* note 4.
[84] *Ibid.*

of the CA is to "cause the establishment of Institutional Biosafety Committees." Tanzania also establishes IBCs in the EMA biosafety regulations, which conduct many of the activities identified earlier for Uganda and Kenya but are also required "to hold discussions on the comparative ecological, economic and social impact of alternatives" to the proposed GMO.[85]

Based on the functions of the IBCs set forth in the biosafety regulatory systems, those committees perform an initial review step in a multistep approval process and conduct self-monitoring of the institution's activities. Is this self-regulation valuable? Should it be trusted when it is done by a private corporation? Do the IBC reviews make experiments with GE organisms safer? Are the preliminary reviews conducted by the IBCs valuable to the NBCs when they review the applications again? Those and other questions should be answered to determine if the IBCs add value to the overall regulatory process and its goal of ensuring safety to humans and the environment. Based on the functions of the government actors in the existing and proposed biosafety regulatory systems for all three countries, most activities conducted by the IBC are repeated by the NBC and/or the CA. When IBCs or other similar institutional committees have been established in developed countries such as the United States, they self-regulate in areas where there is no governmental oversight. For Kenya, Uganda, and Tanzania, however, the IBCs conduct activities that can be duplicated at the national level.

Finally, even if there is merit to IBCs for public institutions, the same may or may not be true for private institutions. For private companies and businesses, there might be little public benefit from the company policing itself, and any initial review they might provide to the NBC might get very little deference. Therefore, it would be valuable to conduct a research study to determine whether IBCs allow the NBC to be more efficient because they defer to the IBC's judgment and oversight, or whether IBCs merely conduct activities that are then repeated by the NBC or the CA.

One way to reduce overlap or duplication but keep IBCs would involve making them the primary regulatory body for contained-use activities and limiting their role for confined field trials and commercial releases. In such a regulatory system, the IBC might be the only body responsible for ensuring that contained-use experiments safely meet the government's protection standards and procedures. No researcher could do a contained experiment without IBC approval. The IBC would submit its approval to the NBC, which would have thirty days to object to the approval. If the NBC did not object, then the IBC approval decision would stand, and the experiment could proceed legally. For confined field trials, which require a review in each country by the NBC, the IBC might play a minor role to ensure that the researcher provides the necessary information to the NBC. It also might work with the researcher to establish plans and procedures to carry out the terms of the issued

[85] *Supra* note 9.

permit. Finally, for commercial releases, when the NBC has an extensive review of food safety, environmental safety, and any other risk issues, the IBC might play no role. Under such a system, duplication would be avoided, and the IBC would be given the primary role in activities involving GE organisms when potential risks and public concern are minimal.

COOPERATION, COORDINATION, AND HARMONISATION OF EA BIOSAFETY REGIMES

Currently, all three countries in this study are in the process of establishing national biosafety regulatory systems. Although those systems are at different stages of development and have different histories, none of them has been finalized. Thus, there is an opportunity to improve the individual systems, coordinate their operations, and even harmonise the procedures and policies among the three countries. What follows is a suggested path forward that provides both for further development by individual countries to finalize their national biosafety systems while simultaneously working collaboratively and cooperatively to coordinate and harmonize those systems so that they are efficient, effective, and make the best use of limited scientific and legal capacity.

Timing of a Regional Effort

Although some people might argue that each country should have its own biosafety regulatory system established and running before it can consider efforts at coordinating with its neighbours, the best time to begin a regional harmonisation effort is before the biosafety regulatory systems are fully functioning. Working on regional harmonisation efforts at the same time that individual countries are working on their own biosafety system allows countries to use their resources efficiently and share their lessons, best practices, and list of things to avoid.

Successful East African community regional harmonisation efforts have occurred before each individual country has established a legally sufficient and fully functional regulatory system. For example, environmental assessment guidelines for shared ecosystems in East Africa were successfully developed before all three countries had the legal authority and regulatory procedures in place to conduct the environmental impact assessments.[86] Those guidelines were approved by the Sectoral Committee on the Environment and Natural Resources of the EAC before being adopted by each country.[87] Therefore, there is precedent for addressing regional environmental issues, even before the necessary national legal and regulatory systems are in place.

[86] The East African Community, Environmental Assessment Guidelines for Shared Ecosystems in East Africa (East African Community: Arusha, Tanzania, 2005), online: http://www.eac.int/environment/index.php?option=com_docman&task=cat_view&gid=66&Itemid=106.

[87] *Supra* note 15.

Potential Areas for Collaboration, Cooperation, Coordination, and Harmonisation

The discussion that follows sets forth several substantive ways in which the three countries could begin a collaborative effort to coordinate and harmonise their biosafety regulatory systems. While implementing all of the suggestions would lead to systems that would be well coordinated and integrated, efforts in this arena could also occur more gradually by establishing some cooperation and coordination initially while waiting for each country to gain more experience with GE organisms, before any extensive harmonisation effort.

Adoption of Codex Standards for Assessing the Food Safety of GE Organisms
Under the Codex Alimentarius Commission, the governments of the world's nations have issued several consensus documents addressing the food safety issues surrounding GE foods. Those documents might form the basis for the food safety risk assessments of GE foods that are conducted in Kenya, Tanzania, and Uganda. Thus, if each country adopted those documents in their biosafety policy and implemented their risk assessment procedures, the region could move toward a harmonised approach for addressing the food safety issues from GE organisms.

Harmonising Applications, Data Submissions, and Government Documents
For each country's approval process, the developer must complete an application form and submit specific data and information about the GE organism. Because of this, it might be beneficial if each country's application form asked for similar information in a similar manner. That would make it easier to fill out the forms and would standardise developers' expectations about what is required of them from neighbouring countries.

Coordinated biosafety regulatory systems also might accept the same data, when it is relevant to the issue being addressed. For example, when conducting a food safety assessment of a GE organism to be commercially released, it may not matter where the laboratory tests for toxicity or allergenicity were conducted. Test results from a neighbouring country or even a developed country that previously analysed the food safety of the GE organism might be acceptable. Other types of data that might be portable from one jurisdiction to the other might be molecular data, evidence of gene stability, or harm to nontarget organisms. Even certain environmental data might be used in more than one jurisdiction if the same environmental conditions and issues exist in both places. For data on certain environmental issues and questions, however, it probably will be necessary to submit data from tests conducted solely in the country for which the release is requested.

One way to help ensure that data are portable from one country to another is for those countries to establish the same field and laboratory standards. Kenya, Tanzania, and Uganda integrated field and laboratory standards for their regulation of seed, which resulted in greater acceptance of data across countries and increased seed availability.[88] A similar collaborative effort by the countries to establish standards for the GE organism risk assessment testing needed in an application would allow relevant data generated in one country to be submitted in the other countries, if these data are relevant to potential risk issues in those countries.

Government-issued documents related to GE organisms, such as notification, permits, and inspection reports, could also be standardised. The East African Community harmonised its sanitary and phytosanitary standards, measures, and procedures and included in that process standard forms to be used in different situations.[89] In fact, one such standardized document already in existence is the permit form for importing LMOs into the country. A similar effort could produce uniform permits and inspection checklists for the different activities involving GE organisms and could also produce common tags and labels for GE organisms that are being moved among countries for research and field testing.

Establishment of Common Regulatory Pathways for Different Activities Involving GE Organisms

As stated earlier, different activities involving GE organisms have different potential risks. Many biosafety regulatory systems provide each activity with risk-appropriate regulatory procedures, so that limited resources are allocated primarily to activities that raise the most potential concern while minimising the costs and time involved in the regulatory process for activities with little potential risk. It could be beneficial if the three countries implemented similar review and approval procedures for four different activities involving GE organisms: contained-use experiments, confined field trials, commercial releases, and GE organisms used only for food, feed, or processing. Establishing risk-based procedures for those categories and integrating those procedures among the three countries could simplify field trials and commercial release activities in the individual countries and reduce the regulatory burden for GE products that might be marketable in more than one country.

CONTAINED-USE EXPERIMENTS. *Contained-use experiments* involve working with GE organisms inside a laboratory or greenhouse that is physically separated from the outside environment. Such physical separation, along with institutionalized procedures for containment, eliminates virtually any risks that might be posed by the GE organism to the environment. Although the three biosafety regulatory systems in Kenya, Tanzania, and Uganda all require review and approval of such activities, the

[88] IJ Minde, *Facilitating Regional Agricultural Trade for Increased Food Security and Incomes: The Case of Rationalization and Harmonization of Seed Policies, Laws, Rules and Procedures in Eastern Africa* (ASARECA-ECAPAPA: Entebbe, Uganda, 2004).

[89] The East African Community, "Harmonized Sanitary and Phytosanitary Standards, Measures, and Procedures" (The East African Community: Arusha, Tanzania, 2004).

potential risks posed by such activities support establishing streamlined, regulatory review procedures that result in quick decisions.

One possible regulatory pathway for contained-use experiments would make the IBC the primary regulator for such activities. Instead of requiring review and approval by the NBC for each experiment, the regulatory system could establish the IBCs as the decision maker, requiring the investigator to submit the applications to the IBC for review and approval. If the IBC approved the submission, they would notify the NBC of their approval and allow the NBC a set period to object to the approval. If there were no objection, the approval would stand, and the experiment would proceed. Thus, the NBC could veto or overrule an IBC decision, but if the NBC did not act, the IBC permit would be the legal permission to proceed.

A second possible regulatory procedure would be to require NBC approval for such applications but provide a legal mechanism to exempt such activities from review and approval once such applications become familiar and routine. The NBC might scrutinize the first few applications, but over time, routine experiments that pose no real risk would be exempt from needing a permit. It might stifle agricultural research, waste precious regulatory resources, and provide little added protection to the environment and human health if contained-use experiments required extensive risk assessments that tried to assess and address speculative or unlikely environmental and food safety risks.

CONFINED FIELD TRIALS. A *confined field trial* is a restricted environmental release of a GE organism under conditions that are designed to prevent the spread of the organism from the field trial site or its persistence in the environment. The trial is usually small-scale (less than one hectare) and conducted for research purposes to evaluate the performance of the organism or to collect data to analyse the safety of the organism.[90] The field trial is considered "confined" because it is conducted under planting conditions that limit the ability for the GE organism to escape from the site. Those conditions might include biologic, physical, geographical, temporal, and/or chemical methods of confinement.

If there is sufficient confinement, a confined field trial poses relatively little risk to human health or the environment because the chance of escape into the food supply or persistence into the environment is small. Thus, the regulatory system's review procedures for such applications should concentrate on analysing the characteristics of the GE organism and matching those characteristics to a testing protocol with multiple layers of confinement. Thus, a risk assessment of the GE organism should be conducted, but it might not be as extensive as the assessment performed for a commercial release, in which the product will be sold to individual people to use without restriction. For a confined field trial, the risk assessment often addresses only risks and not benefits, because the benefits for the potential product are extremely

[90] *Supra* note 11.

speculative at the field trial stage. The assessment of the GE organism to approve a confined field trial application also may not require a food safety assessment. The nature of the field trial – its small scale and extensive confinement – can make the likelihood of impact on the food supply small. Similarly, to the extent that a biosafety system is legally mandated to assess socioeconomic considerations, such issues might not arise for confined field trials.

Tanzania, under Schedule 18 promulgated pursuant to the Plant Protection Act, recognized the need to treat confined field trials differently in the biosafety regulatory system and set up a streamlined process for reviewing and approving such experiments. The other East African nations could copy this system, leading to common procedures for such trials throughout the three countries.

COMMERCIAL RELEASE. An application for the commercial release of a GE organism occurs when field testing is complete and a developer is ready to release the GE organism as a product sold to farmers. At that stage, the GE organism is expected to persist in the environment and interact with humans. Thus, it is this stage in the regulatory process where a full risk assessment is performed to ensure adequate protection of human health and the environment. That risk assessment should address all relevant environmental issues and could include an analysis of the potential benefits of the GE organism. If the GE organism is eaten by humans and/or animals, there should be a food safety assessment that follows the procedures and rules established by the Codex. If a country's biosafety regulatory system requires an assessment of socioeconomic considerations, the place for that analysis would be at the commercial release stage. Any analysis of those considerations, however, must be consistent with that country's international obligations. It therefore might be prudent for such regulatory systems to restrict such an analysis to a small set of well-defined issues that are directly the result of harms to biodiversity.

For commercial release applications, the regulatory system should ensure that the application and risk assessment are thoroughly evaluated. To do this, relevant government offices with expertise in the potential risk issues should review the application and the risk assessment and provide comments to the decision maker. In addition, experts with scientific knowledge who do not work for the government should be consulted for their opinion on the issues raised by the application. The public should also be provided the nonconfidential portions of the application and risk assessment and given an opportunity to provide comments relevant to the issues being decided by the government.

In deciding on the commercial release application, some risk management measures can be imposed with the approval of GE organism for commercial release. The fact that thousands of people might purchase the organism limits the types of measures that can be feasibly implemented, however. In developed countries, the types of risk management measures that have been implemented for commercial products include restrictions on planting the GE organism in certain geographic locations or the establishment of refuges to prevent insect resistance to the GE organism.[91] Finally, whichever decision is made on the application relating to a GE organism, the government should release that decision, as well the basis for the decision, to the public.

GE ORGANISMS USED ONLY FOR FOOD, FEED, OR PROCESSING. The final category for GE organisms relates to those organisms that are to be used solely for food, feed, or processing. An example of a GE organism that falls within that category is GE grain that is imported as food aid. In that instance, the grain is not meant to be sold as seed for farmers to plant but is to be used to feed humans and/or animals. In such a situation, two regulatory pathways are possible. First, the country can allow the GE organism to enter the country based on the risk assessments and approvals that were conducted in the country of export, as allowed by the Protocol. In such a system, no new risk assessment is conducted in the importing country. Alternatively, the country of import can require its own review and approval process. If that path is taken, food safety issues are the primary risks to be addressed. Thus, the data submitted by the applicant and the regulatory process should focus on conducting a detailed food safety assessment consistent with Codex rules. Any risk assessment involving potential environmental risks from the organism's introduction into the environment should be less extensive than that required under a commercial release application, because the number of GE organisms that persist in the environment and reproduce would be a small fraction of that which occurs in a commercial release situation.

REGIONAL RISK ASSESSMENTS FOR GE ORGANISMS. When public or private developers are ready to commercialise a GE organism, that product could be of benefit to farmers not only in the country where it was developed and tested but also in its neighbouring countries. Thus, the three East African countries could benefit from establishing a mechanism to assess GE organisms that might have a regional impact. Such a system could produce a mutually agreed-on risk assessment for a product that would form the basis for individual approval decisions in each country. It could establish uniform risk management measures for a GE organism, which could be important because ecosystems are not confined within national boundaries. Also, such a system could use the limited biosafety capacity in the region efficiently by drawing on scientists from the three countries to conduct one risk assessment on a particular application. Finally, a regional assessment system might result in a single mutually agreed-on decision on the commercial release of the GE organism for all three countries.

91 United States Environmental Protection Agency, "Biopesticides Registration Document – *Bacillis thuringiensis* Plant-Incorporated Protectants" (USEPA: Washington, DC, 2001).

A regional risk assessment process for potential commercial GE organisms with markets in more than one East African country could be a model similar to that established for environmental assessments involving shared ecosystems. The East African Community has developed guidelines to be adopted within each country that address procedures for conducting transboundary environmental assessments in shared ecosystems in East Africa.[92] Under those guidelines, projects that occur in one country but that might have an impact on one or more of the other countries are required to produce a single EIA carried out by the developer. That EIA is submitted to the country of origin of the project as well as the affected countries, and each is given the opportunity to comment on the EIA. Those comments are then taken into account in the decision by the country of origin. If the decision is not satisfactory to any one country, that country can appeal to the EAC Secretariat to resolve any dispute.[93]

If such a system were developed for GE organisms, an application for the commercial release of a GE organism could be submitted to one country, which would then share the application with the other two countries. A regional science advisory board, composed of members of the different country NBCs and technical advisory bodies, could then review the application, conduct the risk assessment, and propose any risk management measures. Each country, as well as the public, then has the opportunity to comment on the review by the regional body, who would finalize their recommendation taking into considerations relevant comments. Following one possible pathway, each country could take its own approval decision under its own legal authority. Such a system would allow for the possibility that the GE organism might be approved in one country but not its neighbour and would also allow individual countries to add their own additional risk management measures. A second option would be to have the legal authorities in the three countries meet and decide, based on the recommendation from the regional body, to jointly approve or deny the application for all the countries. Whether the final decision is left to the individual country or determined jointly, a regional risk assessment procedure could save time and money and make better use of limited expert capacity in the region.

Even if a regional risk assessment system is not established, at minimum, the three East African countries could agree to use the same risk assessment procedures and risk assessment criteria for their individual risk assessments of a GE organism. A similar harmonisation was successfully done with the phytosanitary rules, when the three countries all agreed to use the revised FAO pest risk analysis procedures as the basis for issuing import permits.[94] By using the same risk assessment criteria and procedures for GE organisms, there would be some standardization and predictability for developers in the review and approval process.

92 *Supra* note 89.

93 *Ibid.*

94 Seed Regional Working Group. Harmonization of Seed Policies and Regulations in Eastern Africa. (ASARECA-ECAPAPA: Entebbe, Uganda, 2003).

CONCLUSION

Kenya, Tanzania, and Uganda have been working the past few years to establish national biosafety regulatory systems, in accordance with their own sustainable development objective, so that safe GE organisms can become part of their agricultural system. Although those regulatory systems are not yet completed, the framework for those future systems is set forth in different legal and regulatory documents analysed in this paper. Despite that the proposed biosafety regulatory systems differ in their details, their ultimate goal is the same – ensuring adequate protection of the environment and human health from activities involving GE organisms.

The detailed legal analysis of the current and proposed biosafety regulatory systems in East Africa identified many issues that, if addressed, could improve those systems. Some of the major conclusions and recommendations from the analysis are as follows.

First, food safety is a necessary component of the biosafety regulatory system. Assessing the food safety of GE organisms could be conducted by a food safety regulator under a food safety law, or it could be conducted when the environmental risks are assessed using legal authority granted under a biosafety act. The legal authority and procedures to assess and address potential food safety risks of GE organisms when they are released as commercial products should be spelled out in the national biosafety regulatory system so that they can be understood by all interested stakeholders.

Second, the biosafety regulatory system should set forth the legal standard and criteria on which it will base its decision whether to approve or deny an application. Third, different activities with a GE organism should have different regulatory pathways to an approval that correspond to the potential risks of that activity. Different pathways should be established for contained use, confined field trials, commercial release, and use as food, feed, or processing. Fourth, if socioeconomic considerations are to be included in the assessment process for GE organisms, that process should occur only at the commercial release stage of development. The potential areas for analysis must be set forth explicitly, as well as how the analysis of those factors will affect the approval process. To try to achieve compliance with international obligations, any socioeconomic analysis should be limited to impacts that are closely linked to biodiversity.

Furthermore, GE organisms, and not their products, should be the focus of the national biosafety regulatory system. Products produced from GE organisms should be regulated under a product-specific statute, not within the biosafety regime. In addition, interim measures established in each country have allowed some GE experimentation and begun the process of establishing institutional capacity in the area of biosafety. The knowledge and experience from those systems should be incorporated into the proposed systems to be implemented in the future. Seventh, the role and value of IBCs should be assessed to determine if their actions increase the safety of GE organisms. If IBCs continue to be a part of the national biosafety regulatory systems, countries might consider making them the primary regulators of contained-use experiments. Eighth, while each country is establishing its national biosafety regulatory systems, collaborative efforts to coordinate and harmonise processes and systems should be explored. Areas for cooperation, standardisdation, and harmonisation might include the application forms and data requirements, the different regulatory pathways for different activities with GE organisms, and the risk assessment process. The three countries also might consider establishing a joint procedure that would provide for one risk assessment for a GE organism that is to be commercially released in all three countries.

The issue of biosafety in East Africa will increase in importance in the coming years as local scientists develop GE organisms that directly benefit agricultural interests in that region of the world. To be ready for those developments, Kenya, Tanzania, and Uganda must continue to establish their national biosafety regulatory systems and develop them into functioning bureaucracies, able to plan and defend their countries desire for sustainable development. By reviewing the conclusions in this paper and implementing some of the suggested improvements, those systems can become more responsive to the requirements of good governance: more comprehensive, understandable, workable, and fair.

17

The National Biosafety Regulatory Systems in Asian and Near East Countries

Nizar Mohamed

This chapter examines the experiences of twenty-three countries in Asia in developing their national biosafety frameworks (NBFs) under the joint United Nations Environment Programme – Global Environment Facility (UNEP-GEF) NBF Development project. The development of NBFs is conceptualised as a process of knowledge mapping, allowing each country to produce a NBF tailored to the country's particular situation within a common framework, and in response to the requirements of the Cartagena Protocol on Biosafety[1] (Cartagena Protocol). Based on an analysis of country experiences, the chapter draws out key lessons for future capacity-building activities in biosafety. Lessons learned are that (1) biosafety is a sustainable development issue; (2) NBFs must be responsive to national needs and priorities; (3) the process of developing NBFs must be inclusive, country driven, and strengthen national capacity to ensure sustainability; and (4) regional cooperation is essential for sustainability of NBFs at the national and regional levels.

The signatories to the Cartagena Protocol, recognising the need to enable developing countries to make informed decisions on living modified organisms (LMOs), designated the Global Environment Facility (GEF) as the agency responsible for financing capacity building. In 2000, the GEF Council adopted an initial strategy to assist countries to prepare for the Protocol's entry into force. Under this strategy, the GEF funded the NBF Development Project, implemented by the UNEP, with national projects executed by designated government agencies in each country. This global project's main objective was to prepare countries for the entry into force of the Cartagena Protocol. The project consisted of two major components: the development of NBFs in participating countries, and the promotion of regional and

[1] *Cartagena Protocol on Biosafety to the Convention on Biological Diversity*, 29 January 2000, 2226 UNTS 208 (entered into force 11 September 2003), online: http://bch.cbd.int/protocol/text/.

subregional collaboration and exchanges of experience on biosafety. Its process and outcomes are discussed in Chapter 2 of this book.

The importance of capacity building in biosafety in Asia has been recognised by donor agencies with a proliferation of regional and national capacity building projects on biosafety and the safe use of biotechnology.[2] Many countries in Asia have also recognised the key role of biotechnology and biosafety in promoting sustainable development.[3] This is important, in light of the scope and focus of this book. As of January 2012, thirty-four countries from Asia had ratified the Cartagena Protocol.[4] Of these, most are actively involved in either developing or implementing their NBFs with GEF support. In addition, two countries that have not ratified the Cartagena Protocol, Lebanon and Nepal, have completed the preparation of their NBFs with GEF support and plan to ratify the Cartagena Protocol in the near future.

The twenty-three countries in Asia developing their NBFs under the global project reviewed in this chapter have a variety of different cultures, traditions, systems of governance, regulatory regimes, and languages, providing a challenge to the development of frameworks that conform to the Cartagena Protocol's requirements. This chapter explores these challenges, shows how they have been successfully addressed in these countries, and illustrates that compliance with international legal instruments is possible where there is a will to achieve that objective, focusing both on *process* (i.e., how countries achieved the NBF development process) and *content* (i.e., how countries addressed the choices that faced them on the structure and content of the NBF). Based on an analysis of these experiences, conclusions are presented on some of the lessons learned.

CONCEPTUAL FRAMEWORK

A NBF can be defined as a combination of policy, legal, administrative, and technical instruments that are developed to ensure an adequate level of protection in the field of the safe transfer, handling, and use of LMOs resulting from modern biotechnology, which could have adverse effects on the conservation and sustainable use of

[2] SR Rao, "Issues in Capacity Building: An Overview" (2005) 7(3) Asian Biotechnology and Development Review 1 at 7; E Tamale, "Building Capacities for the Effective Implementation of the Cartagena Protocol" Cartagena Protocol News (16 October 2006), online: http://www.cbd.int/doc/newsletters/bpn/bpn-issue01.pdf.

[3] P Mehta-Bhatt et al, "An overview of Regulation, Perceptions and Priorities for GM Crops in Asia" (2005) 7(3) Asian Biotechnology and Development Review 9 at 24.

[4] These are Azerbaijan, Bangladesh, Bhutan, Cambodia, China, Democratic People's Republic of Korea (DPRK), Georgia, India, Indonesia, Iran, Japan, Jordan, Kazakhstan, Republic of Korea (South Korea), Kyrgyzstan, Lao People's Democratic Republic (Laos), Malaysia, Maldives, Mongolia, Myanmar (Burma), Oman, Philippines, Philippines, Qatar, Saudi Arabia, Sri Lanka, Syrian Arab Republic, Tajikistan, Thailand, Turkey, Turkmenistan, Vietnam, and Yemen.

biological diversity, also considering risks to human health.[5] The development of a NBF can be conceptualised as a process of knowledge mapping,[6] consisting of five steps: (1) a stocktaking exercise to collect information and data on the current status of biosafety and biotechnology within the country, including human and institutional resources; (2) analysis of the data and information with participation by all stakeholders; (3) storage of the data so that it is readily available to stakeholders and can be used in the future to implement the NBF; (4) further analysis and processing of information to determine priorities for the NBF; and (5) preparation of the NBF through a consultative process. A chart showing this process is shown in Figure 1 in Annex VII to this book.

These steps correspond to the five steps in a knowledge-mapping process. The result of this mapping process is a translation of the map into practice through the development of a NBF.[7] Each country has adapted this process of knowledge mapping to its own particular social, political, environmental, and developmental context to produce a NBF tailored to the country's particular situation within a common framework. The common framework can be conceptualized as a knowledge map consisting of five interrelated components, as shown in Annex I to this book. This portrayal of the NBF comes from the definition of a knowledge map as "an association of items of information (e.g., process, network, policy, geography), preferably visual, where the association itself creates new, actionable information."[8]

The concept of the NBF as a knowledge map provides a framework of analysis in discussing the experiences of the countries surveyed in developing their NBFs. The five interdependent components of the NBF provide countries with a guide to setting up the systems necessary not only to comply with the requirements of the Cartagena Protocol but also to ensure that the systems in place are tailored to national needs and priorities for sustainable development.

The framework of analysis for this chapter is thus based on the key elements of an NBF, including (1) the rationale for developing the NBF in the first place; (2) the process used by countries to develop their NBF; (3) the policy framework within which the NBF will operate; (4) the regulatory regime proposed for managing genetically modified organisms (GMOs)[9] (5) the institutional set-up proposed for managing GMOs; (6) how the NBF addresses issues of public participation; (7) how the NBF

5 UNEP-GEF, *Toolkit for Development of National Biosafety Frameworks: Phase 3 – Drafting the NBF: Part I: Formulation of the Regulatory Regime* at 4, online: http://www.unep.org/biosafety/Toolkit.aspx [UNEP-GEF Toolkit].

6 EF Vail, "Knowledge Mapping: Getting Started with Knowledge Management" (1999) 16:1 Information Systems Management 16 at 23.

7 S Ebener et al, "Knowledge mapping as a technique to support knowledge translation" (2006) 84:8 Bulletin of the World Health Organization 636 at 637.

8 *Ibid* at 636.

9 The term *genetically modified organism* (GMO) has been used, but where there is a specific discussion of the Protocol, the term *living modified organism* (LMO) has been used.

addresses socioeconomic issues; and (8) how the NBF addresses the precautionary approach. Finally, this chapter looks at how the twenty-three countries intend to translate the knowledge map into action through implementation of the NBF.

EXPERIENCES IN ASIAN NATIONAL BIOSAFETY FRAMEWORK DEVELOPMENT

Why Develop a NBF?

The countries in Asia that joined the UNEP-GEF NBF Development Project to develop their NBFs did so for several interdependent reasons, including national development priorities and international obligations. In some cases, what started as a response to an international agreement soon became a development issue within the country, as stakeholders began analysing the potential benefits and impacts of GMOs. For example, in Tajikistan, the initial reason for joining the NBF project was the attraction of obtaining funding for capacity building for biodiversity conservation. As the project got under way and discussions with various stakeholders were held, however, it soon became obvious that biosafety was a sustainable development issue. As a result, Tajikistan ratified the Cartagena Protocol in 2003 and has taken a lead role in Central Asia in highlighting the safe use of biotechnology as a sustainable development issue, organising a subregional meeting on biosafety and biotechnology in Dushanbe in 2004. The resolution from that meeting was then considered at the Central Asia Intergovernmental Commission on Sustainable Development at its meeting in Ashgabat, Turkmenistan, in February 2006.

In other countries, the rationale focussed initially on GMOs as a development issue, but this soon brought about an awareness of the need to ratify the Cartagena Protocol as an international instrument to manage transboundary movement of LMOs. For example, in 2001 the Philippines reiterated the government policy of promoting the safe and responsible use of modern biotechnology and its products as one of several means to achieve and sustain food security, equitable access to health services, a sustainable and safe environment, and industry development.[10] The country, as an importer and potential exporter of LMOs, has also recognised the importance of the Cartagena Protocol as an international instrument to manage the transboundary movement of LMOs and ratified the Cartagena Protocol in August 2006.

The Cartagena Protocol

For many countries, at least at the start of the process, the primary reason was to comply with the requirements of the Cartagena Protocol. Of the twenty-three

[10] Philippines, "National Policy Statement on Modern Biotechnology," 16 July 2001, online: http://www.searca.org/bic/info_kits/policies/policy_statement.htm.

countries in this study, more than twenty-one have ratified the Cartagena Protocol, and others have not as yet completed their national processes for ratification.

National Development Priorities

Many countries have recognized biotechnology and biosafety as being integral to their national development planning priorities, particularly for sustainable development, even though the development of the biotechnology sector within these countries is a recent phenomenon, for example, in Tajikistan. Conversely, several countries in Asia began adopting modern genetic modification technologies in the late 1980s and developed regulations to manage GMOs well before the Cartagena Protocol had been finalised. Several of these front-runners in the development of biotechnology in Asia decided to join the NBF project to review and revise their existing systems and make them consistent with the Cartagena Protocol and changes in national priorities; these countries include the Philippines, Iran, Syria, Thailand, and Indonesia.

Many of the Asian countries saw the promotion of the safe use of biotechnology as part of sectoral policies in areas such as agriculture, forestry, and fisheries (e.g., Syria, Indonesia, and Vietnam), as well as for the promotion of science and technology within the country (e.g., Democratic People's Republic of Korea [DPRK], Iran, Laos, and Thailand). Others considered biosafety within the context of biodiversity conservation (e.g., Bhutan and Lebanon), whereas others were also aware of the trade implications of GMOs (e.g., Lebanon, Republic of Korea [South Korea], and Yemen).

How Did Countries Develop Their NBFs?

The guidelines for the process of developing a NBF, provided by the UNEP-GEF NBF Development Project, helped countries to tailor their NBFs to their own situation. Countries used the process of mapping first to establish a baseline for their national status through a stocktaking exercise, and then to decide the priorities for their NBF. Putting this NBF into practice then required the setup and implementation of the five components of the NBF.

This flexible and iterative process helped countries at various stages of development in biotechnology to tailor both the process and the final product to their own specific situations. For example, countries with a high level of biotechnology research and development (R&D), such as the Philippines, could bring together information on the country's experiences with biotechnology research and regulation of GMOs through a process of consultation. The participatory and inclusive process advocated by the NBF project not only helped to bring different stakeholders together but also ensured that the importance of biosafety came to the forefront of the debate, so that environmental agencies as well as those promoting science and technology and agriculture development could play an active role in the decision-making process on GMOs. Other countries, such as Myanmar, which has a relatively low level of activity in biotechnology, could document these activities, as well as the existing laws and regulations that could have an impact on the safe use of GMOs, to be able to make informed decisions on their NBFs. This process of collecting information and analysing that information using national consultants has also helped to build capacity within the country.

The development of the NBF in the Asian countries is best seen as an iterative learning process as countries collect and analyse information on biotechnology and biosafety within the context of their specific development priorities, and refine and develop their ideas about the different components of the NBF. This iterative process leads to a dynamic and evolving map of the NBF, which can change with experience as systems are established and become operational.

The actual NBF produced by each country was also tailored to their own special needs and priorities within the framework provided by the UNEP-GEF Project. Thus, each NBF is different, and there is no standard format or content for the NBFs in Asia. Some countries, such as Bangladesh and Tajikistan, chose to follow the standard format for the NBF provided by the NBF Project team, with NBF chapters on the policy framework, regulatory regime, administrative systems, monitoring and enforcement, and public participation. Others, such as Jordan, the Republic of Korea, and Syria, which used the development of the NBF to set out their priorities for biotechnology and biosafety, decided to include in their NBF an analysis of the status of their biotechnology sector, including both institutional and human resources. This helped to set their proposed systems for the NBF within the national development context and to identify the capacity-building needs for biosafety and biotechnology within the country. Some countries, like Nepal, decided to use the NBFs as a means to educate both policy makers and the public about biosafety and biotechnology. Their NBF therefore included a chapter on technical aspects of biosafety, which explained some of the key scientific areas of risk assessment and management. Several countries used innovative approaches; Vietnam developed its draft NBF as an action plan for biosafety, complementing the Government's National Plan for Biotechnology. Another example of an innovative approach is the Philippines, where the NBF was in the form of an Executive Order (EO) that brought the existing regulations on biosafety under the umbrella of the NBF; the EO was eventually promulgated by the government after signature by the president.

NBF: POLICY CONTEXT

A national policy that addresses biosafety is considered to be an integral part of a NBF.[11] In Asia, the policy context that provides the vision or rationale for the NBF

[11] MA McLean et al, eds, *A Framework for Biosafety Implementation: Report of a Meeting* (The Hague, NL: International Service for National Agricultural Research, 2003), online: http://www.ifpri.org/sites/default/files/pubs/isnararchive/publicat/bioframework.pdf.

has focussed on both the obligations and responsibilities under the Cartagena Protocol, and the potential of biotechnology in promoting the country's development priorities. The areas of the Cartagena Protocol addressed in policies include the main subjects of its focus: transboundary movement;[12] protection of natural and agricultural biodiversity;[13] protection of human health;[14] public participation, education, and awareness;[15] and capacity building for biosafety.[16] Most national policies on biosafety also include the precautionary approach,[17] including the policies of Bhutan, Indonesia, Jordan, Lebanon, Myanmar, Yemen, and Iran.

In addition to the main priorities of the Cartagena Protocol, national biosafety policies address national development issues, such as promotion of national research and development in biotechnology consistent with biosafety (e.g., DPRK, Vietnam, and Iran); agricultural development and food security (e.g., Bhutan, Myanmar, Philippines, and Yemen); promotion of conservation and use of genetic resources for biotechnology (e.g., Bhutan, Lebanon, Jordan, and Laos); trade issues, both in relation to maintaining exports of agricultural products and importation of food and feed products (e.g., Lebanon); sustainable development and Agenda 21[18] (e.g., the Philippines and Indonesia); and labelling requirements that go beyond the Cartagena Protocol (e.g., Indonesia and Vietnam).

NBF: Regulatory Regime

In Asia, nine countries had some form of regulatory regime in place before they started their NBF Project.[19] In these countries, the regulatory instruments for GMOs were developed originally in response to national priorities on biotechnology and biosafety, and were reviewed and revised as part of their NBF Project. The other fourteen countries had no preexisting legal instruments for biosafety.

The various types of primary legal instruments chosen by countries to regulate GMOs depended both on their previous experiences with regulating GMOs and their regulatory setup in relation to areas such as quarantine, biosecurity, science and technology, trade, and food safety. These include *level 1* instruments approved by the legislative branch of government, such as a parliament, congress, legislature, or house of assembly, which are then promulgated with binding effect; and *level 3* instruments created under a delegated authority by an individual or group but they

[12] Cartagena Protocol, *supra* note 1 at arts 1 and 4.
[13] *Ibid* at arts 1 and 4.
[14] *Ibid* at arts 1 and 4.
[15] *Ibid* at art 23.
[16] *Ibid* at art 22.
[17] *Ibid* at art 1.
[18] *Agenda 21: Programme of Action for Sustainable Development*, U.N. GAOR, 46th Sess., Agenda Item 21, UN Doc. A/Conf.151/26 (1992).
[19] See Annex VII, Table 2.

do not need further approval by the legislature before promulgation that is legally binding.

Several countries lacking a preexisting biosafety regime decided to select a level 3 instrument (e.g., Jordan, Lebanon, Maldives, Yemen, and Bhutan). Reasons for this choice included building on existing legal instruments such as an environment or food law to make best use of a functioning system; expediency – as a by-law or decree requires either ministerial or cabinet approval and does not require amendment of an existing law, and therefore parliamentary approval; a by-law or decree can therefore be promulgated fairly quickly, and a country would have a legal basis for regulating GMOs within a short time frame; in a rapidly changing scientific field, countries preferred to work with a legal instrument that could be reviewed and revised easily, leaving their options open for formulating a suitable instrument later if needed.

Other countries without a preexisting biosafety regime decided to adopt a level 1 biosafety law. Their reasons for choosing a full biosafety law included: no suitable existing laws that could serve as a "home" for a level 3 biosafety legal instrument; legislative systems that enabled them to pass the level 1 law relatively quickly, especially with the political support of the government. For example, four of the countries that chose a full biosafety law had the draft bill passed by their parliament or assembly within a year of completion of the NBF (i.e., DPRK, Tajikistan, Kyrgyzstan, and Laos PDR), whereas Cambodia's biosafety law was passed by their National Assembly in 2007. The drafting and approval of a full biosafety law was considered to be a sign of support not only from a government minister but also from Parliament as a whole, strengthening ownership within the country and making it easier to implement the NBF.

Some countries with a preexisting biosafety regime, such as Iran and Thailand, have opted for the option of drafting a national biosafety law, because they had considerable experience with implementing regulations for biosafety. Others, such as the Philippines, Indonesia, Vietnam, and Sri Lanka, decided to review and revise their existing legal instruments to develop a level 3 legal instrument consistent with the Cartagena Protocol but they have all left the option open to develop a level 1 biosafety law during the implementation phase of their NBF.

In all of the countries in the region that have drafted their regulatory instrument, the scope of the regulatory regime is, at minimum, in accordance with the Cartagena Protocol and covers transboundary movement – export, import, and transit – referring either to LMOs (e.g., Jordan, Laos, Kyrgyzstan, Cambodia) or to GMOs (e.g., Bhutan, Maldives, Vietnam). The instruments also address procedures for food and feed products (LMOs-FFP) in accordance with Article 11 of the Cartagena Protocol.

In addition, many of the draft instruments go beyond the Cartagena Protocol in accordance with national priorities to include issues related to the production of GMOs within the country. These include research, contained use, field trials, marketing, and labelling of GMOs (e.g., Philippines, Maldives, Bhutan, and Vietnam).

NBF: Institutional Setup

In the draft NBFs in the region, the proposed institutional setup for the National Competent Authority (NCA) and a National Focal Point (NFP) as required under Article 19 of the Cartagena Protocol includes a single NCA and/or a NFP receiving and processing applications; this is usually in consultation with a National Biosafety Committee or Commission (NBC) and line ministries; for example, Lebanon (Ministry of Environment), Cambodia (Ministry of Environment), and Bhutan (Bhutan Agriculture and Food Regulation Authority). The most common model is more than one NCA, each with sectoral responsibilities. The receipt of applications for GMOs can be either through a single window for applications received by a central coordinating body and sent for processing to the relevant NCA (e.g., in Iran, DPRK, Sri Lanka, and Myanmar); or multiple windows with applications received by the responsible government agencies, with coordination provided by an NBC (e.g., in Jordan, Yemen, the Philippines, and Indonesia).

In the draft NBFs in the region, the proposals for the risk assessment setup usually assign that responsibility to the NCA or overall biosafety body with advice from either an ad hoc scientific advisory body, or an established advisory committee (e.g., Iran, Yemen, and Lebanon). The composition of this advisory body is often determined by the regulatory instrument (e.g., Lebanon and Yemen), and in some cases, NGO representation on advisory committee is allowed (e.g., Iran, Jordan, Yemen, and Bhutan).

The proposals for decision making also include a variety of different models: a multistakeholder body (usually made up of government agencies) for decision making, sometimes chaired by the environment agency (e.g., Jordan, Kyrgyzstan, Iran, Lebanon, and Yemen); decision-making function vested in a single NCA, after consultation with other agencies or with a multistakeholder advisory body (e.g., Bhutan and Cambodia); or the decision-making function is vested in multiple NCAs, after consultation with other agencies and/or with a multistakeholder advisory body (e.g., Indonesia, Philippines, and Vietnam).

NBF: Addressing Public Participation

Countries in Asia have addressed Article 23 of the Cartagena Protocol on public awareness, education, and participation differently, depending on each country's particular social, political, and economic situation.

The main way in which the public has been involved has been through each country's NBF Development Project. The project required each country not only to implement awareness and education activities on biosafety but also required that the process of developing the NBF involved public participation in making key decisions.[20] The main mechanism for participation in the development of NBFs was through the National Coordinating Committee (NCC), established under the

[20] See Annex VII, Figure 1.

project. The NCC was responsible for overall policy guidance for the development of the NBF, and all countries formed such a body. In these countries, the NCC became the overall policy- and decision-making body for GMOs as part of the institutional setup within the NBF. The NCC included relevant government agencies, as well as representatives from other stakeholders in the twenty-three countries.[21]

The main focus of countries in terms of Article 23 during the NBF development was on increasing public awareness about GMOs through workshops, use of media such as television and radio, regular newspaper articles, and publications on biosafety. This was one of the main activities required for each national project, accounting for some twenty percent of the total costs of the project. Given the diversity of languages in Asia, with more than twenty different languages in the twenty-three countries, the emphasis was on publications and awareness in the main national languages. To complement these activities on increasing awareness, all of the twenty-three countries also implemented extensive activities on strengthening public education on GMOs, using both informal and formal means.

Enabling public access to information on GMOs for their effective participation in decision making was done through production of information on GMOs in printed form in an appropriate national language, and by setting up Web sites on biosafety. For example, sixteen of the twenty-three countries set up Web sites (usually as part of the National Environment Agency Web sites) on biosafety or specifically on the NBF project. These were in the main national language of the country. One country, Yemen, even used its site to conduct a survey on whether Yemen should ratify the Cartagena Protocol.

In those countries that have completed their NBFs, the main provisions for promoting participation by stakeholders in biosafety decision making are consistent with Article 23. In most countries, these provisions are enshrined in the biosafety regulatory instrument and emphasize public consultation on GMO-related activities. These include not only applications for permits for importation or environmental release but also regulation of biotechnology research. In some countries, institutional biosafety committees also include representation from civil society (e.g., Iran and the Philippines).

Public input into the decision-making process is usually through mechanisms that provide for comments within a specified time (usually thirty days), and set out in each NBF the procedures described for processing of applications. The decision-making body is also sometimes required to consider public comments in making their decisions on GMOs (e.g., the Philippines, Jordan, Bhutan, Sri Lanka, and Iran).

Information on applications is to be made available to the public through the national Biosafety Clearing-House (BCH), and other means that are easily accessible (e.g., Philippines, Indonesia, Iran, and Yemen). Access to information is usually restricted to nonconfidential information as defined in the regulatory instrument,

[21] See Annex VII, Table 3.

and in conformity with Article 21 of the Cartagena Protocol. Some countries have provided for non-government representation on the biosafety decision-making body: Iran – on institutional and national biosafety committees, private sector and/or public; Jordan – private sector and consumer representatives on the NBC; Philippines – civil society representation on the National Committee on Biosafety of Philippines and on institutional biosafety committees; Lebanon – private sector, NGO representatives on the NBC; and Yemen – Chamber of Commerce and Consumer Society on the NBC.

NBF: Addressing Socioeconomic Issues

In many of the countries in Asia, the socioeconomic impacts of biotechnology are an important consideration; this is reflected in their NBFs, which require the decision-making process to include consideration of socioeconomic issues. These provisions reflect Article 26 of the Cartagena Protocol and might be included in the national policy on biosafety (e.g., Lebanon) or in the regulatory instrument (e.g., Republic of Korea, Yemen, Philippines, and Bhutan). Others, for example, Iran and Jordan, have chosen not to include socioeconomic issues in their NBF.

The following are some examples of the ways in which countries have chosen to include socioeconomic issues in their NBF. In the Philippines, Executive Order 514 states in Article 5.4 (on Socio-economic, Ethical, Cultural and Other Considerations) that, consistent with Article 26 of the Cartagena Protocol, concerned government departments and agencies may take into account socioeconomic considerations arising from the impact of regulated articles on the conservation and sustainable use of biologic diversity, especially with regard to the value of biologic diversity to indigenous and local communities. The draft legal instrument in Yemen states that the decision-making process shall consider an evaluation of socioeconomic risks in parallel with scientific risk assessment, and that biosafety decisions shall take into account issues of poverty alleviation and food security; livelihoods of small farmers, indigenous people, women, and small and micro enterprises (SME); and cultural integrity of the country and communities. In the Republic of Korea, Article 14 (3) of the biosafety law states that the relevant government authority may "prohibit or restrict" the import or production of LMOs that could have "adverse socio-economic effects on the values of the country's biodiversity."

Implementing the NBF: Translating the Map into Action

The process of developing the NBF is a knowledge-mapping process that helps the country to identify not only its needs and priorities for the safe use of biotechnology but also its human, technical, and institutional resources. The next step is to translate this knowledge map (i.e., the NBF) into action by harnessing these resources, and by establishing and strengthening the systems that make up the NBF: the requisite regulatory regime, the systems for handling applications for GMOs, risk assessment and management, monitoring and enforcement, and for public participation.[22] (Figure 2). The final step is the implementation of the NBF itself (i.e., making these systems operational).

For the Asian countries participating in the NBF Development Project, the long-term aim is first to prepare their NBF and then to translate this NBF into practical and workable systems. The process of translation requires government commitment, which has been demonstrated in several ways: approval of the NBF by the government, approval of the biosafety policy in the NBF, promulgation of the biosafety regulatory instrument by the appropriate government body, and allocation of funds for recurrent costs of putting the NBF into action.

Most countries that are Parties to the Cartagena Protocol have turned to the GEF for further assistance to implement their NBF. Provision of funding from the GEF requires a demonstration of government commitment through allocation of financial and other resources to the implementation of the NBF. In Asia, the current status of proposals for implementation is summarised in Annex VII, Table 4.

SOME LESSONS FROM THE DEVELOPMENT OF NBFs IN ASIA

The comparative analysis of the experiences of countries in Asia in developing their NBFs under the UNEP-GEF project highlight some of the key challenges faced by countries in meeting their obligations as Parties to the Cartagena Protocol. These experiences provide lessons that will be relevant to future capacity-building activities in biosafety as these countries begin to implement their NBFs.

Biosafety as a Sustainable Development Issue

Probably the most important lesson emerging from these experiences is that biosafety is a sustainable development issue and that it cannot be considered in isolation from a country's development priorities. Three critical elements can be highlighted. The first is that biosafety policies must address both environmental and social and economic development concerns and not just the conservation of biodiversity, even though the Cartagena Protocol is part of the Convention on Biological Diversity (CBD) regime. For example, Myanmar has formulated a policy on biosafety and biotechnology, which brings together development priorities and the conservation of biodiversity. Second, biosafety laws and administrative systems have to complement and strengthen existing national systems, such as quarantine or environmental impact assessment (EIA) laws, rather than trying to set up a new parallel system.

22 See Annex I.

For example, Bhutan has developed its systems for administration of GMOs within the context of its Food Act (2005); the biosafety regulations will be administered by the Bhutan Food Regulatory Authority under its mandate for food safety. Third, the NBF has to work within the context of national development strategies and plans, and should not be seen as a stand-alone issue. For example, in the Philippines, the NBF has been developed within the overall context of the Philippines Agenda 21, and the country's priorities on development of the biotechnology sector.

Responsiveness to National Needs and Priorities

Recognition of biosafety as a sustainable development issue means that the process of NBF development, and particularly the resultant product (the NBF itself), must be responsive to national needs and priorities. This will promote sustainability of the NBF in several ways.

First, it helps ensure national ownership by grounding biosafety in a country's national needs and priorities. For example, the Nepal NBF is being formulated within the context of the government's desire to promote biotechnology R&D to ensure that the country as a whole benefits from the utilization of its rich genetic biodiversity, ensure that the country's obligations under the Cartagena Protocol are used as an external stimulus to kick-start the NBF process, and stimulate public debate on GMOs, rather than being an end in itself.

Second, the mapping process advocated by the UNEP-GEF Project in developing the NBF helps a country to decide its priorities for biotechnology and biosafety within the context of the overall national plans for sustainable development. For example, Yemen has developed its NBF to be able to harness the benefits of biotechnology to promote food security whilst ensuring that its rich agrobiodiversity is protected.

Third it helps promote political and public support for the NBF, especially when the NBF is seen by policy makers and the public as supporting national development priorities. This is best achieved through an inclusive and participatory process for preparation of the NBF, as well as making full use of available national expertise and building on existing systems. This will help to ensure that all sectors can contribute and that national priorities guide the process of developing the NBF. For example, in Sri Lanka, the development of a biosafety policy as part of the NBF has helped to address public concerns about food safety while ensuring that its active biotechnology R&D programme is maintained.

Fourth, it helps tailor the NBF to the country's needs and priorities rather than imposing a set formula on a country. Within the overall framework provided by the NBF project, the development process for the NBF has been flexible so that countries are able to adapt it to their own situation. For example, the Vietnam NBF has been formulated as an Action Plan on Biosafety to complement the government's Action Plan on Promoting the Biotechnology Sector.

A Country-Driven Process

A key lesson from the NBF Development Project is the importance of a country-driven process in preparing the NBF. The strong emphasis on this principle has meant that the NBFs developed in Asia have a strong sense of national ownership: governments in many of these countries have committed substantial resources to setting up the necessary systems and maintaining them continuously through financial allocations in the national budget for recurrent costs. These national resources not only complement external assistance for capacity building but also help promote sustainability.

Some of the critical elements helping to ensure a country-driven process include having a national "champion" to support the process of developing the NBF as well as to ensure that the results of the project, the components of the NBF, have support from a wide range of stakeholders, including government agencies, the private sector, and civil society. For example, in Tajikistan, the head of the Biodiversity Research Centre, a former Minister in the government, has helped to champion the cause of biosafety not only within the country but also within the Central Asian subregion.

A second element is that NBF projects in each country have used and strengthened national expertise to help develop the NBF and then to operationalise the systems, procedures, and processes. The role of outside experts has been to support and peer review national contributions rather than to take over the development and running of the NBF. For example, national experts from universities, NGOs, and private consultants in Bangladesh have been entirely responsible for developing the Bangladesh NBF, which was then reviewed by an international expert. An important aspect of the use of national experts has been to provide incentives for these experts so that they are able to contribute and know that their contribution is valued, rather than being asked to make their contributions without any financial rewards.

A third element is to provide access to external technical resources to support national resources – this could be in the form of manuals and toolkits, access to training opportunities, study visits, and visits by outside experts. The series of regional and subregional training workshops by the global NBF project has been a key factor in helping to strengthen national capacity. For example, participants who attended the training workshop for Asia on drafting legislation have been responsible for drafting the biosafety legal instruments in several countries, including Myanmar, Nepal, Sri Lanka, Yemen, Iran, and Tajikistan.

Last, each national project has been aimed at building on existing systems in government rather than inventing new ones to develop the NBF. The process of mapping the existing institutions, laws, and resources has helped each country to identify how best to build on these systems to formulate a workable NBF. For example, the extensive mapping process in Myanmar by national experts has helped to ensure that biosafety fits within the existing quarantine and food safety systems in the country.

Regional Cooperation

Another key lesson is that within a region, countries need to work together, particularly in getting their national decision-making systems on GMOs up and running. In other words, regional cooperation is essential for sustainability of the NBF at the national and regional levels. The critical elements for regional cooperation include sharing of resources among countries within the region, including technical facilities, materials, and expertise; sharing of experiences among countries in developing and implementing NBFs (including methodologies, materials, and skills); sharing of information on biosafety among countries and through regional networks, including the BCH; and building and strengthening regional capacity (Regional Centres of Excellence) for biosafety and biotechnology to support national decision making. The importance of regional cooperation is recognised by the NBFs in at least three countries in Asia: Yemen, Indonesia, and Jordan.

One of the key lessons emerging from the projects in Asia, however, is that regional cooperation *must* be country driven[23] and not as a response to an externally imposed agenda. For most countries, therefore, initial attempts at regional cooperation have taken the form of regional meetings to identify potential areas for collaboration. Examples of such meetings during the development project can be seen in Central Asia – two meetings have been held to discuss subregional cooperation in biosafety and biotechnology. These were in Tajikistan in 2004 and in Kyrgyzstan in 2005; and South Asia – a similar meeting was held in Sri Lanka in April 2005 to discuss future areas for cooperation. This was followed by another meeting to be held in Bhutan early in 2007 to discuss regional cooperation in biosafety and biotechnology; Southeast Asia – a meeting of ASEAN countries was held in Manila in June 2004 to discuss potential areas for cooperation in biosafety, focussing on capacity-building needs within the region for implementing NBFs; West Asia – a meeting of West Asian and North African countries was held in September 2007 in Syria to discuss the formulation of a regional project on capacity building for biotechnology and biosafety within this subregion; this project was submitted to GEF for funding at the end of 2008.

As countries start to implement their NBFs, they are beginning to build on these initial discussions to work on real areas for cooperation. For example, Cambodia and Vietnam (both implementing their NBFs) and Thailand (currently finalising its NBF), carried out a joint subregional training on drafting of biosafety secondary regulatory instruments. One of the aims of this training was "To work with their neighbouring countries in developing their national regulations with a view to possible harmonisation in the future."[24]

[23] UNEP-GEF, *Building Biosafety Capacity in Developing Countries: Experiences of the UNEP-GEF Project on Development of National Biosafety Frameworks*, online: http://www.unep.org/biosafety/files/UNEPGEFstudyVersion170605.pdf.

[24] UNEP-GEF, *Report of UNEP/GEF workshop on secondary biosafety legislation in Vietnam* (October 2006).

An Inclusive Approach

Another important lesson is that an inclusive approach that facilitates the involvement of all stakeholders is crucial if the NBF is to be accepted by all Parties within the country. This helps to ensure support for the implementation of the NBF and promote the sustainability of the achievements.

A critical element for an inclusive approach is the active involvement of relevant government agencies in all aspects of the development and implementation of the NBF. These agencies include not only those for the environment but also those for science and technology, agriculture, forestry, fisheries, health, and education. In Iran, for example, the inclusive process of developing the NBF adopted by the Department of Environment has helped to promote dialogue between the environmental NGOs and those involved in biotechnology research. As the Cartagena Protocol comes under the CBD, the entry point for the NBF project in all countries has been the focal point for the Cartagena Protocol, usually the ministries of the environment. The NBF projects have used the multisectoral NCC in each country to ensure the involvement of other stakeholders, which has helped to ensure that the main messages of biosafety are introduced to all stakeholders so that they become aware of the potential benefits of the safe use of biotechnology.

For example, in Vietnam, where the Ministry of Natural Resources and Environment is the focal point for the CBD, the development of the NBF and subsequent efforts to implement it are supported strongly by both the agriculture and the science and technology ministries. This is because of the recognition that nongovernment stakeholders (private sector as well as civil society) have a key role in developing the NBF and in implementing the proposed systems and procedures. The initial emphasis on awareness and education during the NBF project has helped to lay a foundation for the future involvement of stakeholders in decision making, because awareness and education are prerequisites for effective participation. For example, in Yemen, where there was little awareness of GMOs prior to the NBF project, current efforts to implement the NBF have the strong support of civil society, agricultural scientists, the media, and the private sector. The NBF developed through the project includes mechanisms for meaningful participation by stakeholders in decision making. This is usually enshrined in the regulatory instrument, with explicit mechanisms for public participation provided in the systems for handling applications. For example, Iran and the Philippines now have representation by civil society in the national decision-making body as well as in institutional biosafety committees.

The NBFs also seek to provide stakeholders with access to biosafety information in a form that is readily understandable to all stakeholders and uses easily accessible media. The means for access to information in the NBFs of various countries include not only Internet-based approaches such as the BCH, but also more traditional media such as printed materials, radio, television, and public meetings. For example, the

Indonesia NBF requires dissemination of information on biosafety to the public using forms of media that are readily accessible to all stakeholders.

CONCLUSION: "LEARNING BY DOING"

The experiences of the NBF development Project in Asia demonstrate that building capacity for biosafety and the safe use of biotechnology within a country requires an approach based on learning by doing rather than relying on outside expertise to develop the NBF and to put its provisions into practice. This means that in each participating country, it is crucial to strengthen national capacity to ensure sustainability of the NBF, by laying a strong foundation for effective and sustained implementation of the NBF.

The experiences of the countries in Asia also highlight their commitment to biosafety and the Cartagena Protocol: eighteen of the countries in the NBF Development Project are already Parties to the Cartagena Protocol, and another four countries are completing their national procedures for ratification. The number of ratifications, and the preparation of draft NBFs that provide each country with its own "road map" for establishment of the systems necessary to implement its NBF in compliance with the Cartagena Protocol, represent a major advance in the two years since August 2004, when Watanabe et al. stated that many countries in Asia have "inadequate national legal provisions to put the protocol into practice."[25] The NBFs not only provide the necessary legal instruments and other systems for implementation of the Protocol, but the process of preparation of the NBF has started to build national capacity for *effective* implementation of the Protocol's provisions. In the future, this must be sustained through both externally funded and nationally supported capacity-building efforts to meet the challenge of translating the NBF map into practice.

Lessons from the experiences of the NBF Development Project in Asia include recognition that biosafety is a sustainable development issue and that it cannot be considered in isolation from a country's development priorities. The process of NBF development, and particularly the resultant product (the NBF itself), must be responsive to national needs and priorities, and must be country driven to ensure sustainability. Regional cooperation is also essential for sustainability of the NBF at the national and regional levels, but it must be country driven and not in response to an externally imposed agenda. Last, an inclusive approach that facilitates the involvement of all stakeholders is crucial if the NBF is to be accepted by all actors within the country.

[25] KN Watanabe, M Taeb & H Okusu, "Putting Cartagena into Practice" (2004) 22(10) Nature Biotechnology 1207.

18

The Regulatory and Institutional Biosafety Systems in the Americas

Jorge Cabrera Medaglia

The regulatory and institutional frameworks associated with modern biotechnology, and biosafety in particular, are facing changes and are evolving continuously due to an increasing number of national, regional, and international processes. In addition, contrary to what has occurred to other technologies (i.e., agriculture and livestock innovations), the regulatory and public policy background that pervades the development of modern biotechnology is more complex in terms of underlying frameworks, actors, and rules.

This regulatory and public policy background consists of relevant international instruments and processes such as the Cartagena Protocol on Biosafety,[1] Codex Alimentarius Guidelines,[2] the Food and Agriculture Organization (FAO) Commission on Genetic Resources for Food and Agriculture[3] (CGRFA) and its International Treaty on Plant Genetic Resources for Food and Agriculture[4] (ITPGRFA), and the World Trade Organization (WTO) Agreements on Sanitary and Phytosanitary Measures (SPS Agreement) and Technical Barriers to Trade (TBT Agreement). Finally, the legal and institutional context of modern biotechnology, particularly the field of

[1] *Cartagena Protocol on Biosafety to the Convention on Biological Diversity*, 29 January 2000, 2226 UNTS 208, 39 ILM 1027 (entered into force 11 September 2003), online: http://bch.cbd.int/protocol/text/ [*Cartagena Protocol*].

[2] See, Codex Alimentarius, *Guideline for the Conduct of Food Safety Assessment of Foods Derived from Recombinant-DNA Plants*, CAC/GL 45-2003; *Guideline for the Conduct of Food Safety Assessment of Foods Produced Using Recombinant-DNA Microorganisms*, CAC/GL 46-2003; *Guideline for the Conduct of Food Safety Assessment of Foods Derived from Recombinant-DNA Animals*, CAC/GL 68-2008; *Compilation of Codex texts relevant to the labelling of foods derived from modern biotechnology*, CAC/GL 76-2011.

[3] See FAO, "Commission on Genetic Resources for Food and Agriculture," online: http://www.fao.org/nr/cgrfa/.

[4] *International Treaty on Plant Genetic Resources for Food and Agriculture*, FAO Res. 3/2001, 3 November 2001 (entered into force 29 June 2004), online: http://www.fao.org/nr/cgrfa/cgrfa-global/cgrfa-planttr.

This chapter presents information prepared for the Inter-American Institute for Cooperation in Agriculture (IICA). The author thanks Dr. Enrique Alarcón for his comments.

agricultural biosafety, is complex and integrated with numerous institutions involved within a diverse national, regional, and multilateral regulatory framework.

The chapter is divided into three different sections. The first section focuses on the various impacts and challenges faced by legislators and policy makers in the Americas resulting from agricultural biotechnology, from a sustainable development law perspective. More specifically, it addresses key issues such as the variety of biosafety policies and regulatory systems in the region and the difficulty of regulating ever-evolving technologic fields like biotechnology. The second part of the chapter presents a comparative analysis of the many biosafety regulatory and institutional systems in the Americas, exploring aspects such as competent authorities, institutional order, labelling, and traceability. Finally, the third section focuses on outlining emerging trends in biosafety regulation in the Americas and how these trends reflect the important sustainable development objectives and principles enshrined in the Protocol.

THE IMPACTS AND CHALLENGES FACED IN THE FIELD OF AGROBIOTECHNOLOGY

In terms of research and use of living modified organisms (LMOs), Latin America produces approximately 27 percent of the world's transgenic crops, placing it second after North America, which produces another 65 percent.[5] Thus, in total, the continent produces roughly 92 percent of the world's genetically modified crops. Several studies report commercial seeds growing in Argentina, Brazil, Mexico, Uruguay, Honduras, Colombia, Bolivia, and Paraguay,[6] and there are exports of LMOs from winter nurseries[7] in Chile and Costa Rica, where trade in genetically engineered (GE) organisms is not allowed in the domestic market.[8] The main transgenic crops being grown are soybean, maize, cotton, and canola, and the prevailing features are herbicide tolerance, insect resistance (Bt), and stacked genes for the two traits (i.e., a combination of both traits).[9] The region also has several laboratories and centres of excellence working in agricultural biotechnology which are studying a broad range of crops, such as banana, coffee, tomato, rice, sunflower, papaya, yucca, sugar cane,

5 C James, *ISAAA global status of commercialized Biotech/GM crops 2008, executive summary* (International Service for the Acquisition of Agri-biotech Applications [ISAAA], 2009); G Traxler, "The GMO experience in North and South America" (2006) 2 Int Technology and Globalisation 46 (This study analyses the use, adoption, and benefits of Cotton Bt and Soya RR in the region); E Alarcón and R Artunduaga, "América Latina y el Caribe: Nuevas agrobiotecnologías, desafíos, tendencias y consideraciones institucionales" (2005) 2 *Comuniica*, online: http://unpan1.un.org/intradoc/groups/public/documents/icap/unpan045773.pdf.

6 C James, *ibid.*

7 *Winter nursery* refers to the practice of allowing the GMO crops to be planted in the country and subsequently exported without authorization for national consumption. Companies do this to take advantage of the different crop seasons.

8 Alarcón and Artunduaga, *supra* note 5.

9 *Ibid.*

sweet potato, alfalfa, tobacco, pumpkin, flowers, vegetables, fruits, and some timber species.[10]

In terms of regulatory and institutional frameworks, Latin American countries have traditionally tackled biological safety, a broader expression than biosafety, as well as the introduction of invasive species, in regulatory frameworks related to sanitary and phytosanitary measures.[11] Unsurprisingly, certain nations have simply used plant, animal, seed, and pest control safety laws and other regulations as the regulatory framework for LMOs in agriculture.[12]

In other words, countries in the Americas are facing the challenge of regulating modern biotechnology through several different policies and regulatory systems, which are discussed further in this chapter. Such regulatory frameworks are being enacted swiftly[13] because of four crucial factors:

1. Companies' requests for implementation field trials and the growing trade in GE organisms have given rise to the need to adapt, understand, and enact regulatory frameworks.
2. The signature, adoption, and entry into force of the Cartagena Protocol initiated regulatory reform processes as well as the creation of specific biosafety laws aimed at promoting compliance with its provisions. This process has mainly been triggered by Article 34 of the Protocol, which urges Parties to approve mechanisms destined to promote compliance and address cases of noncompliance.[14]
3. Several important projects and initiatives for capacity building, such as the projects financed by the Global Environmental Facility (GEF) and executed by the United Nations Environment Program (UNEP; hereinafter referred to as UNEP-GEF Projects), have aimed to (a) develop regulatory biosafety frameworks, (b) implement regulatory biosafety frameworks, and (c) strengthen the Biosafety Clearing-House (BCH) mechanism.
4. The trade in GE organisms, particularly those intended for direct use as food or feed or for processing, is on the rise.

10 See Red de Cooperación Técnica de Biotecnología Agropecuaria (REDBIO), online: http://www.redbio.org/.

11 R Brañes and O Rey, "Política, Derecho y Administración de la Bioseguridad en América Latina y el Caribe" (Chile: CEPAL, 2001).

12 See Brazil, Law 11.105/05; and Peru, Law n° 27.104.

13 Contrary to biosafety implementation, the adoption and implementation of regulatory frameworks for access to genetic resources has been gradual. See J Cabrera, F Perron-Welch & O Rukundo, *Overview of National and Regional Measures on Access to Genetic Resources and Benefit-Sharing: Challenges and Opportunities in Implementing the Nagoya Protocol, First Edition* (Montreal, Canada: Centre for International Sustainable Development Law, 2011), online: http://cisdl.org/biodiversity-biosafety/public/ABS_Overview_2011.pdf.

14 See Cartagena Protocol, *supra* note 1 at Article 34. See Chapter 11 of this book for more information.

Although the development of regulatory regimes is essentially based on nationally specific practices, legal systems, public management strategies, and local socioeconomic considerations, national regulatory biosafety frameworks generally consider the following elements most of the time: biosafety policy, occasionally as part of a broader biotechnology or biodiversity policy; regulatory policy; a mechanism to handle requests and permits related to use of LMOs, in particular notifications required by the Cartagena Protocol (i.e., an administrative system); a mechanism for monitoring and inspections; and a system that allows public participation and access to information.[15]

One general observation is clear, however. Decision makers are facing an impressive challenge because biotechnology and biosafety fields are evolving at a drastic pace. In this context, overly precise and detailed regulatory frameworks can easily become obsolete in a short period. To avoid constantly having to enact new legislation and policies, developing a regulatory framework that takes the form of a general framework is preferable, while regulating more specific biosafety aspects by way of bylaws and other regulations. This option allows more efficient regulation of future situations but has the effect of conferring the power of regulating specific aspects of biotechnology to the executive branch of government and other administrative institutions, instead of leaving that power with the legislature.[16]

Understanding the regulatory and institutional biosafety system in the region is a challenging task,owing especially to ongoing changes in biosafety regulations and also to the coexistence of other laws that indirectly tackle biosafety issues (e.g. health codes and seeds laws), but it is necessary to promote the safe use of agricultural biotechnology, maximise the benefits resulting from their use, and minimise the risks to the environment and to human health. To understand the complex situation, it is important to note that the lack of biosafety regulatory systems has been identified as the leading cause of the delay to adopt the use of biotechnology in many countries.[17] This lack of biosafety regulatory systems is because significant difficulties, such as the absence of technical and scientific capacity and pressure

[15] Basically, these are the elements recommended by the UNEP-GEF Biosafety Projects. See UNEP-GEF Biosafety Unit, *Guidance towards implementation of National Biosafety Frameworks: Lessons learned from the UNEP demonstration projects* (Geneva, Switzerland: UNEP-GEF, 2008), online: http://www.unep.org/biosafety; M.A. Lean et al, *A conceptual framework for implementing biosafety: linking policy, capacity and regulation* (ISNAR), online: http://www.isnar.cgiar.org; Moisés Burachik and Patricia L. Traynor, "An Analysis of a National Biosafety System: Regulatory Policies and Procedures in Argentina," ISNAR Country Report No 63 (April 2002).

[16] K Garforth, "When Biosafety Becomes Binding" in MC Cordonier Segger & CG Weeramantry, eds, *Sustainable Justice: Reconciling Economic, Social and Environmental Law* (Boston: Martinus Nijhoff Publishers, 2005).

[17] See, J Cohen, "Poorer Nations Turn to Publically Developed GM Crops" (2005) 23(1) Nature Biotechnology (commenting on the difficulties of complying with regulatory processes and presenting eight reasons explaining the slow development of the trade of GE organisms in developing countries – not exclusively in the Americas); see also E Trigo et al, "Agricultural Biotechnology and Rural Development in Latin America and the Caribbean" (Washington, DC: Inter-American Development Bank – Sustainable Development Department, 2000). Other reasons include inadequate intellectual property rights, and inexistent or inefficient seeds handling and identity protection systems.

from groups militating against the use of biotechnology, have hindered the adoption of specific biosafety laws and regulations. As a result of these problems, the cost of complying with biosafety regulations, which are not always easily understood, has become considerable. In one example, the cost of complying with the Brazilian regulatory framework in 2005, for a single processing event including field trials and food security requirements, ranged from $700,000 USD, for virus-resistant papaya, to $4 million USD, for herbicide-resistant soy.

In light of these facts, it can be concluded that the elaboration of biosafety regulatory regimes plays a crucial role in the development and usage of agricultural biotechnology. This conclusion is supported by three different observations: (1) government and society require regulations on the development and use of agricultural biotechnology to ensure the reduction or the elimination of negative impacts on biodiversity and health; (2) research centres and private sector corporations call for regulatory frameworks that establish specific compliance requirements and conditions, which need to be followed to develop and use LMOs; and (3) depending on its legal features, a regulatory framework can result in substantial costs and increased bureaucratic lag regarding the requests for the authorization of LMO-related activities, particularly trade in GE organisms.

BIOSAFETY REGULATORY AND INSTITUTIONAL FRAMEWORKS IN THE AMERICAS

The following section presents an analysis of biosafety frameworks in the Americas.[18] However, it does not provide an exhaustive study of all the frameworks in force.

Regional International Organization for Plant Protection and Animal Health

In 2000, the Regional International Organization for Plant Protection and Animal Health (OIRSA) produced a Regional Guideline on the Safety of Plant Biotechnology.[19] In fact, the guideline is based on the Cartagena Protocol and essentially regulates the same issues as the Protocol. In view of its nature, the guideline also aims to harmonise the various regional laws and practices pertaining to the issue of biosafety.

[18] Several sources of information were used, such as the BINAS website, online: http://binas.unido.org; UNEP-GEF Biosafety Projects' website, online: http://www.unep.org/biosafety/Projects.aspx; REDBIO website, online: http://www.redbio.com; Convention on Biological Diversity website, online: www.cbd.int; ISAAA website, online: http://www.isaaa.org; Alarcón and Artunduaga, *supra* note 5; Trigo et al, *ibid*; Schaper et al, "Organismos Genéticamente Modificados: Su Impacto Socioeconómico en la Agricultura de los Países de la Comunidad Andina" (Chile: MERCOSUR, 2001); national regulatory authorities' websites; specific country studies; and personal communications. Laws and regulations analysed are compiled in Annex VIII to this book.

[19] OIRSA, *Directriz Tecnica Fitosanitaria Regional Sobre Seguridad de la Biotechnologia Agricola Moderna*, online: http://www.oirsa.org/.

Central American Commission for Environment and Development

The Central American Commission for Environment and Development (CCAD), which is the environmental authority of the Central American Integration System (SICA), through its biodiversity program and territorial legal system, prepared a Central American Accord on the Safety of Modern Biotechnology.[20] A series of technical consultations were undertaken and led to the adoption of a draft Accord, which was then approved by the Central American Ministries of Environment in 2002. Although the Central American Accord is based on the Cartagena Protocol on Biosafety, some of its provisions go beyond the scope of the Protocol. In fact, provisions regarding labelling, documentation, liability, contained use, and transit, as well as various other biosafety principles, have been included in the Central American Accord in the attempt to avoid the inconsistencies and ambiguities that resulted from the multilateral negotiations that led to the adoption of the Protocol.

The Inter-American Institute for Cooperation in Agriculture

As requested by the Ministries of Agriculture in 2002, the Inter-American Institute for Cooperation in Agriculture (IICA), in collaboration with the OIRSA and the Tropical Agricultural Research and Higher Education Centre (CATIE), were charged with preparing a regional biosafety regulatory framework, in response to the lack of regulations in certain countries that are receiving, or could receive, requests for field trials or trade in LMOs. A group composed of various agencies and a consultative process, including participation from regional environmental authorities was formed, and they prepared a Model Regulatory Framework for LMOs for Use in Agriculture in Central American Countries. This initiative intends to promote a standard for developing regulations on agricultural biotechnology that can be adapted by each country, according to its particular needs and situation. The hope is that this regulation will be enacted in a similar fashion by the various nations, thereby triggering a process of harmonisation of regulatory frameworks in the region.

OUTLINE OF THE LEADING BIOSAFETY TRENDS IN THE AMERICAS

In terms of regulatory systems, several approaches have been adopted in the Americas since the early 1990s with the aim of ensuring the safe use of modern biotechnology.[21] Some countries decided to apply preexisting plant or animal health regulations to LMO-related issues, or to incorporate biosafety provisions into plant and animal health protection laws (e.g., Costa Rica, Nicaragua, Belize). Others adopted specific biosafety regulations that need to be enacted by parliament (e.g., Mexico, Brazil,

[20] CCAD, *Acuerdo Centroamericano sobre Seguridad de la Biotecnología Moderna*, online: http://www.sica.int/ccad/.

[21] The trends are based on the legal information catalogued in Annex VIII to this book.

Peru, Panama), the executive power (through agreements or resolutions), or public sector institutions (e.g., Bolivia, Colombia, Chile, Honduras, Guatemala, Argentina, Bolivia). Certain countries decided to mention biosafety in their environmental regulations (e.g., El Salvador, Venezuela, Ecuador, and Uruguay), where others apply seeds, pest control, and plant or animal regulations to modern biotechnology, enacting but a few coordination rules and specific provisions associated to biosafety (United States and Canada). Only one country has addressed biosafety from a broader biosafety perspective that incorporates other issues, such as alien species (Cuba).

Although there is no consensus in our society as to the definition of modern biotechnology, there seems to be a general understanding that genetic engineering should be addressed in specific regulatory frameworks. Most regulations in biosafety pertain to the technology involved in the biotechnological process, namely genetic engineering, and not to the final product (except in Canada, where regulations apply to plants with novel traits).

In view of the complexity of biosafety issues, particularly issues of risk assessment and management, and several other issues associated with the introduction of LMOs into the environment, almost all of the countries have established a committee or a national commission to assist decision makers in the development of biosafety regulations. The composition of these committees and commissions varies considerably in the region, mainly with respect to the inclusion of the productive sector, consumers, and nongovernmental organizations (NGOs). The trend points toward the inclusion of all of these organizations, even though such an inclusion can lead to conflicts of interest.[22] Currently, these committees or commissions are for the most part directly connected to the regulatory authorities in place. Nevertheless, independent national commissions (commissions not connected to regulatory authorities) were created in some countries, namely in Panama, Mexico, and Brazil (partially), and were assigned political and coordinative functions, leaving the issue of technical recommendations to be handled by other sector-based authorities established by each country's competent ministry or secretariat.

The original trend of agricultural biosafety regulations focusing on transgenic plants[23] continues to exist in most American countries but is in the process of being replaced by a broader focus that comprises biosafety regulation of animals, microorganisms, fish, and forest species. This progressive expansion of biosafety toward new areas results from the need to regulate research and eventual trade of other types of organisms created through genetic engineering. The Cartagena Protocol, which has a larger scope than simply plants, and the regulatory frameworks

[22] This statement is true not only for the productive sector and universities but also for NGOs advocating against the use of LMOs.

[23] In the early 1990s, most field trials and commercial releases or semicommercial releases of LMOs were of transgenic plants; however, this initial stage has now passed. A growing number of animals, fish, and timber species are subjects of biotechnological research and development.

implementing its provisions which were subsequently developed, brought a gradual increase in the number and variety of organisms and activities being regulated. In addition, health authorities have increased participation in biosafety advisory commissions and committees.

Despite these significant advancements, food safety aspects regarding LMOs have not been tackled consistently in the Americas. Some countries continue to apply the concept of "substantial equivalence," which does not require specific regulation or food labelling, except when genetic engineering affects food composition or current allergenic elements. Furthermore, only a few countries have established specific regulations regarding labelling or traceability, in addition to other more general provisions.

The entry into force of the Cartagena Protocol and pre-existing provisions relating to LMOs in the Convention on Biological Diversity[24] had the effect of introducing ministries of environment to the biosafety debate. For the reasons mentioned previously, the ministries of agriculture have traditionally been the authorities involved in the decision-making process regarding LMOs. Gradually, environmental authorities started demanding and assuming an active role in the regulation of LMOs, because of their mandate to protect the environment.

This trend has been strengthened by the incorporation of biosafety provisions into environmental laws and regulations. For example, biosafety provisions were included in biodiversity regulations in Costa Rica, Venezuela, and Peru. Furthermore, biosafety is mentioned in general environmental laws in countries such as Uruguay, Guyana, El Salvador, Cuba, and Ecuador. In some cases, LMOs are subject even to environmental impact assessments (at least based on a literal interpretation of the pertinent legal provisions).[25] Finally, in countries such as Bolivia and Ecuador, the ministry of environment and related authorities were designated as the competent authorities to regulate biosafety issues. These countries' national biodiversity strategies also allude to the development of national biosafety frameworks by the competent authorities, however, which can generate conflicts between the ministries of environment and agriculture. These two ministries do not necessarily share the same policy or perspective on the role of LMOs and sustainable development, especially when it comes to risks.

The mechanisms for public participation established in the region are limited, at least in the context of specific biosafety regulatory frameworks. Such mechanisms increase considerably, however, when provisions from sector-based environmental regulations, which underline civil participation, are taken into consideration. As

[24] *Convention on Biological Diversity*, 5 June 1992, 1760 UNTS 79, 31 ILM 818 (entered into force 29 December 1993) at arts 8 (g) and 19(3), online: http://www.cbd.int/convention/text/.

[25] In Brazil, environmental impact assessments are determined by way of jurisdictional decision. In other countries, the issue is governed by environmental laws or other rules that establish the procedures for environmental impact assessment. However, it is not clear how this requirement can comply with the biosafety process, which needs to consider more than just environmental impacts.

noted in the Introduction and Chapter 1 of this book, the commitment to facilitate public participation, supported by access to information and access to justice, is an important principle of many international treaties on sustainable development, including the Cartagena Protocol and the Aarhus Convention.

In general, mechanisms for public participation are limited to the opportunity to submit written comments in response to published requests for permits or in the integration of national or sectoral committees or commissions. Although most regulations that are analysed in the current study contain various provisions on public requests to obtain information on potential LMO imports (including documentation, risk assessment and management, biosafety conditions, and penalties), other legal aspects are not clearly addressed in the regional regulations. For instance, the precautionary principle is not mentioned nearly as often as expected, nor are the precise regulations regarding coexistence and liability, which are certainly among the more general legal aspects governing the issue of biosafety. The concepts of labelling, traceability, and identity preservation are mentioned in some regulations, although these concepts are scarcely developed. Provisions regarding LMOs destined for food, feed, and processing (LMOs-FFP) are not entirely comprehensible. In some cases, LMOs do not seem to be subject to any type of administrative control. Nevertheless, this situation can be remedied through the national design of regulatory frameworks that specifically comply with the terms of the Cartagena Protocol and with support from the UNEP-GEF projects.

Most regulations analysed do not comply with the specific requirements of the Cartagena Protocol as far as procedure, in particular Articles 7 to 10.[26] Particular situations involving LMOs, such as procedures for contained use or transit of LMOs or specific categories of LMOs (e.g., LMOs-FFP), are not specifically addressed in national biosafety frameworks. Moreover, concepts and terminology are frequently unclear, and mechanisms to review administrative decisions are generally absent, among other loopholes. Thus, the procedural issue shall surely receive special attention in the future development of national biosafety regulatory systems.

Numerous initiatives for capacity building exist, including regulatory initiatives, which can substantially change the regional biosafety regulatory "map." Proposals of new regulations and laws have been prepared in several countries, and thus particular attention should be paid to the ever-evolving biosafety framework in the region. Of note, English-speaking Caribbean countries do not currently have fully developed regulatory systems, but this situation could change considerably in the near future.

Although there are emerging regulatory provisions regarding biosafety issues in America, the underlying regulatory policy on LMOs remains ambiguous. In other words, comprehensive strategies to develop and use biotechnology seem not to be linked to the development of biosafety regulatory frameworks. There are, however,

26 See also O Rey Santos, "Régimen legal para la seguridad de la biotecnología en América Latina" (México: UNEP, 2005), online: http://www.prodiversitas.bioetica.org/des64.htm.

allusions to the promotion of biotechnology in the biodiversity and biosafety strategies of some countries.

Furthermore, the adoption and ratification of the Cartagena Protocol has been inconsistent throughout the Americas. The Protocol gives rise to relevant concerns in the region, particularly regarding trade between Parties and Non-Parties. These concerns focus on four different aspects: (1) the Non-Party status of important LMO exporting countries such as Canada, the United States, and Argentina; (2) the large amount of international trade (transboundary movement) in genetically engineered products that have been approved for trade in the market (e.g., soy, corn, and canola) within the region; (3) the potential incorporation into the market of new LMOs, which will be approved for trade in the future; (4) the sparse participation of international trade authorities in the LMO approval process (except in countries such as Argentina). Finally, in some countries, such as Bolivia, Colombia, Argentina, Chile, and Brazil, legal disputes concerning the absence of environmental licenses and impact assessments, or the lack of access to information on LMO-related activities, have been brought to the courts.

CONCLUSION

The adoption of biosafety regulatory frameworks in the Americas is a challenging task, because decision makers are faced with various difficulties like ever-evolving technology (which can quickly render specific regulations obsolete) and lack of public support and funding. Legislators and policy makers in the Americas are responding to these various problems with different legislative and policy-based strategies. Certain countries have decided to apply existing legislation pertaining to the environment, plant or animal health, seeds, and pest control to LMO-related issues. Other countries have chosen to enact specific biosafety regulations. In both cases, the resulting biosafety regulatory frameworks are far from being as comprehensive and detailed as required by the Cartagena Protocol. In fact, many of the existing regulatory systems are rather incomplete because they fail to address or develop important issues such as precaution, liability, labelling, traceability, and public participation. These deficiencies can be remedied through careful drafting of regulatory provisions that comply both substantially and procedurally with the Cartagena Protocol on Biosafety, as well as with the support of capacity building projects, such as the UNEP-GEF Projects on Development of NBFs.

PART IV

Case Studies of Domestic Regimes and Sustainable Development

19

National Experiences with Legislative Implementation of the Protocol

Tomme Rosanne Young

This chapter focuses on national legislative implementation of the Cartagena Protocol on Biosafety,[1] providing practical indications of how many developing country Parties are interpreting the Protocol's provisions and requirements. Because of the lack of complete or non-contradicted information, the chapter does not attempt to categorize either the countries or the instruments according to either new or traditional comparative law categories, such as their national legislative system, general legislative approach, or level of involvement in commercial/scientific aspects of LMO development. The third of these factors has been shown to have a significant impact on the nature of each country's implementation of the Cartagena Protocol.[2]

This discussion makes no attempt to be comprehensive, merely focusing on non-scientific elements of national legislation, emphasising particular points on which questions have arisen in national implementation projects and training seminars. The chapter focuses only on particular points that are discussed in national legislation and attempts to avoid speculation about future measures and needs. For these purposes, national implementation is generally described as including (1) a policy or

[1] *Cartagena Protocol on Biosafety to the Convention on Biological Diversity*, 29 January 2000, 2226 UNTS 208, 39 ILM 1027 (entered into force 11 September 2003), online: http://bch.cbd.int/protocol/text/ [*Cartagena Protocol*]. Legislative and policy frameworks can be accessed on the Biosafety Clearing-House (BCH) database, online: http://bch.cbd.int/database/.

[2] See Global Environment Facility (GEF), *Evaluation of GEF's Support to the Cartagena Protocol on Biosafety* (January 2006) at 4.3, online: http://www.thegef.org/gef/node/2140. [*GEF Evaluation*]. For the most part, legislation reviewed was identified through the BCH and the author has attempted to review all instruments posted in the legislative database as well as other instruments provided directly to the author in the course of various unrelated national legislative projects, as well as statements by individual officials from governments and other delegates (the results of interviews, notes of presentations, and other interventions).

general policy approach to LMOs, (2) binding measures (laws, regulations, or other practices), and (3) other domestic legal and administrative systems.

GENERAL POLICY APPROACHES

As noted in several chapters of this book, the negotiators of the Cartagena Protocol could not come to a uniform decision on the level of risk posed by LMOs and the level of national responsibility to restrict and control their use. As noted previously, the Protocol's text contains a few general mandates but does not include an agreed standard regarding the level of risk posed by LMOs. As such, each country must come to its own policy position regarding the value of LMOs, the level of risk posed thereby, and the level of risk that would be acceptable in each case of introduction of an LMO.

At the national level, these determinations are first matters of policy approach and only second matters of legislative framework and regulatory systems. Most of the key factors determining LMO policies are domestic, including public opinion, level of the country's commercial involvement in LMO activities (development, production, and/or marketing), and the social and scientific conclusions regarding LMO safety and/or inability to come to such conclusions and the concomitant need for a precautionary approach. For obvious reasons, many developing countries, even those whose conclusions on the foregoing are clear, must also consider a third issue – the attitude of trading partners that constitute a large part of their market for LMOs. The most significant of these trading partners, for many countries, are the United States and Europe, whose divergent policy and regulatory positions relating to LMOs have been widely documented[3] and are not further explored here, apart from their impact on other countries.

Although popular and policy controversies have been widely discussed in the press and many other forums, they are generally not aired in the Biosafety Clearing-House (BCH). More obvious are the results of national policy choices – national legislative frameworks, some of which include specific bans (full or partial) on LMO introductions, whereas others enable and even promote their development. Among the best descriptions of these policy challenges is found in Argentina's draft national biosafety framework (NBF).

Argentina is a country that has made a policy commitment to LMO development and the production of LMO crops and products since the 1990s.[4] The document

[3] See L Levidow & S Carr, *GM Food on Trial: Opening Up Alternative Futures of Euro-agriculture (Genetics and Society)* (New York: Routledge, 2009); J Candlish, *Genetics, Molecular Biology & the Law* (London, UK: Wildy Simmonds & Hill, 2009); and V Santaniello, RE Evenson, & D Zilberman, *Market Development for Genetically Modified Foods* (New York, New York: Cabi Publishing, 2001). See also the first national reports submitted by the 23 EU countries which are implementing the EU's biosafety regulations, Australia, Japan, and New Zealand, online: http://bch.cbd.int/database/reports/.

[4] For a description of Argentina's commercial and political history relating to LMOs, see Argentina, *Draft National Biosafety Framework*, 2004, BCH Record ID 43463.

posted as its draft NBF constitutes a report of existing legislative provisions and a discussion of other needs and priorities, rather than particular legislative material. As such, it clearly encapsulates the manner in which international policy concerns relate to domestic political decisions in the field of LMOs, noting particularly the comparison between the regulatory approaches:

> Each [country] has adopted a different approach regarding the regulation of biotechnology and/or its derived products. It can be best illustrated based on comparing the United States with Europe, which have completely different views on the subject. The United States has adopted an optimistic position and evaluates new products and technologies based on a "why not?" approach, while the European approach is more pessimistic and it implies predicting the unknown questioning "why?". These philosophical differences have translated into unique regulatory systems that widely differ in terms of the rule origin (product vs. process), the legal framework (existing laws vs. specific laws in the realm of technology) and the decision-making process (institutional vs. political).[5]

This document appropriately describes the basis on which the Protocol can exist amid such divergent national positions:

> Despite the differences in the philosophical and/or political approaches, the scientific and technical approaches are similar in terms of assessing the potential environmental and food safety risks posed by GMOs. Although this consistency may not be reflected in the regulatory decision-making process, the approval rate or the final decision of whether to approve a GMO commercialization, scientists in the different countries are highly consistent in their opinions about the safety of the currently commercialized GMO. This results from the fact that risk assessment is focused on the GMO traits and not the process through which it was obtained. All in all, what is crucial for the impact assessment of potential gene transfers is the gene and the traits conferred, as well as whether it brings about a selective advantage for the recipient organism.[6]

As a consequence, for Argentina it has been possible to place its critical national priorities on domestic political and economic processes relating to LMOs, where:

> [T]wo now very well known restrictions have not been taken into account. On the one hand, the difficulty to make the public accept these products, which has translated into market impacts and constant changes in the approaches to the existing regulatory frameworks. On the other hand, the lack of harmonization in terms of international regulatory systems, creating uncertainty both as to GMO

[5] *Ibid* at 10.

[6] *Ibid.*

> development and market behaviour. Foreseeing such problems, it would have been necessary (and it is still an unresolved matter) to ensure that national and regional institutions were appropriately prepared to start regulating the release and handling of genetically modified organisms, to reach international/regional harmonization in terms of biosafety regulations, and finally, to create social awareness about GMO food and environmental safety characteristics.
>
> We can briefly state that, foreseeing such problems, it would have been necessary (and it is still an unresolved matter) to ensure that national and regional institutions were appropriately prepared to start regulating the release and handling of genetically modified organisms, to reach international/regional harmonisation in terms of biosafety regulations, and finally, to create social awareness about GMO food and environmental safety characteristics.[7]

Beyond these challenges faced by developing countries that already have a high level of LMO involvement, those with less involvement are further challenged by the lack of technically expert personnel in appropriate agencies to implement the Protocol presuming that national policy can make more basic policy choices relating to LMOs.

MANDATORY ELEMENTS: OBLIGATIONS TO OTHER PARTIES

As discussed previously, Protocol provisions imposing direct obligations between countries are strongly mandatory and not generally subject to application of flexibility mechanisms. The Protocol includes only a few commitments of this type, which generally fit into four categories. All Parties specifically commit to the following:

i. *Annex 1 notifications*: To satisfy this commitment, a Party must either ensuring in either case that the exporter is under an obligation to provide accurate information;[8]
   a. make the Annex 1 notifications of the proposed movement of an LMO into another Party when that movement will be or presage the first international introduction of that LMO; or
   b. pass this responsibility firmly to those persons and entities under their jurisdiction who will be taking the action (exporting the LMO);
ii. *Acknowledgement*: To satisfy this commitment, a Party must acknowledge receipt of the notification, providing the data specified in Article 9.2;[9]
iii. *Unintentional transboundary movement*: To satisfy this commitment, a Party must "notify affected or potentially affected States, the Biosafety

[7] *Ibid* at 11.
[8] *Cartagena Protocol*, *supra* note 1 at art 8.
[9] *Ibid* at art 9.

Clearing-House and, where appropriate, relevant international organizations" including the information specified in Article 17.3 in the event of an imminent or actual unintentional transboundary movement of a living modified organism, and coordinate with them, in the event that emergency measures are necessary/possible;[10] and

iv. *Illegal transboundary movement*: To satisfy this commitment, a Party must impose controls to prevent or react to illegal transboundary movement of LMOs, and notify the BCH when they occur.[11]

These obligations have been and continue to be the Cartagena Protocol requirements addressed in the fewest number of national instruments.[12] The following discussions are based on national legislation in the BCH.

EXPORT OF LMOs (NOTIFICATION AND ACKNOWLEDGEMENT)

The Cartagena Protocol provisions requiring notification by the export country or exporter and acknowledgement by the import country are expressed in unequivocal, mandatory terms (e.g., "Each Party shall . . . "). There appears to be little or no justification for application of the flexibility mechanisms to reduce or eliminate either responsibility. Indeed, it would be difficult to implement any part of the Protocol as a binding instrument, if countries had the right simply to ignore their international notification and acknowledgement responsibilities.

The Protocol obliges Parties to impose specific obligations on the LMO-related actions of their own citizens (and/or entities under its jurisdiction), particularly with regard to notifying other Protocol Parties, before exporting any LMOs to recipients in those countries. The Protocol provisions described earlier enabling stricter and more lenient measures would be difficult (perhaps impossible) to apply in cross-border contexts. These cross-border notification requirements rely on the fact that Cartagena Protocol Parties share an understanding of the specific activities and LMOs to which the notification requirement applies. If a country has expanded the coverage of its law, that country cannot expect other countries to adopt similar expansions of their law on notification. Similarly, a country that has decided not to regulate most LMOs cannot expect other countries to acquiesce in this narrower definition.[13] Thus, unless the two or more countries agree on alternative mechanisms

10 *Ibid* at art 17.

11 *Ibid* at art 25.

12 *GEF Evaluation*, *supra* note 2.

13 A discussion of the various challenges faced by states in implementing international provisions calling for legislation with cross-border impact can be found in T Young, *Administrative and Judicial Remedies Available in Countries with Users under their Jurisdiction and in International Agreements*, UN Doc UNEP/CBD/WG-ABS/5/INF/3, reprinted in T Young, Covering ABS: *Addressing the Need for Sectoral, Geographical, Legal and International Integration in the ABS Regime: Papers and Studies of The ABS Project* (Bonn, Germany: IUCN-ELC, 2008).

in a "Bilateral and Multilateral Agreement or Arrangement" entered into pursuant to Article 14, all parties are obliged to address at least the minimum notification and acknowledgement requirements set forth in Articles 8 and 9.

Certainly, either Party may provide the other with more information than the Protocol requires, but it is doubtful that any Party could increase the number or type of notifications that other countries must provide to that Party, or refuse to issue acknowledgements, except on a bilateral/multilateral basis under an Article 14 Agreement.

As a result, it appears that every export country has an absolute obligation to make (or require those under its jurisdiction to make) appropriate notification prior to every "first international introduction" of a non-FFP LMO that is proposed for a particular import country. When the import country imposes AIA and other requirements on all introductions (not only the first international introduction of non-FFPs), it must enforce those provisions domestically and cannot expect or rely on any notification from the exporter or export country or under its laws.

In practice, the Parties' Article 8 and 9 obligation demonstrates the vital role of the BCH in operation of the Protocol. Export countries are obliged only to provide or require notification, when introduction of the material being exported has never been permitted (first international introduction) in the proposed country of import. If a country has not posted any prior decision permitting a particular LMO, then arguably any effort to export that LMO to that country must necessarily be a subject of notification.

Relatively few countries have formally included provisions relating to export of LMOs from their country to some other country of import.[14] One of the few direct legislative provisions implementing this requirement is found in the law of the Republic of Korea:

> One who wishes to export living modified organisms shall notify in advance the Minister of Commerce, Industry, and Energy of the information as prescribed by the Presidential Decree including items, quantity, and the name of the importing country.[15]

In some countries legislation touches the notification requirement only as a demand on exporters from other countries who might bring LMOs into

[14] The legislative database of the BCH does not include search term functions for Articles 8 and 9. Attempts to full-text search using the protocol terms *notification* and *notifier* demonstrated that such searches are limited to the metadata cards, which do not consistently use Protocol terms as such.

[15] Republic of Korea, *Trans-boundary Movements of Living Modified Organisms Act, 2001*, BCH Record ID 9320 (Korean) at art 20 [*Korea LMOs Act*], summarised in and translated to English in Republic of Korea, *Draft National Biosafety Framework (Developed under the UNEP-GEF Biosafety Project)*, BCH Record ID 43444 at 92 and 195 [*Korea Draft NBF*].

the regulating country.[16] An example of this approach to the notification requirement is found in Bangladesh's Biosafety Guidelines, which require that the notifier include "a declaration that the above-mentioned information is factually correct."[17]

Generally, under national law, such a declaration must be factually correct, and if not, the notifier could be punishable by law. For example, legislation in Brazil provides simply that

> For the purposes of Article 8 of the Protocol, the exporter under the Brazilian jurisdiction shall be responsible for notifying the Party of import before the first intentional transboundary movement of a living modified organism that falls within the scope of Article 7.1 of the Protocol. . . . The exporter for the purposes of Article 4 above shall communicate to the appropriate National Competent Authority, in accordance with the scope of its activities, the fulfillment of the notification to the Party of import. . . . All documents submitted with the notification to the Party of import shall be presented with the communication in accordance with Article 4.1 above.[18]

In Bolivia, by contrast, a legislative exemption for a particular variety of LMO soy appears to allow cross-border commerce in that LMO across the without notification to the country of import:

> personas jurídicas o naturales que sean propietarias, arrendatarias, comodatarios de semillas, cultivos o granos de soya genéticamente modificada resistente a glifosato (evento 40–3–2) que se hayan registrado en el marco de lo dispuesto en los artículos 3 y 5, podrán comercializar el producto cosechado dentro y fuera del territorio nacional.[19]

Finally, a very complete notification provision, although not binding law as yet, is found in the Second Draft Law on Biosafety, contained in Cambodia's NBF:

> Any legal or natural person who intends to export LMOs covered by this law from the Kingdom of Cambodia shall notify the national competent authority of the

[16] See Argentina, *Liberación al Medio de Organismos Vegetales Genéticamente Modificados (OVGM)/ Resolución n° 39*, 2003, BCH Record ID 5649 [*Argentina Resolución 39*] which includes a declaration in the application form that binds all applicants, including domestic importers and foreign exporters, under Argentinean law. It is not clear whether such a provision could be enforced in the exporter's country of origin, but it might be enforceable under general interpretative principles that maximise the enforceability of formally executed contractual provisions (Discussed in T Young, *supra* note 13).

[17] Bangladesh, *Biosafety Guidelines of Bangladesh*, 2005, BCH Record ID 43880 at annex 6 [*Biosafety Guidelines of Bangladesh*].

[18] Brazil, *Decree No.6925 of August 6. 2009*, 2009, BCH Record ID 49478 at art 4 [*Brazil Decree 6925*].

[19] Bolivia, *Resolución Multiministerial No 1*, 2005, BCH Record ID 12011 at art. 6 [*Bolivia Resolución Multiministerial*].

> proposed importing State in writing prior to applying to the concerned competent authority of the Kingdom of Cambodia for an export permit.
>
> The concerned competent authority of the Kingdom of Cambodia may issue a permit to export LMOs only after receipt of authorization in writing from the national competent authority of the proposed importing State. The authorization from the proposed importing party must be attached to the export permit and/or other documentation accompanying the consignment.
>
> The exporter shall get from the Ministry of Environment written certification of the accuracy of all information provided about the LMO to be exported. Such certification shall be attached to the export permit and/or other documentation accompanying the consignment.[20]

In another clause, Cambodia's draft also provides an example of the difficulty involved in national implementation of the notification requirement. It specifically requires "[a]ny legal or natural person who wants to import a LMO for direct use as food or feed or for processing" to provide "written certification issued by the Competent National Authority of the State of export that attests to the accuracy of the information provided concerning the LMO to be imported."[21] For many importers, this requirement could ultimately create a major obstacle to LMO importation, because few countries' national officials would be authorized or willing to provide certification of this type. Indeed, in many developed countries, an official might be exceeding his or her authority if he or she attempted to provide one. At the same time, this clause might be intended to address fallout from situations reported in recent years, in which commercial entities were permitted to export products to developing countries, where those products had been banned from sale in the export countries.[22] Arguably, the value of this example is lessened by the general expectation that it will be a long time before persons and entities in Cambodia will be regularly in the business of exporting LMOs. Cambodia also follows the more conventional approach of inserting a (more easily enforceable) provision, which requires persons in Cambodia to provide accurate information.[23]

[20] Cambodia, *Second Draft Law on Biosafety* at art 21 [*Cambodia Draft Law*] in Cambodia, *National Biosafety Framework*, 2004, BCH Record ID 43438 at 36 [*Cambodia NBF*]. See also Ethiopia, *Draft Biosafety Proclamation* at s 13 [*Ethiopia Draft Proclamation*], reprinted in Ethiopia, *National Biosafety Framework (Developed under the UNEP-GEF Biosafety Project)*, 2007, BCH Record ID 43387 at 57–8.

[21] *Cambodia Draft Law*, *ibid* at art 15 in *Cambodia NBF*, *ibid* at 35.

[22] M.Tran, "Branded", *The Guardian*, UK, 1 September 2005, online: http://blogs.guardian.co.uk/businessinsight/archives/2005/09/01/branded.html.

[23] See *Cambodia Draft Law*, *supra* note 20 at art 15, in *Cambodia NBF*, *supra* note 20 at 35, which requires importers of LMOs-FFP to provide written certification issued by the exporting CNA that attests to the accuracy of the information provided.

Intentional and Unintentional Transboundary Releases and Movement of LMOs

Although Article 17.3 is expressed in hortatory terms, the overall responsibility is indisputably mandatory.[24] To date, the BCH includes few clauses addressing these obligations. One reasonable suggestion is that the reticence to legislate on these transboundary notice provisions arises from the current state of uncertainty regarding liability and redress. Given that the contents of the notice are not specified in a mandatory way, most countries will want to craft such notice provisions carefully to avoid providing inappropriate bases for liability claims.

Beyond this, however, many countries do not have any legal need to adopt laws that specify the manner in which they fulfill their diplomatic duties of notification to other countries. These are matters of diplomacy, decided on the basis of the country's commitments but not normally legislated. In this respect, the differences among countries are significant. In some countries, it is de rigueur for government officials at or below the ministerial level to coordinate with their opposite numbers in other countries, but this is by no means true in all countries. In some, intergovernmental contacts must go through channels to the highest diplomatic levels before being transmitted to diplomatic officials of the country receiving notice, where the document filters down to the appropriate officials in an unspecified time.[25] In the former countries, no legislation may be needed to enable the CNA or NFP to give notice directly, but in the latter, legislation may be required to empower an agency or ministry official to contact his opposite number directly.

Legislatively, however, most countries are more concerned about ensuring that the relevant officials are *aware* of unintentional releases of LMOs, in order to know which factors exist that could trigger their intergovernmental reporting obligation. To this end, as discussed later, many national laws focus on user/permit-holder reporting as a tool that provides the government with a source of underlying information to enable/mandate their notifications under Articles 17 and 25. To this end, several countries' legislation requires various members of the LMO sector to report unexpected or unpermitted releases by the applicant or other person responsible, to the relevant government agency.[26] In the Philippines, for example, provisions of this type are distributed throughout biosafety legislation, separately discussing each type of LMO permit:

[24] Article 17(3) of the Cartagena Protocol is an horatory discussion of the contents of the notification communication, whereas the duty to notify in Article 17(1) is mandatory.

[25] See, e.g., C Strulick & N Dragffy, *Enforcement of Wildlife Trade Controls in the EU* (Bonn, Germany: IUCN-ELC, 2001).

[26] *See* Indonesia, *Regulation of the Government of the Republic Indonesia Number 21 Year 2005 on Biosafety of Genetically Engineered Product*, BCH Record ID 11576 at art 27 [*Indonesia Regulation on Biosafety*].

The permittee shall notify the Director of BPI, within the time periods and in the manner specified below, in case of any of the following occurrences:

a. verbally immediately upon discovery, or in writing within twenty four (24) hours, in the event of any accidental or unauthorized release of the regulated article or new information becomes available indicating that the regulated article could pose significant risks to human health and the environment; and
b. in writing as soon as possible, but not to exceed three (3) working days, if the regulated article or associated host organism is found to have characteristics substantially different from those listed in the application for a permit, or suffers from any unusual occurrence (e.g., excessive mortality or morbidity, unanticipated effect on nontarget organisms)[27]

Although clearly beneficial for biosafety officials to maximise their access to critical information about risks and releases (as discussed later), these provisions generally do not transfer a country's ultimate responsibility to notify other countries, for the same diplomatic reasons that control intergovernmental communications more generally.

LEGISLATION OF DOMESTIC SCOPE

On review of the legislation in the BCH, there are few examples of direct application of these provisions among the legislative documents in the BCH. In the BCH, many countries have chosen to post their existing national (and sometimes subnational) legislative instruments on environmental impact assessment, phytosanitary control measures, veterinary laws, seed acts, customs laws marketing laws, and national criminal and administrative laws, in toto, even when they do not specially reference or treat LMOs.[28]

After these are eliminated, a relatively small number of documents in force remain for consideration. This chapter, therefore, does not focus on whether a law is in force, choosing instead to consider all laws posted by developing countries that are actively participating in the Cartagena Protocol processes, supplemented by several other countries. It has considered the various draft documents in the BCH, including the draft NBF documents, as if they are adopted legislation, with no attempt to

[27] Philippines, *Administrative Order No. 8, Series of 2002, Subject: Rules and Regulations for the Importation and Release into the Environment of Plants and Plant Products Derived from the Use of Modern Biotechnology*, BCH Record ID 46538, at ss 6(d)(8), 8(j)(8), 10(g)(2) and 12(f)(7) [*Philippines AO 8*].

[28] Many of the draft NBFs posted support this view, including a list of a broad range of legislative instruments that may apply to LMOs as well as other non-LMO activities. See Bahamas, *Draft National Biosafety Framework (Developed under the UNEP-GEF Biosafety Project)*, BCH Record ID 43464 at 37–9 (annex I) [*Bahamas Draft NBF*].

determine their provenance or the extent to which they have received governmental approval.

The following sections identify key areas in which the Protocol specifically addresses administrative needs, within six overarching categories: the risk analysis/management processes (AIA and Article 11); scoping choices, including stricter measures; other measures relating to matters referenced in the Cartagena Protocol; non-Protocol mechanisms; linkage to other national legal frameworks; and coordination with other international issues. These discussions are based on examples selected from among the (draft and in-force) documents in the BCH.

Risk Analysis and Risk Management

The Protocol's processes for risk analysis and risk management are generally focused on two elements – the legal and the scientific/technical. Some of the most interesting legislation posted relative to the AIA and Article 11 processes focuses on technical elements (standards for risk assessment, and minimum standards for permits and risk control). Apart from these, most of the legislation specifically identified with the AIA or Article 11 processes focusses on provisions required for administrative operation of the Protocol as an international instrument. In some countries, government agencies can be authorised to undertake these tasks without adoption of any legislative instrument. In others, the government must authorise a particular implementing agency or focal point for the actions taken on behalf of the country to be considered binding or legally effective. In some countries, such as Brazil,[29] a large volume of instruments appears to be necessary to create and empower relevant bodies and empower it to adopt the regulations that will actually include the provisions described in Article 10.

The largest volume of instruments in force in the BCH that are tagged as "Intentional Introduction into the Environment (AIA)" are national sanitary and phytosanitary measures, plant protection legislation, veterinary legislation, and seeds legislation, which apply to LMO imports and exports in the same way that they apply to other biological materials or stocks moving between countries or proposed for domestic marketing. In most instruments, the specific process described in Articles 10 to 13, 15, and 16 (i.e., notification, acknowledgement, decision making, and

[29] See Brazil, *Normative Resolution No. 04 – National Technical Biosafety Commission*, BCH Record ID 45561 [*Normative Resolution 4*]; *Normative Resolution no. 05 – Commercial release of GMOs and their derivatives (including risk assessment)*, BCH Record ID 48410 [*Normative Resolution 5*]; *Normative Resolution No. 06 – planned release to the environment of Genetically Modified Organisms (GMO) of plant origin and their derivatives*, BCH Record ID 48411 [*Normative Resolution 6*]; *Communication No. 01 – National Technical Biosafety Commission*, BCH Record ID 45562 [*Communication 1*]; *Communication No. 02 – National Technical Biosafety Commission*, BCH Record ID 45563 [*Communication 2*]; *Brazil Decree 6925*, *supra* note 18.

resubmission) are not stated in legislation, which often relies on pre-Cartagena Protocol laws or practices.[30]

Ultimately, the particular application and decision-making processes set out in Articles 10 to 13 are legislatively simple. A few national laws posted in the BCH would make it even simpler.[31] Countries that have not adopted such processes (including most of those which have posted legislation in force to date) face challenges of capacity and of political controversy, rather than difficulty in understanding and adopting the appropriate regulatory provisions.[32]

Interpreting the Biosafety Framework: "Stricter Measures"

One key question in national and regional biosafety legislative workshops relates to the extent to which regulatory procedures implementing the Cartagena Protocol must track the provisions of the Protocol exactly. It might be difficult to say with certainty that a particular procedural change in a biosafety permit/license statute is stricter or more lenient than the original Cartagena Protocol requirement.

Scope and Coverage

The scope provisions in the Cartagena Protocol do not generally restrict Parties from legislating over other biosafety issues or from adopting other legislation relating to GMOs, including on issues only partly addressed by the Protocol. Review of the BCH indicates that virtually every country has taken advantage of this opportunity, including some element(s) beyond the scope of the Protocol. One of the most common extensions, found in many of the legislative instruments reviewed apply AIA procedures and/or Article 11 procedures to all LMOs imported – not limited to the first international introduction.[33] This extension is generally permitted under Cartagena Protocol Article 12.4. The legislative database includes a variety of different approaches on this point. For example, Argentina's draft NBF applies its AIA/Article 11 procedures to initial or subsequent, international or domestic introductions.[34] Draft legislation in Cambodia would apply the AIA process to (1) the first international importation of a given LMO for contained use or as LMO-FFP; and (2) "[a]ll

[30] See *Biosafety Guidelines of Bangladesh, supra* note 17. See also Austria, *GMO Act (Gentechnikgesetz)*, BCH Record ID 26; Austria, *Ordinance on Public Hearings for GMO Notifications (Anhörungsverordnung)*, BCH Record ID 47146.

[31] See Republic of South Africa, *Genetically Modified Organisms Act, 1997 (Act No 15 of 1997) Regulations*, BCH Record ID 269 at art 3(1) [*SA GMO Regulations*], which requires a risk assessment for all activities involving genetic modification.

[32] Interviews conducted during global expert workshops and other meetings.

[33] See *Cambodia Draft Law, supra* note 20 at art 12; *Ethiopia Draft Proclamation, supra* note 20 at s 3(1).

[34] Argentina, *Revision of Argentina's National Biosafety Framework*, BCH Record ID 43463, annex at 85: "Organismos Genéticamente Modificados (OGM) cuya liberación al ambiente sea consecuencia del uso exclusivo de los mismos en medicina humana."

imports of LMOs into the Kingdom of Cambodia for intentional introduction into the environment."[35]

One reason given for adopting this type of scope-broadening clauses is the desire to avoid trade discrimination charges. As noted in the Republic of Korea,

> According to Article 12 of the Act, living modified organisms produced within Korea shall obtain risk assessment and the evaluation of the risk assessment. By keeping equity between living modified organisms produced in abroad and those in Korea, we may prevent international conflicts and attain the purpose of the Act.[36]

This concern for avoiding claims of discrimination against foreign producers is addressed in many laws. On occasion, this is done indirectly, by requiring that the relevant provisions apply to "all introductions." In other laws, subsequent introductions are discussed differently and can be excused from AIA requirements under other theories.[37] Presumably, even in countries that include subsequent introductions in their AIA and Article 11 legislation, subsequent risk evaluation processes often follow the holdings of former decisions, unless new facts have been discovered. The primary administrative benefits of extension of AIA and Article 11 processes to subsequent introductions would appear to be (1) added assurance that new risk information is promptly included in national decision making and (2) the government's formal awareness of all introductions, for purposes of oversight. As another extension beyond the most basic scope of the Protocol, most countries' specifically apply their AIA procedures, rather than the Article 11 processes, to importation of LMOs for food, feed and processing, as discussed later in this chapter.

In addition, many (most) countries have formally included all LMOs imported for contained use[38] and transit[39] within their national biosafety frameworks (NBFs). In Namibia, this coverage is explicit and detailed:

> This Act is applicable to the following activities involving a GMO or GMO product, namely – (a) the import of a GMO or GMO product; (b) the export of a GMO or

35 *Cambodia Draft Law*, *supra* note 20 at arts 11, 12, and 15.

36 *Korea LMOs Act*, *supra* note 15 at art 12; *Korea Draft NBF*, *supra* note 15 at 89–90 and 189–190.

37 See Argentina, *Tratamiento diferenciado para el otorgamiento de la autorización de comercialización para aquellas acumulaciones provenientes del cruzamiento de parentales que contienen eventos que en forma individual ya tienen autorización para la comercialización en la República Argentina, Resolución 60/2007*, 2007, BCH Record ID 40386.

38 See *Ethiopia Draft Proclamation*, *supra* note 20 at s 11. The Philippines legislation on plants derived from biotechnology appears initially to exclude contained use, but later includes a detailed permit system for such uses. See *Philippines AO 8*, *supra* note 27 at art 2(b) ("This order shall not apply to contained use of a regulated article, which is within the regulatory supervision of NCBP [National Biosafety Committee of the Philippines]") and arts 5–6 (regulatory permit system for contained use of LMOs).

39 See *Cambodia Draft Law*, *supra* note 20 at arts 4 and 12. Another example is found in Cameroon, *Law laying down safety regulations governing modern biotechnology in Cameroon*, 2003, BCH Record ID 10787, at s 20(1) [*Cameroon Safety Regulations*]. See also, *Ethiopia Draft Proclamation*, *supra* note 20 at s 12.

> GMO product; (c) the release into the environment of a GMO or GMO product; (d) the contained use of a GMO or GMO product; (e) the placing on the market of a GMO or GMO product; (f) the transport of a GMO or GMO product, including transport in transit through Namibia; and (g) the use or handling in any other way of a GMO or GMO product.[40]

Commonly, nearly all countries that issue permits regulate them according to a variety of factors, specifically addressing at least three levels of use: laboratory use, field trials, and commercial release (under various names.) Several legislative documents and drafts provide detailed controls for each level of use.[41] Beyond this, several countries add another component – identification of particular LMOs according to risk categories.

One example of such categories involves four levels, primarily applicable to contained uses but also used as a guide for commercialization activities:

> Safety level 1 – Biotechnological projects that are known to present no risks to the community and the environment
>
> Safety level 2 – Biotechnological projects that are known to present minor risks to the community and/or the environment
>
> Safety level 3 – Biotechnological projects that are known to present slight risks to the community and/or the environment
>
> Safety level 4 (1) – Biotechnological projects that are known to present risks or high risk probability for the community and/or the environment[42]

For each risk level, national law frequently sets out specified minimum mitigation measures or "safety levels."[43] The result of the identification of these categories is a matrix of requirements, addressing each risk category for each level of use.

Some countries have chosen to conform to particular aspects of their legislation to the Protocol's coverage limitations,[44] albeit expressing these scoping provisions in different ways. For example, Brazil's normative resolution on commercial release notes that "a derivative the GMO of which has been passed for release shall not

[40] N, *Biosafety Act (Act No. 7 of 2006)*, BCH Record ID 7257 at s 3 [*Namibia Biosafety Act*].

[41] See *Biosafety Guidelines of Bangladesh*, *supra* note 17; *Ethiopia Draft Proclamation*, *supra* note 20 at s 11. *Philippines* AO 8, *supra* note 27, uses four categories: contained use, field testing, propagation, and direct use.

[42] *Cameroon Safety Regulations*, *supra* note 39 at ch III. See also *Biosafety Guidelines of Bangladesh*, *supra* note 17 at 4(1)(3).

[43] *Cameroon Safety Regulations* at ss 6(2)-(3), 8, 15, and 19, as implemented by Cameroon, *Décret No 0737 PM du 31 Mai 2007 d'Application de la loi 2003/006 du 21 Avril 2003 portant régime de sécurité en matière de Biotechnologie Moderne au Cameroun*, 2007, BCH Record ID 40349 at ss 48–49, see also s 6.

[44] *Cambodia NBF*, *supra* note 20 at 10 and 31. See also *Indonesia Regulation on Biosafety*, *supra* note 26 at art 6.

be submitted to CTNBio analysis and issuance of technical opinion."[45] This has been supplemented by a provision that "[a] GMO containing the same genetic construct used in a GMO of a same species, for which a technical opinion favorable to commercial release in Brazil has already been issued, shall be submitted to a simplified analysis for its release, at the discretion of CTNBio."[46]

In some countries, the extensiveness of the scope of national biosafety legislation been a source of controversy quite apart from the contents of the legislation developed. In Chile, for example, the inclusion of other LMO-related matters in the NBF proved to be a basis of controversy, leading to a decision to implement only within the scope of the Cartagena Protocol.[47]

Another coverage concern raised in several documents in the BCH relates to the general focus on the agricultural sector. Although agriculture has been a major factor in LMO development and introduction, concerns have been raised that other sectors such as forest and fishery sectors and particular classes of LMOs, such as microorganisms, have been underconsidered – a serious omission if true, in light of the Cartagena Protocol's origins and focus on biodiversity issues.[48]

Deadlines Different from Those in the Cartagena Protocol

Questions have been raised about laws that call for a "final decision" (i.e., approval or prohibition of the LMO or a statement that more information is needed). The Cartagena Protocol sets an outside limit at 270 days for such decisions, but several countries have specified a much shorter (90-day) deadline,[49] raising the question – is a change to the deadline a "stricter measure"? Although many LMO developers have objected to lengthy deadlines set by the Protocol, it is not clear that a country that shortens those deadlines is doing the applicant any favours. In light of other provisions in the Protocol, a shorter deadline might mean that more applications are denied (or deemed denied from lack of action) under a 90-day deadline rather than 270 days. The denial of a permit normally means that, at best, the applicant must begin the process again, submitting the entire package plus application fee anew. By contrast, an applicant who is informed during an ongoing review process that the government cannot come to a decision for some stated reason might have

45 *Normative Resolution* 5, *supra* note 29, at art 2. Compare with Brazil, *National Biosafety Law (Law 11.105/2005)*, 2005, BCH Record ID 44500 at art 14(5) [*Brazil National Biosafety Law*]: "GMOs whose derivatives have already been approved by the CTNBio will not be submitted to analysis and technical reports by CTNBio."

46 *Ibid* at art 2.

47 Chile, *Bases para el marco nacional de bioseguridad de Chile (NBF posted under the UNEP-GEF Project)*, BCH Record ID 43466 at 26.

48 *Ibid* at 36.

49 See *Cameroon Safety Regulations*, *supra* note 39 at s 31, shortening the final decision time limit to ninety days; *Cambodia Draft Law*, *supra* note 20 at arts 13–14; *Argentina Resolución* 39, *supra* note 16 at 18.

an opportunity to supplement the application in a way that helps the government come to a final decision.

Shorter deadlines could also impact public participation. In many countries, public participation is a major component of national biosafety law and/or policy. For example, where the application period is very short, a governmental agency charged with making the decision is more likely to use the time restriction as a justification for minimising public participatory processes.[50] The Philippines has acknowledged this approach, while using a different method, generally accepting the Protocol deadlines as outside limits but calling on its government to act more promptly where possible:

> The . . . decision-making process shall be conducted in an efficient, coordinated effective, predictable, cost-effective and timely manner. Undue delay shall be avoided without compromising transparency, public participation, public safety, and public welfare.[51]

National legislative draftsmen have generally assumed that shortening Cartagena Protocol-related deadlines constitutes "stricter measures" pursuant to the Protocol. Many experts have noted that Parties could (and probably should) integrate Protocol-related decision-making into their preexisting decision-making processes under other national laws (e.g., phytosanitary and customs). Most of these processes have deadlines that are much shorter than those in Cartagena Protocol Article 10.

LMOs-FFP

The application of AIA processes to LMO-FFPs has been considered by some to be a "stricter measure" in implementing the Protocol.[52] For example, Article 7.2 specifically states that: "Intentional introduction into the environment" in paragraph 1 above, does not refer to living modified organisms intended for direct use as food or feed, or for processing. Article 7 goes on to state that the first transboundary movement of LMOs-FFP shall be governed by Article 11's streamlined processes. A review of the BCH legislative database suggests, however, that most specific biosafety legislation (in force or in draft) applies a single mechanism to all LMOs, essentially requiring LMOs-FFP to go through the full AIA process. This presumes that all of the drafters of such legislation feel that the abbreviated LMOs-FFP permit process (Article 11) is more lenient than the AIA process, and that applying the AIA permit

50 At a 2004 UNEP-GEF Workshop on European experiences with LMOs for developing country delegates in Bonn, Germany, attended by the author, European LMO officials discussed public participation and noted that they limited it to only a few LMO-legislative discussions, because the level of controversy which would arise if the public participated in all permitting decisions would prevent the Agency from acting expeditiously.

51 Philippines, *Executive Order 514: Establishing the National Biosafety Framework*, 2006, BCH Record ID 46541 at paras 2 and 12 [*Philippines EO 514*].

52 *GEF Evaluation*, supra note 2 at 68–9.

process to LMOs-FFP is a stricter measure for purposes of Cartagena Protocol Article 2.4.

Several of the countries with longer histories of LMO involvement have their own provisions about LMO foods and their production, similar to Article 11 – creating separate practices for governmental assessment of foods than would be applied to materials brought into the country for propagation and production.[53] Among these, the Biosafety Guidelines of Bangladesh incorporate the LMOs-FFP provisions by reference.[54] As such, it is one of the few legislative documents, including drafts, to specifically adopt the LMOs-FFP concept in a way that appears actually stricter than the AIA process. In Annex 3 to that document, FFP applications must complete the "Information requirement for import and export of GMOs," including the full text of Cartagena Protocol Annex 1 (information required for the AIA process), and supplement it with the full content of Cartagena Protocol Annex 2 (information required for the LMO-FFP process), as "additional information . . . required for GMOs intended for direct use as food or feed, or for processing." That instrument notes that "While taking a decision on the import of GMOs/LMOs intended for direct use as food or feed, or for processing, the provisions laid down in Article 11 of the Cartagena Protocol shall be followed."[55]

Another approach to this issue, although it does not follow Article 11, is found in Brazil: "Foodstuffs and food ingredients intended for human or animal consumption that contain or are produced from GMOs or their derivatives should contain information to this effect on their labels, according to the implementing legislation."[56] Cameroonian legislation includes many specific requirements beyond the scope of the Cartagena Protocol, with regard to these requirements, however, the law makes a potential (but not yet complete) exception for LMOs-FFP, when it states that "Notwithstanding the above provisions, products based on genetically modified organisms intended for human or animal consumption shall be subject to specific norms determined by special instruments." Although not a specific legislative commitment to apply only Article 11 to LMOs-FFP, this clause enables Cameroon to make that choice, if it so desires.

One of the simplest and most unequivocal examples of special provision for LMOs-FFP (and pharmaceuticals) is found in Ethiopia's draft Biosafety Proclamation, which states, "This Proclamation shall not apply to any transaction of a pharmaceutical for humans, or a commodity for food, feed or processing listed

53 See Indonesia, *Decree of the Minister of Agriculture Number 67/Permentan/OT.140/12/2006 concerning Decree on Conservation and Utilization of Plant Genetic Resources*, BCH Record ID 45153 at art 14 [*Decree on Conservation and Utilization of Plant Genetic Resources*].

54 *Biosafety Guidelines of Bangladesh*, *supra* note 17 at 4.1 through 4.1.2. Annex 7 of the Guidelines incorporate Annex II to the Cartagena Protocol.

55 *Ibid.*

56 *Brazil National Biosafety Law*, *supra* note 45 at art 40.

under any regulations."[57] Cambodia's draft biosafety legislation takes a much more ambiguous approach to LMOs-FFP processes, requiring that:

> The first import into the Kingdom of Cambodia of a LMO for direct use as food or feed or for processing shall be subject to risk assessment and approval by the Ministry of Environment, in consultation with the concerned competent authority, prior to application for the import permit. Any legal or natural person who wants to import a LMO for direct use as food or feed or for processing shall submit to the Ministry of Environment written certification issued by the Competent National Authority of the State of export that attests to the accuracy of the information provided concerning the LMO to be imported.[58]

This clause does not allow the regulated community to make any individual decisions regarding whether appropriate regulation already exists and specifically includes the Cartagena Protocol's Article 11 deadline for BCH notification, but provides no other indication of how Article 11 applies.[59]

The fact that very few countries have formally included an LMOs-FFP process is reflected in the BCH's decisional database, in which only six documents are formally identified as "Decisions on the importation of LMOs for direct use of food, feed and for processing."[60]

Collective and Individual Exceptions to AIA/Article 11 Requirements

The Protocol includes a variety of options by which LMOs may be granted exceptions from administrative requirements. Some of these are collectively decided (by the COP-MOP), others are generically decided at the national level, and in other cases the country may make individual exceptions.[61]

A potentially complex, but still only theoretical inconsistency arises from Article 7.4, in which the Protocol specifically provides that:

> The advance informed agreement procedure shall not apply to the intentional transboundary movement of living modified organisms identified in a decision of the Conference of the Parties serving as the meeting of the Parties to this Protocol as being not likely to have adverse effects on the conservation and sustainable use of biological diversity, taking also into account risks to human health.

[57] *Ethiopia Draft Proclamation*, *supra* note 20 at s 3(2).

[58] *Cambodia NBF*, *supra* note 20 at 36.

[59] *Cambodia Draft Law*, *supra* note 20 at art 16.

[60] Canada, *NMK-89576-1-Superior NewLeaf™ potato*, BCH Record ID 5762 and Mexico, *DAS-24236-5 - Insect-resistant cotton*, BCH Record ID 7047; *MON-Ø1445-2 - Roundup Ready™ cotton*, BCH Record ID 7981; *MON-ØØ531-6 x MON-Ø1445-2 - Roundup Ready™ Bollgard™ Cotton*, BCH Record ID 8010; *DAS-24236-5 x DAS-21Ø23-5 - WideStrike™ cotton*, BCH Record ID 8040; *DAS-21Ø23-5 - Insect-resistant cotton*, BCH Record ID 8076. In compiling this list, the author looked only at the metadata (whether the document was labeled as "LMO-FFP decision" in the BCH list) and did not read all of the documents in the BCH's "Decisions and Communications" database.

[61] Cartagena Protocol, *supra* note 1 at art 13.

A separate clause in Article 7 expressly states itself to be "subject to Articles 5 and 6" (i.e., to the provisions excluding pharmaceuticals and LMOs in transit of for contained use)[62] and in turn Articles 5 and 6 are both expressly nonmandatory at the implementation level (as a result of exceptions in text, as discussed previously). Taken altogether, it is unclear how these provisions would apply if the COP-MOP has listed some LMOs under Article 7.4. At present, the apparent stalemate regarding COP-MOP decision making[63] should operate to minimise the possibility that this issue will actually arise. Otherwise, the relation of such a decision with Article 2.4 might be difficult. If Article 2.4 is transcendent, no decision under Article 7.4 would have any reliable impact. If Article 7.4 prevails, the COP-MOP could ultimately dictate import permissions in all Party countries.

Only two of the countries whose legislation is reviewed appear to have adopted formal provisions for the approval of generalized exceptions to the requirement of risk assessment (although other countries' silence could simply mean that the agency's ability to make such choices is presumed based on other law or existing practice). In the Republic of Korea, the law requires the head of the CNA to

> make available to the public any information, including the list of living modified organisms that have no adverse effects on public health, and the conservation and sustainable use of biological diversity. . . . [O]ne who wishes to import, or produce living modified organisms that are included in the list and made available to the public pursuant to paragraph 1 above, shall report to the head of the competent national authority of such a fact.[64]

The other country that has adopted relevant legislation is Indonesia, which expresses the concept differently: "This government regulation is based on the precautionary approach to contribute to ensuring environmental safety, food safety and or animal feed safety based on an accurate scientific method by considering religious, ethic, socio-cultural and aesthetic norms."[65]

Although the COP-MOP has not acted collectively, a few countries have specifically adopted formal legislation designating particular LMOs to be exempt from the AIA process. Brazil's National Biosafety Law, for example, provides that

> Production and commercialization of genetically modified glyphosate tolerant soybeans that are registered in the National Crop Registry of the Ministry of Agriculture,

62 *Ibid* at art 7(1).

63 Currently, the COP-MOP Rules of Procedure (ROP) suffer from the same dysfunction found in the ROP of the CBD COP. In both documents, the rule on the body's decision-making power is bracketed owing to disagreement about whether decisions may be taken by vote in cases where the body cannot reach consensus. Pending removal of the brackets, the operations of both bodies are governed by general principles of international law, which require the body to make all decisions by consensus, unless it specifically adopts a contrary rule for its own operations.

64 *Korea LMOs Act*, *supra* note 15 at art 15; translated to English in *Korea Draft NBF*, *supra* note 15 at 191

65 *Indonesia Regulation on Biosafety*, *supra* note 26 at art 3.

> Livestock and Food Supply are hereby authorized.... Planting of genetically modified glyphosate tolerant soybeans reserved by farmers for their own use from the 2004/2005 harvest is hereby authorized, but trading of the production as seeds is forbidden.[66]

Similar provisions (similarly focused in scope) have also been adopted in other countries.[67] It is not clear whether these decisions indicate that all importation of these LMOs is clearly permitted, or that the agency can issue such permissions only by adoption of a national law saying so.

Precaution

The concept of precaution, as it is set forth in the Protocol, has been discussed in many fora but continues to be problematic. In earlier Chapters others argue that the commitment to respect the precautionary principle is found in many international treaties on sustainable development, including the Cartagena Protocol.[68] In its implementation at the national level, precaution raises some additional questions about what constitutes a "stricter measure" and whether the Protocol's permissive attitude regarding stricter measures in national implementation applies to its precaution-related provisions. These questions are being raised (and perhaps decided) in the World Trade Organization, as is discussed in a later chapter. In this chapter, it is most appropriate to describe some national provisions relevant to precaution in the biosafety context and leave their analysis for another author.

Many national expressions of precaution re-conceive the concept in various ways. In some cases, these reconceptions suggest that Parties expect to require denial of LMO introductions unless risks can be known and prevented.[69] In Cameroon, this issue is outlined in national biosafety law as follows:

> Risk assessment in any activity related to genetically modified organisms should take into account the precautionary principle, and conducted in a suitable manner, to guarantee the safety of humans, animals and plants, as well as to protect biological diversity and the environment [and]... should take into account expert opinion and guidelines drawn up by relevant international organizations."[70]

More specifically that law notes that "[l]ack of scientific certainty or agreement by scientists should not be interpreted as reflecting an acceptable risk level."[71]

In Indonesia, the law's expression regarding precaution is addressed by an "official explanation," accompanying the law, explaining that the precautionary approach means:

[66] *Brazil National Biosafety Law*, supra note 45 at arts 34–5.
[67] *Bolivia Resolución Multiministerial*, *supra* note 19.
[68] See Introduction and Chapter 1 of this book.
[69] *Bases para el marco nacional de bioseguridad de Chile*, *supra* note 47 at 20.
[70] *Cameroon Safety Regulations*, *supra* note 39 at s 18.
[71] *Ibid* at s 18.

> making decisions on preventive . . . even before conclusive scientific evidences are found with respect to that impact. In this government regulation, the precautionary approach shall be implemented as such, provided that [before an approved application for an LMO introduction] is implemented, a biosafety assessment and or food safety and or animal feed safety (risk assessment and management) will need to be conducted by using an accurate scientific method by considering social, economic, and ethical factors to guarantee that the risk to . . . environment and human health shall be acceptable based on the existing regulation. Consideration in term of religion, ethic, socio culture among others, [shall be addressed, and] genes transformed into [LMO] shall not coming from organism violating certain religious norms accepted in the country, shape or phenotype of [LMO] should equal to its parental in line with prevailing esthetic norm.[72]

In some countries, the concept of precaution relative to biosafety is expressed in quite different terms, possibly on the basis of that concept's expression in other national policy instruments. Zambia's national biosafety policy, for example, states that

> No approval for transfer, use and release of GMO(s) shall be given unless there is firm and sufficient evidence that the GMO(s) or products thereof pose no risk to human and animal health, biological diversity or the environment. Approval shall not be given where there is reason to believe that harm or damage may result, even when there is lack of scientific evidence or certainty.[73]

South Africa states its precaution concept more succinctly, using specific excerpts from the Protocol: "Lack of scientific knowledge or consensus on the safe use of genetically modified organisms shall not be interpreted as indicating a particular level of risk, an acceptable risk or an absence of risk."[74]

Precautionary concepts are also reconceived in Ethiopia's draft biosafety legislation:

> 1. Any government organ shall, in the implementation of this Proclamation, take into account the need for caution, particularly where there is scientific uncertainty about a risk emanating from any transaction. When faced with any uncertainty in any risk, it shall assume that that risk can occur and shall act to prevent or contain it.
> 2. Any authorized person shall, in the implementation of this Proclamation, take due care and implement measures to ensure that all transactions are carried out in conformity with the provisions of this Proclamation.[75]

[72] *Indonesia Regulation on Biosafety*, *supra* note 26 at art 3.

[73] Zambia, *National Biotechnology and Biosafety Policy*, 2003, BCH Record ID 9602 at para 5.1.

[74] See, e.g., *SA GMO Regulations*, *supra* note 31 at art 3(2).

[75] *Ethiopia Draft Proclamation*, *supra* note 20 at s 5.

In the Philippines, specific reference to precaution is expressly made, subject to the Cartagena Protocol and Rio Principles:

> In accordance with Principle 15 of the Rio Declaration of 1992 and the relevant provisions of the Cartagena Protocol on Biosafety, in particular Articles 1, 10 (par. 6) and 11 (par.8), the precautionary approach shall guide biosafety decisions. The principles and elements of this approach are hereby implemented through the decision-making system in the national biosafety framework.[76]

The country's Executive Order 514, *Establishing the National Biosafety Framework 2006*, also makes specific reference to the "scientific basis" for LMO decision making:

> The implementation of the national biosafety framework shall be based on the best available science and knowledge. Such science and knowledge shall be of the highest quality, multidisciplinary, peer-reviewed, and consistent with national standards as they evolve.[77]

The country's national perspective on biosafety is best understood, however, by comparing its primary precautionary provision with the precaution language of Cartagena Protocol, Articles 10(6) and 11(8). In the following quotation, the specific variations between the Philippine law and the specific language of the protocol are underscored:

> lack of scientific certainty *or consensus* due to insufficient relevant scientific information and knowledge regarding the extent of the potential adverse effects of a *genetically* modified organism on *environment, particularly on* the conservation and sustainable use of biological diversity, *and on* human health, shall not prevent *the concerned government departments* from taking *the appropriate* decision to avoid or minimize such potential adverse effects. *In such cases, concerned government department and agencies shall take the necessary action to protect public interest and welfare* [emphasis added].[78]

Addition of the phrase "or consensus" appears to indicate a loosening of the Cartagena Protocol's provisions, although placing "human health" on a par with the biodiversity considerations might have been intended to tighten them. The various national re-conceptions of the precautionary language in the Protocol suggest that precaution continues to be controversial within the Cartagena Protocol.

Other Provisions Having an Impact on the Cartagena Protocol

Not surprisingly, it is not the regulatory process of the Cartagena Protocol that poses the greatest challenge to national legislators. Much more difficult in many cases is the interaction between new biosafety laws and other particular provisions,

76 *Philippines EO 514, supra* note 51 at para 2.6.
77 *Ibid* at para 2.4.
78 *Ibid* at para 5.1.

necessitating national legislative action that might apply to LMOs in the same way that they apply to non-LMOs. Based on review of the BCH legislation, focused especially on new draft NBFs prepared under the recent UNEP-GEF project, coordination with these laws could be more challenging than simply writing a new administrative framework using the Cartagena Protocol as a template. These issues could pose ongoing concerns when the Cartagena Protocol's programme for oversight of biosafety begins to function as a commonly used mechanism of transborder trade, commerce, and transportation. The following discussions consider nine of the most obvious issues and emphasise some of the challenges to be faced.

Appeal of Permit Decisions

For most countries, national law governing all permits, licenses, and administrative approvals already includes the concept of appeals by individual persons or entities whose applications have not been approved, and/or by other persons who are affected or aggrieved by the permit decision. It is not surprising, then, that appeal provisions are often mentioned or included in national biosafety legislation and draft legislation.[79]

Appeals provisions are often limiting. It is common to state the basis on which an appeal may be requested, thereby effectively limiting the group of people who may appeal. In Cambodia's draft legislation, for example, appeal by a rejected applicant is only possible where "the applicant . . . can prove that the risk to biological diversity and human health [from the activity for which a permit was sought and denied] is minimal."[80] This clause illustrates two concepts. First, it appears to limit appeal rights – permitting only unsuccessful applicants to appeal (excluding appeals by those who oppose decisions granting the permit.) Conversely, however, this clause calls on the appellant to prove that his or her proposed activity involves "minimal risk." The challenge would be extremely difficult in certain countries, however it might not be particularly challenging to Cambodia's courts and parties.

South Africa's law contains a relatively typical approach regarding who is permitted to appeal, apparently enabling appeals by both applicant and opponent:

> A person who feels aggrieved by any decision or action taken by the Council, the registrar or an inspector in terms of this Act may, within the period and in the manner prescribed and upon the payment of the prescribed fee, appeal against such decision or action to the Minister, who shall appoint an appeal board for the purpose of the appeal concerned. . . .

79 See *Cambodia NBF*, *supra* note 20 at 10.

80 *Ibid.*

> The decision of an appeal board, together with the reasons therefore, shall be reduced to writing, and copies thereof shall be furnished to the Minister, whereupon the Minister may take such further action as he or she may deem necessary.[81]

The appeal provision in the law of the Republic of Korea also appears broadly available to the opponents of a given permit:

1. In the event where an individual does not agree on a decision . . . that person may request the head of the related central government authority for a re-deliberation as prescribed by the Presidential Decree.
2. Where a re-deliberation is requested, the head of the related central government authority shall make a decision through consultation with the head of the competent national authority and examination of the Biosafety committee pursuant to Article 31.[82]

This right of appeal is also somewhat open ended under national law. As noted in the country's draft NBF: "In case where the applicant will not accept the decision made by the head of the related central government Authority, he/she can lodge a suit with the Committee of Administrative Litigation under the Prime Minister's department or lodge an administrative suit."[83]

The Philippines' provisions for appeal are relatively straightforward, reflecting consistent national practice on this issue:

> Any person whose permit has been revoked or has been denied a permit or whose petition for delisting has been denied by the Director of BPI, may appeal the decision in writing to the Secretary of the Department within fifteen (15) days after receiving the written notification of the revocation or denial. The appeal shall state all the facts and reasons upon which the appellant bases his appeal. The Secretary of the Department shall grant or deny the appeal in writing within sixty (60) days from receipt of the appeal.[84]

Legislation on this point in Namibia is also relatively detailed, identifying a very specific procedure with expressly stated time limits and decisional powers.[85]

Permit Reopeners and Rescission

The Protocol recognises one possibility under which each permit or other biosafety decision may be reopened and reconsidered. Article 12 specifically notes that the country approving an application for introduction of an LMO might find it necessary to reopen that decision, "at any time, in light of new scientific information

[81] Republic of South Africa, *Genetically Modified Organisms Act (No. 15 of 1997)*, BCH Record ID 255, at s 19(1) and (6) [*SA GMO Act*]. See also *SA GMO Regulations*, *supra* note 31 at s 9.

[82] *Korea LMOs Act*, *supra* note 15 at art 18; *Korea Draft NBF*, *supra* note 15 at 91 and 195.

[83] *Korea Draft NBF*, *ibid* at 91.

[84] *Philippines AO 8*, *supra* note 27 at s 18.

[85] *Namibia Biosafety Act*, *supra* note 40 at s 46.

on potential adverse effects on the conservation and sustainable use of biological diversity, taking also into account the risks to human health" and to change that decision (either by rescission or by imposing additional mitigation measures).

In many countries, national legislation addresses the possibility of reopening the permit decisions of all types. These provisions may apply to (or apply in addition to) the processes developed in NBFs. In other countries, each law that creates an administrative framework must also detail its procedural rules. Among such rules, the most important and challenging are the rules relating to the government's powers of unilateral revision, rescission, or amendment of a preexisting license, permit, or other provision. As a result, many permit-reopening provisions are relatively national biosafety detailed and specific. They are also strongly linked to national laws and organic laws, which enumerate the legal rights of citizens. In the Republic of Korea, for example, provisions for the reopening and rescission of LMO approval include very specific standards on which these actions must be based:

> The head of the related central government authority may revoke an approval of import, or production of living modified organisms granted pursuant to Article 8 and 9, or 12, in case where such living modified organisms fall into any of the following subparagraphs. However, an approval for those that fall into (1) or (2) must be revoked:
>
> 1) Where it is found that the living modified organisms approved for import or production have, or are believed to have, adverse effects on public health, and the conservation and sustainable use of biological diversity,
> 2) Where the approval is received by fraud, or other wrongful means,
> 3) Where the living modified organisms are used for a different purpose, from which they were approved,
> 4) Where this Act, or an order, or a disposition issued under this Act, is violated, and
> 5) Where it is prescribed by the Presidential Decree.
>
> In the case where an approval has been revoked, pursuant to paragraph 1 above, the head of the related central government authority shall promptly notify the head of the competent national authority of such a fact.[86]

Bangladesh provides another example of a rescission clause – one that demonstrates the fact that rescission can be only partial. This example also demonstrates an important element that must be included in a partial rescission – a determination about which rules must apply where a particular provision is rescinded, but the overall permit remains in force:

> The [National Technical Committee on Biosafety] may, at any time, rescind the certification of any host-vector system. If certification of host-vector system is rescinded,

[86] *Korea LMOs Act*, *supra* note 15 at art 17; *Korea Draft NBF*, *supra* note 15 at 10, 91 and 193–4.

NCB will instruct investigators to transfer cloned DNA into a different system or use the clones at a higher physical containment level unless the NCB determines that the already constructed clones have adequate biological containment.[87]

In the Philippines, detailed revocation procedures are specified for each different type of permit, specifying a variety of grounds for such revocation, including "(1) Revocation of the [agency] endorsement. (2) Provision of false information in the Application to import for Contained Use; (3) Mis-declaration of shipment, (4) Violation of relevant SPS or biosafety rules and regulations or any conditions imposed in the Permit to Import for Contained Use; (5) Refusal to allow inspection . . . or (6) New technical information becomes available." A somewhat different list applies for revocation of permission for field testing, and still different ones apply to LMO propagation and their "direct use" as LMOs-FFP.[88]

Other laws have addressed this issue in similar ways, specifying grounds for the reopening permits, including revocation due to nonpayment of an annual fee.[89] Few countries impose no administrative penalties in their biosafety legislation, but presumably these countries rely on other national legislation governing administrative actions to empower these actions. In addition, as noted later, legal and administrative penalties are closely related to the legislative obligation to report any information discovered or heard regarding LMO releases or risks.

Transport and Documentation

The issue of LMO labelling, and specifically whether and when labels are to be required with regard to the transport and handling of LMOs, has presented a challenge for international processes under the Cartagena Protocol, owing in part to the prominence of labelling issues in WTO documents and decision making. The Protocol expresses mandatory obligations of the Parties about these issues, to the extent that they apply to transboundary movements of LMOs. Extension of handling, transport, packaging, and identification of LMOs to domestic movement and packaging of LMOs would not appear to be problematic, in light of the Cartagena Protocol's provisions regarding stricter measures. The more frequently aired concerns in connection with transport and handling, and relevant documentation relate to the WTO and its controls on national measures that indirectly restrain external trade. As a result of these concerns, the Cartagena Protocol COP-MOP's process of developing and adopting "detailed requirements for [LMO labelling], including specification of their identity and any unique identification" and determining "the need for and modalities of developing standards with regard to identification, handling, packaging and transport practices," have been relatively difficult, extending well beyond the time limit specified in Article 18(2)(a).

[87] *Biosafety Guidelines of Bangladesh, supra* note 17 at 4.2.2(iv).
[88] *Philippines AO 8, supra* note 27 at ss 6(F), 8(P), 10(K), and 12(I).
[89] *Namibia Biosafety Act, supra* note 40 at s 33–6.

Most of the BCH documents referencing labelling issues recognise this and refer to developing "international standards" in this area.[90] Pending resolution of this issue at the international level, however, these issues have had an ongoing impact on many countries, both Cartagena Protocol Parties and countries that are not Parties but are active participants in the LMO sector. Countries that have not adopted labelling provisions or have adopted language that parrots or compares with Cartagena Protocol article 18(2) are numerous.[91] Possibly, this signals that the country intends to implement whichever labelling measures finally agreed on by the COP-MOP. Others have indicated that their level of commitment to the Protocol has been affected by these issues, which have delayed both their entry into the Protocol as well as its legislative implementation.[92]

One country whose domestic regulatory framework directly discusses labelling is the Republic of Korea, whose framework predates adoption of the Protocol and apparently generated its own labelling requirements and procedures:

> 1. One who develops, produces, or imports living modified organisms shall identify them accordingly, or any information including the type, which will be marked on the packaging or on the container as prescribed by the Presidential Decree. . . .
> 2. No one shall provide false information, or alter, or delete the information required by paragraph 1 above at his own discretion.
> 3. Methods of the identification, required by paragraph 1 above, and other matters where necessary, shall be made available to the public by the head of the related central government authority.
>
> . . .
>
> 1. One who undertakes activities, including the import or export of living modified organisms, shall respect the guidelines on handling and administration, as prescribed by the Presidential Decree, including transportation, which is to be done in a sealed container.
> 2. Methods of handling and administration, pursuant to paragraph 1 above, and other matters where necessary, shall be made available to the public by the head of the related central government authority.[93]

This law's other key provisions relative to labelling are somewhat more direct. For example, that law strictly bans all import and sales of a GM product that is mixed with a substance that did not undergo safety assessment tests, with strict penalties for violation.[94]

90 See *Bahamas Draft NBF*, *supra* note 28 at para 8.2.3.

91 See *Cambodia Draft Law*, *supra* note 20 at art 24; *Ethiopia Draft Proclamation*, *supra* note 20 at s 23; and *Namibia Biosafety Act*, *supra* note 40 at s 32. The Parties have addressed transportation issues under Article 18 at each meeting of the COP-MOP; see Chapter 6 in this book for further information.

92 *Bases para el marco nacional de bioseguridad de Chile*, *supra* note 47 at 18.

93 *Korea Draft NBF*, *supra* note 15 at 10 and 197–8.

94 Republic of Korea, *Food Sanitation Act*, 2002 at art 4, summarised in *Korea Draft NBF*, *ibid* at 56–8.

A somewhat different approach is found in Cameroon's Biosafety Law, which specifies domestic labelling requirements in detail, without waiting to see how the COP-MOP deliberations are resolved. Relevant clauses in that law provide that

> Any genetically modified organisms or products thereof intended for intentional release or marketing on the national territory shall be packaged and labeled in order to safeguard ethical and cultural values, and to avoid risks to human and animal health. . . . All the genetically modified organisms perfected and marketed on the national territory shall be packaged and labeled by the producer and sender as follows: "Product based on genetically modified organisms," or "contains genetically modified organisms," in compliance with other supplementary norms defined by the competent national administration in collaboration with the other authorities involved. The following information shall be specified:
>
> - distinctive marks of the model or specifications of packaging, irrespective of the container, generally used by the manufacturer of packages;
> - packaging with marks indicating content, donor and consignor;
> - labels with specific colours corresponding to dangerous contents.
>
> Moreover, the consignor shall fill in and sign two copies of a manifest. The said manifest shall attest to the respect, by the sender, of the requirements of the advance informed agreement.[95]

Related to this issue, draft legislation in Cambodia requires that the letter of approval of the Ministry of Environment must be attached to the import permit for any LMO.[96]

Although this law would not ineluctably prevent the use of a single permission for multiple importations, it does suggest the possible need for national legislative measures and customs regulations addressing this possibility. This link – between LMO labelling and customs controls – has been directly addressed in very few biosafety laws, with Ethiopian law providing the main example:

> Every person in possession of any genetically modified organism or product thereof as part of his or her personal effects or baggage shall, on arrival or departure, declare such possession to the Customs Officer on duty at the port of entry or exit. . . . Where a Customs Officer suspects that any person is in possession of a genetically modified organism or product thereof for which there is no valid permit, she/he shall require the said person to surrender such organism or product thereof. . . . 3. Any genetically modified organism or product thereof surrendered to a Customs Officer pursuant to sub-article 2 shall be sent to the Authority immediately.[97]

95 *Cameroon Safety Regulations*, *supra* note 39 at s 49.
96 *Cambodia Draft Law*, *supra* note 20 at art 17.
97 *Ethiopia Draft Proclamation*, *supra* note 20 at s 27.

Further provisions specifically empower seizure of LMOs brought through customs in violation of LMO law.[98]

The regulation of LMOs in transit suggests another concept that is relevant in this context. Typically, LMOs that are in transit through a country need not submit to the full AIA or Article 11 process in that country, because such application can be challenging from a practical standpoint. Few of the laws that generically apply biosafety requirements to LMOs in transit speak to this point. One that does, albeit indirectly, is found in the Republic of Korea, which provides that

> [o]ne who wishes to transport living modified organisms through the country, in order to export them to other countries, shall report to the head of the competent national authority with regards to the information including items, quantity, the name of the exporting country, and name of the importing country as prescribed by the Presidential Decree.[99]

Ultimately, resolution of the labelling issues could wait on international law and the ability of global experts and tribunals to address a more challenging question – the relationship of two separate international instruments among sovereigns. Many authors and experts appear to base their analysis of labelling and other controls on movement of LMOs on the assumption that the WTO is preeminent, and "trumps" all other international instruments. This result is far from certain as a matter of law, particularly when particular controls are internationally agreed among sovereigns as well (as in the Cartagena Protocol).[100]

EIA and Risk Assessment

Although efforts were made to separate this concept from EIA, discussions of risk assessment in the Protocol have inevitably resulted in significant attention to EIA laws and their relation to the Cartagena Protocol. For this reason, several countries have posted their basic national EIA legislation in the BCH. Unfortunately, in many cases these laws are limited in application and according to their terms are specifically not applicable to aquacultural, forest, agricultural activities, or any other activities that involve plants, animals, or microbial specimens, given that most countries' EIA legislation applies, by its terms, to only a specific list of activities.[101] Without in-country consultation, one might not necessarily be able to determine whether unwritten practices or other legal mechanisms are in use, which would enable national EIA law to meet the country's Cartagena Protocol risk assessment

[98] *Ibid* at s 28.

[99] *Korea LMOs Act*, *supra* note 15 at art 21; *Korea Draft NBF*, *supra* note 15 at 195.

[100] An excellent discussion of the relationship between the Cartagena Protocol and the WTO is found in Mackenzie et al, *An Explanatory Guide to the Cartagena Protocol on Biosafety* (Gland: IUCN-ELC, 2003) at paras 844–914.

[101] World Bank, *Environmental Assessment Sourcebook Vol. 1: Policies, Procedures, and Cross-Sectoral Issues*, World Bank Technical Paper No 139 (Washington, DC: World Bank, 1991).

obligations. By posting of EIA legislation, however, the countries might be indicating that they will perform risk assessments only for activities covered by national EIA law.

Some countries national EIA laws do specifically consider some LMO/biosafety issues. In South Africa, for example, EIA regulations specifically list the following among its schedule of "Activities which may have a Substantial Detrimental effect on the Environment": "5) The release of any organism outside its natural area of distribution that is to be used for biological pest control; and 6) The genetic modification of any organism with the purpose of fundamentally changing the inherent characteristics of that organism."[102]

Although this law potentially excludes LMOs with different purposes, this clause rather clearly suggested that some proposed LMO activities (LMOs[103] released for pest control and LMO development in contained-use activities) must be subject to an EIA. For these LMO activities, there remains only the question of whether scrutiny under South Africa's EIA legislation satisfies the requirements of risk assessment under the AIA and Article 11 processes. Other activities listed, which could also have some relevance for LMOs, include

> 3) The concentration of livestock, aquatic organisms, poultry and game in confined structures for the purpose of commercial production, including aquaculture and mariculture.
>
> 4) The intensive husbandry of, or importation of, any plant or animal that has been declared a weed or an invasive alien species.
>
> ...
>
> 7) The reclamation of land, including wetlands, below the high-water mark of the sea, and in inland waters.
>
> 8) The disposal of waste as defined in section 20 of the Act, excluding domestic waste, but including the establishment, expansion, upgrading or closure of facilities for all waste, ashes and building rubble.

In addition, LMO users and developers must be curious whether these clauses will be interpreted as requiring an EIA for every new contained use or introduction, or if they will be excused where the particular LMO or event has been imported for

[102] Republic of South Africa, *Environment Conservation Act, 1989, Regulations: The Identification under Section 21 of activities which may have a substantial detrimental effect on the environment*, BCH Record ID 253, online at: http://www.acts.co.za/enviro/index.htm. See also description of Republic of Korea, *Guidelines for the Environmental Risk Assessment of GM Agricultural Products*, 2001 in *Korea Draft NBF*, *supra* note 15 at 69–75.

[103] In most national legislative definitions, an LMO is not considered to have a "natural area of distribution" and normally fits within national controls on "alien" species introductions, unless the law specifies otherwise. See T Young, *National and Regional Legislation for Promotion and Support to the Prevention, Control and Eradication of Invasive Species* (Washington, DC: World Bank, 2006).

contained use and/or introduced previously. The Philippines has taken this issue into specific account, noting in relation to its national policy and framework's treatment of risk assessment, as follows:

> The application of the EIA System to biosafety decisions shall be determined by concerned departments and agencies subject to the requirements of law and the standards set by [national biosafety body] where applicable under the matter.[104]

Confidentiality and Trade Secrecy

Implementation of the confidentiality provisions in the Protocol presents a relatively complex legal challenge, which differs from country to country. This issue could have a potentially serious impact on the commercial operability of the Protocol. Discussions in workshops and other expert meetings have identified several particular problems related to confidentiality, most especially the significant divergence among countries, and the fact that some exporters think that relevant officials in the import countries cannot be relied on to maintain and protect the applicant's trade secrets, even if they grant confidentiality. Fearing that confidential material outside of their control will be lost, despite legal provisions for confidentiality, many exporters have insisted that they will withhold that information, regardless of confidentiality laws.[105] Once trade secrets have been presented, we should remember, they can no longer be guaranteed to be "secret." At least one person (the government official who evaluated the application to decide if confidentiality provisions shall apply) will have seen the "secret" and might (intentionally or unintentionally) pass it to others. Persons and entities in the LMO sector must rely on governmental protection and liability clauses as commercial protection, in the event that confidential information is unduly disclosed.

Several documents in the BCH database address confidential information.[106] Although many of these generally parrot the relevant clauses of the Cartagena Protocol, a few national laws and practices are more specific. The variation among these laws demonstrates the basic differences among countries in approach to the rights of commercial entities.

[104] *Philippines EO 514, supra* note 51 at para 5.3.

[105] Names and national affiliations withheld.

[106] See e.g., *Cambodia Draft Law, supra* note 20 at arts 27–30; Former Yugoslav Republic of Macedonia, *Law on GMO (Official Gazette of RM 35/08)*, BCH Record ID 100989; and Latvia, *Procedures for the release into the environment or placing on the market of genetically modified organisms, the procedures for monitoring and issuance of a permit, as well as the procedures for providing information regarding circulation of genetically modified organisms and public involvement in the decision taking process*, BCH Record ID 48915. See also *Argentina Resolución 39, supra* note 16 at ss 26–31 and Argentina, *Solicitud para la concesión de permisos para Experimentación y/o Liberación al Medio de Microorganismos Genéticamente Modificados y/o sus productos para aplicaciones en animales./ Resolución No 656*, 2004, BCH Record ID 11812.

One example of a confidentiality provision in biosafety legislation comes from Brazil:

> [The Commission] shall take the necessary measures to safeguard the confidential information of commercial interest identified by the applicant and considered as such by [the Commission], provided such information do not affect private or collective interests protected by the Brazilian Constitution.
>
> In order to maintain the confidentiality mentioned in this Article 8, the applicant shall forward to the President of [the Commission] an express and well grounded request containing specifications on the information to be kept confidential.
> . . .
>
> The request may be dismissed through a well grounded order, which may be subject to appeal to [the Commission's] plenary meeting, and confidentiality shall be maintained until a final decision is reached to the contrary. . . .
>
> The applicant may . . . withdraw its proposal in case the request for confidentiality is definitely dismissed and, in this case, [the Commission] may not disclose the information for which confidentiality has been sought. Registration and monitoring bodies and entities may request access to certain confidential information deemed indispensable to the body or entity in a petition explaining the request and indicating the agent that may have access to the confidential information. Registration bodies and entities may not disclose any confidential information.[107]

The last paragraph of the foregoing excerpt discusses the issues of greatest concern to the applicant. They note that it is not sufficient to state, "[the Commission] may not disclose the information for which confidentiality has been sought." If the government cannot provide sufficient measures to reimburse applicants whose secrets are lost, and to penalize officials who cause that loss, then the adoption of confidentiality protections is merely "so much ink on paper."

The extent of protection awarded varies greatly from country to country, and its value cannot be determined on the basis of "paper" evaluation but may require in-depth consultation with the relevant officials. Still, the difference between them could be instructive. For example, South Africa's domestic biosafety framework includes the full range of provisions comparable to Cartagena Protocol Article 21. Its confidentiality provision begins, however, by stating four exceptions to confidentiality that are not limited by agency decision – that is, a person may determine for him- or herself that a particular item is not protected:

> No person shall disclose any information acquired by him or her through the exercise of his or her powers or the performance of his or her duties in terms of this Act, except:
>
> a. insofar as it is necessary for the proper application of the provisions of this Act;

[107] *Normative Resolution 6*, *supra* note 29 at 8–9.

b. for the purposes of any legal proceedings under this Act,
c. when ordered to do so by any competent court; or
d. if he or she is authorised to do so by the Minister.[108]

A later clause imposes a duty on persons who obtain secret data to protect that data: "If, for whatever reasons, the applicant withdraws an application, any party who has knowledge of the details of the application must respect the confidentiality of the information supplied."[109]

Within a similar scope, the Republic of Korea has also adopted relatively detailed provisions, including inter alia, the following:

> The heads of the . . . relevant . . . authorit[ies and] agenc[ies] . . . shall take necessary measures to ensure the confidentiality of the information on living modified organisms, so that it is not subject to theft, leakage, or distortion. . . .
>
> [Those] authorities . . . shall under no circumstance use the information on living modified organisms for commercial purpose, or provide it to a third party, unless prescribed by the Presidential Decree. . . .
>
> In the case where such information is provided to a third party, the heads of the information-handling authorities may have restrictions on the necessary provisions regarding purposes or uses, or request to take necessary measures to protect confidentiality. . . .
>
> One who works, or used to work, for an authority handling information, both an officer and an employee, shall under no circumstance use the information that he has obtained in the course of his service for wrongful purpose, including disclosure or provision to a third party for possible use.[110]

Proposed legislation in Ethiopia would adopt similar provisions, presuming that only legislation is needed to fulfil this obligation:

> Every applicant shall have a right to notify the Authority specifying the information to be treated as confidential. . . . 2. The Authority shall protect information which is not essential for biosafety and which it determines as confidential following a request to that effect made in writing by the applicant. . . . If the applicant withdraws the application before approval, the authority shall respect the confidentiality of the information except for that part referred to under the provisions of sub-Article 3 of this Article.[111]

In countries in which biosafety protection is politically important, concerns of public access to information are often pitted against the commercial and scientific concern for protecting trade secrets. The confidentiality provisions in Namibia apply

[108] *SA GMO Act*, *supra* note 81 at s 18(1).
[109] *Ibid* at art 18(4).
[110] *Korea LMOs Act*, *supra* note 15 at arts 28–30; *Korea Draft NBF*, *supra* note 15 at 76–8, 96–8, and 200–1.
[111] *Ethiopia Draft Proclamation*, *supra* note 20 at s 26.

to this dichotomy, giving an indication of the bases that might be used in denying confidentiality:

> The Council may refuse to declare that information is commercially confidential if the Council is satisfied that the public interest in disclosure of the information outweighs the prejudice that disclosure would cause to any person.[112]

That law also indicates the lower standard of protection that will be given in the event that confidentiality is denied: "If the Council refuses an application under subsection (1) in relation to any information, the information is to be treated as commercially confidential until the determination of any appeal against such refusal under section 46 or the period for lodging such an appeal has expired."[113]

To balance these provisions, however, that law also imposes a legal restraint (similar to those described previously, but somewhat less intense) on government officials who gain awareness of the confidential information:

> [A] person who – (a) has knowledge of commercially confidential information; (b) has acquired that knowledge because of performing duties or functions under this Act or any other law; and (c) knows that the information is commercially confidential information, must not disclose the information.
>
> [The foregoing] does not apply to commercially confidential information which is disclosed – (a) to a member or the Registrar or to any staff member or other person engaged in the performance of any function under this Act and the disclosure is made within the ordinary scope of the duties of the person making the disclosure; (b) by order of a court; or (c) with the consent of the person who applied to have the information declared as commercially confidential commercial information. . . . A person who contravenes [the above] commits an offence and is liable to a fine . . . or to imprisonment . . . , or to both such fine and such imprisonment.
>
> In this section: . . . "disclose," in relation to information, means give or communicate in any way.[114]

Note that this provision applies only to "commercially confidential information" (possibly limited to information the confidentiality of which has been approved under the earlier provision) about which the person disclosing it can be proved to have known the confidentiality of the material.

Legislation in the Philippines provides another clear indication of both sides of the heretofore unresolved conflict between the applicant's desire for confidentiality and the public's desire for transparency:

> In all applications for approvals, whether domestic or foreign, concerned departments and agencies shall ensure that it has [*sic*] procedures and regulations to determine and protect confidential information; [*sic*] Provided, however, that the

[112] *Namibia Biosafety Act*, *supra* note 40 at s 43(4).
[113] *Ibid* at ss 43(7) and 44.
[114] *Ibid* at ss 43(7) and 45.

concerned agencies may refuse declaring the confidentiality of such information if it is necessary to enable stakeholders to effectively conduct a scientific risk assessment.[115]

In some countries, the countries' general administrative/commercial legislation might already exist and apply so broadly that it protects or fails to protect information provided to biosafety agencies. Ultimately, the difficulty involved in confidentiality legislation is not the drafting but the need to reconcile conflicting interests. Regardless of how expressed, however, the "real" challenge of confidentiality legislation is the actual protection of the disclosed information, specifically, finding a mechanism that ensures that individual officials with access to files containing confidential records have a disincentive for violating confidentiality. For many countries, governmental employees are low paid, and there is no national culture discouraging legal violation. Such countries will continue to face difficulties with imposing LMO disclosure requirements until some mechanism can be developed to aid with this, possibly at the international level.

Costs and Conduct of AIA or other Assessment

Virtually all of the developing country legislation and drafts pass the duty to pay for (and sometimes, to conduct) the risk assessment to the applicant.[116] The clue provided in Cartagena Protocol Article 15.3 ("The cost of risk assessment shall be borne by the notifier if the Party of import so requires") has apparently been well heeded.

For countries that want (and are able) to maintain primary responsibility for the assessment, the most direct approach may be both the easiest and the least misunderstood. For example, Indonesia's biosafety legislation provides that "[t]he expenses for evaluation and technical assessment, and the laboratory, greenhouse, and confined field tests . . . shall be charged to the applicant."[117]

The additional legal challenges that can arise when the applicant has conducted the AIA himself generally have not been considered. These include the possibility that the AIA will be scientifically unsound or deceptive. One interesting provision in proposed legislation for Ethiopia would state that "[a]n approved risk assessment report shall remain valid for the period specified by the Authority during its approval."[118] It is not clear in this document who is responsible for ensuring validity,

[115] *Philippines EO 514*, *supra* note 51 at para 6.2. In the Philippines, public participation has long been important to national and local governance, even at highly technical levels.

[116] See *Ethiopia Draft Proclamation*, *supra* note 20 at s 9. See also *Namibia Biosafety Act*, *supra* note 40 at s 23.

[117] Indonesia, *Joint Decree of the Minister of Agriculture, the Minister of Forestry And Estate Crops, the Minister of Health, and the State Minister of Food and Horticulture Concerning Biosafety and Food Safety of Genetically Engineered Agricultural Products*, BCH Record ID 9591 at art 40(3) [*Joint Decree on Biosafety and Food Safety*].

[118] *Ethiopia Draft Proclamation*, *supra* note 20 at s 14(2).

and whether the creator of the risk assessment is bound to update it throughout that period.[119]

Technical Standards

For full implementation of the biosafety objective of imposing standards of biosafety on LMO activities and ensuring that potential risks are minimised, it is generally thought that technical standards regarding those activities must be formally understood and adopted/applied in any country in which such activities are proposed.[120] For developing countries, apart from those with more intensive involvement in LMO development, production, and use, however, both the adoption and the application of such standards may pose nearly insurmountable challenges. It might be possible (perhaps advisable) for a legislating country with limited LMO technical capacity to rely on technical decisions of other countries having an attitude/policy regarding LMO risk similar to that of the legislating country. In that case, they can perhaps adopt or adapt detailed technical standards have been adopted in those other countries of similar opinion. Two developing countries with detailed technical standards are the Philippines and Bangladesh.[121]

Harm, Violation and Responsibility

The issues of "liability and redress" have been particularly challenging in the context of the Cartagena Protocol. As a legal and practical matter, however, these issues are linked to a variety of other concepts, including "offenses and penalties" and "obligations to report." The following discussions describe how various focus countries have addressed these two issues.

Liability and Legal Redress

It is perhaps not surprising that national Cartagena Protocol implementing legislation has not resulted in significant innovation or other progress in the areas of liability and redress, given that international negotiations only recently concluded and the challenges they face in identifying the standards and other bases of such liability.[122] Liability questions relating to biosafety are multifaceted, but some of the key concerns are expressed in the Republic of Korea's draft NBF:

[119] Another example of provisions assigning some or all assessment costs to the applicant is *Indonesia Regulation on Biosafety*, *supra* note 26 at arts 19–20.

[120] See UNEP-GEF, *Toolkit for Development of National Biosafety Frameworks*, online: http://www.unep.org/biosafety/Toolkit.aspx; especially "Phase 3 – Drafting the NBF, Part II: Designing the administrative systems for handling applications and notification."

[121] *Philippines EO 514*, *supra* note 51 at s 3; *Biosafety Guidelines of Bangladesh*, *supra* note 17 at chs III, IV, and annexes.

[122] *Cartagena Protocol*, *supra* note 1 at art 27. See *Nagoya – Kuala Lumpur Supplementary Protocol on Liability and Redress to the Cartagena Protocol on Biosafety*, online: http://bch.cbd.int/protocol/NKL_text.shtml. Among the challenges still being faced are uncertainty about the evidence necessary to document a type/level of injury or damage that is substantial enough to be subject to legal redress, and the challenge in developing legal measures that can be implemented in developed (strict-rule-of-law) countries.

> When GM products are transferred from the developing company, (i.e., the producer or the distributor) to the food processor who utilizes the ingredient and finally to the consumer as the final food product, the issue of who is responsible for ensuring safety arises. It is the manufacturing company or representative company which is able to submit the data for evaluation.
>
> At this phase when there is no domestic production, the distributor of the GM foods is not the developing company but the food distributor or domestic manufacturer. Therefore the domestic importer or food manufacturer must take the responsibility for the mixing between the GM foods and the conventional foods. To eliminate this problem, a responsible attitude towards safety evaluation on the part of the developer is important [continuing with discussion of the need for a "Post-monitoring plan"].[123]

In many cases, biosafety legislation appears to presume that the country's general liability law and practices will apply in cases of biosafety-related harm. The law of Cameroon takes this approach directly, stating, "The user of genetically modified organisms shall be liable for any damage caused by the deliberate or accidental release of such organisms."[124]

South Africa's law addresses liability in terms of a "user pays" approach that is frequently found in national environmental laws – by enunciating the duty of the user first and then discussing the concomitant liability:

> Users shall ensure that appropriate measures are taken to avoid an adverse impact on the environment which may arise from the use of genetically modified organisms.
>
> . . .
>
> The liability for damage caused by the use or release of a genetically modified organism shall be borne by the user concerned: Provided that when such an organism was in the possession of an inspector as set out in section 15(4), the user concerned at the time of such use or release shall not be held liable for any damage unless such user foresaw or should have foreseen such damage and could or should have prevented the damage but failed to take reasonable action to prevent such damage.

Relatively few national biosafety laws identify LMO use as an activity subject to strict liability.[125] In common law countries, if LMO activities are essentially given the status of an "ultrahazardous activity," persons who engage in those activities, even when acting with permission, could be "strictly liable" – liable for all injury and damage caused by the activity, even if they can show that they used due care and did not violate any duty owed the plaintiff.[126] Where national law does not consider

[123] *Korea Draft NBF*, *supra* note 15 at 64–5.

[124] *Cameroon Safety Regulations*, *supra* note 39 at s 37.

[125] See, e.g., Brazil, *Decree 5591/2005, regulating Law 11.105/2005*, 2005, BCH Record ID 11777, at title VIII, art 68. [*Brazil Decree 5591*]: "Without prejudice to the application of the penalties provided for . . . parties responsible for damage to the environment and to third parties have joint and several liability for full compensation or reparation of such damage, regardless of fault."

[126] See, e.g., Restatement (Second) of Torts, ss 519, *et seq.* (American Law Institute, 1979).

LMO activities to be ultrahazardous, the user is held to a "reasonable" standard – liable for damage that he or she foresaw or should have foreseen and failed to take reasonable action to prevent. Liability then might be "joint and several" – any person involved in the activity is liable for the entire amount, unless he or she can obtain a share from other responsible parties.[127] In many countries, liability issues are contentious, in some cases preventing the adoption of legislation and/or forcing legislators to leave the issue unresolved.[128]

National legislative provisions on liability in proposed Ethiopian legislation provide a clear example of the application of strict liability in biosafety legislation:

> Approval for any transaction by the Authority does not exonerate the applicant from liability. . . . A person who is engaged in any transaction shall be strictly liable for any harm caused by such a genetically modified organism or products thereof. The harm shall be fully compensated. . . . Without prejudice to sub-Article 2 of this Article, exemption from liability shall be granted only when it is verified that it is the victim himself or a third person for whom the proponent is not responsible that has caused the damage. . . . Liability shall attach to the person responsible for the transaction, which results in the damage, injury or loss as well as to the provider, supplier or developer of the genetically modified organism or product thereof.[129]

This draft further enunciates principles of joint and several liability, as well as duties to reinstate, rehabilitate, or clean up any ecosystem that is damaged, as well as recompense for any injury or harm suffered, authorising a broad range of plaintiffs to bring civil suits.[130]

As described in some BCH literature, liability and redress provisions are both complicated and important, becoming nearly impossible to satisfy when the goal is to find a single international solution.[131] The draft NBF submitted for the Republic of Korea noted,

> There are a lot of problems in compensating for the damage resulting from LMOs from the general budget. The mechanism that he who imports or exports LMOs takes out an insurance and settles the matter does not fit the basic principle of compensation for the damage resulting from them, since he who takes out an insurance can compensate whilst he who does not cannot compensate enough for the damage, and after all government should compensate for the damage.[132]

That document and many others have expressed doubts about whether these matters can be resolved in the foreseeable future.

[127] *Ibid.*
[128] See *Bases para el marco nacional de bioseguridad de Chile, supra* note 47 at 36.
[129] *Ethiopia Draft Proclamation, supra* note 20 at ss 29-30.
[130] *Ibid.*
[131] *Korea Draft NBF, supra* note 15 at 101–2.
[132] *Ibid.*

Offenses and Penalties

National law in all countries includes some penalty provisions. In some countries, biosafety legislation relies on that existing law,[133] whereas in others the biosafety framework itself contains numerous, specific, and separate statements of specific offenses, penalties, and other forms of redress that apply in the case of LMO violations.[134]

Some consideration of the penalty provisions in biosafety law can be instructive. For example, Cameroon's biosafety law includes provisions potentially affecting the importer's/user's liability:

> In the event of accidental release of genetically modified organisms resulting in an adverse impact on human, animal or plant health or on the biodiversity and the environment, and that could have been prevented in accordance with the criteria laid down by the competent national administration or the [national biosafety agency], the user, persons or institutions informed of such accidental release shall immediately pass over such information to the competent national administration, indicating the place where the said release occurred, detail [*sic*] information on the actions taken and the authorities notified.[135]

Other countries base their penalty legislation on a specific list of duties or offenses.[136] Brazilian legislation, for example, lists twenty-seven administrative offenses that are subject to penalties, including "Carrying out activities or projects involving GMOs and their derivatives . . . in scientific research, in technological development and in industrial production" either "as a self-employed individual" or "without authorization from [the authorized agency] or in breach of its rules." Other administrative offenses include impose a penalty on the importer or other in-country user of LMOs for their failure to bind "the legal entity that finances or sponsors activities . . . involving GMO and derived products."[137] This offense has a clear objective of ensuring that the country has the legal ability to take action against at least one person or entity having "deep pockets" – able to take financial responsibility for offenses leading to serious harm or significant levels of financial liability. Other such offenses include failure to maintain records and/or to submit required reports about GMO-related activities and accidents as required by law; destruction or discarding or other release GMOs into the environment without permission and/or in violation of relevant rules; violating the rules and permits governing contained use,

[133] See e.g., *Philippines EO 514*, *supra* note 51 at s 6. *Decree 5591*, *supra* note 125 at title VIII, art 68.

[134] The identification of illegal activities relating to GMOs and biosafety is straightforward: many countries list objective activities that will be considered crimes and/or misdemeanors and specify particular penalties. See *Korea LMOs Act*, *supra* note 15 at arts 38–44; translated in *Korea Draft NBF*, *supra* note 15 at 205–10. See also *Ethiopia Draft Proclamation*, *supra* note 20 at ss 31 and 32. See also *Namibia Biosafety Act*, *supra* note 40 at s 47; and *Brazil National Biosafety Law*, *supra* note 45 at title VIII, arts 24–9.

[135] *Cameroon Safety Regulations*, *supra* note 39 at s 28(1).

[136] See *Brazil National Biosafety Law*, *supra* note 45 at art 7.

[137] *Ibid* at art 6; *Decree 5591*, *supra* note 125 at title VIII, art 69.

commercial release of GMOs, and/or natural or recombinant DNA/RNA; "[u]se, trade, register, patent or license genetic technologies that restrict use"; and failure of a private institution involved in GMO research to set up an internal commission to address GMO issues and otherwise to comply with rules governing such institutions and rules on worker safety and accident response.[138] This list includes general offenses as well, including "[g]enetic engineering on living organisms in breach of the rules of this Decree," and "[p]roduction, storage, transportation, commercialization, importation or exportation of GMOs and their derivatives, without authorization or in breach of the rules . . . and the registration and oversight bodies and agencies."[139]

Sanctions and penalties vary widely across the BCH, both in type and severity. In Brazil, sanctions enunciated in the event that an offence is committed and the offender apprehended include several different potential legal actions:

> [W]arning; fine; seizure of the GMO and its derivatives; suspension of sales of GMOs and its derivatives; embargo of the activity; partial or full interdiction of the establishment, activity or undertaking; suspension of the registration, license or authorization; cancellation of the registration, license or authorization; loss or restriction of fiscal incentives and benefits granted by the government; loss or suspension of right to receive financing from official credit establishments; state intervention in the business; [and] prohibition of entering into contracts with the Public Administration, for a period of up to five (5) years.[140]

Penalty and sanction application often vary in severity according to whether the offense was petty, serious, or very serious, and the Decree includes multipliers and other requirements for repeat and continuing offenders.[141]

In many cases, particularly when biosafety legislation takes the form of regulations, guidelines, and other administrative instruments, it may not be legally possible to adopt separate offenses, penalties, or rights of redress.[142] In these situations, however, the legislation can provide special insight into the way that sanctions should be imposed on those who violate their permits and appropriate regulations. One example is found in the provisions of the Bangladesh Guidelines. Under the heading "Sanctions against violation of Biosafety guidelines" they discuss all types of sanctions that can be included in guidelines and recommend further application of general national criminal, civil, and administrative law:

> In addition to the revocation of the project approval, any violation of the provisions of this guideline or the concealment or withholding by the proponent of any

[138] *Ibid.*

[139] *Ibid.*

[140] *Ibid* at title VII, art 21; *Decree 5591, supra* note 125 at title VIII, art 70.

[141] *Brazil National Biosafety Law, ibid* at arts 22–23; *Decree 5591/2005, supra* note 125 at title VIII, arts 71–9.

[142] In many countries, the power to issue regulations sits with the ministry or agency. Often this power does not include the power to specify constitutional matters such as penalties and legal rights.

> information necessary to evaluate risks to human health or the environment shall be penalized by the concerned ministry by stopping the work immediately and forfeiting the government grants/funds.
>
> Further, any incentives that may have been granted the proponent or institution for contributing to advanced scientific or technological research and development will be withheld. These penalties are exclusive of any other penalties tenable by existing law. It means that until and unless specific acts and rules pertaining to Biosafety of GMO/LMO is [*sic*] in place, any activity undertaken in violation of Biosafety guidelines shall also be treated under the penal provisions of existing law, including, but not limited to, civil, criminal and administrative liabilities for gross negligence.
>
> The [National Technical Committee on Biosafety] will inform the government on all issues pertaining to violations of these guidelines by any organization. The Ministry will, if necessary, issue public statement [*sic*] on any such violation to caution other organizations.[143]

Another penalty/liability approach focuses on the duty to take remedial action. This approach is found in an Indonesian instrument that considers another type of offense: "When [an LMO] causes biosafety and food safety harm, then in that case the person or legal entity who or which has obtained approval for the utilization of [that LMO] is obligated to participate in the control and overcoming [*sic*]."[144]

In some countries, penalty provisions also address the failings of government officials. For example the draft biosafety law in Cambodia provides that: "Any environmental inspection official or agent who is negligent, fails to pay attention to, or fails to comply with rules and regulation of the Ministry of Environment or conspires with a violator or facilitates the commission of violation, shall be subject to administrative sanctions or faces prosecution before the court."[145]

One recurring penalty-related challenge for many developing countries relates to the size and type of penalties assessed, seeking to ensure that developed country violators will be penalized in a manner that provides an appropriate deterrent, whereas those from developing countries are not subjected to an inappropriately high fine. The penalty provisions in the biosafety law of Cameroon address this challenge by specifically authorising officials to engage in "settlement" negotiations regarding LMO-related penalties:

> The competent national administration in charge of biosafety shall have full powers to work out a compromise. To this end, the accused must refer the matter to the authority concerned.... The amount of money paid as settlement shall be determined in consultation with the authority in charge of finance. The said amount shall not be less than the minimum amount of the corresponding penal fine.... Under

143 *Biosafety Guidelines of Bangladesh*, *supra* note 17 at 3.3.
144 *Joint Decree on Biosafety and Food Safety*, *supra* note 117 at art 43.
145 *Cambodia Draft Law*, *supra* note 20 at art 41.

> pain of nullity, the settlement procedure shall be carried out before any possible court proceeding. . . . The method of collecting and allocating the proceeds of the settlement shall be determined by regulation.[146]

Finally, two of the most unique offense/penalty provisions in the BCH are found in Ethiopia's draft biosafety legislation:

- Any transaction of a genetically modified organism or product thereof with the intention of causing harm to human health, biological diversity, the environment, or property, or with the intention of any other hostile act, is prohibited.
- Any transaction of a genetically modified organism or products thereof with the intention of peaceful application which causes significant risk to human health, biological diversity, the environment, or property which has not been given an Advance Informed Agreement in writing by the Authority is prohibited.[147]

That instrument focuses its primary provisions around *transactions* – a term that is defined to include "any research and development, import, export, transit, handling, release, contained use, transport, placing on the market, use as a pharmaceutical for humans or animals, use as food, feed or for processing of any genetically modified organism or products thereof." No transaction would be permitted unless it complies with the Proclamation.[148]

Obligations to Report

One important aspect of liability provisions is their relation to reporting obligations. As most national environmental officials know, the most fruitful source of information on environmental risks is frequently the permit holder, local resident, or private sector, who becomes aware of newly discovered risks long before they are broadly known to the government. Some countries recognise this by imposing a duty on permit holders to disclose new information regarding risks when it becomes known to them:

> In the event new information becomes available indicating that the regulated article could pose significant risks to human health and the environment, the applicant shall on its own report to [locally relevant agency] and immediately take measures necessary to protect human health and the environment.[149]

[146] *Cameroon Safety Regulations*, *supra* note 39 at s 65.

[147] *Ethiopia Draft Proclamation*, *supra* note 20 at s 6.

[148] *Ethiopia Draft Proclamation*, *supra* note 20 at ss 1(21) and 7. A similar approach is taken in Namibia, whose law focuses on "dealing with GMO or GMO product" (*deal with* being defined as carrying out any activity defined in section 3 with a GMO or a GMO product); see *Namibia Biosafety Act*, *supra* note 40 at ss 1 and 20.

[149] This is one of many such examples; this exerpt is from *Philippines* AO 8, *supra* note 27 at s 12(F)(8).

This obligation is described in Brazil more specifically:

> The following activities are mandatory:
> I – Investigation of accidents taking place during genetic engineering research and projects and submitting a report thereon to the competent authority within a period of 5 (five) days starting from the date of the event;
> II – Immediately notify [the Commission] and public health, farming and environmental protection authorities of any accident that may lead to the dissemination of GMOs and their derivatives;
> III – Adoption of the means necessary to fully inform [the Commission], public health, environment and farming authorities, the community and employees of the institution or company on the risks to which they may be exposed, as well as on procedures to be followed in the case of accidents with GMOs.[150]

In Cambodia, draft legislation would place an affirmative duty on the user, importer, researcher, and/or introducer to give information leading to reopening or rescission of this permit. In particular, permit holders are obligated to "report immediately to the Ministry of Environment when the operator becomes aware of new scientific information indicating that these activities or operations and/or the LMOs involved may adversely affect the conservation and sustainable use of biodiversity, taking into account risks to human health."[151]

Reporting requirements can be immensely useful to the regulating agency, as a practical matter. Where they function well, they increase the agency's knowledge of actual events, conditions, and discoveries, improving their ability to do their job. There is one other characteristic of reporting laws, however, that can inhibit their usefulness. In many countries, the permit holder who provides information under a reporting obligation ultimately can be required to take action, pay a penalty, or assume liability for the action, condition, event, or risk that they reported. Reporting thus might be decidedly harmful to him or her, legally and financially.

The law of Cameroon also indirectly refers to the application of national tort, civil responsibility, and other liability provisions, by stating, "The fact that the competent national administration is informed shall in no way relieve the user of any obligation binding him, by virtue of ordinary law or duty, to give notice to persons likely to be affected."[152]

As a consequence, agencies might be unable to rely on reporting as a source of safety information without legislating special protection for self-reporting permit-holders. A few countries have specifically considered this issue, apparently opting for information over liability, including a specific provision regarding the user's freedom from responsibility in the event of an accidental release within the country,

[150] *Brazil National Biosafety Law*, *supra* note 45 at art 7.
[151] *Cambodia Draft Law* at art 5 in *Cambodia NBF*, *supra* note 20 at 33.
[152] *Cameroon Safety Regulations*, *supra* note 39 at s 28(2).

as long as he or she reports the event. One example is found in South Africa, which has biosafety regulations that specify the responsibilities of "the user concerned":

> to ensure that – (a) the registrar is notified immediately..., or as soon as possible thereafter,... and (b) all appropriate short-term, medium-term and long-term measures are taken to avoid or mitigate any adverse impact of such accident on the environment and human health.[153]

Public Participation and Awareness

Nearly every BCH instrument includes specific reference to public participation and awareness.[154] In practice, a few developed countries have suggested or implemented limitations of national processes for public participation to legislative and regulatory development processes (i.e., not required for individual permit approvals), out of concern that these requirements will engender undue delays.[155] The BCH, and particularly the Parties' commitments to post certain types of data, may result in legal consequences relating to confidentiality and the authority of individual persons or groups to post documents in the name of the country. This issue has not yet been addressed in any of the legislation reviewed. It is expected that all countries are well aware of national legislative and administrative practices in setting up agencies, boards, and commissions. The large volume of the BCH includes such legislation, often in detail, and is not further discussed here. Experience indicates that such programs are easily developed by each country to reflect its own national rules and practices, when the country finds a way to surmount the funding and technical challenges that are often obstacles to effective public participation.

Other Mechanisms

A few countries have specifically noted the need for legislative mechanisms to implement the Cartagena Protocol, which are not mandated by, or even mentioned in, the Protocol. A few of the most frequently arising mechanisms are described below.

Emergency Planning and Management

One mechanism that is specifically contained in several countries' biosafety legislation is the adoption of an emergency response plan to address the possibility of an accidental release of LMOs. One such provision is found in Cameroon's biosafety law, which requires that

> [b]efore introducing any genetically modified organisms or any activity related thereto into an open environment, appropriate measures and emergency response plans shall be put in place to properly manage accidents.... Response strategies

153 SA GMO *Regulations*, *supra* note 31 at s 7.

154 See, e.g. *Bahamas Draft NBF*, *supra* note 28 at paras 2.3.5–2.3.7 and 6.0; *Brazil Decree* 5591, supra note 125 at s VII.

155 UNEP-GEF Workshop on European experiences with LMOs, for developing country delegates, Bonn, Germany 2004.

> and emergency plans shall be implemented by all persons involved in the production, manipulation and marketing of genetically modified organisms, in collaboration with the competent services, in order to properly manage emergency cases resulting from the deliberate or accidental release of genetically modified organisms and products thereof in their possession. . . . In the event of a disaster or of imminent danger resulting from the deliberate or accidental release of genetically modified organisms representing a threat to human, animal or plant health, biodiversity and the environment, the competent national administration shall inform the authorities responsible for disaster management and the services involved. The said administration shall also provide advice on the appropriate emergency response strategies. . . . The competent national administration may suspend the activity, the importation/exportation of genetically modified organisms concerned, pending an investigation on the causes of the accident.[156]

Similar, less detailed provisions are found in other national legislation which provides a duty of every person or entity operating under an LMO permit or license to "ensure that" actions taken under that permission are "carried out in conformity with this law . . . by, among other things . . . [d]eveloping a risk management strategy,"[157] in particular where "LMOs are deemed to have had or possibly have had significant adverse effects on the public health and the conservation and sustainable use of biological diversity."[158] Other components of risk management that are also sometimes specifically required of users, importers, introducers, and others may include the following:

- Establishing mechanisms for internal monitoring of safety
- Taking measures to prevent an unintentional introduction of LMOs into the environment
- To respond to and mitigate any harm to biodiversity and human health when unintentional introduction into the environment occurs

Namibia's Biosafety Law includes a similar provision, leaving the need for such a plan to the discretion of the authorities:

> Before any release or contained use of a GMO or GMO product is made or carried out, the Council may require the applicant to – (a) prepare an emergency plan for the protection of human or animal health or the environment outside the area of release or contained use in the event of an unintentional release and ensure that the appropriate emergency services are aware of the potential hazards and informed thereof in writing; (b) provide information on safety measures and procedures to be adopted in the case of an unintentional release to persons likely to be affected by the release.[159]

156 *Cameroon Safety Regulations*, *supra* note 39 at s 37(1).
157 *Cambodia Draft Law* at art 5 in *Cambodia NBF*, *supra* note 20 at 32.
158 *Korea LMOs Act*, *supra* note 15 at art. 27; *Korea Draft NBF*, *supra* note 15 at 96 and 198.
159 *Namibia Biosafety Act*, *supra* note 40 at s 40(1).

This provision is supported by further detail regarding the steps that the permit holder must take in the event that an emergency release occurs.[160]

Currently, there are significant international discussions, projects, and processes focused on disaster prevention, minimisation, and management, which can be relevant and useful in this process.[161] These combine consideration of both natural and human-caused disasters, and focus on local or regional vulnerabilities, as well as climate-change adaptation, food security, and other issues. Biosafety and LMO issues can figure in national emergency/disaster planning of this type at several levels and in many different ways.

User/Facility Certification and Registry

Another innovation that is found in more than one country involves certification or registration of facilities, individual people, and/or entities involved in LMO research, development, production, or commercialisation within the country, rather than relying solely on individual permit/approval with regard to specific LMOs and/or introductions. These requirements, although particularly valuable for national implementation, are not among the registration mechanisms effectively required under the Cartagena Protocol or recorded in the BCH. The Cartagena Protocol system relies on the BCH to tell exporters and export countries whether a particular LMO has already been introduced in the import country (requiring export notification). Once the Cartagena Protocol is internationally functional, the minimal requirement imposed on countries of export (and through them, LMO exporters) will be notification of LMOs of which the import country has not allowed any prior international introduction. At the same time, most countries' ability to license facilities and certain types of technical activities have generally been recognized.[162]

A level of oversight of LMO-involved users, researchers, developers, producers, and commercialisers within each country, although completely unnecessary to the international mechanism, is particularly valuable to national implementation. As a result, countries have adopted different approaches to enable such oversight. For example, Brazil's biosafety legislation provides for the issuance of a Certificate of Quality in Biosafety to particular users, introducers, and developers by the country's biosafety commission. Such a certificate is issued "for development of activities with GMOs and their derivatives in laboratories, institutions or companies" and is copied to the relevant oversight authorities.[163]

Several countries have created registers and registries of LMO users and introducers. In some cases, the registry focuses on LMO users and developers that are

[160] *Ibid* at ss 41–2.

[161] The preeminent source of international deliberations and technical assistance on this topic is the UN's International Strategy for Disaster Reduction, online: http://www.unisdr.org/eng and http://www.preventionweb.net.

[162] The WTO's observations on "trade in services" and LMO controls have been less hostile than when addressing direct restrictions on LMO introductions.

[163] *Brazil National Biosafety Law*, *supra* note 45 at arts 14(XI) and 32.

otherwise outside the country's Cartagena Protocol framework. For example, Bolivian law calls for a registry of all users and producers (including agriculturists) of certain types of LMO soy that have been declared in that country to be specifically exempt from AIA and LMOs-FFP processes.[164] In this case, registration is essentially a legal declaration to comply with specified requirements, and the registrant is subject to sanctions if those requirements are breached.[165] The Cameroonian law also includes a registry of "LMO importers and commercial agents,"[166] whereas South Africa's law calls for registration at the level of "facilities." Under the latter, such registration applies to "any place where organisms are being genetically modified under conditions of contained use."[167] The certificate of each facility shall be essentially a passport for facilities. The document must be maintained on the facilities and provided to inspectors on request. This approach is also found in Namibia[168] and Indonesia.[169]

The Republic of Korea takes a similar approach, calling for licensing of anyone "who wishes to develop LMOs or to establish and operate a research facility dedicated to tests involving LMOs," and specifically authorising revocation of such licenses, "where the said development or tests were found to have or possibly have adverse effects on public health and the conservation of biological diversity."[170] The Korean draft NBF explains this approach:

> It is important to ensure safety of a research facility, but if it is too strictly regulated, it could impede the development of the bio-industry. Since test and research of LMOs bring about bio-technology, too strict a regulation on research facilities may result in restricting the research & development of biotech itself. There are no provisions on research facilities in the Protocol. The approval for the establishment and operation of research facility has no direct relationship with the transboundary movements of LMOs. . . . It is necessary to identify the safety on research facilities of LMOs in the domestic Act. LMOs are developed and experimented innumerably in research facilities. In doing so, it is inevitable to produce unintended LMOs more than intended LMOs. Unintended LMOs are released to the environment as waste or through wastewater. LMOs that are discarded and released unintentionally in the course of research may survive and do unintentional damage to nature. The aim of regulating a research facility for LMOs is to prevent adverse effects resulting from such a situation.[171]

[164] *Bolivia Resolución Multiministerial*, *supra* note 19 at arts 2–3. See also declaration regarding glyphosate-resistant soy (event 40-3-2) in Bolivia, *Resolución Administrativa No 044/2005*, 2005, BCH Record ID 12010.

[165] *Ibid*.

[166] *Cameroon Safety Regulations*, *supra* note 39 at s 50.

[167] *SA GMO Regulations*, *supra* note 31 at s 4.

[168] *Namibia Biosafety Act*, *supra* note 40 at ss 27–29 and 39(1)(b).

[169] *Regulation of Indonesia on Biosafety*, *supra* note 26 at arts 20 and 35.

[170] *Korea LMOs Act*, *supra* note 15 at arts 22–23; *Korea Draft NBF*, *supra* note 15 at 195–196.

[171] *Korea LMOs Act*, *supra* note 15 at arts 23–24; *Korea Draft NBF*, *supra* note 15 at 94–95 and 154–55.

Most of the recommendations expressed by Korea's draft NBF have focused on improvement of this facility-registration system, rather than particular matters specified in the Protocol.[172]

Conflicts of Interest

Throughout the LMO and biosafety discussions, the question of individual interests has been raised as a means of forestalling final acceptance of particular decisions on LMO-related matters, but has not been addressed in the Protocol. In national legislation, provisions that seek to ensure that an LMO-related decision is not challenged on a claim of "conflict of interest" have been rare. Ultimately, the qualifications most frequently considered in determining whether a person is being objective or should be seen as "interested" relate to educational and experiential characteristics, most readily available within LMO development industries and institutions. At the same time, affiliation with those institutions alone is sometimes cited as a justification for dismissing a particular expert statement.[173]

In light of the general level of controversy relating to LMOs, it is somewhat surprising that few Cartagena Protocol implementing laws raise the question of decision-maker conflict of interest. As argued by others in this book, good governance can be a principle of international law on sustainable development.[174] This arises from the existence of other laws within many countries that already apply to prevent such conflicts. One biosafety law containing a conflict provision is South Africa's. It requires that "a person appointed to the Committee shall immediately recuse himself or herself as a member of the Committee if a subject matter is in issue in which he or she has any direct or indirect interest or if, for any other reason, there is or there is likely to be a conflict of interest as a result of his or her participation in the proceedings of the Committee."[175] This interesting provision suggests that national cases and regulations could offer useful insights into how to characterize a "conflict of interest" in the context of biosafety.

Such characterization might be more challenging for countries in which "subsequent imports and introductions" (i.e., after the first international introduction) will be given different treatment or exemption, regardless of whether a different person or entity is seeking to import or introduce the species. Under many national commercial laws addressing conflicts of interest, the question of whether the recused official's interest is in the particular applicant or the sector is often raised. Namibia's biosafety law includes more specificity on this point:

[172] *Korea Draft NBF*, *ibid* at 155–158.

[173] The challenge of finding staff that would be perceived as "unbiased" was faced courageously by the UNEP-GEF Biosafety Project. Its success in achieving that goal was challenged, but its endeavour cannot but be admired.

[174] See Introduction of this book.

[175] *SA GMO Act*, supra note 81 at s 13. A similar provision is included in s 19(2) in reference to boards hearing appeals of LMO decisions.

> If a member of the Council or of a committee has a direct or indirect financial or other interest in any matter which is the subject of consideration by the Council or the committee and which may cause a conflict of interests in the proper performance of his or her duties as such a member, he or she must disclose that interest to the Council or to the committee as soon as is practicable after the relevant facts come to his or her knowledge.... A person who has such an interest... (a) if he or she is present at a meeting of the Council or the committee at which the matter is to be considered, must disclose the nature of his or her interest to the meeting immediately before the matter is considered; or (b) if he or she is aware that the matter is to be considered at a meeting of the Council or the committee at which he or she does not intend to be present, must disclose the nature of his or her interest to the chairperson of the Council or the committee before the meeting is held.... A member of the Council or of a committee who has an interest... must not (a) be present during any deliberation; or (b) take part in any decision, of the Council or the committee in relation to the matter in question.... A disclosure by a member under this section must be recorded in the minutes of the relevant meeting of the Council or committee.... A member who knowingly fails to comply... commits an offence.[176]

Linkage to Other Laws and Legal Frameworks

The BCH legislative database includes many "other laws" that are encoded with specific Cartagena Protocol related headings (often "all functions pursuant to the Cartagena Protocol on Biosafety") but that are best described as "general laws applicable to all commercial and/or agricultural enterprises." The laws so posted include national or subnational EIA legislation, hazardous substance controls, worker safety provisions, and other instruments, without any particular reference to LMO and/or biosafety matters. In fact, the clear scope of the legislation often appears to exclude biosafety and/or species introduction issues completely.[177] In a few instances, however, national law in these areas specifically addresses LMO issues, as discussed later. These laws pose a more complex challenge that could arise in some countries that expect to address various aspects of LMOs under different authorities. In such countries, a permit given by another agency (i.e., other than the agency enforcing the LMO/biosafety law) can be treated as a final decision on import, so that LMO importers may seek such permits when they have doubts about their ability to obtain permission under the national biosafety legislation, or when time is tight and that process is expected to take too long. Such challenges arise rather frequently in countries in which ministerial and administrative authority is a basis for promotion

[176] *Namibia Biosafety Act*, *supra* note 40 at s 14.

[177] A detailed summary of national laws that could apply to LMO developers, producers, and marketers is found in Bangladesh, *Draft National Biosafety Framework*, 2006, BCH Record ID 43436 at 25–27 and 66–82.

or increased budgetary attention. In some countries, the potential for linkage challenges is significant, as in Chile, where ten different legal bodies can be involved in biosafety-related decision making.[178]

General Legal Issues

For purposes of dealing with LMO export countries and exporters, the most important point to include in national Cartagena Protocol implementing legislation should be that legislation's relationship to other laws of the country. The BCH contains examples of some of the most common legislative methods for addressing the possibility that another national law might be or appear to be relevant to LMO movements, uses, and/or introductions.

One such example is found in the Republic of Korea, which expressly denies any preemptive impact of its biosafety law on other laws:

> This Act shall apply to all activities, including import or export of living modified organisms, unless however there are specific provisions by other laws regarding their handling and safe administration.[179]

Later, this law appears to take a contrary position, stating that

> Living modified organisms whose import is approved, or a report on such an import is received by the head of the related central government authority pursuant to Articles 8 and 9, shall be deemed as approved for import by the Minister of Commerce, Industry, and Energy pursuant to Article 14 paragraph 2 of the Foreign Trade Act.
>
> [and that] Living modified organisms in which the importer or exporter has reported or notified to the head of the competent national authority pursuant to Article 15, paragraph 3 or Article 20, shall be deemed as approved for import or export by the Minister of Commerce, Industry, and Energy pursuant to Article14 paragraph 2 of the Foreign Trade Act.[180]

Another approach to legislative integration (or the lack thereof) that is commonly found in other countries, however, is inclusion of a *supremacy clause* – a self-supporting statement that this legislation shall prevail over other national laws to the extent that they both require or address a particular issue inconsistently. Cambodia's draft law includes two such provisions. The first is a relatively common form of supremacy clause: "Any provisions that are contrary to this law shall be considered as null."[181]

[178] *Bases para el marco nacional de bioseguridad de Chile, supra* note 47 at 37.
[179] *Korea LMOs Act, supra* note 15; *Korea Draft NBF, supra* note 15 at 86 and 184. See also *Philippines AO 8, supra* note 27 at 12(E) and *Joint Decree on Biosafety and Food Safety, supra* note 117 at 37.
[180] *Korea LMOs Act*, supra note 15 at art 16; *Korea Draft NBF, supra* note 15 at 191.
[181] *Cambodia NBF, supra* note 20 at 31 and 41; *Cambodia Draft Law, supra* note 20 at art 45.

The second type of supremacy clause is less direct and calls on the Ministry of Environment to "notify the Biosafety Clearing-House that this law shall apply with respect to any import of LMOs into the Kingdom of Cambodia."[182] In essence, this second approach seeks to provide a kind of estoppel by recording the preeminence of this law in an international forum in which Cambodia participates. At present, the Cartagena Protocol's legal elements (i.e., liability and compliance) have not been used in a way that would make such estoppel legally functional, but it could become so.[183]

Alien and/or Invasive Species

One cross-framework issue that has frequently been linked to LMO and biosafety frameworks is the national law governing the introduction of alien species and the eradication of such species when they are discovered to be "invasive." Interviewees noted that in many countries, national law on "alien species" would technically include LMOs (which are not native anywhere) unless LMOs are specifically exempt from that law's coverage. These issues have arisen primarily in island nations (one of the largest groupings of countries within the UN family), where invasive species pose a particular threat.[184] For many of these countries, national concern about LMOs is limited; their primary import/export policies are focused on specific trading partners and their needs. In some Small Island Developing States (SIDS), national agencies and legislation relating to alien and invasive species were specifically identified as base institutions and documents on which Cartagena Protocol implementation was to be grounded.

The linkage to invasive species issues is thought to be a tool for limiting or minimising LMO introductions; however, when the country's approach to introduction of alien species is permissive, this linkage can operate to reduce the number of LMOs that must go through the biosafety process. One important example is found in the Philippines, whose national law on "plant products derived from modern biotechnology" specifically links invasiveness issues and LMO-related biosafety:

> This order covers the importation or release into the environment of: 1. Any plant which has been altered or produced through the use of modern biotechnology if the donor organism, host organism, vector or vector agent belongs to any of the genera or taxa classified by [the Philippine Bureau of Plant Industry] as meeting the definition of plant pest or is a medium for introduction of noxious weeds; or any plant product altered or produced through the use of modern biotechnology which

[182] *Cambodia NBF*, *ibid* at 31 and 41; *Cambodia Draft Law*, *ibid* at art 43.

[183] See Chapter 11 of this book for more information on the Cartagena Protocol's Compliance Committee.

[184] See e.g. *Bahamas Draft NBF*, *supra* note 28 at paras 1.3, 2.2.3, 2.4 and 4.2.1–5. The proposals for new legislation centre on the term "unwanted organisms."

> may pose scientific risks to human health and the environment based on available scientific and technical information.[185]

This provision is specifically limited to LMOs derived from species that have been formally designated as "plant pests" and/or "noxious weeds," also including other plants developed through the use of modern biotechnology, "which may pose significant risks." Clearly, the law requires some reason to think that the LMO is dangerous to apply.

The combination of alien species and LMO concerns have sometimes also included integration of national law governing "access to genetic resources and equitable sharing of the benefits arising from their utilisation" or "ABS" and also might be tied to national programmes on climate change and adaptation.[186] As yet, the BCH and COP-MOP do not include any direct decisions relating to or resolving this potential overlap.[187]

Trade and Discrimination

Concerns about the application of international trade law to LMOs have loomed large in national legislative processes relating to biosafety. Relatively few countries have taken this issue into their national law, however. One that has done so is the Republic of Korea, whose primary biosafety law's approach to international trade law concerns has been described as follows:

> According to Article 12 of the [Trans-boundary Movements of Living Modified Organisms] Act, living modified organisms produced within Korea shall obtain risk assessment and the evaluation of the risk assessment. *By keeping equity between living modified organisms produced in abroad [sic] and those in Korea, we may prevent international conflicts and attain the purpose of the Act.*[188] [emphasis added]

Another approach taken by the Philippines effectively subordinates biosafety processes to international trade and environmental law:

> In accordance with international law, the national biosafety framework shall be implemented in a manner consistent with and mutually supportive of the international obligations of the Philippine, in particular its obligations under international trade and environmental law. Multilateral, regional and bilateral cooperation in implementing the national biosafety framework, in particular its sections on capacity building and financial resources shall be encouraged.[189]

[185] *Philippines* AO 8, supra note 27 at s 2.

[186] See *Bahamas Draft NBF, supra* note 28 at paras 2.2.4 and 4.2.5.

[187] Most SIDS (except those that are highly involved in the LMO sector) have posted, at most, a draft NBF in the BCH.

[188] See *Korea Draft NBF, supra* note 15 at 89–90. Article 12 requires those interested in producing LMOs to obtain approval from the relevant government authority.

[189] *Philippines* EO 514, *supra* note 51 at para 2.11.

Codex Alimentarius

In 2000, the Codex Alimentarius (a joint body under the UN Food and Agriculture Organisation and the World Health Organization) recognised the potential value of its general process focused on development of internationally accepted food-safety standards in regard to biosafety. To internationalise the movement of "biosafe" LMOs, including especially GM food, the Codex created a special team mandated to establish safety protocols for genetically modified foods. The basic product of its work was produced in June 2003.[190] Many countries are expected to set up organisations for the supervision of GM foods based on the safety protocols established by the Codex. A similar team under the OECD was created in 1999, with the goal of further spreading the application of the "concept of substantial equivalence" across the world.[191]

For many countries, the Codex and its standards constitute a major element of national commercial law, providing a basis for controlling and supporting production of foods and plants for international trade.[192] Standards of the type mentioned earlier have been seen as a major tool enabling countries to comply with the Cartagena Protocol, while engaging in agriculture and other biological production activities.

LMOs and Waste Disposal

A few countries have specifically discussed the concept of LMOs as "hazardous waste," indicating, sometimes indirectly, how and to which extent national regulations of hazardous waste will apply in the event that LMOs are disposed of or unintentionally released. One example, found in the biosafety law of Cameroon, states that:

> The management of waste resulting from research and development, the manipulation and marketing of genetically modified organisms shall be in compliance with the provisions of the laws in force. . . . Waste and contaminated effluents containing genetically modified organisms shall be inactivated using approved means before the final discharge. Waste disposal shall be in conformity with the laws in force. Gas and other toxic emissions originating from facilities which use genetically modified organisms shall be treated before being released into the environment.[193]

[190] *Report of the Seventh Session of the CODEX ad hoc Intergovernmental Task Force on Foods Derived from Biotechnology*, UN Doc FAO/WHO ALINORM 08/31/34.

[191] The outputs of OECD's Task Force for the Safety of Novel Foods and Feeds are online: http://www.oecd.org/document/63/0,2340,en_2649_34391_1905919_1_1_1_1,00.html.

[192] T Young et al, *Independent Evaluation of FAO Corporate Strategy B-1 (International instruments concerning food, agriculture, fisheries and forestry, and the production, safe use and fair exchange of agricultural, fishery and forestry goods)* (Rome: FAO-PBEE, 2008) at 46–52.

[193] *Cameroon Safety Regulations*, *supra* note 39 at s 38–39.

Biosafety regulations in South Africa, presumably referring to LMO-related wastes, specifically discuss "waste disposal" issues as an affirmative responsibility that can extend beyond the user:

> (1) *Any person who possesses or controls waste* shall manage and dispose of such waste so that the waste does not have any negative impact on the environment and human health.
> (2) The person mentioned in subregulation (1), shall comply with all relevant national, provincial and local authority legislation in his or her management and the disposal of the waste in his or her possession or under his or her control.[194] [emphasis added]

The draft biosafety legislation in Ethiopia addresses this issue with regard to imported material:

> Where imported genetically modified organisms or products thereof remain unclaimed at the port of entry for more than two weeks, the Authority, together with the Customs Authority or other concerned government body, may take action to dispose of them in an environmentally sound manner.[195]

Bans on Import/Production

The possibility that a country or its agency may formally ban or prohibit LMOs has been the subject of broader legal analysis elsewhere. Although the issue remains open, national legislative drafters continue to consider the possibility that its biosafety process might identify one or more LMOs whose introduction or other use or movement in the country should be prohibited. One example is found in the Republic of Korea, where the law provides that:

> The head of the related central government authority may prohibit or restrict import or production of living modified organisms that fall into any of the followings subparagraphs:
>
> 1) That has, or is worried to have, adverse effects on public health and the conservation and sustainable use of biological diversity,
> 2) That which is produced through breeding with living modified organisms, and also fall into subparagraph 1) above, and
> 3) That has, or is worried to have, adverse socio-economic effects on the value of the country's biological diversity.[196]

[194] SA GMO *Regulations*, *supra* note 31 at s 9. Similar provisions are found in *Korea LMOs Act*, *supra* note 15 at art 19; *Korea Draft NBF*, *supra* note 15 at 91–2 and 194–5.
[195] *Ethiopia Draft Proclamation*, *supra* note 20 at s 7(4).
[196] *Korea LMOs Act*, *supra* note 15 at art 14; *Korea Draft NBF*, *supra* note 15 at 90 and 191.

In its defence against claims that such a prohibition would constitute trade discrimination, the Republic of Korea has specifically applied its biosafety law to both domestic and foreign-developed LMOs.

Coordination with Other Legal Frameworks

Most countries with posted legislation in the BCH appear to have included several laws that apply generally to biologically derived products, activities, and similar materials that move across national borders. Relatively few of these laws include any special provision for LMOs or other modern biotechnology.

One country that has made specific LMO provisions in a variety of such laws is the Republic of Korea, which addresses biosafety in its Korea Food & Drug Administration (KFDA) Regulation, Inspection Guidelines on Risk Assessment Documents for GM Foods and Additives (KFDA Notice), Sanitation Act (2002, as amended) and Ministry of Maritime Affairs & Fisheries (MOMAF) Regulation: Notice on Marking Items and Methods for Genetically Modified Fisheries Products (Notice of the MOMAF 2002), as well as its Biotechnology Support Act (2001).[197] Each of the relevant agencies has specific risk assessment and risk management responsibilities and is responsible to develop regulations and/or guidelines for that purpose. This approach is intended to address the necessity to broaden national legislative focus beyond agricultural activities: "Most of the LMOs that are imported into Korea are agricultural products. Therefore, risk evaluation of GM crops plays an important part in the risk management of LMOs in Korea."[198]

Pharmaceuticals and Medical Research

The Protocol's language about pharmaceuticals does not necessarily exclude all pharmaceuticals from its coverage, being limited to "pharmaceuticals for humans that are addressed by other relevant international agreements or organizations."[199] Countries that have followed the Protocol's lead in this area have generally used exactly the same language.[200] Neither the Protocol nor any of these laws has as yet given much guidance regarding this limitation, neither defining the scope of the term "pharmaceuticals" nor listing the "relevant agreements." Some of the most influential commentators to address the Cartagena Protocol have suggested that appropriate standards and definitions can be adduced, stating that

> LMOs that are pharmaceuticals for humans are principally genetically engineered vaccines (e.g., micro-organisms genetically modified to transmit the hepatitis B vaccine). In order to be exempt, such LMOs must be addressed by other relevant

[197] *Korea Draft NBF*, *ibid* at 14–17.
[198] *Ibid* at 67.
[199] *Cartagena Protocol*, *supra* note 1 at art 5.
[200] *Cambodia Draft Law*, *supra* note 20 at art 1; *Cambodia NBF*, *supra* note 20 at 31.

international agreements or organizations. The principal relevant international organization in this area appears to be the World Health Organization.[201]

In Cameroon, the law specifies the country's decision to extend its coverage to pharmaceuticals in furtherance of Cartagena Protocol Article 5 extends to particular categories of LMO use but does not consider whether they are regulated internationally:

> Recombinant-DNA vaccines and other pharmaceutical products manufactured through genetic modification and marketed on the national territory shall be subjected to the same safety norms provided for in this law. . . . Recombinant-DNA products and other imported pharmaceutical products shall be quarantined at entry ports until samples which shall be tested by the competent national administration shall prove that the said products are not dangerous, before they are placed on the market. In the absence of any proof of danger, the competent national administration shall, in collaboration with the other services involved, take the responsibility to authorise the release of the products. Consequently, the manufacturer shall set up strategies and ensure the follow-up of the products, in order to fully guarantee their safety to human and animal health as well as to the environment.[202]

In other countries, discussions have specifically considered the questions of whether this exception should apply to traditional and herbal medicines, and whether products designed to address concerns such as baldness, weight loss, wrinkles, and skin pigmentation can (or should) be considered "pharmaceuticals" for this purpose.

Access and Benefit Sharing

Another interlinkage that is sometimes included in national policy documents and some other biosafety instruments is access to genetic resources and equitable sharing of the benefits arising from their use (also known as "access and benefit-sharing" or ABS) under CBD Article 15. This issue is often linked to the rights of the holders of indigenous and traditional knowledge.[203] ABS is generally focused on the "other end" of the LMO process – the identification of genetically valuable species, which ultimately might be developed into various products of genetic modification (e.g., new agricultural varieties, medicines, cosmetics). Many of the countries most involved in or interested in ABS have linked the two frameworks, whether rhetorically or actually. It is not yet clear whether and/or how this linkage will be expressed in national legislation. To date, discussions of ABS in NBFs have only described

[201] Mackenzie et al, *supra* note 100 at para 237.

[202] *Cameroon Safety Regulations*, *supra* note 39 at s 51–2.

[203] See, *e.g.*, *Bahamas Draft NBF*, *supra* note 28 at 2.2.4–2.2.5, 2.3.2 and 8.2.5. Bolivia has posted ABS legislation in the BCH: *Reglamento de la Decision 391 del Acuerdo de Cartagena y de Bioseguridad*, 1997, BCH Record ID 309.

the concept; however, in most countries, legislation for both ABS and biosafety are being developed, and so the relationship could develop in future.

The national legislation in Cameroon, without fully discussing the relationship between the concepts, states the following: "With regard to genetically modified organisms perfected on the basis of genetic resources taken from the national heritage, the provisions of the regulations in force relating to access to genetic resources and sharing of benefits shall be applied mutatis mutandis."[204] This provision addresses only a limited number of potential ABS/Cartagena Protocol combinations – those that involve a foreign user gaining access to Cameroonian genetic resources, developing a commercially valuable GMO on the basis of them, and then proposing to export those products back to Cameroon.

In Chile, the relation between Cartagena Protocol implementation and ABS is seen as somewhat mutually exclusive, leading to recommendations that future work should focus on Chile's own genetic resources and not technologically enhanced materials:

> Que el impulso de la biotecnología, tema a ser considerado a futuro, conforme a la regulación que se propone, por instancias nuevas tales como el Comité de Regulaciones en Biotecnología y el Foro Biotecnológico, se canalice no sólo hacia el área agrícola, silvícola o acuícola, con el foco colocado en organismos transgénicos, que son especies comerciales e introducidas, sino también hacia el aprovechamiento de los recursos genéticos nativos, sin necesariamente recurrir a la transgenia. En la búsqueda de soluciones biotecnológicas, se puede conocer y valorar nuestro propio patrimonio genético, a la vez que desarrollar productos comerciales o industriales a partir de ellos.[205]

Similar provisions are found in several other instruments in the BCH.[206]

Other Provisions of Interest

Several other interesting clauses have been identified through the course of this research but are not so common in application that they are addressed in one of the earlier sections. The following are among the most interesting as issues for legal study and discussion.

Embryonic Stem Cells

Although this issue remains controversial in many countries, only one of the focus countries has linked it to biosafety. Brazil's primary biosafety legislation is designed around the concept of genetic manipulation and the use of genetic material. As

[204] *Cameroon Safety Regulations*, *supra* note 39 at s 54.

[205] *Bases para el marco nacional de bioseguridad de Chile*, *supra* note 47 at 27.

[206] See, e.g., *Decree on Conservation and Utilization of Plant Genetic Resources*, *supra* note 53.

such, it specifically includes detailed provisions about scientific and other research using stem cells.[207]

Innovations in Biosafety Administrative Frameworks

Brazil's biosafety framework imposes an additional affirmative administrative responsibility on "[e]very institution that uses genetic engineering techniques and methods or carries out research using GMOs and their derivatives." Specifically, each such institution "must set up an Internal Biosafety Committee, and also designate a responsible technical officer for each specific project."[208]

Post Facto Application of Biosafety Law to Previously Approved LMO Activities

Draft legislation in Cambodia includes a provision that might be impossible for developed country agencies to adopt. It would provide that "[t]his law shall also apply to LMOs related activities or operations covered by this law, which have been undertaken prior to the date on which this act shall come into effect."[209] A similar concern seems to have arisen in the UNEP-GEF project in Chile and was described as follows:

> La jerarquía de las normas y disposiciones enfocadas hacia la bioseguridad no supera, en general, el ámbito administrativo, lo que importa un cierto grado de inestabilidad jurídica, por tratarse de disposiciones que, llegado el caso, pueden perder vigencia más rápido que aquellas que emanan de leyes propiamente dichas. Esto no debiera ser óbice para que, con las debidas facultades y dentro de las pertinentes atribuciones, se dicten resoluciones específicas, para cubrir situaciones especiales no suficientemente abordadas por la norma legal y que pudieran requerir de una solución inmediata y ad hoc.[210]

Delisting Particular LMOs

Philippines legislation specifically focuses on "Delisting of Regulated Article" as opposed to the more common approach of identifying particular LMOs or activities that are exempt from the Cartagena Protocol processes. Delisting will occur only when "based on the nature of a regulated article and its use, the regulated article will not pose any significant risks to human health and the environment."[211]

CONCLUSION

This chapter has canvassed national legislative implementation of the Cartagena Protocol on Biosafety, providing practical indications of how many developing

[207] *Brazil Decree 6925*, *supra* note 18 at art 4.

[208] *Brazil Decree 5591*, *supra* note 125 at title VI, arts 61–62. See also *Normative Resolution 6*, *supra* note 29 at art 5.

[209] *Cambodia Draft Law*, *supra* note 20 at art 42.

[210] *Bases para el marco nacional de bioseguridad de Chile*, *supra* note 47 at 27.

[211] *Philippines AO 8*, *supra* note 27 at s 13.

country Parties are interpreting the Protocol's provisions and requirements. The level of involvement in commercial/scientific aspects of LMO development has a significant impact on the nature of each country's implementation of the Cartagena Protocol. In this chapter, national implementation was generally described as including (1) a policy or general policy approach to LMOs, (2) binding measures (laws, regulations, or other practices), and (3) other domestic legal and administrative systems. Evidence has indicated that many countries are advancing in this respect.

20

The Costa Rican Legal Framework on Agricultural Genetically Modified Organisms

Jorge Cabrera Medaglia

This chapter explains the main provisions of the Costa Rican legal framework regarding agricultural genetically modified organisms (GMOs) from the perspective of implementing the sustainable development objective, principles, and provisions in the *Cartagena Protocol on Biosafety*.[1] The chapter begins with a brief description of the governing body on environmental issues and then delves into special legislation related to GMOs. This section also reviews the regulatory framework, including the *Phytosanitary Protection Law*,[2] *Biodiversity Law*,[3] *Organic Law of the Environment*,[4] and *Law for Development, Promotion, and Encouragement of Organic Farming*,[5] among others. Particular claims made under the Administrative Environmental Tribunal (TAA) are described as well. Finally, this chapter provides some conclusions and recommendations.

The Costa Rican governing body for environmental issues is designated as the Ministry of Environment, Energy and Telecommunications (MINAET).[6] MINAET, as established by its governing law and the *Organic Law of the Environment*, must formulate, plan, and implement natural resources, energy, and environmental protection policies, as well as maintain the administration, control, promotion, and development of these areas. Article 2 of its governing law sets out the following functions:

[1] *Cartagena Protocol on Biosafety to the Convention on Biological Diversity*, 29 January 2000, 2226 UNTS 208, 39 ILM 1027 (entered into force 11 September 2003), online: http://bch.cbd.int/protocol/text/[*Cartagena Protocol*].

[2] *Ley* No 7.664 – *Ley de protección fitosanitaria*, La Gaceta No 83, 2 May 1997 [*Phytosanitary Protection Law*].

[3] *Ley* No 7.788 – *Ley de biodiversidad*, La Gaceta No 101, 27 May 1998 [*Biodiversity Law*].

[4] *Ley* No 7.554 – *Ley Orgánica del Medio Ambiente*, La Gaceta No 215, 13 November 1995 [*Organic Law of the Environment*].

[5] *Ley* No 8.591 – *Ley de desarrollo, promoción y fomento de la actividad agropecuaria orgánica*, La Gaceta No 155, 14 August 2007 [*Law for Development, Promotion, and Encouragement of Organic Farming*].

[6] *Ley* No 7.152 – *Ley Orgánica del Ministerio del Ambiente, Energía y Telecomunicaciones*, La Gaceta No 117, 21 June 1990 [*Organic Law of the Ministry of Environment, Energy and Telecommunications*].

- Encourage the development of natural resources, energy, and mining,
- Promote and manage legislation on conservation and sustainable use of natural resources, to obtain sustained development in this field and to ensure compliance with the law,
- Establish, by executive decree, binding rules and regulations regarding rational use and protection of natural resources, energy, and mining,
- Promote and manage legislation on exploration, distribution, protection, handling, and processing of natural resources within its competence, and ensure compliance with the law,
- Process and approve permits and concessions within its competence,
- Promote the signing of international treaties, conventions, and agreements, and represent the government on matters within its competence, and
- Develop and promote environmental educational programs at all educational levels and for the general public.

In the Costa Rican legal framework, the main rules in this area are the *Phytosanitary Protection Law* and the *Phytosanitary Regulations*,[7] and amendments. Before the promulgation of the *Phytosanitary Protection Law*, however, Costa Rica already had a legislative regime that regulated this issue, the *Plant Health Law*[8] and related legislation.

THE PHYTOSANITARY PROTECTION LAW AND RELATED REGULATIONS

Governing Bodies

Article 5 of the *Phytosanitary Protection Law* establishes the functions and obligations of the State Phytosanitary Service (SFE).[9] Section II of the *Phytosanitary Protection Law* clearly regulates GMOs. Article 20 of the *Phytosanitary Protection Law* created the Biosafety Technical Commission, to give advice on biotechnology matters to the SFE. The law also provides that individual persons or legal corporations that import, research, experiment with, transport, release into the environment, multiply, or commercialise transgenic plants or products thereof for agricultural use, whether produced inside or outside the country, shall obtain prior authorization from the SFE. GMOs are subject to the regulations, standards, and procedures laid out

7 *Decreto No 26.921/MAG – Reglamento a la Ley de protección fitosanitaria*, La Gaceta No 98, May 1998 [*Phytosanitary Regulations*].

8 *Ley* No 6248 – *Ley de Sanidad Vegetal*, La Gaceta No 101, 29 May 1978 [*Plant Health Law*].

9 *Decreto No 35.668/MAG – Reglamento de la estructura organizativa del Servicio Fitosanitario del Estado*, La Gaceta No 32, 16 February 2010, s 34 establishes the organisational structure of the State's Phytosanitary Service. It establishes a Safety and Environmental National Office with a Safety Regulation and Environment Department, and a Biotechnology and Biosafety Office within this department.

in Article 41. The SFE may modify or revoke any authorization granted based on technical criteria,[10] as well as prohibit the transfer, research, experimentation, release into the environment, multiplication, and commercialisation of these organisms for the purpose of protecting agriculture, the environment, and human and animal health. Even though the *Phytosanitary Protection Law* regulates agricultural matters, there is an explicit concern for the protection of human health.

The *Phytosanitary Regulations* establish the functions and composition of the Biosafety Technical Commission (BTC) in Articles 111–6. The BTC was first created by Executive Decree in 1996 and designed to work under the supervision of the Ministry of Agriculture and Livestock.[11] The BTC subsequently gained independent legal status through the *Phytosanitary Protection Law*, which established that it would become an advisory office on biotechnology issues to the SFE.

The BTC's functions are to:

1. Advise public institutions on biosafety matters.
2. Advise the SFE and National Seed Office on the establishment and implementation of measures and technical procedures, and draft executive orders and regulations needed to regulate the importation, movement, experimentation, environmental release, multiplication, marketing, and use of transgenic plants and other GMOs.
3. Provide advice to official institutions responsible for issuing permits to import, test, mobilize, release into the environment, multiply, and sell plants or other GMOs.
4. Promote the dissemination of information and training on biosafety issues.[12]

The Biosafety Technical Commission's responsibilities go beyond giving advice in the agricultural field. For example, Article 111, clause 3 states that the commission may provide technical criteria to other institutions in the field of genetic engineering. Clauses 2 and 3 also refer to plants or other organisms without establishing limitations, as the *Phytosanitary Protection Law* does when it refers only to the agricultural field.

The BTC's composition was amended by Executive Decree No. 31.946 of July 9, 2004, as follows:

1. A representative of the Ministry of Science and Technology
2. Two representatives of the Ministry of Agriculture
3. Two representatives of the Ministry of Environment and Energy
4. A representative of the National Seed Office

[10] *Phytosanitary Protection Law*, *supra* note 2, s 42.

[11] *Decreto No 25.219/MAG/MICIT – Crea el Comité Técnico Nacional Asesor de Bioseguridad*, La Gaceta No 116, 19 June 1996.

[12] *Phytosanitary Regulations*, *supra* note 7, s 111.

5. Two representatives of the National Academy of Sciences
6. A representative of the Federation for Environmental Conservation
7. A representative of the Biodiversity Coordination Network[13]

The BTC has a president (who must be the representative of the SFE), a vice president, and a secretary.[14] Members of the Commission hold their positions for two-year terms and can be reelected. The institution that names the representative is free to remove him or her.[15] Decisions can be made by a simple majority of members. The legal quorum for meetings should be at least five members, two of whom should be government officials.[16]

The BTC seeks to gather people with established scientific experience and acumen, and credentials in scientific fields. For example, two of its members are from the National Academy of Sciences. There are no representatives from the Ministry of Public Health, however, despite the mention of human health considerations (an issue still unresolved) in the Cartagena Protocol.

Authorization to Work with GMOs

The *Phytosanitary Regulations* sets forth legal requirements for the release or trade of GMOs or products thereof for agricultural purposes, regulations regarding certification of release into the environment, and the procedure for granting such certification.[17] Authorization is granted on the understanding that commercialisation of modified organisms as food is not allowed. Thus, any natural or legal person who wishes to release into the environment or trade transgenic material or its products must have a phytosanitary certificate and must fulfil phytosanitary legal requirements for trade.

The Costa Rican legal framework requires the use of special forms known as "BIO" for release, transport, and so forth of GMOs. The information required in these BIO questionnaires is the same as that required in Annex 1 of the Cartagena Protocol on Biosafety. Questionnaire BIO-4 requires that any person who wishes to move GMOs in Costa Rica should notify the SFE. The SFE, jointly with the Biosafety Technical Commission, then sets up requirements and measures for the transportation of the transgenic material.[18]

To obtain the certificate of release into the environment, the interested party must submit an original BIO-2 questionnaire and two copies thereof to the SFE. The SFE will issue an approval or rejection of the request within thirty calendar

[13] *Ibid*, s 112.
[14] *Ibid*, s 113.
[15] *Ibid*, s 114.
[16] *Ibid*, s 115.
[17] *Ibid*, ss 117–8.
[18] *Ibid*, s 117.

days. If the information is incomplete, the service will request the presentation of missing information within sixty working days. In some exceptional cases, under the petition of the interested party, the SFE can extend this term. Once the information is complete, the SFE has sixty calendar days to submit its decision. The BTC will review the questionnaire, and once it give its approval, the service will grant the certification of release into the environment, in accordance with the BIO-3 questionnaire.

Both the certificate of release into the environment and the phytosanitary certificate for trade issued by these authorities are valid only for environmental release, trade, transport, research, or the reproduction of transgenic material. These certificates do not include the right to commercialise transgenic food products in Costa Rica. All field trials and demonstrations conducted in the country need the permission of the SFE and are under its supervision.[19]

Registration is required for all natural or legal persons who carry out activities specified in Article 40 of the law, as well as for those who wish to manipulate transgenic material or other organisms, and for any projects that include genetic manipulation of such materials.[20] Also required is a letter of commitment in which these persons assume responsibility for the treatment or destruction of the products (once the project ends) to avoid pollution[21] and for labelling the products in accordance with the BIO-5 Questionnaire.

The regulation sets out rules for location and storage of these products[22] and the supervision of testing fields. The National Seed Office, with the coordination of the service, must supervise location, storage, and testing field supervision in the case of genetically modified seeds.[23] Monthly and final reports are required regarding the characteristics and behaviour of the transgenic products and any emergencies that might have occurred.[24] Accidental releases must be reported to the SFE, which will take measures on the matter.[25]

The regulations establish rules for unexpected changes (e.g., when diseases or unforeseen environmental effects emerge), which should be reported within twenty-four hours after the event;[26] rules for internal transfers, in accordance with BIO-1 Questionnaire;[27] and an obligation to manage the packaging materials that go along with the organisms properly, to prevent dissemination.[28]

19 *Ibid*, s 118.
20 *Ibid*, s 119.
21 *Ibid*, s 120.
22 *Ibid*, s 121.
23 *Ibid*, s 123.
24 *Ibid*, s 124.
25 *Ibid*, s 125.
26 *Ibid*, s 126.
27 *Ibid*, s 127.
28 *Ibid*, s 128.

All legal persons with authorization shall inform the SFE of the arrival date of the transgenic product at its final destination or of failure to import.[29] In the case of a commercial product, the product must be labelled as genetically modified so that consumers are aware of its contents.[30] Because no commercial products have been introduced under this article, it is thus far undeveloped. Information provided for the records shall be confidential.[31] The request of authorisation must be published in a national newspaper and paid for by the applicant. The publication should contain a simple description of the project. Project approval must be published in the official journal, which also must be paid for by the applicant and must contain a simple description of the processes and their impacts.[32]

Any person who desires to introduce genetically modified products into the country for agricultural purposes should first inform the SFE. In the case of violation of this rule, sanctions established in Article 74 will apply. Article 74 reads, "Whoever, maliciously, causes harm to agriculture, the environment or human and animal health, imports, releases in the environment or commercializes transgenic vegetables or other [GMOs] or their products, biological control agents or other type or organisms for agricultural use, shall be sentenced to a prison term of three to ten years." This statement of criminal liability is of prime importance to potential users of GMOs. In addition, the *Biosafety Audit Regulations*[33] establishes a legal framework for agricultural biosafety audits that may be performed by third parties if they are duly accredited to the Ministry of Agriculture and Livestock.

Costa Rica's legal framework on genetically modified agricultural products is relatively modern. Issues such as labelling and publications of questionnaires are not common in other Central American legislation. Some aspects, however, are not covered, such as:

a) The legislation does not define the term *products* mentioned in the law, and therefore we cannot know the scope of it.
b) The legislation is less developed in the case of commercialisation of GMO and their products.
c) The applicability of the legal framework to GMOs for food, feed, and processing is uncertain.
d) Labelling is mentioned but not properly developed in the legislation.

Members of the Commission representing nongovernmental organizations (NGOs) have not yet signed the confidentiality agreement, although all other members

29 *Ibid*, s 129.
30 *Ibid*, s 131.
31 *Ibid*, s 132.
32 *Ibid*, s 133.
33 *Decreto No 32.486/MAG Reglamento de Auditorías en Bioseguridad Agrícola del Ministerio de Agricultura y Ganadería de Costa Rica*, La Gaceta No 142, 22 July 2005 [*Biosafety Audit Regulations*].

have. Protection of commercial secrets can be covered under the scope of the *Undisclosed Information Law*,[34] however, and this applies to members of NGOs. Moreover, the competent authorities' understanding is that as the Commission's role is an advisory office, it may decide the extent of the information given to them.

In addition, the *Undisclosed Information Law* and its amendments protect natural or legal persons' industrial and commercial secrets. Undisclosed information must meet the requirements of Article 2: that it be secret, that reasonable and proportional measures be adopted to keep it secret, and that it have commercial value. Applications related to transgenic plants, which usually meet these requirements, are protected by the law. The law forbids the use and disclosure of this information without the consent of the information holder or by any person who, because of his or her work or position, has access to information that is undisclosed and expressly deemed confidential. This obligation continues after the information holder leaves his or her position. Noncompliance with this law can lead to criminal sanctions. In the case of other competitors, unfair competition rules from the *Law for Protection of Competition and Effective Consumer Defence*[35] may be applied.

PROVISIONS OF THE BIODIVERSITY LAW

The *Biodiversity Law* sets out five provisions regarding GMOs in Chapter III under the title "Guarantees of Environmental Safety." These provisions establish that, to prevent and avoid damage, present and future, to human, animal, or plant health, or to the integrity of the ecosystems, the regulation of the law will set out the mechanisms and procedures for access to the elements of biodiversity for the purposes of research, development, production, application, release, or the introduction of exotic or GMOs into the environment.[36] The State is obligated to avoid any risk or danger that threatens the permanence of ecosystems. It should also prevent, reduce, or restore environmental damage that threatens life or deteriorates its quality. The civil liability of titleholders or people responsible for the management of the GMOs and any damage caused is set out in the *Organic Law of the Environment*, the Civil Code, and other applicable laws. Criminal responsibility is set out in the existing legal regulations.[37]

Any person who proposes to use GMOs created inside or outside Costa Rica in the agricultural sector for import, export, experimentation, research, transport, release into the environment, reproduction, or commercialisation must obtain prior

[34] *Ley No 7975 – Ley de Información No Divulgada*, La Gaceta No 12, 18 January 2000 [*Undisclosed Information Law*].

[35] *Ley No 7472 – Ley de Promoción de la Competencia y Defensa Efectiva del Consumidor*, s 17 in particular. La Gaceta No 14, 19 January 1995 at s 17 in particular.

[36] *Biodiversity Law*, *supra* note 3, s 44.

[37] *Ibid*, s 45.

permission from the SFE. All natural or legal persons, national or foreign, who carry out genetic manipulation are obliged to register with the Technical Office of the National Commission for Biodiversity Management.[38]

Any person may participate in the permitting process and give observations and submit documents in writing. They may also request the repeal or revision of any given permit.[39] Based on technical, scientific, or security criteria, the technical office may modify or repeal any given permit. In the face of imminent danger, unforeseeable situations or failure to comply with official arrangements, the office may retain, confiscate, destroy, or return the GMOs.[40]

The *Biodiversity Law* differs in some important ways from the provisions of the *Phytosanitary Protection Law*:

(a) It extends the scope of the protection to exotic species.
(b) It establishes that resolutions of the Biosafety Commission are binding.
(c) It gives the National Commission for Biodiversity Management (CONAGEBIO), which depends on the Ministry of Environment, Energy, and Telecommunications, the power to revoke permissions issued by the Service.[41]
(d) It requires registration by any natural or legal person performing genetic manipulation under the Technical Office.

The authority issued to a committee with less experience and expertise in the biotechnology safety field seems inadequate. The role of the registration issued under the technical office is not clear enough in accordance with regulations of the *Phytosanitary Protection Law*. Because the technical office started its operations in 2002 and because its primary role is related to access to genetic resources and benefit sharing, that office has not assumed responsibility for registration. The *Biodiversity Regulations*,[42] which address access to genetic and biochemical resources, have no provisions on biotechnology. Public officers from the technical office have indicated that the office is planning to jointly use the public registry of the Ministry of Agriculture and Livestock. Finally, it must be noted that MINAET is the focal point of the Convention on Biological Diversity (CBD).

[38] *Ibid*, s 46.
[39] *Ibid*, s 47.
[40] *Ibid*, s 48.
[41] The *Biodiversity Law* contains a contradiction, because although section 14 gives the CONAGEBIO the authority to revoke permissions and establishes that on this matter the administrative process shall be exhaustive, article 48 gives this same authority to the technical office.
[42] *Decreto No 34.433/MINAE – Reglamento a la Ley de biodiversidad*, La Gaceta No 68, 8 April 2008, s 96, transitory provision IX establishes that to develop the General Regulation of Chapter III of Biodiversity Law on guarantees of environmental safety, the CONAGEBIO should be supported by the Phytosanitary Protection Service and with experts in this area [*Biodiversity Regulations*].

PROVISIONS RELATED TO ENVIRONMENTAL IMPACT ASSESSMENT

According to the *Organic Law of the Environment*, any human activity that alters or destroys elements of the environment or generates residues or toxic or hazardous materials requires an environmental impact assessment (EIA), which must be executed by the National Technical Environmental Secretariat. Prior approval from this body is required to initiate activities, works, or projects. The laws and regulations indicate which activities, works, or projects require an EIA.[43] Subsequently, the *Organic Law of the Environment* regulates aspects such as the approval process, guarantees, and mandatory resolutions of the National Technical Environmental Secretariat, among others.

The *General Regulations on the EIA Procedure*[44] and amendments set forth which activities must undergo an EIA and which should follow other procedures. This depends on the specific conditions of each project. These assessments are classified in four categories: A, B1, B2, and C, depending on the possibilities of environmental impact. The regulation also establishes procedures and special documentation to obtain an Environmental Feasibility License, which is required to initiate activities.

Chapter VII of the *Biodiversity Law* refers exclusively to the EIA of the proposed project, which at the discretion of the commission's technical office shall be requested to undertake an EIA when it considers that they can affect biodiversity.[45] Table 1 in Annex IX of this book sets forth the guidelines for EIA, environmental hearings, and international notifications. In the particular case of GMOs, no legislation explicitly requires an EIA.

PROVISIONS OF SEED LAW NO. 6289

The *Seed Law*[46] created the National Seed Office as an institution connected to the Ministry of Agriculture and Livestock. The *Seed Law* is governed by the *Seed Law Regulations*.[47] Because one of the functions of the National Seed Office is the promotion and protection, improvement, control, and use of higher-quality seeds, rules and specific mechanisms necessary for circulation and trade are continually updated.[48] The Office's main objective is to promote and organise the production and use of superior quality seeds, as well as to intervene in all stages of the process.[49] The law applies to those species of seeds that are useful to mankind.[50]

43 *Organic Law of the Environment*, *supra* note 4, s 17.

44 *Decreto No 31.849/MINAE/S/MOPT/MAG/MEIC – Reglamento General sobre los Procedimientos de Evaluación de Impacto Ambiental (EIA)*, La Gaceta No 125, 28 June 2004.

45 *Biodiversity Law*, *supra* note 3, s 92.

46 *Ley No 6289 – Ley de Semillas*, La Gaceta No 7, 10 January 1979 [*Seed Law*].

47 *Decreto No 12.907-A – Reglamento a Ley de Semillas*, 31 October 1989.

48 *Seed Law*, *supra* note 46, s 1.

49 *Ibid*, s 2.

50 *Ibid*, s 3.

Regarding biosafety matters, some aspects of its mandate should be taken into account. First, the National Seed Office participates, as indicated previously, in the control of the release into the field of genetically modified seeds. Second, any natural or legal person who imports or exports seeds must register the importation or exportation of those seeds, including GMO seeds. Strictly speaking, however, approval is not an actual authorisation. To date, in coordination with the National Seed Office, hundreds of projects involving the field trials of transgenic products have been approved, as indicated in Tables 2 and 3 in Annex IX of this book.[51]

PROVISIONS OF THE ORGANIC FARMING LAW

The *Law for Development, Promotion, and Encouragement of Organic Farming* contains many regulations that refer to GMOs. These regulations are specifically designed to prevent genetic contamination.[52]

Article 21 in particular says that:

> [t]he State, through the Ministry of Agriculture and Livestock, shall determine by regulation the requirements and procedures that must be followed to prevent genetic contamination of local genetic resources by genetically modified organisms. The Ministry will execute the necessary control mechanisms to ensure compliance with the law, and will define and implement measures and will take actions to protect organic crops, such as containment areas and management plans, among others.

Public officials who fail to carry out the necessary controls to prevent a farm dedicated to organic farming from being contaminated by GMOs shall, jointly with the State, take responsibility for the damage caused, according to the law of General Public Administration.

Article 22 refers to the protection of organic production from the risk of contamination by GMOs and provides that:

> [w]ithout prejudice of controls established under the [*Phytosanitary Protection Law*], permits for planting, reproducing, exchanging and multiplying genetically modified organisms shall be granted by the Ministry of Agriculture, through the competent authority. To grant such permits and in the event that there is reasonable doubt, based on scientific and technical criteria, of the possibility of adverse effects that may be caused by the transgenic material in the application to organic crops in the area, the state should request the producer who asked permission to plant transgenic seeds for sufficient technical evidence to minimize the risk of such effects; this evidence will be technically evaluated for the authorization of the

[51] See also the website for the State Phytosanitary Service, online: http://www.sfe.go.cr/.

[52] There are precedents of claims in Canada (Class Actions) undertaken by organic farmers against genetically modified seed products on the basis of, among other reasons, the contamination of organic crops. See Chapter 21 for a more detailed discussion.

permit. As input to establish the necessary technical criteria, the procedure for granting the permit must comply with a non-binding consultation made by the authorities with individuals and organizations engaged in organic production and living in the area, and that have registered under the Ministry of Agriculture and Livestock.

In cases in which organic farms, or farms in transition to organic production, are exposed to a threat of contamination by GMOs, the Ministry of Agriculture must apply protective measures such as adequate physical barriers, containment areas and plans to manage, protect, and ensure the integrity of the area, and also to supervise the implementation of these measures. In all cases, if there is contamination, it must be documented in the records of the farm, and the organic farmer will communicate immediately with the certifier. Contaminated products should be identified and separated.

If there is proof of unauthorized production of transgenic crops in areas near organic production, officials of the Ministry of Agriculture should immediately take the measures necessary to prevent contamination of organic crops and the environment. They must obtain, within the due process framework, evidentiary material for potential lawsuits. For this purpose, the Ministry of Agriculture has the agency established in Article 42 of the [*Phytosanitary Protection Law*]. In such cases, the Ministry of Agriculture shall perform the necessary studies in order to dismiss or determine the damage caused to organic production.

The *Phytosanitary Protection Law* also prohibits the use, production, and testing of GMOs or transgenic organisms in organic farming.[53] Finally, under Article 33, it is considered to be a specific offence, punishable by a prison term of one to three years, to:

a) Use, with deceit or for profit, genetically modified organisms or their products in organic farms duly certified or in transition to become organic, in cases where a more serious crime is not committed.
b) Plant or produce genetically modified organisms without appropriate permission in areas which are devoted to organic farming as defined by the Ministry of Agriculture and Livestock in cases where a more serious crime is not committed.

The *Law for Development, Promotion, and Encouragement of Organic Farming* is regulated by the *Organic Farming Regulations*.[54] Chapter X of the decree sets forth protection rules related to GMOs, such as access and use of native seeds, pollution liability, measures to avoid pollution, measures to be taken in case of pollution, and measures to be taken in case of the detection of unauthorized GMOs.

53 *Phytosanitary Protection Law*, *supra* note 2 at art 31.

54 *Decreto No 35.242/MAG/H/MEIC – Reglamento de la Ley No 8.591, Ley de desarrollo, promoción y fomento de la actividad agropecuaria orgánica*, La Gaceta No 107, 4 June 2009.

LABELLING, TRACEABILITY, AND LIABILITY

There is no law that regulates labelling and traceability of GMOs, but a draft law was published in the official journal on 26 August 2003.[55] The bill aims to regulate labelling and traceability of genetically modified products to be used for human consumption and agricultural activities.[56] It defines which products are considered GMOs.[57] The bill does not apply in cases in which the content of GMOs does not exceed 0.9 percent of the food ingredients or products, considering them individually; if the food has only one ingredient; when the presence of the material is caused by an accident; or when it is technically unavoidable.[58] The bill's main objectives are to inform consumers through labelling, monitoring of GMOs in all stages of the process, and simplifying the recall of the products.[59] It contains legal principles such as precaution, consumer interest protection, and nonapplication of the principle of substantial equivalence.[60]

The bill establishes the power of the Ministry of Public Health, the Ministry of Agriculture and Livestock, and the Ministry of Economy, Trade, and Industry, in these matters.[61] In case of product registration, the applicant must submit, among other information, risk analysis, as well as information on sampling, identification, and detection methods, and safety and security measures.[62] The bill establishes a traceability system. This process requires that copies of the documentation shall be recorded in a Registry of Transactions. In all trade and distribution phases, the operator must give the information to the recipient.[63] To deliver the product to the end consumer, the labels and advertising of the product must state, where applicable: (1) that the product is a genetically modified organism; (2) that the product contains GMOs; or (3) that the product has been produced from GMOs.[64]

Offences are within the jurisdiction of the Ministry of Public Health, the Ministry of Agriculture and Livestock, and the National Consumer Commission.[65] With regard to liability, Article 31 of the *Phytosanitary Protection Law* sets forth responsibility for harm and damages: "Those who perform research or experiments, mobilize, release into the environment, import, export, multiply and commercialize vegetables or the organisms or products referred to in article 41, shall be responsible for damages caused to agriculture, the environment, human and animal health." The

55 *Proyecto de ley No 15.342 – Ley Sobre la Informaicon y la Trazabilidad de los Organismos Modificados Geneticamente*, La Gaceta No 163, 26 August 2003.

56 *Ibid*, s 1.

57 *Ibid*, s 2.

58 *Ibid*, s 3.

59 *Ibid*, s 4.

60 *Ibid*, s 5.

61 *Ibid*, s 7 *et seq*.

62 *Ibid*, s 14.

63 *Ibid*, s 6.

64 *Ibid*, ss 19–20.

65 *Ibid*, s 30.

Organic Law of the Environment refers to liability for damage in Article 98, stating: "the damage or pollution to the environment can occur because of an act or omission, and can be attributed to natural or legal persons who commit those activities." Article 101 states that without prejudgment of other responsibilities, natural or legal persons responsible for noncompliance with this law and laws that regulate the environment and biodiversity will have civil liability and will be responsible for the damage caused. Joint business owners and directors will be responsible for actions or omissions that cause damage.

Based on the above provisions, the Administrative Environmental Tribunal (TAA), a decentralized bureau of the Ministry of Environment, has determined that strict liability applies in terms of environmental damage. In the case of damage to the environment caused by GMOs, these principles should be applied, establishing a strict liability standard.

CLAIMS MADE BEFORE THE ADMINISTRATIVE ENVIRONMENTAL TRIBUNAL

On 27 February 2008, three complaints were made before the TAA by the representative of BASF COSTA RICA S.A. against Mr. Nelson Marchena and Ms. Marianela Villegas Bermudez,[66] Mr. Franklin Hidalgo Briceño and Miriam Badilla Quesada,[67] and Mr. Hansel Quiros Abellan.[68] The claims are based on the following facts: Through conventional breeding techniques, the company developed a production system called "Clearfield Arroz." This production system is divided into three parts: (1) the Clearfield Certified Seed, (2) the Clearfield Authorized Herbicide, KIFIX 70WG, and (3) the Guardian Program. The company claims that the accused persons planted these seeds without authorisation. Although these cases do not involve GMOs as such, their outcomes will illuminate the state of liability for environmental damage.

CONCLUSION

The *Phytosanitary Protection Law* and *Phytosanitary Regulations* develop the legal framework regarding GMOs in Costa Rica. BIO Questionnaires establish the specific information that should be submitted by the applicants. The existence of an advisory committee composed of members of the universities, through the National Academy of Sciences, allows the State Phytosanitary Service to have scientific support for their decisions. Up to this point the legal framework has been sufficient to process the applications submitted to the service. However, some consideration of GMOs should

66 Tribunal Ambiental Administrativo, Case No 50–08–02 TAA.
67 Tribunal Ambiental Administrativo, Case No 52–08–03 TAA.
68 Tribunal Ambiental Administrativo, Case No 52–08–01 TAA.

be included in the *General Public Health Law*,[69] and it is important to incorporate the Ministry of Health within the National Biosafety Committee.

The Costa Rican agricultural biosafety legislation, especially in the field of seeds, provides important references to public participation (through publication of applications and approvals), labelling, liability for damages, and other issues that are not very common in the legislation of other Central American countries. The legislation referred to previously focuses on GMOs for use in agriculture. The interpretation of the scope of these terms could extend the jurisdiction of the SFE to the assessment of GMOs used for food and processing, but practical situations have prevented this from happening. Documentation on the Ministry of Agriculture and Livestock's website points out that in the case of trade of GMOs, additional permission from the biotechnology management is required, and in this case the Biosafety National Commission would intervene. Authorities of the service also have the power to allow or prohibit the importation of plants such as vegetables and grains that are intended for consumption or industrial processing.

In practice, however, the legal regime for GMOs for food, feed, or processing is not yet clear. Control mechanisms of GMOs for animal or human consumption (beyond quarantines) are not in place. In the case of processed food, the Costa Rican legislation does not regulate cases in which food has been made with GMOs, because the safety of the products is ensured through sworn statements and origin certifications, especially in the case of products originating from the United States and the European Union. These foods could become health risks that have not been evaluated locally because these products can come from countries where labelling is not required.

To date, the Ministry of Environment and the Biosafety Commission have not issued any formal criteria about the environmental impact of GMOs on biodiversity, nor on whether these organisms require EIAs. This would be redundant, because process assessment and risk management undertaken by the Ministry of Agriculture and Livestock already exist in this field. Finally, clear and science-based rules on labelling of GMOs and products containing them should be established to inform and protect consumers.

[69] *Ley No 5395 – Ley General de Salud*, La Gaceta No 222, 24 November 1973.

21

Innovations in Biosafety Law in New Zealand

Frederic Perron-Welch

New Zealand approaches biosafety (known as *biosecurity*) from the perspective of a biodiverse and relatively isolated island state with many endemic species. There are also a number of introduced species, some of which have become invasive and threaten local flaura and fauna, others of which now form the basis of New Zealand's agricultural sector. From the perspective of sustainable development law, the country can be described as having adopted an approach of keeping its options open about biotechnology. This means that New Zealanders are not turning their backs on the potential benefits of biotechnology, but at the same time they are proceeding carefully, minimising and managing risks and encouraging the coexistence of different forms of agriculture.[1]

REGULATORY SYSTEM

Animals like cattle and sheep were introduced to New Zealand in the nineteenth century and quickly became important contributors to the country's economy. Because they are introduced animals, the country is free of many of the pests like scrapie and foot and mouth disease that plague other parts of the world. New Zealand has worked hard to stop these pests from affecting its herds and has a long history of

[1] The information in this section is drawn from Biosecurity Council, *Tiakina Aotearos Protect New Zealand: The Biosecurity Strategy for New Zealand* (2003); Royal Commission on Genetic Modification, *Report of the Royal Commission on Genetic Modification* (2001), online: http://www.gmcommission.govt.nz/RCGM/rcgm_report.html.

This chapter is the result of ongoing CISDL research on biosafety decision making and international sustainable development law. See Martin Endicott, Christine Frison, and Kathryn Garforth with Marie-Claire Cordonier Segger, Jorge Cabrera, and Sylvestre Manga, *Innovations in Biosafety Law: A CISDL Working Paper* (Montreal: Centre for International Sustainable Development Law, 2005). The author would like to thank Alexandra Keenan, LL.B. (Dalhousie) for her assistance in further developing this chapter.

measures designed to protect species from the arrival of pests and disease. In 1993, New Zealand passed the *Biosecurity Act*,[2] which is described as the world's first legislation designed to protect both native and nonnative species from the harmful effects of exotic pests and diseases.[3] The Act underwent a substantial review in 2010 and proposed amendments were approved in the areas of border risk management, readiness and response, and enforcement and compliance.[4]

In 1996, New Zealand passed the *Hazardous Substances and New Organisms Act*[5] (HSNO Act), which is meant to combine the management of hazardous substances and new organisms in one comprehensive piece of legislation. The purpose of the Act "is to protect the environment, and the health and safety of people and communities, by preventing or managing the adverse effects of hazardous substances and new organisms."[6] The country signed the Cartagena Protocol in May 2000 and ratified the agreement in February 2005. The HSNO Act is the primary instrument through which requirements of the Protocol are met.

An important development in the New Zealand regulatory system arose from the Royal Commission on Genetic Modification. This was a government-commissioned group that studied the issue of genetic engineering from May 2000 to July 2001. After the Commission released its report, the government instituted a two-year moratorium on applications to release genetically modified organisms (GMOs) to be able to implement the report's findings. This included releasing a biosecurity strategy in August 2003 and amending the HSNO Act in October 2003. The two-year moratorium expired on October 29, 2003. MAF Biosecurity New Zealand (MAFBNZ), the division of the Ministry of Agriculture and Forestry responsible for enforcing the HSNO Act's provisions on "new organisms," released a strategic five-year plan for 2007 to 2012.[7]

STRATEGIC VISION

The stated vision of MAFBNZ's strategic plan is that "New Zealanders, our unique, natural resources, our plants and animals are all kept safe and secure from damaging pests and diseases."[8] The plan takes its strategic vision from the 2003 Biosecurity

2 Biosecurity Act 1993, Public Act 1993 No 95.

3 Except where otherwise indicated, the information in this section is drawn from the Biosecurity Strategy, *ibid*; and New Zealand, Ministry for the Environment (MFE), *Your Guide to the Hazardous Substances and New Organisms Act* (Wellington, NZ: MFE, 2001), online: http://www.mfe.govt.nz/publications/hazardous/guide-to-hsno-act-jul01.html.

4 New Zealand, Office of the Minister for Biosecurity, "Biosecurity Act 1993: Approval for Amendments" (2010), online: http://www.biosecurity.govt.nz/biosec/pol/biosecurity-act-review.

5 *Hazardous Substances and New Organisms Act*, Act 30 of 1996 (HSNO Act).

6 *Ibid*, s 4.

7 New Zealand, *Strategic Plan* (MAFBNZ), online: http://www.biosecurity.govt.nz/biosec/org/plan.

8 *Ibid* at 2.

Strategy and envisions a broad leadership role for MAF, covering "economic, social, cultural, health and environmental outcomes."[9]

In the strategic plan, MAFBNZ committed to achieving seven goals: to maintain an effective biosecurity system, to make timely and informed decisions, to improve the effectiveness of border interventions, to improve biosecurity outcomes through collaboration among organisations, to involve all stakeholders in taking responsibility for biosecurity, to foresee and prepare for emerging risks, and to create an atmosphere that attracts and retains quality employees.[10] Specific actions include closing loopholes by identifying areas of accountability for MAFBNZ and partner agencies; establishing decision-making frameworks based on scientific principles, Maori values, and other stakeholder interests; as well as prioritising interventions and using a new business model to improve institutional effectiveness. Efforts will be made to promote greater industry involvement in both decision making and resourcing. Compliance and enforcement will also improve through a shift from primarily inspection-based border interventions to a more targeted approach. Many of these changes are reflected in the approved amendments to the Biosecurity Act.

PERMITTING MECHANISM

The HSNO Act is the primary piece of legislation regulating the introduction of new organisms into New Zealand. Under the Act, *new organisms* are defined to include species that were not present in New Zealand prior to 29 July 1998 and genetically modified species, among others.[11] To release a new organism into New Zealand, a proponent must receive the approval of the New Zealand Environmental Risk Management Authority (ERMA). In some cases, parallel approvals under other pieces of legislation, like the *Biosecurity Act* or the *Medicines Act*,[12] are also required.[13]

The approval process, which takes place at the proponent's expense, begins with the proponent filing an application with ERMA. An application should include a clear and comprehensive identification of the foreseeable risks, costs, and benefits of the organism in question. This means an assessment of risks that is appropriate to their significance (i.e., the higher the risk, the more detailed the assessment that is required); an assessment of how significant risks can be managed; and the potential adverse effects of a new organism on the environment and public health, including social and cultural effects. Applicants must show that it is possible to manage the environmental and health risks of a new organism effectively in containment before it will be allowed into New Zealand. ERMA bases its decision on the introduction of

9 *Ibid* at 6.

10 *Ibid* at 12.

11 HSNO Act, *supra* note 5, s 2.

12 Medicines Act 1981, Public Act 1981 No 118.

13 Except where otherwise indicated, the information in this section is drawn from the *Guide to the HSNO Act*, *supra* note 3.

a new organism largely on the evaluation of the risks, costs, and benefits contained in the application. Applicants must also consult relevant stakeholders, including Maori, as treaty partners, when preparing an application. ERMA takes into account the adequacy of the consultation between the applicant and affected stakeholders when assessing and deciding on applications. In addition, ERMA has a dedicated Maori focus group and receives input from a statutory advisory committee, the Nga Kaihautu Tikanga Taiao, on Maori perspectives, including Maori approaches to risk management.[14]

Different types of applications pass through different processing pathways within ERMA. There are four types of applications: standard notified applications, standard non-notified applications, rapid assessments, and ministerial call-ins. Standard notified applications are applications that must always be publicly notified and include applications to import a new organism for release (with certain exceptions), release a new organism from containment, and field test a GMO. Standard non-notified applications do not require public notification and include applications to import or develop in containment any new organism, including a GMO, or to field test a new organism that is not a GMO. ERMA retains the discretion, however, to notify these applications publicly if there is likely to be significant public interest.

Rapid assessments are used for applications that are considered low risk, including the importation of a new organism that is not a GMO. The application must meet specific low-risk criteria to fall into this category. Such low-risk assessments may be delegated to institutional biological safety committees at institutions such as universities and Crown research facilities.[15] Finally, ministerial call-ins involve the Minister for the Environment calling in an application that he or she thinks is of particular national or international importance. ERMA processes the application, conducts an inquiry, and prepares a report of recommendations to the Minister, who then makes the decision and notifies the public.

In consultation with a wide range of stakeholders, ERMA has developed and adopted a methodology for decision making under the HSNO Act. The criteria on which the methodology is based are specified in Part V of the HSNO Act. More specifically, ERMA must evaluate risks, costs, and, where applicable, benefits, based on the nature and, characteristics of the substance or organism; the applicant's assessment, and where applicable, proposals for managing the risks concerned; any submissions received; the reviews prepared by ERMA New Zealand or an appointed external expert; and matters arising at a hearing (if notification is public). When deciding on an application, ERMA also decides if any controls should be placed on the new organism. All decisions by ERMA are publicly notified. ERMA may

14 New Zealand, Environmental Risk Management Authority, "Nga Kaihautu Tikanga Taiao," online: http://www.ermanz.govt.nz/tehautu/ngakaihautu/index.html.

15 New Zealand, Environmental Risk Management Authority, "Institutional Biological Safety Committees and Low-Risk Genetically Modified Organism Decision Making," online: http://www.ermanz.govt.nz/no/applications/ibsc/index.html.

decline an application for containment when minimum standards are not met, adverse effects outweigh positive effects, not enough information to assess adverse effects is available, or the organism cannot be contained adequately.

The HSNO Act mandates the use of the precautionary approach, stating that "all persons exercising functions, powers and duties under this Act . . . shall take into account the need for caution in managing adverse effects where there is scientific and technical uncertainty about those effects."[16] The precautionary approach is not yet well established in New Zealand and can carry different weight depending on the context, even within the HSNO Act.[17] Amid this ambiguity there is a fierce debate over how the tension between scientific knowledge and other values should be resolved. Many of ERMA's membership and executive staff are scientists. Despite the opportunities for public input, outlined later, many members of the public, and in particular the Maori, criticise the idea of scientists making technical decisions on issues involving value judgments about social, environmental, and other concerns.[18]

The HSNO Act does provide for appeals of ERMA's decisions, but only on questions of law and not on technical or other assessment matters. Applicants may reapply for approval of a new organism if the initial application is declined. This is most likely to occur when significant new information on the environmental risks of the organism has come to light or when improved means of managing the risks have been developed. The Act also allows reassessment of new organisms approved for containment if certain criteria are met. These criteria include the availability of significant new information about the risks of a new organism, the availability of another substance with similar or improved beneficial effects and reduced adverse effects, and a significant change in the quantity of the substance being used or how it is being used. A reassessment involves reexamining the risks and benefits of a new organism and can result in ERMA's changing the conditions placed on an approval or even completely withdrawing approval. For a reassessment to proceed, ERMA must first decide whether there are grounds for a reassessment. The party seeking the reassessment must then prepare an application, which must include much of the same information as that required in the initial application for approval.

INFORMATION SYSTEM

The application and approval process for a new organism allows for significant public input.[19] As mentioned previously, the public must always be notified about standard notified applications and also might be notified about some standard non-notified applications. ERMA's public notification procedure includes publishing

[16] *HSNO Act*, *supra* note 5, s 7.

[17] Mitsuhiko Takahashi, "Are the Kiwis Taking a Leap? – Learning from the Biosecurity Policy of New Zealand" (2005) 24 Temple Journal of Science, Technology & Environmental Law 461 at 471.

[18] *Ibid* at 473.

[19] *Guide to the HSNO Act*, *supra* note 3.

notices in newspapers, on its Web site, and in *The Bulletin*, its official record. After public notification, people have thirty days to make submissions to ERMA on the application. Submissions can relate to various parts of ERMA's decision-making criteria, including the technical information in the application, cultural or ethical values, or other consultations with stakeholders.

Public participation in the New Zealand system includes public hearings on individual applications. ERMA will arrange a public hearing for a notified application if it considers such a hearing necessary or if either the applicant or a submitter requests a hearing. The purpose of a hearing is to allow all parties to be heard and for ERMA to understand the full range of issues at stake before making a decision. Most hearings are conducted by a committee of ERMA members, but committees may also include nonmembers with expertise in a particular area. All submitters have the right to be heard at a hearing. They can request a hearing in their application but may be required to assume the costs of the hearing if ERMA considers their grounds to be unfounded or frivolous. Submitters may attend the hearing or call a witness but are not obliged to do either. If they attend, they may be questioned by the chair and potentially by the other parties as well.

Finally, when an application undergoes public notification, ERMA keeps commercially sensitive information confidential, in accordance with New Zealand's Official Information Act. ERMA's prime objective, however, is to protect the environment and the health and safety of people and communities. Information is kept confidential only when it does not affect the validity and integrity of any information about possible risks and adverse effects of new organisms.

MONITORING AND INSPECTIONS, ENFORCEMENT, AND COMPLIANCE

Part VII of the HSNO Act outlines the provisions on compliance, inspection, enforcement, liabilities, and defences. Compliance includes both voluntary compliance through self-responsibility, codes of practice, industry standards and guidelines, and mandatory compliance through inspections, enforcement, and compliance with the controls on an approval. ERMA is responsible for monitoring and inspections. These duties include monitoring compliance with and enforcement of the Act; enquiring into any incident or emergency involving a new organism; advising and assisting the Minister for the Environment on international issues and the Act's effectiveness; reporting to Parliament annually on incidents caused by inadequate management of new organisms, and the extent to which the Act has contributed to the health and safety of people and the environment.[20]

ERMA also coordinates enforcement, which is carried out by seven central agencies and all territorial authorities (city and district councils). Each agency enforces the area for which its existing legislation gives it responsibility. Powers of

[20] *Ibid.*

enforcement for officers include powers of entry for inspection and the power to obtain search warrants. Enforcement officers can seize property, take samples, and collect evidence when these are linked to a breach of the Act. No enforcement mechanisms apply to new organisms once they have been approved for release, because approvals for release do not have any conditions attached.

Section 109 of the HSNO Act spells out the offences under the Act. Any person is allowed to report information on any of the listed offences. Penalties for the offences include fines, imprisonment, and other penalties, such as ordering an offender to remedy or mitigate the adverse effects of their noncompliance at their own cost or requiring that a new organism be destroyed. Liability under the Act is strict for most offences; the prosecution need not prove that the offender intended to commit an offence, merely that one occurred. All people responsible for an offence can be held liable for it.

The Act also provides for defences to liability. Possible defences include that even though an offence was committed, it was reasonable and necessary to protect life or health, prevent serious damage to property, or avoid an actual or likely adverse effect on the environment, and that the conduct was reasonable in the circumstances and all reasonable steps were taken to remedy or mitigate the effects. Proof of the exercise of due diligence is also a defence under the Act.

CONCLUSION

New Zealand has crafted a rigorous and innovative response to the threats that exotic organisms, including GMOs, pose to its unique ecosystems and to its economy. The strategic plan, with its inclusion of economic, social, environmental, and other factors, strongly reflects the principles of sustainable development. The financial and evidentiary burdens placed on proponents of new organisms, as well as the strict liability imposed on many violations of the HSNO Act, demonstrate a robust way to implement the Cartagena Protocol, using the "polluter-pays" principle to force proponents to internalise their costs. At the same time it is responsive to different circumstances, with different levels of scrutiny applied to lower-risk products. At the same time, the country's choice to apply a cost-benefit analysis in approving new organisms poses a challenge. It makes the uncertainty regarding the precautionary principle highly relevant, because the costs and benefits of emerging biotechnology might be unknown. The ongoing debate among scientists, the public, and the Maori, a part of the public who largely mistrust genetic engineering, is likely to keep a spotlight on these questions, especially given the ability of the public to participate in ERMA's assessments. New Zealand's history, economy, geography, and political relationship with the Maori make it unique, but it could also prove a useful case study for policy makers elsewhere.

22

Liability and Redress in Canadian Case Law

Hoffman v. Monsanto Canada Inc.

Kathryn Garforth and Paige Ainslie

The case of *Hoffman v. Monsanto Canada Inc.*[1] involved a group of Saskatchewan organic farmers bringing a motion for certification as a class to pursue class action litigation against both Monsanto Canada Inc. ("Monsanto") and Bayer CropScience Inc. ("Bayer"). The farmers sought redress for alleged damage caused to them by Monsanto and Bayer's genetically modified varieties of canola. The farmers were denied certification by both the Saskatchewan Court of Queen's Bench and the Saskatchewan Court of Appeal. They sought leave to appeal to the Supreme Court of Canada, but this too was denied. A review of the arguments and findings is worthwhile, because the case is one of the first of its kind and could have potentially far-reaching implications both within Canada and abroad.

The dispute concerned the plaintiffs' application for certification as a class, and therefore the decisions address this issue rather than the potential liability of Monsanto and Bayer for the damage alleged against them. That said, the Saskatchewan *Class Actions Act*[2] requires a court to assess the pleadings and be satisfied that they disclose a cause of action as one step in deciding whether to certify the class. The courts thus engaged in a preliminary analysis of the claims, and this analysis is our focus here, rather than the discussion of the other criteria that must be met for a class to be certified.

THE FACTS AT ISSUE

In the late 1980s and early 1990s, both Monsanto and AgrEvo Canada (the predecessor to Bayer) developed varieties of genetically modified (GM) canola. Both

[1] *Hoffman v Monsanto Canada Inc.* 2005 SKQB 225 [*Hoffman*], 2007 SKCA 47, [*Hoffman Appeal*], leave to appeal to SCC refused, 2007 CanLII 55334.

[2] *Class Actions Act*, SS 2001, c C-12.01.

companies received approval from Agriculture and Agri-Food Canada[3] (AAFC) for the unconfined release of their varieties in 1995.[4] The varieties, known as Roundup Ready and LibertyLink canola, respectively, were made commercially available the following year and are now widely grown in western Canada.[5]

The plaintiff farmers had three major complaints against Monsanto and Bayer. The first and primary complaint was that the release of Monsanto and Bayer's GM varieties of canola has resulted in the spread of the plants and the modified genes that they contain, such that it is now impossible to grow organic canola.[6] The spread, through cross-pollination and volunteer plants, of the modified canola genes and plants into the canola fields and crops of the organic farmers has, it was alleged, made it impossible for the farmers to comply with the prohibitions on genetically modified organisms (GMOs) in organic farming. The second complaint was that even organic farmers who are not attempting to grow canola suffer from the appearance of volunteer Roundup Ready and/or LibertyLink canola plants in their fields.[7] The plaintiffs claimed for past and future cleanup costs incurred by organic farmers in response to the volunteers. The final complaint concerned the creation and subsequent abandonment of an identity preservation program (IPP) by the defendants. The IPP was introduced by the defendants in the mid-1990s, at the same time they first made their modified canola varieties commercially available. The purpose of the IPP was to segregate GM canola from conventional canola for export markets. The plaintiffs alleged that the abandonment of the IPP resulted in the loss of the European market for all Canadian canola.[8]

THE DECISIONS OF THE COURTS ON THE CAUSES OF ACTION

Court of Queen's Bench for Saskatchewan

The three complaints against Monsanto and Bayer were broken down into six causes of action: negligence, the rule in *Rylands v. Fletcher*,[9] nuisance, trespass,

[3] Agriculture and Agri-Food Canada was the federal government department responsible for assessing the applications at the time when Monsanto and AgrEvo applied for approval for unconfined release of their modified varieties of canola. The situation changed in 1997 when the Canadian Food Inspection Agency was created for this purpose.

[4] *Hoffman, supra* note 1 at paras 8–9.

[5] *Ibid* at paras 8–10.

[6] "Amended, Amended Statement of Claim" RSSC No 67 (2002) at paras 26–7, online: http://www.saskorganic.com/oapf/pdf/amended-claim.pdf [*Statement of Claim*].

[7] *Ibid* at para 27(a).

[8] *Ibid* at para 36(a).

[9] *Rylands v Fletcher* (1866), L.R. 1 Ex 265; (1868), LR 3 HL 330.

The Environmental Management and Protection Act[10] of Saskatchewan, and *The Environmental Assessment Act*[11] of Saskatchewan. Each of these is discussed in turn in the next sections. In her 177-page decision, Justice Smith of the Court of Queen's Bench for Saskatchewan – the court of first instance – adopted a standard of assessing whether it was "plain and obvious" that the claims under each of these causes of action would fail at trial.

Negligence

The plaintiff farmers alleged that there were two different duties of care owed to them by Monsanto and Bayer that were breached. These were:

- A duty to ensure that GM canola would not infiltrate and contaminate farmland where it was not intended to be grown,[12] or at least the defendants "ought to have warned growers purchasing their products of cross-pollination, and advised them of farming practices designed to limit the spread of the gene"[13]
- A duty not to negligently undertake the development of export rules to ensure continued access to foreign markets, and to maintain an adequate identity preservation program (IPP) "to preserve the European canola export market where most of the organic canola produced in Canada was sold"[14]

The damage to the farmers arising from the breach of these alleged duties "derive[d] from damages and loss of revenues caused by: (a) loss of canola as a crop to be used within their regular rotations; (b) loss of opportunity to participate in the certified organic canola market; (c) past and future cleanup costs caused by Roundup Ready or LibertyLink canola volunteers growing on the fields of organic farmers."[15] In her decision, Justice Smith addressed the two alleged duties separately.

Duty to Ensure that GM Canola Would Not Infiltrate and Contaminate Farmland

As the plaintiffs conceded that the duty of care they were asserting was novel, Justice Smith turned to the test from *Anns v. Merton London Borough Council*[16] to determine whether this new duty of care should be recognised: "The *Anns* test is . . . said to be a two-pronged test, first to determine whether a *prima facie* duty of care arises, and

10 *The Environmental Management and Protection Act*, SS 1983–84, c E-10.2 ["EMPA"]. This law has been repealed and replaced by *The Environmental Management and Protection Act, 2002*, SS 2002, c E-10.21 ["EMPA, 2002"].

11 *The Environmental Assessment Act*, SS 1979–80, c E-10.1 ["EAA"].

12 *Statement of Claim*, *supra* note 6 at para 34.

13 *Ibid* at para 35.

14 *Ibid* at para 36(a).

15 *Ibid* at para 44.

16 *Anns v Merton London Borough Council*, [1978] AC 728 (HL) [*Anns*]. While the *Anns* test was adopted in Canada and has been modified by subsequent judgments, it is worthwhile pointing out that it is no longer good law in the United Kingdom; see, in particular, *Murphy v Brentwood District Council*, [1991] 1 AC 398 (HL).

second whether there are any policy considerations that ought reduce or limit the scope of the duty."[17] The first part of the test involves determining whether the plaintiffs' pleadings allege "reasonably foreseeable harm and relational proximity sufficient to establish a *prima facie* duty of care."[18]

Justice Smith found that the farmers' allegations were sufficient to support the general claim that the "adventitious presence"[19] of GM canola where it was not intended to be grown was foreseeable.[20] She was less certain, however, that the farmers' claim supported the allegation that the loss and damage being claimed was foreseeable. She pointed to the fact that the organic standards in place at the time did not, for the most part, prohibit the use of GMOs in organic agriculture.[21] Nonetheless, she was prepared to find that the plaintiffs' pleadings were sufficient to support the allegation or could be so amended.[22] More critically, Justice Smith found that the plaintiffs did not allege there to be any relationship between themselves and the defendants, and so their pleadings did not support the relational proximity needed to establish a *prima facie* duty of care.[23]

In addition, Justice Smith found that the plaintiffs failed on the second part of the *Anns* test – policy considerations. She cited two policy considerations against applying the alleged duty of care to the defendants. The first was the issue of the government approval of the unconfined release of the defendants' varieties of GM canola: "The imposition by the courts of a duty of care not to release these substances into the environment would therefore appear to be in conflict with express governmental policy."[24] Second, the judge characterized the plaintiffs' claim as one for pure economic loss. She stated that "[i]n effect, the alleged damage is not of physical harm to the plaintiffs' crops, but arises from the alleged inability to meet the requirements of organic certifiers or of foreign markets for organic canola."[25] Common law is generally reluctant to impose a duty of care to avoid

[17] *Hoffman*, *supra* note 1 at para 54.

[18] *Ibid* at para 60.

[19] Smith J adopted the term *adventitious presence* as a more neutral term advocated by the defendants instead of the term *contamination* used by the plaintiffs, *Hoffman*, *supra* note 1 at para 11. In their memorandum on their application for leave to appeal to the Court of Appeal for Saskatchewan, the plaintiffs stated that Justice Smith's adoption of the term *adventitious presence* as defined by the defendants illustrates that she "started off on the wrong foot" and "prejudge[d] the outcome of an action where it is alleged that there has been negligent and systemic contamination of the environment by GMOs." See "Memorandum of law on behalf of the prospective appellants" in the Court of Appeal for Saskatchewan, Case No 1148, online: http://www.saskorganic.com/oapf/pdf/Memo_Appeal_Application.pdf ["Plaintiffs' memo"] at para 20. See also Heather McLeod-Kilmurray, "*Hoffman v. Monsanto*: Courts, Class Actions, and Perceptions of the Problem of GM Drift" (2007) 27 Bulletin of Science, Technology & Society 188 at 197.

[20] *Hoffman*, *supra* note 1 at para 63.

[21] *Ibid* at para 64–5.

[22] *Ibid* at para 66.

[23] *Ibid* at para 67.

[24] *Ibid* at para 71.

[25] *Ibid* at para 72.

causing foreseeable pure economic loss, and Justice Smith found that imposing a duty here would expose the defendants to indeterminate liability.[26]

Duty Concerning the Identity Preservation Program

The plaintiffs amended their statement of claim to include an alleged breach of an alleged duty of care not to negligently undertake the development of an IPP. Justice Smith found that for such a duty to arise, the plaintiffs must have somehow detrimentally relied on the defendants' undertaking. Justice Smith concluded that the plaintiffs failed to plead detrimental reliance, and so this element of the cause of action in negligence could not be sustained.[27]

The Rule in *Rylands v. Fletcher*

The rule in *Rylands v. Fletcher*, also known as strict liability, was said to require for its imposition that: "(i) the defendant has made a non-natural use of its land; (ii) the defendant brought onto his land something which was likely to do mischief if it escaped; (iii) the substance in question escaped; and (iv) damage was caused to the plaintiff's property or person as a result of the escape."[28]

The court did not rule on whether GM canola was a dangerous substance or whether field trials of GM canola were an unnatural use of land. Rather, Justice Smith found that it could not be reasonably argued "that the commercial release and sale of Roundup Ready and LibertyLink canola seed constituted an 'escape' of a substance, dangerous or otherwise, from property owned or controlled by the defendants in the sense of 'escape' required by the rule in *Rylands v. Fletcher*."[29] She therefore was not satisfied that the plaintiffs' pleadings disclosed a cause of action.

Nuisance

The court distinguished between physical and amenity losses.[30] Justice Smith found that the plaintiffs' claims could be interpreted as including both types of nuisance and, regardless, "there is authority for the proposition that no action can be brought by a plaintiff who is unduly reactive to the defendant's conduct because he is carrying on a business or operation that is particularly sensitive to the kind of intervention that is in question."[31] She did not discuss this point further nor cite the authority to which she was referring; however, the issue of the sensitivity of organic farmers to GM production was raised as a potential barrier to a claim in nuisance in the British case of *R. v. Secretary of State for the Environment ex parte Watson*.[32]

[26] *Ibid* at paras 73, 77–80.
[27] *Ibid* at para 88.
[28] *Ibid* at para 91.
[29] *Ibid* at para 97.
[30] *Ibid* at para 100.
[31] *Ibid* at para 104.
[32] *R v Secretary of State for the Environment ex parte Watson* (1998), [1999] Env LR 310 (CA).

Monsanto and Bayer made two counterarguments to the plaintiffs' claim in nuisance. They questioned whether the harm claimed fell within the scope of nuisance, and if so, whether they could be held liable in nuisance for the harm alleged.[33] On the first point, the defendants argued that the damage alleged by the plaintiffs was not caused by the release of GM canola:

> [B]ut by the actions of third parties who have promulgated the standards affected by the inevitable adventitious presence of GM canola and by the decisions of individual organic farmers to seek to adhere to those standards. Implicit in this argument is the claim that the adventitious presence of GM canola is not inherently harmful to crops or to land. The defendants would on this basis distinguish cases that have imposed liability, for example, for the spread of weeds or for the drift of herbicides, both of which have harmed or endangered the physical well-being of growing crops.[34]

The defendants argued further that the alleged injury or interference was not sufficiently "unreasonable" or "substantial" to sustain the claim. They stated that growing GM canola is a "usual and ordinary" activity and that pollen flow is a natural phenomenon. Justice Smith ruled that the analogy to contamination of land by weeds was too close to be dismissed and had to be assessed by a trial judge because of all the evidence.[35]

On the second point, Monsanto and Bayer argued "that they cannot be held liable unless the alleged nuisance emanated from land they occupied or controlled."[36] The court disagreed and found that "[a]nyone who actively creates a nuisance whether or not in occupation of the land from which it emanates can be liable and this liability continues so long as the offensive condition remains regardless of his ability to abate it and stop the harm."[37] Justice Smith went on, however, characterising the defendants' "true objection" as being one of the impossibility of finding harm as a result of merely marketing or selling GM canola: "The adventitious presence of canola in the crops and on the land of organic farmers required the intervention of neighbouring farmers who cultivated GM canola.... While the 'release' of GM varieties of canola by the defendants may have been a necessary condition for the occurrence of the harm alleged, it was far from sufficient, in itself."[38]

Justice Smith found that holding a manufacturer or inventor liable in nuisance for damage caused by the use of its product would be very sweeping. She relied on U.S. law to conclude that when the activity complained of is the activity of one who is not in occupation or control of adjoining land, and no independent malfeasance

33 *Hoffman*, *supra* note 1 at para 105.
34 *Ibid* at para 106.
35 *Ibid* at para 108.
36 *Ibid* at para 112.
37 *Ibid* at para 113.
38 *Ibid* at para 114.

is alleged, then, "at the very least, direct causation of the damage alleged must be alleged."[39] She found that the plaintiffs did not allege direct causation, and therefore it would not be possible to support a claim "that the defendants substantially caused the nuisance alleged."[40] Her decision that this claim could not stand was qualified by the possibility of causes of action against the defendants, lying in the Environmental Management and Protection Act and the Environmental Assessment Act, which support "the necessary allegation of malfeasance" and would thus preclude striking the claim of nuisance at this stage.[41]

Trespass

The court described the requirements for sustaining a claim in trespass as follows: "the plaintiffs must establish an unauthorized entry upon another's land of a physical object or physical contact with land, and the defendants' act or interference must be voluntary and intentional or negligent."[42] The dispute centred on whether direct interference by the defendants was required for the action in trespass.

The plaintiffs argued that directness should not be required when, as was alleged here, the spread of the genes to where they were not wanted was foreseeable. Furthermore, they argued that the genes are Monsanto and Bayer's property, making this case analogous to the stray bull cases "which hold an owner strictly liable for damage caused by a bull which strays onto a neighbour's land."[43] Justice Smith disagreed on both points. She found that much more than "natural and inevitable forces" are required between the marketing of GM canola and its spread to the plaintiffs' land.[44] She also stated that stray bull cases are not trespass cases:

> The imposition of strict liability for the consequences of stray bulls is clearly a policy decision intended to place a heavy onus on the owners and possessors of bulls to keep these animals confined and under control. Although the plaintiff claims that the defendants "own" their GM canola gene (a claim that the defendants say is a misunderstanding of the nature of their interest under patent law), it can point to no similar public policy that would have, in effect, placed an onus on the defendants not to have commercially released GM canola, for the plaintiffs' claim is that, once GM canola was commercially released, cross-pollination of conventional canola crops was natural and inevitable.[45]

Justice Smith thus concluded that the action in trespass could not be sustained.

39 *Ibid* at para 122.
40 *Ibid* at para 122.
41 *Ibid* at para 123.
42 *Ibid* at para 127.
43 *Ibid* at para 129.
44 *Ibid* at para 131.
45 *Ibid* at para 132.

The Environmental Management and Protection Act

The plaintiffs' statement of claim asserts a cause of action under both the original *Environmental Management and Protection Act* (EMPA), as well as the new *Environmental Management and Protection Act* (EMPA, 2002) that replaced the original legislation in 2002. Justice Smith considered the two pieces of legislation separately.

The EMPA

The court found that to establish a claim for liability under section 13(3) of the original EMPA, the farmers needed to allege and prove:

1. that the GM canola varieties of the defendants are "pollutants" within the meaning of the Act
2. that immediately before the first discharge of the "pollutant", the defendants were either the "owners" of the pollutants or "persons having control" of the pollutants
3. that the "pollutant" has been "discharged"
4. that the discharge of the pollutant has been into the "environment"
5. that the discharge has caused loss or damage to the plaintiffs[46]

The plaintiffs made each of these allegations in a general manner but argued that delving into the meaning of the terms in section 13(3) was a matter to be left for trial. Specifically, the plaintiffs argued that it was not necessary for them to allege or prove that GM canola is harmful or unsafe "except to the extent of alleging that the adventitious presence of canola has caused damage or loss to the plaintiffs in respect of loss of the market for organically grown canola and the costs related to removing volunteer canola plants from their fields."[47] They also indicated, however, that if the court did not agree, they were prepared to amend their statement of claim to allege that GM canola is inherently unsafe.[48] The court was not satisfied that the pleadings as they stood disclosed a cause of action under section 13 of the EMPA, but Smith concluded that if the statement of claim were amended as proposed, it was no longer plain and obvious that the claim that GM canola is a pollutant could not succeed.[49]

Justice Smith went on, however, to examine whether GM canola could be said to have been discharged into the environment in the meaning of the Act, and if so, whether Monsanto and Bayer could be said to have been the owners of the pollutant or the persons having control of the pollutant immediately before the first discharge. She found that simply marketing the GM canola seed to be used by farmers did not fall within the definition of discharge nor did the dismantling of the IPP program.[50] On the issue of whether Monsanto or Bayer could be considered

46 *Ibid* at para 138.
47 *Ibid* at para 146.
48 *Ibid* at para 146.
49 *Ibid* at para 150.
50 *Ibid* at para 154.

the owners of or the person having control of the pollutant, the farmers argued that the companies' patents and technology use agreements meant that the companies retained ownership and control of the modified genes. Justice Smith disagreed:

> In my respectful view it is not reasonably arguable that ownership of a patent in the modified gene and enforcement of patent rights through "technology user agreements" are sufficient to constitute "ownership" or "control" of the "pollutant" (GM canola seed and resulting pollen) after the seed is sold to farmers and cultivated by them, as these words are used in the Act.[51]

She thus concluded that a claim under section 13 of the EMPA could not be sustained.

The 2002 EMPA

The 2002 EMPA is significantly different from the original EMPA. Section 15(3)(a)(i) of the new Act gives any person a right to compensation from "the person responsible for a discharge for loss or damage incurred as a result of . . . the discharge of a substance." Under section 15(1), "loss or damage" is defined to include personal injury, loss of life, loss of use or enjoyment of property, and pecuniary loss, including loss of income. Section 15(3)(a)(i) does not require the discharge of the substance to cause adverse effects – although this requirement is included in other sections of the Act – and the definition of substance does not require a substance to have harmful or unsafe attributes.[52]

In interpreting the civil liability provisions in the EMPA, 2002, the court described their potential scope as "staggering."[53] Justice Smith was inclined to interpret section 15(3)(a)(i) in light of the criminal or quasi-criminal liability standards in section 4 of the EMPA, 2002. These require that the discharge of the substance may cause or is causing adverse effects. Furthermore, the standards in section 4 do not apply to discharges authorised by government. Nonetheless, on the plain wording of section 15, Justice Smith could not conclude that it was plain and obvious that the plaintiffs' claim would fail at trial.[54] The claim was, however, limited to only those losses that were alleged to have been sustained since 2002, when the Act came into effect.

The Environmental Assessment Act

The plaintiffs contended that the testing and unconfined release of GM canola into the Saskatchewan environment was a "development" under the *Environmental Assessment Act* (EAA). The defendants were thus required to conduct and submit an environmental impact assessment under the Act and receive the necessary ministerial approval prior to proceeding with the testing and release. Because the defendants

[51] *Ibid* at para 157.
[52] Definition of *substance* in s 2(bb) of the EMPA, 2002, *supra* note 10.
[53] *Hoffman*, *supra* note 1 at para 163.
[54] *Ibid* at para 168.

failed to do this, the plaintiffs contended that the defendants should be liable under section 23 of the Act for the loss or damage suffered by the plaintiffs.[55]

"Development" is defined in the EAA as

> any project, operation or activity or any alteration or expansion of any project, operation or activity which is likely to: . . .
>
> (iv) cause widespread public concern because of potential environmental changes; [or]
> (v) involve a new technology that is concerned with resource utilization and that may induce significant environmental change[.][56]

Justice Smith turned to government documents and the case law to help interpret the definition of "development" in the Act. She concluded that the testing and unconfined release of GM canola was not a "development" under the terms of subclause (v) because, although it might involve "new technology," it was not concerned with resource use.[57] She did find, however, that the defendants' activities could constitute a development under subclause (iv), and so it was not plain and obvious that the plaintiffs' claim on this ground would fail at trial.[58]

Justice Smith also explored the issue of the defendants' being liable for the "loss and damage" of the plaintiffs under section 23 of the Act. There is no definition of the term *loss and damage* in the legislation, and it has not been subject to judicial interpretation. She was not certain that it would cover the pure economic loss claimed by the farmers but ruled that it was not plain and obvious that this claim and the claim for loss and damage arising from the adventitious presence of volunteer GM canola plants would fail at trial.[59] Accordingly, the cause of action under the EAA was sustained.

In summary, Justice Smith upheld two of the statutory causes of action with the possibility that the 2002 EMPA and the EAA could support a cause of action in negligence. Ultimately, however, these possible statutory causes of action were largely moot. In the remainder of the decision, she examined the four other elements that must be satisfied for a class to be certified, that is, the existence of an identifiable class, the claims of the class members raising common issues, a class action being the preferable procedure, and whether there is a suitable representative plaintiff. These do not relate to the substance of the liability claims and so are not discussed here, but it is of note that Justice Smith ruled that the proposed class failed to meet the requirements of each of these other elements.

55 Statement of claim, *supra* note 6 at para 41.
56 EAA, *supra* note 11 at s 2(d).
57 *Hoffman*, *supra* note 1 at para 190.
58 *Ibid* at para 191–2.
59 *Ibid* at para 193.

The Court of Appeal for Saskatchewan

In 2005, the farmers sought and were granted leave to appeal Justice Smith's decision to the Court of Appeal for Saskatchewan. The ruling of Justice Cameron on behalf of the Court of Appeal was released in May 2007.[60]

The text of the Court of Appeal's decision is significantly shorter than that of the Court of Queen's Bench, and approximately one half of it reviews the appropriate standard on which to assess whether the plaintiffs' pleadings disclosed a cause of action. The Court of Appeal's ruling changes the standard from the "plain and obvious" test used by Justice Smith to one in which the representative plaintiff must satisfy a judge "that the class has an apparently authentic or genuine cause of action, . . . that there exists a plausible basis in principle and presumed fact for supposing the defendants could be held liable."[61] The impact of this change on the Court of Appeal's ruling is unclear because, after having elaborated the test, Justice Cameron assessed whether Justice Smith had erred in any of her conclusions that it was plain and obvious that the causes of action would fail at trial and, as is explored in more detail later, he generally agreed with her throughout. The change in standard could have broader implications for the ability to obtain certification for environmental class actions and class actions in general, but an exploration of this matter is beyond the scope of this chapter.[62]

Having determined the appropriate test to apply, the Court turned its attention to the causes of action. The plaintiffs did not appeal Justice Smith's ruling on strict liability, and Monsanto and Bayer did not cross-appeal her ruling on the causes of action under the 2002 EMPA and the EAA but nonetheless contested it in their factums. The Court of Appeal chose to consider these two causes of action in its ruling, resulting in a decision in which they considered negligence, nuisance, and trespass together and the three statutory causes of action – EMPA, EMPA 2002, and the EAA – together as a second group. Their analysis of the causes of action takes a mere five pages.

Negligence, Nuisance, and Trespass

The Court of Appeal offered little in the way of new analysis of these torts. Justice Cameron supported Justice Smith's findings on negligence, agreeing "that the pleadings did not disclose a relationship of sufficient proximity to give rise to a *prima facie* duty of care of the nature of that pleaded and, even [if] it were otherwise, there are sound policy reasons for negating the duty in view of the fact the federal government had approved the unconfined release of LibertyLink and Roundup

[60] *Hoffman Appeal*, *supra* note 1.

[61] *Ibid* at para 53.

[62] For more, see "Memorandum of argument" in the Supreme Court of Canada, File No 32135 at 17–18, online: http://www.saskorganic.com/oapf/pdf/Argument-Jul-25-07.pdf; McLeod-Kilmurray, *supra* note 19.

Ready."[63] On nuisance, the Court again agreed with Justice Cameron and found that the defendants did not occupy or control the land from which the alleged nuisance emanated. Justice Cameron did not discuss how the companies' intellectual property rights might give them the necessary control to found an action in nuisance, despite the plaintiffs' having argued this point in their memo on appeal. Finally, Justice Cameron also cited with approval Justice Smith's ruling on trespass, that even with a "liberalized view of the requirement for direct interference, the cause of action in trespass still lacked a proper foundation."[64]

The Statutory Causes of Action

The Court of Appeal supported Justice Smith's findings that the plaintiffs' pleadings did not disclose a cause of action under the EMPA and did not discuss the matter any further. It then turned to the 2002 EMPA and ruled that Justice Smith's finding of a possible cause of action under this statute was "unrealistic."[65] The Court of Appeal thought that unconfined release of the companies' GM varieties of canola was authorised under an act of Parliament and so was exempt from the prohibition on discharging a substance into the environment, as contained in section 4 of the EMPA, 2002. Thus eliminated, the only potential cause of action that remained was that under the EAA. The Court was sceptical that the field testing and release of the GM canola varieties could cause widespread public concern – the ground on which Justice Smith had upheld this claim. The Court ruled: "Still, there is more to the matter than this, given the remaining indicia of what constitutes a development within the meaning of the *Act*, so it may be something could be made of this cause of action. In light of what follows, however, we do not find it necessary to determine the matter."[66] This rather lazy assessment of the cause of action suggests that Justice Smith's ruling stands, and it is possible that the testing and unconfined release of GM canola could constitute a "development" under the EAA.

Having dismissed all the causes of action with the possible exception of the EAA, the Court of Appeal considered the other requirements that must be met for a class to be certified. They agreed with Smith's rulings on these points, found that the class did not meet the requirements for certification, and the appeal was dismissed. As stated previously, the farmers sought and were denied leave to appeal to the Supreme Court of Canada. As per its usual practice, the Supreme Court did not provide any reasons for its denial of leave to appeal. The farmers have stated that this is not the end of the road and that they will consider other legal options.[67]

63 *Hoffman Appeal*, *supra* note 1 at para 59.
64 *Ibid* at para 64.
65 *Ibid* at para 67.
66 *Ibid* at para 70.
67 Letter from the OAPF Committee (13 December 2007), online: http://www.saskorganic.com/oapf/pdf/13Dec07-letter.pdf.

Larry Hoffman and Dale Beaudoin, the two proposed representative plaintiffs for the class action, however, have subsequently ruled out bringing individual claims against the companies.[68]

UNDERLYING CONCERNS

The *Hoffman* case was born from the Organic Agriculture Protection Fund (OAPF), which was launched in 2001 by the Saskatchewan Organic Directorate to pay for what it viewed to be a precedent-setting case. As detailed earlier, the action rested on the assertion that genetically modified canola has contaminated the fields of organic farmers, resulting in great economic damage. There are several broader issues underlying this action, however, that go beyond economics and that were not put directly before the courts. Simply stated, the case pitted the proponents of biotechnology against its rejecters. Although many view designer seeds as a boon to farming (by 2003, approximately 70 percent of all canola grown in western Canada was either Roundup Ready or LibertyLink canola),[69] some farmers see them as an affront to traditional farming practices and are concerned about the wholesale adoption of GM crops. This section examines these concerns in three categories: the issue of safety, the appropriate balance of rights and responsibilities, and access to information.

Safety in an Economic Veil

The arguments in this case were largely veiled in the language of economics rather than considerations of health and safety. Beneath the veil, however, the issue of safety lay at the heart of the whole dispute. In the context of assessing the cause of action in negligence, Justice Smith remarked: "There is no allegation that GM canola is unhealthy or causes detrimental physical problems to humans or plant life."[70] What Justice Smith did not mention, however, are the very reasons why organic farmers and organic certification standards have rejected GMOs in the first place. These reasons are based, at least in part, on lingering questions over the potential for GM crops and foods to cause harm to humans and the environment. With the case never having been heard on its merits, the costs of these risks continue to be borne by farmers rather than biotechnology companies.

The plaintiffs' hesitation to litigate the safety issue directly is understandable, however, in light of the practical considerations at play. Foremost among these is

68 Saskatchewan Organic Directorate – Organic Agriculture Protection Fund, Media release, "Individual action not the way to go: organic farmers" (16 April 2008), online: http://www.saskorganic.com/oapf/.

69 *Hoffman, supra* note 1 at para 40.

70 *Ibid* at para 72.

the difficulty that the plaintiffs would have faced in proving any assertions that the GM canola varieties are unsafe. As recognized by Justice Smith:

> Indeed, any allegation that GM canola is inherently harmful or dangerous in any respect, or at least that any possible or potential inherently harmful quality was known to the defendants at the time that GM canola was commercially introduced in Canada, would seem to be inconsistent with the express pleading of the AAFC decision documents described above.[71]

Any attempt by the plaintiffs to assert that the AAFC erred in its findings that the modified canola was safe for unconfined release into the environment would have been an uphill battle, to say the least. The plaintiffs would have borne the burden of having to prove these assertions and probably would have been faced with contradictory evidence from the defendants' experts and their own. In addition, the defendants have by far the deeper pockets, which would have allowed them to hire more experts with greater scientific and courtroom expertise. The plaintiffs would also have faced the challenge of gaining access to information on the safety of the GM canola, as is discussed later.

A further reason for the plaintiffs' avoidance of the safety issue was likely their recognition that a court would be loath to engage in a determination of the safe or hazardous nature of GM canola. With the highly controversial nature of GMOs and biotechnology, a court would be very reluctant to engage in a scientific rather than a legal exercise and replace government assessment of the canola varieties with its own. Undertaking the economic exercise of assessing monetary damages, however, is a much more comfortable exercise for a court.

All this said, the plaintiffs' expressed a surprising willingness to contest the safety of the defendants' canola varieties, particularly in their application for leave to appeal to the Saskatchewan Court of Appeal and also in their application for leave to appeal to the Supreme Court. In arguing to the Saskatchewan Court of Appeal that their appeal had merit, the plaintiffs described a perceived error by Justice Smith in her interpretation of their claims: "While the Learned Chambers Judge was correct that the Plaintiffs' position was, for the bulk of their claims, that they did not have to prove that the genetic modifications belonging to the Defendants were inherently harmful. It is incorrect to say that the Plaintiffs' position was that these modifications were therefore environmentally benign."[72] The plaintiffs thought that a ruling from Justice Smith from an earlier hearing meant that the safety issue was to be left for trial. In her decision on certification, however, Justice Smith quoted with approval the AAFC's regulatory documents that approved the release of the defendants' GM canola varieties: "In effect, the Learned Chambers Judge banned the Plaintiffs from filing an affidavit challenging the safety of GMOs because 'such an extremely complex and heated scientific controversy' could be determined only at trial. Then

[71] *Ibid* at para 23.

[72] Plaintiffs' memo, *supra* note 19 at para 22.

she subsequently determined the scientific controversy in the Defendants' favour because the Plaintiffs did not challenge the safety of GMOs."[73] The plaintiffs again asserted this point in their memo to the Supreme Court, describing it as premature at that stage of the action for "the courts below to dismiss the action on the basis of federal regulatory approval, given that the regulatory approval can be demonstrated not to be foolproof, particularly as failing to adopt a precautionary approach or to employ adaptive management once these organisms are released."[74]

The farmers also appeared to raise a new dimension of the safety issue in their arguments in seeking leave to appeal to the Supreme Court, namely, the issue of precaution. There are two aspects to this dimension: a perceived lack of precaution in the Canadian regulatory system for the unconfined release of plants with novel traits and the lack of precaution in the decisions of the courts when they denied the action certification as a class.

Perhaps the most widely quoted definition of the precautionary principle is that from Principle 15 of the 1992 Rio Declaration on Environment and Development: "Where there are threats of serious or irreversible damage, lack of full scientific certainty shall not be used as a reason for postponing cost-effective measures to prevent environmental degradation." The Supreme Court of Canada endorsed a very similar phrasing of the principle in 114957 *Canada Ltée (Spraytech, Société d'arrosage) v. Hudson (Town)*.[75] The principle has also been incorporated into several pieces of federal legislation, most notably in this context the *Canadian Environmental Protection Act, 1999*[76] (CEPA, 1999).

As mentioned, the plaintiffs asserted that the federal regulatory approval for the GM canola varieties in question failed to adopt a precautionary approach. This raises the question of whether the government has an obligation to adopt a precautionary approach in its regulatory process. The argument is there to be made that the government does have such an obligation – its application is required by CEPA 1999, which regulates in part the products of biotechnology. The Privy Council Office has also developed "A Framework for the Application of Precaution in Science-based Decision Making about Risk."[77]

The argument for precaution in judicial decision making can be interpreted as supporting the view that questions of safety lie at the heart of this action. Applying the definition of precaution from Principle 15 as quoted earlier, the "cost-effective

73 *Ibid* at para 28.

74 *Hoffman Appeal, supra* note 1 at para 39. The plaintiffs switched between the use of the terms *precautionary principle* and *precautionary approach* and appear to treat the two as synonymous. Although the terms are not necessarily synonymous in international law, we will treat them as such here for the sake of argument, simplicity, and clarity.

75 114957 *Canada Ltée (Spraytech, Société d'arrosage) v Hudson (Town)*, 2001 SCC 40 at paras 31–32.

76 *Canadian Environmental Protection Act, 1999*, SC 1999, c 33 [CEPA, 1999].

77 *Ibid*, s 2(1)(a) and Part 6; Government of Canada, "A Framework for the Application of Precaution in Science-Based Decision Making about Risk," online: http://www.bcp-pco.gc.ca/docs/information/publications/precaution/precaution_e.pdf.

measure" to be taken would be the certification of the proposed class as a class action. There appears no reason why the judiciary cannot apply the precautionary principle when it is making decisions on measures concerning the protection of human health or the environment. The main stumbling block could be that the precautionary principle is rooted in a desire to *prevent* harm to the environment, whereas here the farmers were claiming for damages that had allegedly already occurred. Even this obstacle could be overcome, however, when it is considered that one of the objectives of class action litigation is behaviour modification. Although precautionary decision making would be too late to protect the plaintiffs from the damage they allegedly had already suffered, if certification of the class would have resulted in changes in the defendants' behaviour such that the alleged types of damage no longer took place, then it could indeed be decision making that prevents harm.

McLeod-Kilmurray takes a similar view in summarising the scientific uncertainty in relation to GMOs and the legal uncertainty faced by a court in deciding whether to certify a class without having had the benefit of hearing all the evidence that would be presented at trial: "This combination of scientific and legal uncertainty makes the Precautionary Principle doubly relevant in the *Hoffman* case and in many environmental class actions. The Precautionary Principle might lead judges to err on the side of caution and grant certification where there is uncertainty on some of the branches of the certification test. It is always possible to decertify later in the case if the court considers it appropriate."[78]

Whether a court will be willing to lift the economic veil and grapple with the health and safety concerns of GMOs remains to be seen. The Court of Appeal foreshadowed its reluctance to consider the environmental aspects of the dispute when it refused leave to intervene in the appeal, as requested by two environmental organizations. In its decision on leave to intervene, Justice Cameron, writing for the Court, commented that the subject matter of the plaintiffs' pleadings and causes of action "adds the flavour of environmental litigation to the case generally, but that is not the focus of the case on appeal."[79] The Court did not address the farmers' arguments on the health and safety of GMOs in its decision on the appeal and again cited with approval the federal government's approval of the unconfined release of the GM canola varieties.

Advantages and Burdens, Rights and Responsibilities

The opportunity to gain intellectual property protection – and patents in particular – over living organisms and parts thereof has grown dramatically over the past twenty-five years. In Canada, higher life forms are not patentable per se,[80] but in *Monsanto Canada Inc. v. Schmeiser*,[81] the Supreme Court interpreted a patent on a modified

78 McLeod-Lilmurray, *supra* note 19 at 197.

79 *Hoffman Appeal, supra* note 1 at para 7.

80 *Harvard College v Canada (Commissioner of Patents)*, [2002] 4 SCR 45, 2002 SCC 76.

81 *Monsanto Canada Inc v Schmeiser*, [2004] 1 SCR 902, 2004 SCC 34 [*Schmeiser*].

gene and the cells that contain it in such an expansive manner that the effective result of the decision is to extend patent rights to higher life forms.

Monsanto, in particular, actively defends its patent rights, as illustrated by the *Schmeiser* case, but also by the company's technology use agreement (TUA). Any farmer wishing to purchase and grow Roundup Ready canola seed must sign a TUA. The agreement licenses the farmer to use Monsanto's patented technology, that is, the genetically modified gene. The agreement forbids, however, the farmer from saving seed from the GM plants to plant the following year. The American version of the contract, at least, also explicitly vests ownership of the technology with Monsanto and not with the farmer.[82]

The patent rights associated with GMOs have several consequences. First is the issue of the innocent infringer. The Supreme Court in *Schmeiser* made it quite clear that they did not consider Mr. Schmeiser to be an innocent infringer.[83] It did suggest, however, that there might be a defence against allegations of patent infringement for victims of volunteer GM crops. The question of what would constitute an innocent infringer was not canvassed in *Schmeiser*.

As a result, the decision from *Schmeiser* places all of the burdens on the farmer. It is the farmer who must monitor his or her fields for volunteer GM canola plants or the spread of the introduced genes. If and when the patented genes appear, it is the farmer who must call the company to come and remove the offending plants. If the farmer does not monitor his or her fields or does not call the company after having detected the genes, he or she faces the threat of patent infringement litigation. If sued, it is the farmer who must rebut the presumption of use through an innocent bystander defence that is said to exist but has not been defined. The *Hoffman* case can be understood as an attempt also to place burdens on companies that have developed and profited from the GM varieties and require them to face some of the costs for the consequences of their technologies.

The second consequence arising from the patents on genes is the impact on seed-saving practices and the ability of farmers to maintain their independence from corporate agriculture if they so desire. The impact on seed saving takes multiple forms. First of all, as a consequence of *Schmeiser*, if farmers are aware or should be

82 In their memo for leave to appeal to the Saskatchewan Court of Appeal, the plaintiffs asserted that the TUA "clearly provided that ownership of the gene would remain with Monsanto," Plaintiffs' memo, *supra* note 19 at para 51 of Schedule "A". The authors have found American versions of the contract that explicitly vest ownership of the technology with Monsanto, but Canadian versions of the TUA do not appear to contain this clause; for the American TUA, see: "2008 Monsanto Technology/Stewardship Agreement," online: http://www.monsanto.com/SiteCollectionDocuments/tug_sample.pdf. The ownership clause is contained in the first bullet point under the heading "Grower Receives from Monsanto" and states in part that "Monsanto retains ownership of the Monsanto Technologies including the genes (e.g., the Roundup Ready gene) and the gene technologies." For the Canadian TUA, see "2005 Monsanto TUA – English" (draft 20 April 2004), on file with author. Neither of these TUAs are for Roundup Ready canola, and it is possible that the plaintiffs were relying on a different version of the Canadian contract.

83 *Schmeiser*, *supra* note 81 at para 95.

aware of the presence of patented genes in their crop, they would be ill-advised to save seeds from these fields to plant the following year. If they do save and plant the seeds, they run the real risk of not being innocent bystanders and facing patent infringement litigation, as did Schmeiser.

The second form concerns the prohibition on saving seed that is contained in the TUA as well as proposals that would restrict the ability of farmers to save seeds from plants protected by Plant Breeders' Rights, another form of intellectual property. These restrictions are seen by many farmers to be an infringement of farmers' age-old right to save seed.[84]

Finally, industry's response to the pressure to guard against cross-pollination is to introduce "terminator technology" – also known more formally as *genetic use restriction technologies* (GURTs) – into their seeds. This technology renders the second generation of designer seeds infertile and ensures against the proliferation of cross-pollinated seeds for the benefit of organic and other non-GM using farmers. It also, however, helps to enforce the companies' intellectual property rights by making moot the ability of farmers to save seeds. After all, what use is it to save seeds containing patented genes if those seeds are sterile? It is also possible that biotechnology companies will introduce GURTs into their seeds in an effort to minimize their exposure to liability.[85]

The concern over the right to save seed is one piece of a larger puzzle in which organic farmers feel threatened by the encroaching private industry and fear the available avenues of avoidance may be dwindling. They worry that one day, through crop contamination or merely succumbing to the path of least resistance, all farmers might be growing GM crops and consequently might be subject to the conditions of the biotechnology industry. These farmers see the erosion of the traditional publicly owned and minimally restrictive seed market and fear that the result of the industry's

[84] The concept of farmers' rights – including the right to save seed – has developed in response to the expansion of intellectual property protection over living organisms. In this regard, Canada is a party to the *International Treaty on Plant Genetic Resources for Food and Agriculture*, FAO Res. 3/2003, 3 November 2001 (entered into force 29 June 2004), online: http://www.planttreaty.org. The Treaty is the first to include provisions on farmers' rights in the operational text of an international agreement (see Article 9), but it is still largely left to national governments to give content to these rights and take measures to protect and promote them, "as appropriate." At the very least, however, the Treaty offers a boost to farmers' arguments against practices that will undermine their legal ability in Canada to save seeds.

[85] Jeremy deBeer, "Rights and Responsibilities in the Agricultural Biotechnology Industry" (2006) [unpublished] at 32–33. The Parties to the *Convention on Biological Diversity* (5 June 1992), 31 ILM 818 (entered into force 29 December 1993) have agreed to what is often considered to be a *de facto* moratorium on the development of GURTs, pending further research and data; see: COP Decision V/5, "Agricultural biological diversity: review of phase I of the programme of work and adoption of a multi-year work programme." See in particular section III, paragraph 23, which contains the *de facto* moratorium, and paragraphs 20, 22, and 25, which call for more research. This moratorium was renewed at the eighth COP held in Brazil in March 2006, see COP Decision VIII/23, "Agricultural biodiversity," at section C, para 1. Para 2(b) of the decision also encourages further research within the mandate of Decision V/5, section III.

transformation will be to reduce farmers to mere "renters of corporate seeds."[86] Until now, organic farmers have focussed their efforts on pressuring government to adopt measures to protect traditional farming practices and to reject legislative amendments that would afford greater protection to the biotechnology industry. *Hoffman* seems to represent a switch in strategy, with organic farmers now seeking to make industry itself partially responsible for preventing the elimination of organic farming.

The final consequence arising from the patents on genes is in fact more of a question: if the patentees have rights over the patented genes in the genetically modified canola, should they not also bear some responsibility for these genes? As mentioned earlier, this argument has multiple facets. Some suggest that the companies should bear the responsibility for the consequences of their patented genes because they are in a much better position than government or individual farmers to guard against foreseeable losses, including pure economic loss. A 2003 report issued by Agriculture and Agri-Food Canada states that "[t]he Canadian seed industry generates $700 million in combined domestic and export sales every year. Nearly three quarters of the seed produced in Canada is used for domestic purposes."[87] In addition to their economic advantages, breeders, being intimately familiar with their own products, are also the party most likely to be aware of associated risks and available precautionary measures. It logically follows that they would also be in the best position to communicate this information to users of the product. Indeed, the role of access to information is a point to which we shall return.

The second facet to this responsibility argument juxtaposes the outcome from *Schmeiser* with the outcome from *Hoffman*: "Monsanto can exert unprecedented levels of control over things it could not patent, whilst simultaneously being able to deny that it has any control *over the same product* in the context of the common law or statue. This is an unacceptable incongruity."[88] In their statement of claim, the farmers alleged that by virtue of their patents and TUAs, the companies maintained control and ownership of their patented technologies. As described in this citation, Justice Smith largely dismissed these arguments. The plaintiffs reasserted them in their memorandum filed as part of their request for leave to appeal to the Saskatchewan Court of Appeal:

> The pleading that Monsanto at all times owned the gene in issue should not be overlooked. A higher level of responsibility attaches to property that one owns. It is almost comical to compare the position that Monsanto took before the Federal Court, Federal Court of Appeal and Supreme Court of Canada in *Monsanto*

[86] Susan Safyan, "Old MacDonald Had a Seed" (November 2005) 277 Alive Magazine.

[87] Agriculture and Agri-Food Canada, "Canada's Agriculture, Food and Beverage Industry; Canada's Seed Industry" (23 December 2003).

[88] Martin Phillipson, "Giving Away the Farm? The Rights and Obligations of Biotechnology Multinationals: Canadian Developments" (2005) 16 The King's College Law Journal 362 at 372.

> *Canada Inc. v. Schmeiser* . . . as compared to the submissions that Monsanto Canada made in the present case.
>
> In *Monsanto Canada Inc. v. Schmeiser* . . . Monsanto brought a patent infringement action against a Saskatchewan farmer who it alleged was growing canola that contained its genetic modification without a licence. The "unconfined release" of the genetic modification reflected in the Canadian decision documents did not result in a waiver of Monsanto's patent rights. Furthermore, Monsanto submitted that the use of the genetic modification was tightly controlled by it such that it never gave up its ownership rights over the genetic modification. . . . In the present case, by contrast, Monsanto has taken the position that the dissemination of its genetic modification is so widespread and so out-of-control that they should not be held responsible for it. At para. 135 of its Memorandum opposing certification, Monsanto argued:
>
> > [t]here is no way to put the genie back into the bottle. GM canola is now part of the environment. Our federal government authorized and approved that.[89]

Although the farmers' arguments in *Hoffman* were not necessarily consistent with the farmer's (i.e., Schmeiser's) arguments in *Schmeiser* (and it is worthwhile pointing out that both the farmers in *Hoffman* and *Schmeiser* have been represented by the same counsel), the thrust of their assertion was that the broad patent rights awarded to Monsanto by the Supreme Court in *Schmeiser* should also give rise to responsibilities. A refinement on this argument is that it is not so much the patent rights per se that could give rise to liability; rather it is the control that the patent gives the patentee over the invention that could ground a cause of action in nuisance, trespass, strict liability, and/or the statutory obligations.[90] If patent owners have the expansive rights to control their inventions, as awarded by the Supreme Court to Monsanto in *Schmeiser*, then they also have the corresponding duty to prevent the exercise of this control from causing harm to others.[91] When the exercise of their control does cause harm, they should bear liability for the consequent damage.[92]

If the companies cannot be held responsible, then the *Hoffman* case will build on the precedent set by *Schmeiser* and add yet another burden to the list of those that the farmer must bear – the burden of the damage. Indeed, such a decision in fact extends the companies' control powers. One outcome from *Schmeiser* was that Monsanto's property rights as a patent owner trumped Schmeiser's property rights to his seeds.[93] The effect of *Schmeiser* is "to confer shared ownership by virtue of the

[89] Plaintiffs' memo, supra note 19 at paras 34–6. See also Plaintiffs' memo at paras 50–3 of Schedule "A" and their memo on leave to appeal to the Supreme Court, *supra* note 62 especially at paras 30–3.

[90] Jeremy deBeer, "The Rights and Responsibilities of Biotech Patent Owners" (2007) 40 University of British Columbia Law Review 343 at 367–71.

[91] *Ibid* at 361–2, citing AM Honoré, "Ownership" in AG Guest, ed, *Oxford Essays in Jurisprudence* (Oxford: Oxford University Press, 1961) 107 and others.

[92] *Ibid* at 27–8.

[93] Jeremy deBeer, "Reconciling Property Rights in Plants" (2005) 8 Journal of World Intellectual Property 5 at 13.

inseparable integration of tangible and intangible property. The infusion of every seed or speck of pollen into the crops of someone else renders Monsanto a co-owner of every plant thereby affected."[94] With *Hoffman*, the control goes much further. No longer is it just control over how plants containing a genetically modified gene may be used, but that control is extended over the very choice of farming practices:

> Were the courts to refuse a remedy for the unconsented genetic alteration of a claimant's produce by cross-pollination from nearby GM crops, this would render his property right contingent. He may choose to farm organically, but his right to do what he likes on his own land would in effect become subject to the state's right to determine that that right be protected only if its exercise pursues an economic function approved by the state. In not protecting it, the state would be elevating the property right of the GM producer over those of neighbouring non-GM producers, a position which involves a tacit endorsement of the economic priority to be given to GM production. In this regard the regulatory regime for authorising GM releases also has a potentially key role to play in allocating and legitimating property rights.[95]

Choosing organic farming really would become no choice at all; organic farming would be possible only to the extent that the farmers could keep the GM genes owned by others off their property – an apparent impossibility at least in the case of canola. The consequence would be a perpetuation of "information feudalism," characterised by an unequal redistribution of property rights and favouring the technology's proponents at the expense of its detractors.[96]

Access to Information

One of the other important aspects of the *Hoffman* case that has perhaps been less discussed is the issue of access to information. As mentioned previously, this issue goes very much to the merits of the liability claims: who is best placed to bear responsibility for the consequences caused by the GM canola? One answer is that the companies are most likely to be aware of the risks associated with their products and the available precautionary measures to avoid or mitigate these risks. As the holders of this information, they are not only best placed to communicate

94 Bruce Ziff, "Travels with My Plant: *Monsanto v. Schmeiser* Revisited" (2005) 2 University of Ottawa Law and Technology Journal 493 at 503, online: http://www.uoltj.ca/articles/vol2.2/2005.2.2.uoltj.Ziff.493-509.pdf.

95 Christopher P Rodgers, "Liability for the Release of GMOs into the Environment: Exploring the Boundaries of Nuisance" (2003) 62 Cambridge Law Journal 371 at 380. See also Jack Ralph Kloppenburg Jr, *First the Seed: The Political Economy of Plant Biotechnology*, 2d ed (Madison, WI: University of Wisconsin Press, 2004) at 282, where he characterises the commodification of the seed through, in part, intellectual property rights, as "a vehicle for capital to gain an unprecedented degree of control over the shape of the crop production process as a whole."

96 Peter Drahos and John Braithwaite, *Information Feudalism: Who Owns the Knowledge Economy?* (New York: New Press, 2002), see in particular pp. 198–9. See also McLeod-Kilmurray, who states, "The risk of GMOs is not individual. They affect the ecosystem itself, the organic option itself," *supra* note 19 at 195.

information to users but also to bear responsibility for the consequences that these new varieties could cause.

A second angle to this issue is the challenge that the plaintiffs would have faced in gaining access to the information held by the defendants on the Roundup Ready and LibertyLink canola varieties, to assert their argument on the inherent dangers posed by these varieties. Biotechnology companies – including the defendants – have been notoriously reluctant to divulge even the most basic information on the health and safety testing for their biotechnology products. The federal government has aided and abetted this process by refusing to disclose information that should be available to the public under the terms of the *Access to Information Act*.[97] Indeed, it was only after the plaintiffs filed suit to gain access to large portions of an AAFC study on seed lot contamination that the government capitulated and disclosed the full report.[98] During the course of the proceedings in *Hoffman*, the plaintiffs served a Notice to Produce Documents on the defendants in relation to the health and safety data, and Monsanto responded with a motion requesting a confidentiality order.[99] The application was adjourned pending the outcome of the certification process, and with certification having been denied, the issue is now moot. Had the plaintiffs gained access to the information, however, they would still have faced the challenge of sifting through the more than 7,700 pages, whereas the defendants were already intimately familiar with the contents.[100]

On appeal to the Saskatchewan Court of Appeal, the plaintiffs argued that access to information was one reason why a class action was the preferable procedure for the dispute. They stated that being in a class action would give the plaintiffs the clout to request the voluminous health and safety data held by the companies: "It is doubtful that Monsanto would be offering to produce such documents to somebody bringing a $2,000 claim in the Small Claims Court of Saskatchewan. The photocopy costs alone would exceed any recovery achieved."[101] They also considered their success in forcing the government to disclose the seed lot contamination report as evidence

97 *Access to Information Act*, RSC 1985, c A-1. For a discussion of the Canadian government's application of the Act in the context of biotechnology, see Kathryn Garforth, "Balancing Industry Confidentiality with the Public Right of Access: The Case of Biotechnology in Canada" (CISDL Working Paper, September 2005) at 7–16, online: http://cisdl.org/biodiversity-biosafety/public/Balancing_Confidentiality_with_Access.pdf. Other reports have also found deficiencies with the Canadian public's access to information on GMOs; see, for example, The Royal Society of Canada (RSC), *Elements of Precaution: Recommendations for the Regulation of Food Biotechnology in Canada* (Ottawa: RSC, 2001) at 212–4 and recommendation 9.2, online: http://www.rsc.ca//files/publications/expert_panels/foodbiotechnology/GMreportEN.pdf; Peter Andrée & Lucy Sharratt, *Genetically Modified Organisms and Precaution: Is the Canadian Government Implementing the Royal Society of Canada's Recommendations?* (Ottawa: The Polaris Institute, 2004) at 11–13.

98 Plaintiffs' memo, *supra* note 19 at para 11 of Schedule "C"; "GM canola spreading: Agriculture Canada report" *CBC News* (28 June 2002).

99 Plaintiffs' memo, *ibid* at para 9 of Schedule "C."

100 Plaintiffs' memo, *ibid* at paras 8–9 of Schedule "C."

101 Plaintiffs' memo, *ibid* at para 11 of Schedule "C."

of the clout that a class can bring.[102] The Court of Appeal did not consider the issue of access to information in upholding Justice Smith's ruling that a class action was not the preferable procedure.

Finally, the decision by Justice Smith could represent an important precedent in public access to and participation in the regulatory approval process for GM plants. Justice Smith ruled that the defendants' unconfined release of their GM canola varieties could constitute a "development" under the terms of the EAA. In a wider context, this ruling could mean that when biotechnology companies – or indeed anyone – plan the unconfined release of a new genetically modified variety of plant in Saskatchewan, they will first need to conduct an environmental impact assessment. Environmental impact assessments under the Saskatchewan legislation can include the public disclosure of information on the planned "development" and a public information meeting. Similar provisions exist in the environmental impact assessment legislation of the other two prairie provinces.[103] The EAA also gives the minister the power to limit the disclosure of information that relates to a development.[104] As already described, it would appear as though the Court of Appeal's consideration of the plaintiffs' pleadings under the EAA has not disturbed Justice Smith's ruling on this point.

The plaintiffs used this argument to describe why a class action suit was the preferable procedure for their action. They stated that a declaration that the testing and development of GMOs was a "development" under the EAA that "would compel the Defendants to submit their engineered genes to a public environmental scrutiny rather than the behind closed doors approach they have been allowed to use with the federal government's regulatory bodies."[105] In discussing what the farmers were really seeking through their litigation, McLeod-Kilmurray writes: "The claim under the Environmental Assessment Act indicates that an essential part of what the farmers are seeking is greater oversight, participation, and accountability in making these decisions about the safety and impact of these highly contentious seeds, not solely cash recovery."[106] Of course, Justice Smith's ruling was not a determination that the defendants' activities are, in fact, a development under the EAA, but simply a finding that it was not plain and obvious that the plaintiffs' claim to this effect would have failed at trial. Nonetheless, her ruling could help to open what has traditionally been a very closed regulatory process.

[102] *Ibid.*

[103] In Manitoba: *Environment Act*, CCSM, c E125. See the definition of *development* in s 1(2), especially para (h) and the provisions on public hearings in s 10-12. In Alberta: *Environmental Protection and Enhancement Act*, RSA 2000, c E-12, see the factors for consideration in deciding whether to order an environmental impact assessment in s 44(3), especially paragraph (c) and the requirement to publish an environmental impact assessment report in s 52.

[104] EAA, supra note 11 at s 7.

[105] Plaintiffs' memo, *supra* note 19 at para 17 of Schedule "C." They reasserted this point in their memo on leave to appeal to the Supreme Court, *supra* note 62 at para 53.

[106] McLeod-Kilmurray, *supra* note 19 at 195.

CONCLUSION

In Canada, where biosafety law is less developed and the Cartagena Protocol has not yet been signed and ratified, jurisprudence offers the best indication of the current state of legal developments. The farmers in *Hoffman* were unsuccessful in their attempt to gain certification as a class to bring a class action suit on behalf of the organic farmers of Saskatchewan against Monsanto and Bayer. Although the case did not result in any determinations of liability – or lack thereof – against the companies, the implications of the case on environmental, social, and economic interests are clear. To that end, the dispute offers ample illustration of the principles of sustainable development law. Perhaps most particularly, the case highlights the principle of integration and interrelationship of social, economic, and environmental objectives. As described earlier, the farmers' claims were largely couched in the language of economics, but their dispute with biotechnology went beyond than this. They are organic farmers out of concern for the impacts that conventional and industrial agriculture have on the environment and the communities where they live, as well as for their economic well-being as farmers. Whether implicit or explicit, there were a multitude of social, economic, and environmental values integrated into the farmers' claims. Litigation, however, is an extremely blunt instrument for trying to defend these values.

Organic farmers see Canadian farming practices heading down a new road, and not knowing where it leads, they worry about eliminating the option to turn back. Traditional farming practices and the seeds that farmers plant have evolved over a long time. The introduction and rapid spread of genetically modified canola and their introduced genes have already brought significant changes to the Canadian prairies in the ten years that they have been in use. If government and industry are not going to "proceed with caution," organic farmers at least want to be protected in their decision to do so.

23

The Use of GMOs in Chile and the Protection of Indigenous Culture

Konstantia Koutouki and Paula Honorato Marin

The use of genetically modified organisms[1] (GMOs) in agriculture is controversial in most countries. The exponential growth in the number of acres devoted to GMO plants and the rapid increase in the level of technology used have led many to the conclusion that perhaps regulation of this area needs to be examined to determine its sufficiency and efficacy. Keeping up with biotechnology has always been a difficult task for the law; however, these days the consequences of such unfettered development of biotech products could prove irreversible. In fact, "use of Roundup [for GMOs] or its generic equivalent, glyphosate, has skyrocketed to the point that weeds are rapidly becoming resistant to the chemical. That is rendering the technology less useful, requiring farmers to start using additional herbicides, some of them more toxic than glyphosate."[2] Therefore, before legislators had the chance to debate and regulate a technology properly, it already could be on its way to becoming obsolete. Of course, the impact that it has already had cannot be ignored.

The rate at which biotechnologies such as GMOs are evolving presents an important challenge for governments. The clearest example of how difficult this issue can be is found in the European Union (EU). In July 2010 the New York Times reported that:

> [a]fter decades of pushing nations to surrender more power to Brussels, the European Union is about to throw in the towel on one highly contentious issue: genetically modified foods. On Tuesday, the European Commission will formally propose giving back to national and local governments the freedom to decide whether to grow crops that many Europeans still call Frankenfoods. The new policy is aimed at overcoming a stalemate that has severely curtailed the market for biotech seeds

[1] This term is used to refer to transgenic organisms because these organisms are always genetically modified, but GMOs are not always transgenic organisms. The difference lies in whether the modification was a product of genetic engineering.

[2] Andrew Pollack, "Study Says Overuse Threatens Gains from Modified Crops," New York Times, April 13, 2010.

in Europe for years. Only two crops, produced by Monsanto and B.A.S.F., are sold for cultivation here.[3]

In 2009, the International Service for the Acquisition of Agri-biotech Applications (ISAAA) listed the top producers of GMO products as follows:

> USA (64.0 million hectares), Brazil (21.4), Argentina (21.3), India (8.4), Canada (8.2), China (3.7), Paraguay (2.2) and South Africa (2.1). The balance of 2.7 million hectares was grown by the following 17 countries, listed in decreasing order of hectarage: Uruguay, Bolivia, Philippines, Australia, Burkina Faso, Spain, Mexico, Chile, Colombia, Honduras, Czech Republic, Portugal, Romania, Poland, Costa Rica, Egypt, and Slovakia.[4]

At a global level, current challenges such as lack of access to sufficient food supply are exacerbated by poverty, which affects developing countries to a much larger degree that developed ones.[5] For countries such as Chile this problem is aggravated by the fact that although they are developing countries, Chile and others want to be on the fast track to economic development. GMOs are particularly attractive to countries likc Chile that depend heavily on agricultural sales to fuel growth and economic development. Despite this perceived advantage, the use of GMOs has been the subject of controversy due to of possible health and environmental risks, including antibiotic resistance, associated risks in the consumption of products derived from animals fed GMO, toxicity, and allergies,[6] but also owing to the likely dependence of local farmers on seed from foreign multinationals and the monopoly of the latter over the former.[7]

Furthermore, the use of GMOs can have repercussions in other seemingly unrelated fields, such as culture. In this respect, countries like Chile are faced with potential damage to indigenous culture, which is firmly rooted in the natural environment. The indigenous people in Chile have a strong relationship with the

3 James Kanter, "Europe's New Approach to Biotech Food," New York Times, July 7, 2010. The article states that, "in the latest example of the persistent differences, countries failed for a third time to break a deadlock over whether to allow imports of six varieties of bioengineered corn for food and feed made by Monsanto, Dow AgroSciences, Pioneer and Syngenta."

4 ISAAA, "Highlights of 'Global Status of Commercialized Biotech/GM Crops: 2009'," online: http://www.isaaa.org/resources/publications/briefs/41/highlights/default.asp.

5 There is, to some extent, a degree of self-nomination for "developing countries." "Least developed countries" are defined according to a United Nations list. The World Trade Organization (WTO) has no definition for "developed countries" or "developing countries." Member countries self-designate as being "developed" or "developing." Other countries can challenge the decision by a member to use provisions meant for developing countries, however. See WTO, "Quels sont les pays en développement à l'OMC?", online: http://www.wto.org/french/tratop_f/devel_f/d1who_f.htm.

6 See Raúl Brañes and Orlando Rey, "Política, derecho y administración de la bioseguridad en América Latina y el Caribe" (México, DF: CEPAL, PNUMA & ALDA, 1999), UN Doc UNEP/LAC-IGWG.XII/TD.1 at 8-9.

7 See Chidi Oguamanam, "Agro-Biodiversity and Food Security: Biotechnology and Traditional Agricultural Practices at the Periphery of International Intellectual Property Regime Complex" (2007) Mich St L Rev 215.

surrounding ecosystem.[8] According to the Mapuche, the use of GMOs could be harmful to their culture, which reflects ancient practices, in which "man must not bring about a transformation in nature since there exists a dependent relationship between all of its components."[9]

This chapter examines existing Chilean legislation to identify laws that could be used to protect the rights of indigenous peoples in Chile from possible harm suffered from the presence of GMOs in their lands.

GMOs IN CHILEAN FOOD PRODUCTION

As mentioned previously, Chile has a significant GMO food production industry. Chile also has "been an international leader and paradigmatic case of neoliberal law and economics since the late 1970s ... Chile is also a paradigmatic case of national economic dependence on the export of natural resources, making the country especially dependent on ecosystem goods and services in the future."[10] Liberal trade policies have been a cornerstone of Chilean economic development for many years. For example, the U.S.-Chile free-trade agreement (FTA) eliminated tariffs on 90 percent of U.S. exports to Chile and 95 percent of Chilean exports to the United States, a significant proportion when compared with other FTAs.[11] Chile has concluded many free-trade agreements with various trade partners, including the United States, Canada, and the EU. To be fair, there are environmental provisions in all of these agreements; however, as Kirby points out, "allowing state-to-state dispute resolution on environmental issues is not sufficient to actually increase enforcement because States tend to rely on mutual nonenforcement when there are no other consequences."[12]

The use of GMOs in food production is part of Chile's economic growth strategy. In fact, Chile is part of the Miami Group, which is composed of GMO exporting nations such as Canada, the United States, Australia, and others. During the negotiations for the Protocol, this group "rejected any approach that would allow protectionist trade barriers to "masquerade" as environmental protection. They sought

8 Katherine Inostroza Romero, "Le rapport intégral entre l'être humain et la nature à partir de la cosmovision ancestral mapuche: Une étude exploratoire au Lac Lleu Lleu, VIIIème Région" at 182, online: http://ressources-cla.univ-fcomte.fr/gerflint/chili3/katherine.pdf.

9 José Lopez Arellano, "Pour une gestion éthique des OGM: Les représentations véhiculées dans la culture amérindienne du Québec en ce qui trait à l'alimentation, aux organismes génétiquement modifies (OGM) et aux transformations que l'humain peut apporter à la nature" (Québec: Commission de l'éthique de la science et de la technologie, 2002) at 2.

10 Carl J Bauer, "Dams and Markets: Rivers and Electric Power In Chile" (2009) 49 Nat Resources J 583 at 591.

11 Guillermo Gabriel Zorogastua, "Peru's Too Expensive – I'll Get My Cheese From Chile: The Agricultural Market Access Provisions of the U.S.-Chile FTA and the U.S.-Peru TPA" (2007) 16 J Transnat'l L & Pol'y 337 at 339.

12 Rachel T Kirby, "Giving Power to the People: Comparing the Environmental Provisions of Chile's Free Trade Agreements with Canada and the United States" (2009) 10 Sustainable Dev L & Pol'y 65 at 76.

'to limit the possibility of referring to the precautionary principle and socio-economic considerations in decision-making,' while emphasizing risk assessments and sound science as proper foundations for any determination."[13]

The current Chilean administration has breathed new life into the GMO industry in Chile, which goes beyond agricultural foods to include transgenic fish like salmon and other animal products.[14] According to the "Invest in Chile" website:

> Chile has taken significant steps to incorporate the biotechnology on its economy. Through a powerful series of initiatives, including a new regulatory framework and the economic incentives, Chile seeks to develop a sector of the biotechnology worldwide. The biotechnology in Chile contains already a great variety of sectors related to natural resources, including agriculture, livestock, biofuels, fishing, aquaculture, forestry and the industrial biotechnology. Chile also encourages foreign investment in R&D and is working lively in the formation of its human capital in the promotion of clinical investigation [*sic*].[15]

Given Chile's economic strategy, the perspective of the current administration on GMO utilisation, and Chile's lack of ratification of the Protocol, it is important to examine Chile's existing legal framework and see how much and which kind of protection can be found for indigenous culture and the natural environment in which they reside.

GMOs AND INDIGENOUS CULTURE

The relationship between the Chilean government and the various indigenous peoples that reside with in Chilean territory has been adversarial at best.[16] The liberal trade policies mentioned in the previous section have had an impact on indigenous communities in Chile. According to Barrera-Hernández, the tribulations of indigenous communities "are exacerbated in developing countries, like Chile and others in Latin America, where neo-liberal development policies have made the promotion of accelerated, market-driven economic development the governments' single most important goal and raison d'être."[17] To many indigenous peoples of Chile, the use of

13 Maria Julia Oliva, "On Biosafety and the Agreement on Sanitary and Phytosanitary Measures: What Will Decisions Regarding GMOs Have to Be Based on?" (2002) 13 Int'l Legal Persp 22 at 25.

14 USDA Foreign Agricultural Service, "Agricultural Biotechnology Annual – Chile" (2009) GAIN Report No CI9012, online: http://gain.fas.usda.gov.

15 CORFO, "Invest in Chile – Biotechnology Applied to the Food Industry," online: http://www.investchile.com/opportunities/food_industry/biotechnology_applied_to_the_food_industry.

16 See Noah Bialostozky, "The Misuse of Terrorism Prosecution in Chile: The Need for Discrete Consideration of Minority and Indigenous Group Treatment in Rule of Law Analyses" (2007) 6 Nw UJ Int'l Hum Rts 81.

17 Lila Barrera-Hernández, "Indigenous Peoples, Human Rights and Natural Resource Development: Chile's Mapuche Peoples and the Right to Water" (2005) 11 Ann Surv Int'l & Comp L 1 at 1.

GMOs represents a modification of traditional production methods,[18] calling into question the role of traditional knowledge in farming and food production. In Chile the indigenous population is sizable:

> [B]ased on the 2002 Census, approximately 700,000 households, or 4.6 percent of the total population, identify themselves as indigenous. The largest indigenous group is the Mapuche people (approximately 95 percent of all indigenous people in Chile), which is concentrated in the south and which is in turn subdivided into various indigenous territorial groups (the Huilliche, Lafkenche, Pehuenche and Nagche). A sizeable contingent of Mapuche people also lives in the metropolitan area of Santiago, where they are classed as "urban Mapuche;" their characteristics and problems differ significantly from those of the Mapuche in the south. The Aymara (7.1 percent of all indigenous people in Chile), Atacameño (3.04 percent), Quechua (0.89 percent), Colla (0.46 percent), and Diaguita peoples live in the north of the country. The Rapa Nui people – nowadays down to about 2,000 persons – live on Easter Island, and a small number of Kawashkar and Yagán remain in the far south.[19]

There is international and national legislation that could protect Chile's indigenous population. Ideally, a synergy between the two levels of legislation would provide these communities with the support necessary to further their socioeconomic development.

The International Labour Organization (ILO) C169 *Indigenous and Tribal Peoples Convention*[20] is Chile's only ratified international treaty concerned with the rights of indigenous peoples. The importance of the ILO for the indigenous population in Chile was illustrated recently. The country's Supreme Court in a "unanimous decision invoked ILO Convention 169 and could have far reaching consequences for Chile's mining industry. Chile's Supreme Court last week handed down a landmark ruling on indigenous water rights in a case that pitted Region I Aymara communities against Agua Mineral Chusmiza, a company seeking the rights to bottle and sell freshwater from a source used historically by Aymara indigenous residents."[21] Indeed, because of the ratification of the ILP 169 Convention, Chile is now obligated to "consult indigenous communities whenever considering legislative or administrative measures which may directly affect them. This also means that

18 Thierry Raffin, "Les plantes génétiquement modifiées dans les PVD: entre discours et réalité" (2006) 118 Révue Tiers Monde 705.

19 Project for Indigenous Self-Determination, "Indigenous Peoples in Chile," online: http://indigenousnews.org/indigenous-peoples/chile/.

20 Better known as Convention 169 of the International Labour Organization (ILO). This convention was adopted at the ILO General Conference June 27, 1989, ratified by the Chilean Senate in March 2008, and published as law October 15, 2008. The law was approved and incorporated into legislation in full, without any reservation. *Chile ratifica integramente el Convenio 169 de la OIT*, Ministerio de Planification, Gobierno de Chile, online: http://www.mideplan.cl/index.php?option=com_content&view=article&id=2625&Itemid=50.

21 "Chile's Supreme Court Upholds Indigenous Water Use Rights," online: http://int.piplinks.org/chile%E2%80%99s-supreme-court-upholds-indigenous-water-use-rights.

private investors are required to consult these communities on future investment projects."[22]

The ratification of ILO Convention 169 is a major step forward for Chile's indigenous population. Access to an international forum reinforces their capacity to call attention to issues concerning their rights within Chile. Given the outcome of the Agua Mineral Chusmiza case, one can conceive the application of ILO Convention 169 to GMO impacts on indigenous culture and socioeconomic welfare. The *United Nations Declaration on the Rights of Indigenous Peoples* is more difficult to incorporate into internal legislation in Chile owing to possible conflicts with the Chilean legal structure. Concepts such as collective rights[23] and collective ownership, are not well known in the Chilean legal culture. Used in conjunction with ILO Convention 169, however, the UN Declaration could further reinforce international legal protection for indigenous people in Chile faced with the presence of GMOs on their lands.

Ley 19.253 "*Ley Indigena*"[24] recognizes the existence of indigenous peoples who conserve their own culture and have lived on Chilean territory since the pre-Columbian period. In this sense, the State must recognize them as the "origin" of the Chilean nation, and hence there is recognition of cultural pluralism.[25] There exists no recognition of "indigenous peoples" as defined in international law, however; the law makes reference to ethnic groups. From this point of view, the *Ley Indigena* narrower[26] when compared with clauses found in international treaties. Also, *Ley Indigena* does not recognize collective rights[27] and contrary to what is expressed in ILO Convention 169,[28] it does not recognize indigenous rights to natural resources either.[29] Additionally, the law does not mention treaties that have been concluded between the Spanish crown and the Mapuche or other indigenous peoples in Chile, which represent an important legal tool for indigenous organisations.[30]

22 Export Development Canada, "Chile Country Overview", online: http://www.edc.ca/EN/CountryInfo/Documents/Chile.pdf.

23 Because of the application of traditional indigenous law, there could be a conflict between aboriginal human rights as defined by the political constitution of Chile and as defined by international law. See José Zalaquett Daher, "La Declaracion de Naciones Unidas sobre los derechos de los pueblos indigenas," anuario 04/5, Seccion Internacional, Universidad de Chile at 144.

24 *Ley* 19.253, *Ley Indigena* (5 October 1993), Ministerio de Planificacion y de Cooperacion, online: http://www.leychile.cl/Navegar?idNorma=30620.

25 Guillaume Boccara et Ingrid Seguel-Boccara, "*Politicas indigenas en Chile (siglos XIX-XX) de la asimilacion al pluralism, El caso Mapuche*," Nuevo Mundo Mundos nuevos, Biblioteca de Autores del Centro (BAC), 2005, online: http://nuevomundo.revues.org/594#quotation.

26 Fernando Casanueva, "*Indios malos en tierras buenas. Vision y concepcion del Mapuche segun las élites chilenas del siglo XIX*," in Jorge Pinto Rodriguez, ed, Modernización, Inmigración y Mundo Indígena. Chile y la Araucanía en el Siglo XIX (Chile: Universidad de la Frontera, 1998), at 102.

27 Michelle Velasquez, "*Pueblo Mapuche y Derechos Humanos en Chile*," Universidad Alberto Hurtado, documentos, at 14 and 15, online: http://etica.uahurtado.cl/documentos/pueblomapuchederechoshumanos.pdf.

28 See Article 15 of ILO Convention No 169.

29 Velasquez, *supra* note 27 at 17.

30 Norbert Rouland, *Droit des minorités et des peuples autochtones* (Paris: Presses Universitaires de France, 1996) at 347.

In this sense, the *Ley Indigena* does not follow developments in international law concerning indigenous people's rights.

Finally, the Mapuche and other indigenous cultures are seen as constituting part of Chile's national "heritage."[31] Hence, it is established that there is an obligation on the State in particular, as well as society in general, to promote, respect, and protect indigenous culture. To this end, suitable measures must be adopted to preserve their lands and ensure that the use of these lands correspond to the principles of sustainable development.[32] Although the *Ley Indigena* was a watered-down version of the initial demands of the indigenous people in Chile, it represents a solid legal base that they can use to further develop their rights in Chile, as well as a basis for bringing grievances to the international arena on many issues, including possible repercussions from GMO use in Chile.

Ley 19.253 Corporación Nacional de Desarrollo Indígena (CONADI) has as its objective the coordination of all activities and plans established by indigenous law,[33] as well as encouraging active participation of indigenous peoples and communities in mainstream society, especially that of indigenous women.[34] CONADI's upper management is composed of various members, eight of whom must be representatives from indigenous communities,[35] thus ensuring the participation of these communities within the organization.[36] The group is expected to fund projects that contribute to the socioeconomic advancement of indigenous communities.

The provisions of this law appears to be less assimilationist than the older trend of Chilean laws on the subject,[37] because this law recognises that indigenous peoples and indigenous culture must be preserved and encouraged. Nevertheless, the concept of indigenous identity is not precisely defined; the context of this term has yet to be determined.[38] Nonetheless, groups such as the Mapuche have been recognised as a unique ancestral population with their own culture and worldview. This means that their specificity and their right to participate in national life have been established. The State is responsible for the protection of any rights attributed to indigenous peoples,[39] including the protection of their lands and the environment

31 *Ley 19.253*, *supra*, s 7.
32 *Ibid*, s 1(4).
33 *Ibid*, s 39.
34 *Ibid*.
35 *Ibid*, s 41.
36 *Ibid*, s 10(5).
37 Alejandro Saavedra Parra, "*Estado chileno y legislación indígena*" in Armando Marileo, *¿Modernización o sabiduría en tierra mapuche?* (Santiago: Librería San Pablo, 1995) at 171.
38 Rouland, *supra* note 30 at 360–1.
39 Discrimination of aboriginal people and their culture is prohibited according to Article 8 of Indigenous Law, subject to fines. It would be more useful, however, to investigate causes of discrimination to avoid them in the future, and not simply penalise them. Boccara and Seguel-Boccara, *supra* note 28.

in which they live, as well as establishing authorities and organisations charged with ensuring that these standards continue.[40]

Decreto-1[41] specifically addresses development activities in the Araucanía region,[42] a region with a large indigenous population, in particular the Mapuche. The program focussed specifically on the preservation and restoration of the natural environment as well as the sustainable development of the indigenous communities that depend on the natural resources of the area. Furthermore, the emphasis placed on this region by DTR-1 goes far in acknowledging Chile's cultural pluralism. *Decreto 395 Aprueba reglamento sobre el fondo de tierras y aguas indigenas* subsidises the purchase of additional lands for communities affected by land scarcity, and finances mechanisms to permit the solution of land conflicts and the provision of water. According to Human Rights Watch, "up to 2002 . . . CONADI's land and water fund assigned resources to increase Mapuche land by about 50,000 hectares, benefiting about 4,617 families. It also secured Mapuche tenancy of 120,000 additional hectares of ancestral lands already long occupied by Mapuche, benefiting 3,697 families. Despite these important achievements, the still unsatisfied need for land of the Mapuche is estimated to be at least 150,000 hectares."[43]

Decreto 75, Apruebu reglamento para la clasificacion de especies silvestres,[44] a result of *Ley 19.300*,[45] requires the classification of wild species, according to their need for conservation. To date there have been four phases of classification. The first led to the classification of thirty-three species in different categories.[46] The second process led to the categorisation of 71 species, whereas the third allowed for the classification of 61, and the fourth, 133 species. A fifth and sixth phase are currently in process. *Decreto 977, Reglamento Sanitario de los alimentos* requires that all activities involving the modification of food or raw materials destined for human or animal consumption be approved by the health minister to meet the technical standards

40 For example: Government intercultural programs in fields of education and health, Hugo Ormeño et Jorge Osses, "*Nueva legislación sobre indígenas en Chile,*" in *Cuadernos de la Realidad Nacional*, No. 14. Santiago, Universidad Católica de Chile, 1972, at 18 and 19.

41 DTO-1: *Promulga El Convenio Entre La Ministerio De Planificacion Y (Mideplan) De Financiacion Especifico Del Proyecto Denominado "Recuperacion Ambiental Y Desarrollo Socio-Productivo En Areas Campesinas E Indigenas De La IX Region De La Araucania"* 2002, Ministerio de Relaciones Exteriores, online: http://www.bcn.cl/lc/tinterna/tratados_pdf/tratado_mp809.pdf.

42 Jorge Pinto Rodríguez, "*La población de la Araucanía en el siglo XX, Crecimiento y distribución especial,*" Temuco, Chile: Departamento de Ciencias Sociales, Ediciones Universidad de La Frontera, (2009) at 219 *Revista Historia*, No. 43, vol., enero-junio 2010, Santiago, online: http://www.scielo.cl/scielo.php?script=sci_arttext&pid=S0717-71942010000100018&lng=en&nrm=iso.

43 Human Rights Watch, "Undue Process: Terrorism Trials, Military Courts, and the Mapuche in Southern Chile" (2004), online: http://www.hrw.org/en/reports/2004/10/26/undue-process.

44 *Decreto 75, Aprueba reglamento para la clasificacion de especies silvestres,*" Ministerio Secretaria General de la Presidencia, BCN Ley Chile, online: http://www.conama.cl/transparencia/transparencia2009/normativadoc/DECRETO75.pdf.

45 *Bases Generales del Medio Ambiente*. This legislation regulates all major activities that are likely have an environmental impact and maintains in force all previous environmental regulations.

46 Comision Nacional del Medioambiente (CONAMA), "*Especies Clasificacion segun Estado de Conservacion,*" online: http://www.conama.cl/clasificacionespecies/index2.htm.

established by the international scientific community. The standard, *Resolucion exenta no 1.523*,[47] is an example. It governs the introduction of living modified organisms (LMOs), produced in or outside Chile, into the environment.[48] Modified organisms in transit to Chile must meet the phytosanitary demands established in each case.[49] Similarly, the resolution requires that before entering the territory, every LMO must obtain authorisation following a risk analysis.[50] The introduction of these organisms must be preceded by a notification from the originating country to the receiving one.

Ley 20249, Indigenous Coastal Protection, "establishes the coastal marine space of native peoples. This law amends current legislation to recognize and safeguard customary use of these areas by indigenous communities that are closely linked to the coastal area, thus enabling them to maintain their traditions and natural resource use."[51] Chile has established projects aimed at creating protected spheres. For example, the *Sistema Regional de Areas Protegidas* (SIRAP)[52] or even the *Sistema Regional de Areas Protegidas para la Conservacion y el Uso Sustentable del bosque lluvioso templado Valdiviano*.[53] These programs aim to implement regional development while conserving long-term biodiversity. The Chilean strategy is to achieve efficient management of protected spheres, leading to sustainable development. In this process, nongovernmental organisations, private enterprises, indigenous groups, and the public sector are all encouraged to participate. The creation of financial institutions, training programs, and environmental education allows for the expansion of protected spheres and the sustainable use of resources. In this sense, we have noticed a correspondence between Chilean provision and requirements found in the Convention on Biological Diversity and the International Treaty on Plant Genetic Resources for Food and Agriculture.[54]

47 "*Resolucion exenta No 1523, Establece normas para la internacion y introduccion al medioambiente de organismos vegetales vivos modificados de propagación*," 2001, Ministerio de Agricultura; Servicio Agricola y Ganadera; Direction Nacional, online: http://www.bvsde.paho.org/bvsacd/cd38/Chile/R1523-01.pdf.

48 *Ibid*, s 2.

49 *Ibid*, s 3.

50 *Ibid*, s 4.

51 Robert V Kemper, "The Indigenous Peoples of Chile: A Study of Human Rights for Three Indigenous Groups in Chile," online: http://faculty.smu.edu/rkemper/anth_4309-6309/Team_Report_Chile.ppt.

52 CONAMA, "*Sistema Regional de Areas Protegidas*" (SIRAP), online: http://www.conama.cl/portal/1301/fo-article-46149.pdf.

53 CONAMA, "*Sistema Regional de Areas Protegidas para la Conservacion y el Uso Sustentable del bosque lluvioso templado Valdiviano*"; *Programa de las Naciones Unidas para el Desarrollo*, "*Ficha de proyecto-GEF – Bosque Valdiviano*," online: http://www.pnud.cl/proyectos/fichas/bosque-valdiviano.asp.

54 *International Treaty on Plant Genetic Resources for Food and Agriculture*, 3 November 2001, FAO Res 3/2001 (entered into force 29 June 2004), ss 8-10 & 17, online: http://www.fao.org/nr/cgrfa/cgrfa-global/cgrfa-planttr/. *Agreement on the Application of Sanitary and Phytosanitary Measures*, being part of Annex IA to the *Agreement Establishing the World Trade Organization*, 15 April 1994, 33 ILM 1144, ss 2, 6, 7 & 18, online: http://www.wto.org/english/tratop_e/sps_e/spsagr_e.htm.

CONCLUSION

The expansion of genetically modified and transgenic organisms is evident in most regions of the globe. The debate as to the safety of such organisms, especially in food products, continues unabated. Despite possible biosafety issues, many countries see GMOs as key elements to increased production and hence economic growth. For developing countries, such a scenario is very alluring. Countries, such as Chile, see the use of GMOs as an integral part of their economic strategy as they are able to tap into a growing market. The revenue generated from these exports helps to fuel these countries' economic development.

Concurrently, however, Chile is beginning to mend its relationship with the indigenous peoples present in its territory. After centuries of hostility and attempts to assimilate or even eradicate these populations, the national government is taking steps to provide legal protection for these communities though the enactment of national legislation and the ratification of international conventions.

Despite efforts to integrate indigenous people into mainstream Chilean society, the indigenous peoples of Chile maintain their autonomy, working relentlessly to preserve their ways and culture. The use of GMOs threatens this objective, given its potential to contaminate indigenous land. Such contamination compromises traditional indigenous methods of cultivation as well as the well-being of plants native to these areas. These plants are not only a source of nutrition for these poor communities but also part of the health care system for many indigenous people. Furthermore, many of these plants have an intrinsic cultural and religious significance that cannot be understood outside the realm of a particular people. In fact, Shallat writes:

> Like conventional medicine, the diagnosis starts with a urine exam. In this case, however, no lab equipment is involved. Don Manuel Lincovil swirls a small glass jar of urine, holding it up to the light to detect patterns in the swirling sediment. From this observation, he determines whether the pain the patient is suffering from has a physical, psychological or supernatural origin. Don Manuel is a machi, or Mapuche healer, practicing in Santiago, far from his homeland in the south. Increasingly machis and yatiris (Aymara healers, from the north) are being incorporated into Chile's health care system. The Ministry of Health hopes this resurgence of traditional medicine – where medicine men with drums practice alongside physicians in lab coats and where new pharmacy chains tout ancestral remedies – will improve health care for the disadvantaged Mapuche and other indigenous groups. From a ruka (a thatched, dirt-floored hut) behind a conventional clinic, Don Manuel uses invocations and herbal infusions, music, trance and techniques similar to reflexology and aromatherapy. He treats lifestyle ills, such as obesity, diabetes, hypertension and respiratory illness that increasingly affect urban Mapuches.[55]

[55] Lezak Shallat, "Alternative Medicine: Chile's Holistic Approach to Health Care" (2010) Global Post, online: http://www.globalpost.com/dispatch/chile/101012/mapuche-indigenous-medicine-healthcare?page=0,1.

The legislative examples discussed in this chapter represent possible avenues for indigenous people to assert their rights should they wish to do so when faced with GMO contamination of their lands. Given that Chile has not ratified the Protocol, this piecemeal approach provides a basis on which to build a legal structure, to handle possible negative outcomes from the use of GMOs in agriculture and other areas. Even as it stands, a coherent application of these laws can go far in protecting the rights of indigenous peoples as well as the integrity of the lands and the authenticity of the natural organisms found there.

Given the precarious position of both GMO safety and indigenous cultural preservation in Chilean law, existing laws should serve as building blocks to truly innovative legislation that considers the precautionary principle when it comes to the regulation of GMOs, as well as the necessity to preserve and maintain Chile's indigenous heritage. Integration of international legal instruments such as the UN Declaration on the Rights of Indigenous Peoples and the Cartagena Protocol on Biosafety would be a tremendous start.

PART V

Global Policy Trends in Biosafety

24

Sustainable Development, Biosafety, and International Law

Frederic Perron-Welch, Marie-Claire Cordonier Segger, Christine Frison, and Jorge Cabrera Medaglia

Living within the environment's carrying capacity and managing the use of natural resources to meet the needs of current and future generations are tasks that societies have faced throughout history.[1] The modern concept of *sustainable development* is used to describe the efforts of States to improve the quality of life of their citizens in a way that can be maintained over the long term. It emerged in modern international relations to bridge the differing views of developing and developed States over how to achieve economic development while protecting the environment and ensuring the long-term survival of peoples.[2] In areas where impacts of economic growth are diffuse, global, and ultimately cumulative over time (i.e., biodiversity loss), it is more difficult to find common starting points from which to develop agreements, and the language of sustainable development has helped countries come to agreement. Sustainable development has thus become a key objective of many important international economic, environmental, and social treaties and regimes today.[3] Although finding one universal definition of *sustainable* that is appropriate for all sectors, cultures, and regions is not a straightforward process, in each area of law and policy that aims to promote sustainable development, progress is slowly being made.[4]

DEVELOPMENT PARADIGMS PRIOR TO STOCKHOLM

By the late eighteenth and early nineteenth century, scholarly debates recognised the existence of essential links between environment and development, and proposed

[1] See Separate Opinion of Vice-President Weeramantry in *Case Concerning the Gabčíkovo-Nagymaros Project (Hungary/Slovakia)*, (1997) ICJ Rep 7; M-C Cordonier Segger and A Khalfan, eds, *Sustainable Development Law: Principles, Practices and Prospects* (Oxford: Oxford University Press, 2004) at 15.

[2] M-C Cordonier Segger, "The Role of International Forums in the Advancement of Sustainable Development" (2009) 10 Journal of Sustainable Development Law & Policy at 4.

[3] *Ibid.*

[4] *Ibid.*

consequent environmental constraints to development.[5] Throughout, it remained unclear to which extent these issues were of international concern, and how international legal rules might govern the balance between environmental protection and economic development. Equity issues, human rights, and other social concerns were only barely touched upon at that time.[6]

Development and the environment remained distinct issues until well into the twentieth century. After World War II, the United Nations Economic and Social Council (ECOSOC) focussed its energies on promoting full employment and economic stability, as well as economic development, in two respective subcommissions.[7] The development agenda was primarily driven by postwar programmes aimed at the economic reconstruction of areas devastated by war, but it also sought to comprehend long-term economic development, focussing on the "inadequately developed" parts of the world, to promote the fullest and most effective utilisation of natural resources, labour, and capital, raising the level of consumption. They aimed to do so by increasing coal exports and trade in mining gear, reestablishing agricultural production, rehabilitating transport, manufacturing new equipment and machinery, reviving and expanding international trade, and increasing the production of electrical power.[8]

However, because of the unsustainable use and destruction of natural resources during the War, and their importance to reconstruction, the international community came to recognise that the sustainable use of natural resources was a serious issue facing humanity.[9] In 1948, the United Nations Educational, Scientific and Cultural Organization (UNESCO) hosted the first International Conference on the Protection of Nature, which led to the founding of the International Union for the Protection of Nature (IUPN), later renamed the International Union for Conservation of Nature (IUCN).[10] The IUPN Constitution first defined protection of nature as "the preservation of the entire world biotic community, or man's natural environment, which includes the earth's renewable natural resources of which it is composed, and on which rests the foundation of human civilization." It also remarks upon the vital importance of soils, water, forests, wildlife and wilderness areas for economic, social, educational and cultural reasons, and notes the impact

5 See, e.g., TR Malthus, *An essay on the principle of population*; D Ricardo, *Principles of Political Economy and Taxation* (Amherst, NY: Prometheus Books, 1996) at 47.

6 Cordonier Segger & Khalfan, *supra* note 1 at 16.

7 In 1946, ECOSOC established two subsidiary bodies, the Sub-Commission on Employment and Economic Stability, and the Sub-Commission on Economic Development. See "Economic and Social Council" (1947) 1(1) International Organization 103–115, online: http://www.jstor.org/stable/2703522.

8 *Ibid.*

9 *Conservation and utilization of resources*, UN Doc E/RES/32(IV) of 28 March 1947.

10 *International Union for the Protection of Nature* (Brussels: Hayez, 1948), online: http://www.iucn.org/dbtw-wpd/edocs/1948-001.pdf; P. van Heijinsbergen, *International Legal Protection of Wild Flora and Fauna* (Amsterdam: IOS Press, 1997) at 39.

on human well-being of natural resource depletion, and asserts that this trend may be reversed if people are awakened in time to a full realization of their dependence upon exhaustible natural resources and recognize the need for their protection, restoration, and wise and informed administration for the future peace, progress and prosperity of mankind.[11]

The following year, ECOSOC held the UN Scientific Conference on the Conservation and Utilization of Natural Resources (UNSCCUR), noting conservation's potential to contribute to peace and help underdeveloped or resource depleted regions.[12] The objective of UNSCCUR was to further develop scientific knowledge on the conservation and utilisation of forests, water supplies, soil, crops, pasturage and livestock, fish and wildlife, fuel and energy, and mineral resources.[13] Participants discussed three major themes: the influence of natural resources on civilisation and the influence of forms of civilisation on natural resources, the interrelationship between resources and international tensions, and educational methods designed to encourage the protection and more efficient use of resources.[14]

The work of the ECOSOC, UNESCO, and IUCN in the postwar years began bearing fruit in the mid-1950s and early 1960s. The IUCN was quickly recognised as pioneering necessary scientific research on "the conservation and wise use of the entire world biotic community, or man's natural environment, which includes the earth's renewable natural resources of which it is composed, and on which rests the foundation of human civilization."[15] The General Assembly issued a declaration that recognised the sovereign right of States to use and exploit natural wealth and resources freely in pursuit of their own progress and development in 1952.[16] This was followed up by a Commission on Permanent Sovereignty over Natural Resources established in 1958 to survey the status of "this constituent of the right to self determination, with recommendations . . . for its strengthening, and [with due regard] paid to the rights and duties of States under international law and to the importance of encouraging international co-operation in the economic development of underdeveloped countries."[17] The report of the commission led to a string of increasingly

[11] Preamble to *Constitution of the International Union for the Protection of Nature* in *International Union for the Protection of Nature, ibid.*

[12] UNESCO, *The Scientific Conference on Resource Conservation and Utilization* (UNSCCUR), UN Doc. UNESCO/NS/UNR/1 at paras. 1–3; see also *United Nations Scientific Conference on the Conservation and Utilization of Resources*, UN Docs. E/RES/109(VI), E/RES/141(VII) and E/RES/271(X).

[13] *The Proceedings of the United Nations Scientific Conference on the Conservation and Utilization of Resources*, 17 August–6 September 1949, Lake Success, New York, UN Doc E/Conf 7/2.

[14] *Ibid.* at 6.

[15] HJ Coolidge, "International Union for the Protection of Nature" (May 28, 1954) 119 Science Magazine 3A.

[16] *Right to exploit freely natural wealth and resources*, UN Doc A/RES/626(VII) at para 1.

[17] *Recommendations concerning international respect of the right of the peoples and nations to self-determination*, UN Doc A/RES/1314(XIII) at para 1.

broad declarations over the next decade on the rights and duties of States on natural resource use and economic development.[18]

In 1962, UNESCO held a General Conference on the utilisation of natural resources, resulting in the *Recommendation Concerning the Safeguarding of Beauty and Character of Landscapes and Sites*.[19] The resolution recognised that protective measures for safeguarding landscapes and natural sites must consider the needs of communities, their evolution, and the rapid development of technical progress.[20] In late 1962, the UNGA adopted a related resolution on "Economic Development and the Conservation of Nature."[21] The Resolution noted the work of UNESCO and agreed that early measures should be taken to preserve natural resources, flora, and fauna, and that these measures should occur simultaneously with economic development. This was proposed on the basis that preservation could be of considerable importance to the further economic development of countries and of benefit to their populations, to the extent that economic development in developing countries could jeopardise their natural resources and flora and fauna, sometimes irreplaceably, if such development takes place without due attention to conservation and restoration.[22] Early action was hoped for, especially in developing countries, directed toward preserving, reserving, enriching, and making rational use of natural resources and increasing productivity; observing existing international conventions and treaties on the preservation of the world's fauna and flora; introducing effective domestic legislation directed at eliminating the wasteful exploitation of soil, rivers, and flora and fauna; and taking appropriate steps to prevent the pollution of natural resources and to protect landscapes.[23] Finally, the General Assembly called on the Secretary General, specialised UN agencies, and other interested organizations to cooperate and give technical assistance to developing countries, at their request, for the conservation and restoration of natural resources and flora and fauna.[24]

In 1968, ECOSOC recommended to the General Assembly that it convene a conference on problems of the human environment in recognition of the continuing and accelerating impairment of environmental quality, and effects on the condition of people, such as their physical, mental, and social well-being, dignity, and enjoyment of basic human rights, in both developing and developed countries.[25]

[18] See *Concerted action for economic development of economically less developed countries*, 1960, UN Doc A/RES/1515(XV); *Permanent sovereignty over natural resources*, 1962, UN Doc A/RES/1720(XVI); *Permanent sovereignty over natural resources*, 1963, UN Doc A/RES/1803(XVII); *Permanent sovereignty over natural resources*, 1967, UN Doc A/RES/2158(XXI).

[19] *Recommendation Concerning the Safeguarding of Beauty and Character of Landscapes and Sites*, UNESCO General Conference XII, Paris, France, 11 December 1962.

[20] *Ibid* at s 12.

[21] *Economic Development and the Conservation of Nature*, UN Doc A/RES/1831 (XVII) at 21.

[22] *Ibid*.

[23] *Ibid* at para 1.

[24] *Ibid* at para 2.

[25] *Question of Convening an International Conference on Problems of Human Environment*, UN Doc A/RES/1346(XLV).

This conference was first to link environmental issues with human well-being, a relationship now recognised as underpinning sustainable development, and was necessary due to the profound changes in the relationship between humans and the environment resulting from scientific and technological developments.[26] In 1969, the Declaration on Social Progress and Development recognised social progress and development as the common concern of the international community, which should supplement national efforts to raise the living standards of peoples.[27]

INTERNATIONAL POLICY MAKING ON SUSTAINABLE DEVELOPMENT

The 1972 UN Conference on the Human Environment (Stockholm Conference) was the crossroads at which development that integrates economic, environmental, and social justice issues was first discussed and recognised as a necessity.[28] The main outcomes were the Declaration of the Conference on the Human Environment (Stockholm Declaration),[29] which put forward a common outlook and principles to lead the international community in preserving and enhancing the human environment, and the Action Plan for the Human Environment (Stockholm Action Plan).[30]

The twenty-six principles embodied in the Stockholm Declaration were a political response to the numerous ecological crises threatening the planet at the time that demanded worldwide attention and effort.[31] They place humans squarely at the centre of the response to these crises, noting that "[t]o defend and improve the human environment for present and future generations has become an imperative goal for mankind – a goal to be pursued together with, and in harmony with, the established and fundamental goals of peace and of worldwide economic and social development."[32] They are an early manifestation of international sustainable development law and policy.

Key principles include the right to freedom, equality, and adequate conditions of life in an environment of a quality that permits a life of dignity and well-being, and the responsibility to protect and improve the environment for present and future generations;[33] the need to safeguard earth's natural resources for the benefit of present and future generations;[34] the need to maintain, restore, or improve earth's capacity to produce vital renewable resources;[35] and that economic and social development is essential to providing a favourable living and working environment and for creating

26 *Ibid.*

27 *Declaration on Social Progress and Development*, UN Doc A/RES/2542(XXIV) at Article 9.

28 M Orellana, "Introduction" (2009) 10(1) Sustainable Development Law & Policy 2.

29 *Stockholm Declaration on the Human Environment*, UN Doc A/C.48/14 (1972).

30 *Action Plan*, UN Doc A/CONF.48/14/Rev.1.

31 Cordonier Segger & Khalfan, *supra* note 1 at 17.

32 *Stockholm Declaration*, *supra* note 29 at Part I, para 6.

33 *Ibid* at Principle 1.

34 *Ibid* at Principle 2.

35 *Ibid* at Principle 3.

conditions that are necessary for improvement of quality of life.[36] The Stockholm Action Plan also recognises the early international consensus on the importance of preserving the world's genetic resources for the purposes of conservation, evaluation, and utilization,[37] with particular emphasis placed on the conservation of plant, animal, and microorganism germplasm.[38]

Another significant result of the Conference was the launch of the UN Environment Programme (UNEP),[39] which has played a key role in developing new treaties, including the Convention on Biological Diversity. In 1975, the Governing Council of UNEP declared that "environmental management implies sustainable development of all countries, aimed at meeting basic human needs without transgressing the outer limits set to man's endeavours by the biosphere."[40] From this point on, sustainable development as a sociopolitical concept took on a growing role in international debates. It appeared as a key pillar of the 1980 IUCN World Conservation Strategy (WCS), which emphasized the "integration of conservation and development to ensure that modifications to the planet do indeed secure the survival and well-being of all people" and was followed by the book *Building a Sustainable Society* in 1981.[41] In 1982, the UN General Assembly provided further support for the general principles and attitudes of environmental conservation by adopting the World Charter for Nature,[42] which represented international convergence on some core elements of international environmental law and policy.[43]

The concept of sustainable development discussed from 1972 to 1982 recognised the need to ensure that economic and social development became more equitable and sustainable, especially for developing countries. The perception was that many of the most urgent international problems related to the global environment, but that these problems were intricately related to development strategies. It was also noted that economic growth alone would not solve these issues, especially for the poor. Public pressure increased, mainly in developed countries, for new laws and policies to resolve the problem.[44]

In 1983, the General Assembly established the World Commission on the Environment and Development (WCED) to investigate these issues and prepare a report on "environment and the global problematique to the year 2000 and beyond."[45] In 1986, the UN Declaration on the Right to Development defined development as "a comprehensive, economic, social and cultural process which aims at the constant

36 *Ibid* at Principle 8.
37 *Action Plan, supra* note 30, Recommendation 39.
38 *Ibid*, Recommendation 43.
39 UNGA Resolution 2997 (XXVII).
40 UNEP Governing Council, Document 20 (III) Programme policy and implementation at para 9(b). UN Doc UNEP/GC/3/20?
41 LR Brown, *Building a Sustainable Society* (New York: WW Norton & Company, 1981).
42 *World Charter for Nature*, UN Doc A/RES/37/7.
43 Cordonier Segger and Khalfan, *supra* note 1 at 17.
44 *Ibid* at 17–18.
45 *Process of Preparation of the Environmental Perspective to the Year 2000 and Beyond*, UN Doc A/RES/38/161.

improvement and well-being of the entire population and of all individuals on the basis of their active, free and meaningful participation in development and in the fair distribution of the benefits resulting therefrom."[46] The joint term *sustainable development* gained widespread acceptance as a result of the report of the 1987 of the World Commission on Environment and Development, "Our Common Future"[47] (also called the Brundtland Report). The Brundtland Report contains the most generally accepted definition of sustainable development, being "development that meets the needs of the present without compromising the ability of future generations to meet their own needs."[48] It was taken into consideration by the UNGA in establishing an "Environmental Perspective to the Year 2000 and Beyond."[49]

In response to the Brundtland Report, the UN General Assembly resolved to hold the 1992 United Nations Conference on Environment and Development (Rio Conference) to "elaborate strategies and measures to halt and reverse the effects of environmental degradation," listing nine areas "of major concern in maintaining the quality of the Earth's environment and especially in achieving environmentally sound and sustainable development in all countries."[50] At the Rio Conference, governments officially adopted sustainable development as the development paradigm.[51] Specific outcomes included the Rio Declaration,[52] a short consensus statement agreed on by the Heads of State assembled in Rio that affirms the focus on human development that is central to the concept of sustainable development and lays out a series of principles to help achieve sustainable development (including Principle 27);[53] Agenda 21, which is annexed to the Declaration; and the opening for signature of three international treaties, including the *Convention on Biological Diversity* (CBD).[54]

Agenda 21 complements the Rio Declaration by providing an 800-page "blueprint" to halt and reverse the effects of environmental degradation and to promote sustainable development in all countries.[55] As a means of implementation of sustainable development, Agenda 21 noted the need for international action to codify and develop

46 *Declaration on the Right to Development*, UN Doc A/RES/41/128.
47 World Commission on Environment and Development, *Our Common Future* (Oxford: Oxford University Press, 1987), Cordonier Segger & Khalfan, *supra* note 1 at 15.
48 *Our Common Future*, *ibid* at 43.
49 *Environmental Perspective to the Year 2000 and Beyond*, UN Doc A/RES/42/186.
50 *United Nations Conference on Environment and Development*, UN Doc A/RES/44/228.
51 Orellana, *supra* note 28 at 2.
52 *Rio Declaration on Environment and Development*, Annex 2, Report of the UN Conference on Environment and Development Vol. I (13 June 1992) UN Doc A/CONF.151/26 (Vol I); *Agenda 21*, Annex 2, Report of the UN Conference on Environment and Development Vol. I (13 June 1992), UN Doc A/CONF.151/26 (Vol I).
53 M-C Cordonier Segger, "Sustainable Development in International Law" at 357 in D Armstrong, ed, *Routledge Handbook of International Law* (New York: Routledge, 2008).
54 *Convention on Biological Diversity*, 5 June 1992, 1760 UNTS 79, 143; 31 ILM 1004 (entered into force 29 December 1993).
55 NA Robinson, ed, "Agenda 21: Earth's Action Plan," IUCN Environmental Policy & Law Paper No. 27 (1993).

"international law on sustainable development."[56] The Rio Conference contributed significantly to a global understanding of the concept of sustainable development. However, most of the particulars on sustainable development goals and standards, as well as commitments to achieve them, remained to be worked out through the treaties signed there, further intergovernmental negotiations, and the new institutions recommended by the conference such as the UN Commission for Sustainable Development (CSD).[57] The global treaties that were signed at the Rio Conference are one of the principal areas in which standards, rules, and regimes clarify the content of a commitment to sustainable development.[58]

In 1997, the UNGA convened a Special Session on Sustainable Development (Rio+5) to review and appraise the implementation of Agenda 21.[59] The outcome recognised that time is of the essence in meeting the challenges of sustainable development and pledged to continue to work together to accelerate the implementation of Agenda 21. States further agreed that "it is necessary to continue the progressive development and, as and when appropriate, codification of international law related to sustainable development. Relevant bodies in which such tasks are being undertaken should cooperate and coordinate in this regard."[60] In the same year, the UNGA approved the Agenda for Development.[61]

During its 55th session in 2000, the General Assembly adopted both the UN Millennium Declaration,[62] as well as a resolution convening the World Summit on Sustainable Development (WSSD), a ten-year review of progress achieved since 1992 in the implementation of Agenda 21.[63] The WSSD was held in Johannesburg from 26 August to 4 September 2002 and heralded the conclusive adoption of a broader concept of sustainable development. Outcomes include the Johannesburg Declaration, in which participating States assumed "a collective responsibility to advance and strengthen the interdependent and mutually reinforcing pillars of sustainable development, economic development, social development, and environmental protection at the local, national, regional, and global levels," and the Johannesburg Plan of Implementation, in which States committed "to undertaking concrete actions and measures at all levels and to enhancing international cooperation."[64] In Johannesburg the social pillar of sustainable development was emphasized and

56 See *Agenda 21*, *supra* note 52 at paras 39.1–39.10.

57 Cordonier Segger & Khalfan, *supra* note 1 at 21.

58 PW Birnie, AE Boyle, & C Redgwell. *International Law and the Environment*, 3rd ed. (Oxford: Oxford University Press, 2009) at 124.

59 *Report of the United Nations Conference on Environment and Development*, UN Doc A/RES/47/190; *Special session for the purpose of an overall review and appraisal of the implementation of Agenda 21*, UN Doc A/RES/51/181.

60 *Programme for the Further Implementation of Agenda 21*, UN Doc A/RES/S-19/2 at para 109.

61 *Agenda for Development*, UN Doc A/RES/51/240.

62 *Millennium Declaration*, UN Doc A/RES/55/2.

63 *Ten-year review of progress achieved in the implementation of the outcome of the United Nations Conference on Environment and Development*, A/RES/55/199.

64 *Johannesburg Plan of Implementation*, UN Doc A/CONF.199/20.

confirmed, especially with regard to the important health, social, and human rights in sustainable development.

In 2012, the UN Conference on Sustainable Development (UNCSD) or "Rio+20" took place in Brazil. The conference focused on renewing political commitment to sustainable development, assess progress toward internationally agreed-on goals on sustainable development, and addressing new and emerging challenges. The Declaration, entitled 'The Future We Want', focused on ways to advance a green economy in the context of sustainable development and poverty eradication, and on the further developments necessary to strengthen the international institutional framework for sustainable development.[65] Preparatory Committees were held in 2010 and 2011, during which states debated the successes and failures in implementing the major sustainable development summit outcomes and the major themes underlying the UN Conference on Sustainable Development.[66]

As has been canvassed elsewhere, States have also made sweeping commitments to sustainable development as an objective, through international, regional, and bilateral environmental, economic, and development treaty law.[67] This includes the CBD and Cartagena Protocol, the Treaty on European Union,[68] WTO Agreement,[69] UN Framework Convention on Climate Change,[70] UN Convention to Combat Desertification,[71] and International Treaty on Plant Genetic Resources for Food and Agriculture,[72] as well as the Millennium Declaration, among others.[73]

Finally, principles of sustainable development law have also been argued before and applied by international tribunals. The principles can be found in the decisions of the International Court of Justice, Permanent Court of Arbitration, and World Trade Organization (WTO), some of which are discussed below.

SUSTAINABLE DEVELOPMENT LAW OF RELEVANCE TO BIOSAFETY

As discussed in Chapter 1, the field of biosafety emerged concurrently with the development of modern biotechnology. When the science advanced to a point at which

65 *Implementation of Agenda 21, the Programme for the Further Implementation of Agenda 21 and the outcomes of the World Summit on Sustainable Development*, UN Doc A/RES/64/236 at para 20.

66 See Co-Chairs' Summary, First Preparatory Committee Meeting for the UN Conference on Sustainable Development, 2012 (20 May 2010); and *Progress to date and remaining gaps in the implementation of the outcomes of the major summits in the area of sustainable development, as well as an analysis of the themes of the Conference*, A/CONF.216/PC/2.

67 M-C Cordonier Segger, *supra* note 53 at 355.

68 *Treaty on European Union*, 31 ILM (1992) 247: 39, 42. See also K von Moltke, *The Maastricht Treaty and the Winnipeg Principles on Trade and Sustainable Development* (Winnipeg, MB: International Institute for Sustainable Development, 1995) at 9.

69 *Agreement establishing the World Trade Organization*, 33 ILM (1994) 15: 42, 44, 199, 201.

70 *UN Framework Convention on Climate Change*, 31 ILM (1992) 849.

71 *UN Convention to Combat Desertification in Those Countries Experiencing Drought and/or Desertification, Particularly in Africa*, 33 ILM (1994) 1328.

72 *International Treaty on Plant Genetic Resources for Food and Agriculture*, FAO Resolution 3/2001, (2006) 2400 UNTS 379, November 3, 2001 (entered into force June 29, 2004).

73 *Millennium Declaration*, UN Doc A/RES/55/2.

living modified organisms (LMOs) were being released into the environment, regulatory regimes emerged of necessity. For a time, biotechnology – and thus biosafety – remained the exclusive preserve of the developed world. Once the potential economic value of genetic resources became apparent, the developing world seized on biotechnology as a potential tool for development.

This stance led to several outcomes, notably Article 19 of the CBD on the "Handling of Biotechnology and Distribution of its Benefits." Biotechnology is also referenced in Article 16 of the CBD, "Access to and Transfer of Technology" and underlies the provisions of Article 15, "Access to Genetic Resources." The need for biosafety at a domestic level is also made clear in Article 8, "In-situ Conservation," at paragraph (g).

The detailed requirements of Articles 8(g), 16, and 19 – including technology transfer; developing country participation in biotechnology research; access to the results and benefits of biotechnology; consideration of a protocol on the safe transfer, handling, and use of LMOs; and information sharing on adverse impacts – combined with the comprehensive guidance on biotechnology (including biosafety) contained in Agenda 21,[74] transformed biosafety from a scientific challenge to an economic, environmental, and social one, and by extension a matter of importance for sustainable development. Because of the concatenation of issues and fears related to biotechnology and biosafety in the 1990s, there was a need for international rules of engagement to protect those States with little science, less information, and few resources to provide a way forward for biotechnology to be used safely for sustainable development. The result was the Cartagena Protocol, which forms part of a broader set of sustainable development principles and international law of relevance to biosafety.

The regulation of modern biotechnology in international law is a cross-cutting effort that requires recognition in various fora and instruments. The technology is not suitable for a single and all-encompassing international agreement administered by a single international organisation, which means that the development of international law on biotechnology necessarily takes place in multiple fora.[75] Regulation of modern biotechnology requires a carefully balanced allocation of rules on both international and domestic levels. The role of international law is limited in this sense and should focus on key framework issues relevant to the protection of human rights, consumers, and the environment in the international trade of goods and services relating to biotechnology. Allocation of regulatory powers should take into account current multilayered governance systems, assigning appropriate tasks to appropriate decision makers.[76]

[74] See *Agenda 21*, Section II ("conservation and management of resources for development") and Chapter 16 ("environmentally sound use of biotechnology").

[75] *The International Law on Biotechnology*, Resolution 5/2010 of the 74th Conference of the International Law Association, held in The Hague, The Netherlands, August 15–20, 2010 at para 1.

[76] *Ibid* at para 4.

Principles of international law are required to govern where global priorities and norms relating to biotechnology and sustainable development intersect. This is particularly true given that the domestic and international rules addressing biotechnology are highly fragmented, and efforts to achieve greater coherence must be made.[77] There must be a greater effort to develop legal principles and techniques to coordinate and, where necessary, integrate international, social, economic, and environmental regimes – including the design, strengthening, and implementation of laws, instruments, and policies to address current biosafety challenges.[78]

Key Concepts of International Law

The classical framework of international law posits that relations between independent States are governed by binding rules of law, emanating from their sovereign will, expressed in treaties, or in customs generally accepted as binding. Today's far broader public international law framework can be seen as a set of rules for an overarching transnational legal process that increasingly seeks to achieve global public interest goals. This is partly due to greater flexibility on the part of States and the recognition of transnational organizations, including non-governmental and intergovernmental organisations which assist in making, implementing, and monitoring international law. Today, international law provides normative frameworks and procedures for coordinating behaviour, controlling conflict, facilitating cooperation, and establishing common values.[79]

Treaty law is the most common source of international law, and like a contract, binds only those States that become "Party" to the treaty. Treaties can include specific rules or can simply spell out a framework with general objectives and principles. The Cartagena Protocol is an example of the former type of treaty, whereas the CBD falls into the latter category. Some treaties, including the CBD and Cartagena Protocol, require regular meetings of a "Conference of the Parties" to make decisions about treaty implementation. A smaller number, which also includes the CBD and Cartagena Protocol, designate supervisory bodies to monitor compliance. The decisions of these bodies are not usually formally binding on the Parties but are of significant weight and can influence the conduct of States.[80]

Customary law comes into being through the behaviour of States based on norms generally accepted as binding. To form customary law, State practice must be consistent and widespread, and there must also be evidence that States act as they do because they believe that they are bound by these norms (*opinio juris*). These customary norms, whether recognised or in the process of being recognised (*lex ferenda*) are normally called *principles*. Those principles that are consistently used

[77] *Ibid* at Preamble.

[78] Cordonier Segger and Khalfan, eds, *supra* note 1 at 367–8. See also Jutta Brunnée and Stephen J Toope, *Legitimacy and Legality in International Law* (Cambridge, UK: Cambridge University Press, 2010).

[79] *Ibid* at 10–11.

[80] *Ibid* at 8–9; see also Chapter 10.

and reinforced across various contexts can acquire the status of "peremptory" norms of international law (*jus cogens*), which prevail over other norms. Other obligations under customary law are owed to the international community as a whole, not just to a particular, identifiable State (*erga omnes* rules), because of the interest of all members in matters of global concern.[81]

These sources of international law are supplemented both by general principles of law common to the major legal systems and also judicial decisions and teachings of publicists.[82] Much of international law, whether customary or agreed on, reflects principles analogous to those of the major legal systems of the world, often derived from historical practice or a more remote common origin. When there has been insufficient practice by States to give a particular principle the status of customary law, and the principle is not governed by an international agreement, the International Court of Justice (ICJ) or other tribunals can resort to these general principles as a source to develop international law in special circumstances.[83]

Soft law refers to an immense body of legally nonbinding or incompletely binding norms, the most important of which are guidelines, resolutions, declarations, and recommendations that are made by parties to an international agreement in the course of its implementation.[84] Further examples include General Assembly Resolutions, Declarations of Principles adopted by the United Nations or by *ad hoc* conferences, treaty provisions agreed by consensus among the majority of States but not yet in force, and the proposals of the International Law Commission.[85] Soft law is not a source of international law, but certain documents can constitute evidence of emerging customary international law, act as tools to interpret treaties and custom, and serve as templates for generating the text of treaty law. Soft law has indisputable moral force and can demonstrate *opinio juris* or existing customary law as well as providing guidance to States in their behaviour.[86] Compliance with such precepts can shape legitimate expectations and demonstrate good faith. Pertinent examples include the Rio Declaration, Agenda 21, and the 2002 World Summit for Sustainable Development Johannesburg Plan of Implementation.[87]

The principle of *lex specialis derogat legi generali* holds that a special rule of law prevails over a general rule, so that, for example, treaty rules between States as *lex specialis* have priority as against general rules of customary law between the same States.[88] As *lex specialis*, the Cartagena Protocol is the most relevant law on biosafety.

81 *Ibid* at 9.

82 *Statute of the International Court of Justice*, 26 June 1945, TS No 933, 59 Stat 1055, at Article 38(1); Birnie, Boyle, and Redgwell, *supra* note 58, at 15; and Malcolm N Shaw, *International Law*, 5th ed. (Cambridge: Cambridge University Press, 2003) at 66–7.

83 Cordonier Segger and Khalfan, *supra* note 1 at 9.

84 *Ibid* at 9–10.

85 Birnie, Boyle, and Redgwell, *supra* note 58, at 15; Shaw, *ibid* at 107–14.

86 Birnie, Boyle, and Redgwell, *supra* note 58, at 15.

87 Cordonier Segger & Khalfan, *supra* note 1 at 10.

88 Shaw, *supra* note 82 at 116.

It provides sui generis international mechanisms for the biosafety regime, such as the advance informed agreement (AIA) procedure, provisions relating to LMOs for food, feed, and processing (LMOs-FFP), and risk assessment and management. As they are tested and start to work, they can be further developed, contributing to sustainable development. The Protocol strikes a balance, leaving the policy choice to countries, balancing risks with opportunities to move forward with the development and use of modern biotechnology.

The relationship of different international agreements is defined by the rules of customary international law, including in particular those recognized by Article 31(3)(c) of the Vienna Convention on the Law of Treaties (VCLT). It does not presume the priority of one set of rules over another set of rules in different instruments. All are of equal standing and importance. Application and interpretation of pertinent treaty rules should seek to bring about coherence and to avoid conflict among different rules. This should also apply even where the ratifications of the respective treaties are not fully identical.[89] In interfacing the Cartagena Protocol and WTO law, general international rules on interpretation, and more specifically, the application of principles of coexistence and coherence, that is, the "principle of not adding to or not diminishing the rights and obligations" and the concept of "no hierarchy, mutual supportiveness and deference."[90] Conflicts between treaties are resolved in accordance with Articles 30 and 41 of the VCLT and depend on the intentions of the Parties and the *lex specialis* rule. A treaty must also be interpreted and applied in the light of other rules of international law, including new norms of environmental law. An understanding of customary international law and general principles is thus essential, even when the law being interpreted is treaty based.[91]

Relevant Principles

Sustainable development at a minimum requires the integration of environmental concerns in development decision making. As noted in the Introduction to this book, the 1992 Rio Declaration from the UN Conference on Environment and Development (UNCED) states that "[i]n order to achieve sustainable development, environmental protection shall constitute an integral part of the development process and cannot be considered in isolation from it."[92] In the 2002 Johannesburg World Summit on Sustainable Development, this principle was arguably broadened to recognise that human rights and social development priorities also constitute an

[89] *The International Law on Biotechnology*, *supra* note 75 at para 5.
[90] *Ibid* at para 36.
[91] Birnie, Boyle, and Redgwell, *supra* note 58 at 109.
[92] Rio Declaration, Principle 4.

integral part of this balance.[93] Sustainable development must be implemented at all sectors of society and governance.[94] To respect this principle, States can seek to resolve overlaps or perceived conflicts between economic, social, and environmental concerns either through the activation of existing institutions or the establishment of new ones that can balance the competing goals.[95]

Equity and Respect for Human Rights

Existing standards of human rights protection fully apply to biotechnology. Human rights are of particular importance in shaping appropriate procedures in the field of biotechnology and must be taken into account by all levels of governance in shaping rules and regulations relating to biotechnology. The principle of human dignity should be taken as the paramount criterion in this regard.[96]

International human rights law is the most relevant body of social law for sustainable development owing to its wide range of legally and nonlegally binding obligations relating to civil, political, social, economic, and cultural rights.[97] Because of this fact, international human rights law is increasingly being brought into play in sustainable development debates.[98] Given that human rights law is drafted in broad terms that are not limited to any particular sphere of activity, it should apply as much to environmental and economic issues as to political and social issues.[99]

The broad scope of human rights law and discourse, and its imperative character and subversive nature, has caused resistance to the mainstreaming of human rights in international sustainable development law and policy. This push-back does not negate the utility of a rights-based approach but should rather promote a more rigorous debate on how human rights principles should be applied.[100]

Respect for human rights principles is pertinent to biosafety because international human rights law embodies a preexisting imperative on States to provide key rights like the right to political participation, the right to free expression, the right to an adequate standard of living, the right to equality of social welfare and alleviation of poverty, equality rights for women and minorities, the right to self-determination, and the right to development. Furthermore, human rights principles have even been

93 Cordonier Segger and Khalfan, *supra* note 1 at 103.

94 New Delhi Declaration, Principle 7.2.

95 *Ibid* at Principle 7.3.

96 *The International Law on Biotechnology*, *supra* note 75 at para 7.

97 Cordonier Segger and Khalfan, *supra* note 1 at 70. The foremost expression of these rights is the so-called International Bill of Human Rights, which includes the *Universal Declaration of Human Rights*, the *International Covenant on Economic, Social and Cultural Rights*, the *International Covenant on Civil and Political Rights*, and two optional Protocols.

98 *Ibid* at 201.

99 *Ibid* at 70–1 and 203.

100 *Ibid* at 220.

used to advocate the right to a healthy environment, an extension that has gained recognition at the regional and national level.[101]

These assertions are supported by the International Law Association's *New Delhi Declaration of Principles of International Law Relating to Sustainable Development*, which asserts that the realisation of the international bill of human rights is a central piece to the pursuit of sustainable development.[102] Since the World Summit on Sustainable Development in 2002, there has been growing recognition of the connection between environmental protection and human rights.[103] The 2009 *High Level Expert Meeting on the New Future of Human Rights and Environment: Moving the Global Agenda Forward* reviewed and furthered the debate over the linkages between human rights and the environment.[104]

Prevention of Serious Transboundary Environmental Harm and State Liability

One of the most long-standing principles of international environmental law stems from the 1938 *Trail Smelter Arbitration*,[105] which held that no State has the right to use or permit the use of its territory in such a manner as to cause injury in or to the territory of another or the properties of persons therein, when the case is of serious consequence and the injury is established by clear and convincing evidence.[106] The principle was reiterated in the Stockholm Declaration and Rio Declaration, and has been codified in the work of the International Law Commission in the *Draft Articles on Prevention of Transboundary Harm from Hazardous Activities, 2001*.[107]

Integration of Environmental Protection and Social and Economic Development

The *Iron Rhine* arbitral award affirmed that environmental law and the law on development are mutually reinforcing, integral concepts that require that where development could cause significant harm to the environment, there is a duty to prevent or at least mitigate such harm. Notably, it was also found that the principle applies not only in sovereign activities but also in activities undertaken in implementation of specific treaties between the Parties to those treaties.[108]

101 *Ibid* at 71–3.
102 *New Delhi Declaration of Principles of International Law Relating to Sustainable Development*, Preamble.
103 *Human rights and the environment as part of sustainable development – Report of the Secretary General*, UN Doc E/CN.4/2005/96. See also Lynda Collins, "Are We There Yet? The Right to Environment in International and European Law" 3(2) Journal of Sustainable Development Law and Policy 121.
104 High Level Expert Meeting on the New Future of Human Rights and Environment: Moving the Global Agenda Forward, Online: http://www.unep.org/environmentalgovernance/.
105 *Trail Smelter Arbitration (United States v. Canada)*, 3 R. Int'l Arb. Awards 1911 (1938), reprinted in 33 AJIL 182 (1939), 3 R. Int'l Arb. Awards 1938 (1941), reprinted in 35 AJIL 684 (1941).
106 Cordonier Segger and Khalfan, *supra* note 1 at 79.
107 Official Records of the General Assembly, Fifty-sixth Session, Supplement No. 10 (A/56/10).
108 *Award in the Arbitration regarding the Iron Rhine ("Ijzeren Rijn") Railway between the Kingdom of Belgium and the Kingdom of the Netherlands, decision of 24 May 2005* (2005), 27 R.I.A.A. 35 at para 59 (Permanent Court of Arbitration).

Principles of precaution, good governance, and public participation (supported by access to information and justice), which can be viewed as *lex ferenda*, are also reflected in the Cartagena Protocol on Biosafety. These principles are discussed in different chapters of this book. In Chapter 4, for instance, a link is highlighted between the principle of precaution and the risk assessment and risk management mechanisms of the Protocol. The decision-making procedures explained in Chapter 5, and the public participation obligations enshrined in the Protocol as discussed in Chapter 7, concur clearly with the principles of good governance and public participation. As has been shown in Chapters 13 to 17, these principles of international law on sustainable development are also reflected in regional, national, and even subnational biosafety regulations and cooperation mechanisms, and in the decisions of courts and tribunals to a certain extent. Indeed, through their adoption in treaties such as the Cartagena Protocol, these principles could be in the process of "crystallising" in international law, and contribute to a further understanding of how countries can deliver on commitments to sustainable development in specific treaty regimes.

Other general principles of international law, including state responsibility, fully apply in the field of biotechnology. Such principles can complement the Cartagena Protocol on Biosafety and new Nagoya – Kuala Lumpur Supplementary Protocol on Liability and Redress.[109]

Relevant Treaty Law

The World Trade Organization

The provisions of the Cartagena Protocol and the WTO Agreement[110] overlap because they both contain rules that pertain to international trade in LMOs. The most pertinent WTO rules include GATT Article XX(b),[111] the Agreement on the Application of Sanitary and Phytosanitary Measures[112] (SPS Agreement), the Agreement on Trade-Related Aspects of Intellectual Property Rights[113] (TRIPS), and the

[109] The International Law on Biotechnology, Resolution 5/2010 of the 74th Conference of the International Law Association, held in The Hague, The Netherlands, August 15–20, 2010 at para 30 [Hague Recommendations on International Law on Biotechnology].

[110] *Agreement Establishing the World Trade Organization*, 15 April 1994, 33 ILM 144.

[111] *General Agreement on Tariffs and Trade* 1994, 15 April 1994, being part of Annex IA to the *Agreement Establishing the World Trade Organization*, 15 April 1994, 33 ILM 144. Article XX(b) grants a general exemption to trade measures relating to human, animal, or plant life, subject to the requirement that such measures are not applied in a manner that would constitute a means of arbitrary or unjustifiable discrimination between countries where the same conditions prevail, or a disguised restriction on international trade.

[112] *Agreement on the Application of Sanitary and Phytosanitary Measures*, 15 April 1994, being part of Annex IA to the *Agreement Establishing the World Trade Organization*, 15 April 1994, 33 ILM 1144.

[113] *Agreement on Trade-Related Aspects of Intellectual Property Rights*, April 15 1994, 33 ILM 1197, being part of Annex IC to the *Agreement Establishing the World Trade Organization*, 15 April 1994, 33 ILM 1144.

Agreement on Technical Barriers to Trade[114] (TBT Agreement). The CBD Parties negotiating the Protocol were aware of this overlap and appear to have sought to design the Protocol in a way that would avoid conflicts with Parties' existing commitments under the WTO.[115]

Parties to the Cartagena Protocol that are also WTO members need to take both regimes into consideration when regulating the intentional transboundary movement of LMOs in a manner that promotes sustainable development.[116] The Cartagena Protocol makes this clear in its opening statement by promoting the mutually supportive implementation of trade and environment agreements with a view to achieving sustainable development and emphasising that the Protocol does not imply a change in the rights and obligations of a Party under any existing agreements, nor should it be subordinated to other international agreements.[117]

The objectives of the two regimes are not inherently incompatible, but they are clearly distinct. Notably, there are instances in which provisions of the Cartagena Protocol could require, authorise, or justify measures that might impede the flow of trade in LMOs between WTO members, thus running counter to the WTO's objective to liberalise trade.[118] For example, differences could arise over the extent to which the concept of precaution can be invoked in each treaty, as well as the extent to which socioeconomic considerations can be considered in decision making.[119]

Although there is some overlap between the treaties, the design of the Cartagena Protocol limits its impact on international trade. The provisions that have the greatest potential impact on trade are limited to the narrowest category of products. The AIA and risk assessment procedures apply only to the first intentional transboundary movement of LMOs for intentional introduction into the environment. By contrast, the category of LMOs that likely represents the largest percentage of international trade, LMOs-FFP, is subject to less stringent measures under the Protocol, although transboundary movements of such LMOs might still be subject to similar domestic regulations of the Party of import.[120]

Of note, not all Parties to the Protocol are WTO members, and there are also WTO members that are not Parties to the Protocol. Such countries might view WTO rules as being the only rules that apply to their trade in LMOs with Parties to the Protocol. If Parties to the Protocol become more familiar with the potential interactions between the Cartagena Protocol and the WTO, they can be more effective in ensuring that

[114] *Agreement on Technical Barriers to Trade*, 15 April 1994, being part of Annex IA to the *Agreement Establishing the World Trade Organization*, 15 April 1994, 33 ILM 1144.

[115] Mackenzie et al, *An Explanatory Guide to the Cartagena Protocol* (Cambridge, UK: IUCN, 2003) at 226.

[116] Mackenzie, et al, *supra* note 115 at 225.

[117] *Cartagena Protocol*, *supra* note 9 at Preamble.

[118] Mackenzie et al, *supra* note 115 at 225.

[119] Cordonier Segger and Khalfan, *supra* note 1 at 348.

[120] Mackenzie et al, *supra* note 115 at 227.

both regimes are implemented in a "mutually supportive" manner. The Protocol also governs some transboundary movements of LMOs that are unrelated to international trade, such as the noncommercial transboundary movement of laboratory specimens under contained-use provisions, and the unintentional transboundary movement of LMOs.[121]

The 2001 Doha Ministerial Conference agreed to negotiate on the relationship between WTO rules and multilateral environmental agreements (MEAs), in particular those that contain specific trade obligations.[122] Those negotiations have not come to a conclusion, but cooperation and exchange do take place between the WTO and the Secretariat for the Convention on Biological Diversity and its Cartagena Protocol.[123] In absence of binding rules, the Hague Recommendations on International Law on Biotechnology of the International Law Association provide some guidance on the matter. First, the Hague Recommendations suggest that risk analysis methods and procedures in the SPS Agreement should be aligned to the Cartagena Protocol, based upon a clear distinction between risk assessment and risk management, and complemented by risk communication.[124] Secondly, the Hague Recommendations underline that WTO Members should ensure that their sanitary or phytosanitary measures regarding GMOs are based on an international assessment of the risks to human, animal, or plant life or health, as appropriate to the circumstances, and based upon a determination by respective international organizations and accredited specialized scientific research institutions.[125]

The SPS Agreement identifies three institutions that are in charge of international standardisation. These institutions are the Codex Alimentarius Commission, the International Plant Protection Convention (IPPC), and the World Organization for Animal Health (OIE).

The Codex Alimentarius Commission is an international organisation created by the World Health Organization (WHO) and the Food and Agriculture Organization (FAO) that is in charge of establishing regulations related to food security.[126] Codex Alimentarius standards help to ensure fair trade practices in food trade, according to WTO rules. The Codex Alimentarius Commission has established an Ad Hoc Intergovernmental Task Force on Food Derived from Biotechnology to handle issues associated with food obtained through biotechnological processes, particularly food used for health and nutrition purposes.

In July 2003, the Codex Alimentarius Commission held a meeting during which three risk assessment standards for food resulting from biotechnology were approved. These standards establish risk assessment principles for food derived from modern

[121] *Ibid.*

[122] WTO, *Ministerial Declaration*, WTO Doc. WT/MIN(01)/DEC/.

[123] See "Existing Forms of Cooperation and Information Exchange Between UNEP/MEAs and the WTO," UN Doc TN/TE/S/2/Rev.2.

[124] Hague Recommendations on International Law on Biotechnology at para 22.

[125] Hague Recommendations on International Law on Biotechnology at para 23.

[126] Online: www.codexalimentarius.net.

biotechnology. The principles refer to the concept of *tracing*, which is a risk assessment tool whose meaning is the subject of an important debate. In fact, the United States considers this concept to be different from traceability and follows only the previous and subsequent link in the LMO movement chain. The approved standards are the following:

1. Principles for the Risk Analysis of Foods Derived from Modern Biotechnology
2. Guideline for the Conduct of Food Safety Assessment of Foods Derived from Recombinant-DNA Plants
3. Guideline for the Conduct of Food Safety Assessment of Foods Produced Using Recombinant-DNA Microorganisms

The Ad Hoc Intergovernmental Task Force on Food Derived from Biotechnology is currently working on a Draft Guideline for the Conduct of Food Safety Assessment of Foods Derived from Recombinant-DNA Animals as well as on two draft annexes pertaining respectively to Food Safety Assessment of Foods Derived from Recombinant-DNA Plants Modified for Nutritional or Health Benefits and Food Safety Assessment in Situations of Low-level Presence of Recombinant-DNA Plant Material in Food.

Likewise, the Codex Committee on Food Labelling is working on draft Guidelines for the Labelling of Food and Food Ingredients Obtained through Certain Techniques of Genetic Modification/Genetic Engineering, as well as on a draft Amendment to the General Standard for the Labelling of Prepackaged Foods (to address the issue of genetically modified foods). Finally, in 2007, the Codex Alimentarius Commission adopted the Working Principles for Risk Analysis for Food Safety for Application by Governments that were proposed by the Codex Committee on General Principles. These principles allude to the precautionary approach.

The goal of the IPPC is to prevent the international spread and introduction of pests of plants and plant products, and to promote appropriate measures for their control. The IPPC was reviewed in 1997 and entered into force in 2005. A commission working group studied plant pest risks associated to LMO issues and an international standard, the ISPM No. 11 pest risk analysis for quarantine pests, including analysis of environmental risk and LMOs. This standard intends to protect plants and ecosystems from LMO-related risks. According to the text, such protection measures should be cost effective, nondiscriminatory, feasible, and should not limit basic trade needs.[127]

The World Organization for Animal Health, also known as the Office International des Epizooties (OIE), is an intergovernmental organisation created by the International Agreement signed on January 25, 1924. OIE's mission is to ensure the transparency of animal health conditions worldwide. Member countries promise to declare animal diseases detected on their territory. In addition, the OIE is in charge

[127] For more information, see www.ippc.int.

of safeguarding international trade through the elaboration of health regulations destined to be applied to international transboundary movements of animals and animal products.

International Treaty on Plant Genetic Resources for Food and Agriculture

The International Treaty on Plant Genetic Resources for Food and Agriculture (ITPGRFA)[128] was created under the auspices of the United Nations Food and Agriculture Organization (FAO) in 2001.[129] Its objectives are "the conservation and sustainable use of plant genetic resources for food and agriculture and the fair and equitable sharing of benefits arising out of their use, in harmony with the Convention on Biological Diversity, for sustainable agriculture and food security."[130] The ITPGRFA has created the first multilaterally managed global gene pool of the world's major crops and forages: the Multilateral System of Access and Benefit-Sharing (MLS).[131] The MLS represents more than 1.1 million samples of plant genetic material that Contracting Parties govern collectively and multilaterally for the sake of the poor and the hungry.[132] Through this gene pool, the 127 Parties control – and are responsible for – the basis of more than 80 percent of the world's food from plants, and our most important tool for adapting to climate change in agriculture in years to come.[133]

The Treaty is in harmony with the CBD.[134] The ITPGRFA is an original instrument that implements the CBD's objectives and its sustainable development principles multilaterally, thereby reinforcing the links between its social, economic, and environmental dimensions.[135] Contracting Parties to the Treaty have focussed on

[128] *International Treaty on Plant Genetic Resources for Food and Agriculture*, FAO Resolution 3/2003, November 3, 2001 (entered into force June 29, 2004), United Nations Treaty Series I-43345, registered on December 13, 2006. To this date, there are 127 contracting parties [2010]. Online: http://www.planttreaty.org/.

[129] For a detailed account of the history of the negotiation and implementation of the Treaty from the points of view of its major stakeholders, see C Frison, F López F and JT Esquinas-Alcázar, eds, *Plant Genetic Resources and Food Security: Stakeholder Perspectives on the International Treaty on Plant Genetic Resources for Food and Agriculture* (Earthscan, FAO and Bioversity International, 2011) at Chapters 1 and 10.

[130] *ITPGRFA* at Art 1.1.

[131] For more detailed analysis, see G. Moore and W. Tymowski, *Explanatory Guide to the International Treaty on Plant Genetic Resources for Food and Agriculture* (IUCN Environmental Policy and Law paper No.57, 2005).

[132] G Tansey and T Rajotte, eds, *The Future Control of Food* (London: Earthscan, 2008) at 115–40.

[133] K Garforth and C Frison, "Occasional Paper 2, Key issues for the relationship between the Convention on Biological Diversity & the International Treaty on Plant Genetic Resources for Food and Agriculture" (Quaker International Affairs Programme, July 2007), online: http://biogov.uclouvain.be/staff/frison/OP2-Final_000.pdf.

[134] C Gerstetter et al, "The International Treaty on Plant Genetic Resources for Food and Agriculture within the Current Legal Regime Complex on Plant Genetic Resources" (2007) 10 Journal of World Intellectual Property, at 259.

[135] C Frison, "The Principles of Sustainable Development in the Context of the Implementation of the International Treaty on Plant Genetic Resources for Food and Agriculture" (2006) 2(2) Journal of Sustainable Development Law and Policy at 155.

creating a common multilateral exchange mechanism for plant genetic resources for food and agriculture (PGRFA).[136] This mechanism seeks to place every stakeholder on the same level, and facilitate the exchange of material. It also seeks to promote technology transfer, information sharing, and capacity building.[137] By providing a secure basis for the pooling of PGRFA, and a common framework for their conservation and sustainable use, the Treaty directly implements sustainable development principles.[138] Technology transfer for PGRFA is an important component of the benefit-sharing mechanism of the Treaty.[139] Furthermore, the improvement of the Global Information System on PGRFA is a way to enhance access to samples, in line with Article 17 of the Treaty platforms for cooperation, so that the whole can be larger than any one input. The Treaty makes such a strategy explicit, for example in Article 17, where the Global Information System will be built "based on existing information systems."[140] The Parties to the Protocol could explore these components of the MLS as sources of inspiration in the design of their national biosafety frameworks, or in the establishment of regional collaborations.

The Farmers' Rights recognised in the ITPGRFA are vested in the "past, present and future contributions of farmers in all regions of the world, particularly those in centres of origin and diversity, in conserving, improving and making available these resources."[141] These rights include the right to "save, use, exchange and sell farm-saved seed and other propagating material."[142] Farmers' rights clearly embody sustainable development principles, by promoting the social and economic development of farmers, taking into account the need to conserve and sustainably utilise resources from the environment.[143] These rights are relevant to the biosafety debate,[144] particularly at the national level, where Parties to both instruments have to implement these obligations. Indeed, the Treaty does not mention LMOs, but by

[136] *Ibid.*

[137] C Frison et al, "Intellectual Property and Facilitated Access to Genetic Resources under the International Treaty on Plant Genetic Resources for Food and Agriculture" (2010) 1 European Intellectual Property Review, online: http://biogov.uclouvain.be/staff/frison/christine.html; see also K Garforth and C Frison, *supra* note 130; and M Halewood and K Nnadozie, "Giving Priority to the Commons: The International Treaty on Plant Genetic Resources for Food and Agriculture" in G Tansey and T Rajotte eds, *The Future Control of Food – A Guide to International Negotiations and Rules on Intellectual Property, Biodiversity and Food Security* (Earthscan 2008) at 115.

[138] G Tansey and T Rajotte, *ibid.*

[139] *International Treaty on Plant Genetic Resources for Food and Agriculture* at Article 13.2(b).

[140] C Frison, F López, and JT Esquinas-Alcázar, "Chapter 20: General Conclusion" in C Frison, F López, and JT Esquinas-Alcázar, eds, *Plant Genetic Resources and Food Security: Stakeholder Perspectives on the International Treaty on Plant Genetic Resources for Food and Agriculture* (Earthscan, FAO and Bioversity International 2011) at 257.

[141] *International Treaty on Plant Genetic Resources for Food and Agriculture*, at Preamble, para 7.

[142] *Ibid* at art 9.

[143] R Andersen, "Agrobiodiversity and Farmers Rights in Developing Countries" Conference Papers – International Studies Association 2008; R Andersen et al, *The Plant Treaty and Farmers' Rights: Implementation Issues for South Asia* (South Asia Watch on Trade, Economics & Environment 2009).

[144] Cordonier Segger and Khalfan, *supra* note 1 at 350.

allowing researchers to develop and patent new products (based on plant material originating from the MLS), the Treaty implicitly recognises the use of LMOs. This therefore could have an impact on Farmers' Rights, if the new patented products hinder these rights.

The ITPGRFA's multilateral approach to capacity building and technology transfer is a tool that could be explored and applied to biosafety. Contributors to this book have expressed the need for innovative tools to help countries to design and implement their NBFs efficiently at the national level and at the regional and global level. The need for a system to provide common problem examination tools and concerted solutions at the regional level has been highlighted several times throughout this book.

Finally, the Treaty is becoming a model in other international processes. Other UN bodies are looking at the Treaty as their reference in crafting customised multilateral systems, in particular because it provides innovative tools that are designed in a sustainable development perspective. Examples include WHO and its process on influenza virus sharing and benefit sharing, and the UN Convention on the Law of the Sea and its process on the genetic resources of the areas beyond national jurisdiction (e.g., the deep seabed).

Aarhus Convention

The Aarhus Convention on Access to Information, Public Participation in Decision-making, and Access to Justice in Environmental Matters contains specific guidelines on public participation in LMO-related decision-making processes. The Convention entered into force in 2001. Although it was enacted by the United Nations Economic Commission for Europe, the convention is open to other nations. The Convention was amended in 2005 to devise more precise provisions on the deliberate release of GMOs, but the amendment has not yet entered into force. The amendment will enter into force once it has been ratified by at least three-quarters of the Parties. In September 2007, the amendment had been ratified by only four countries.

The Strategic Plan of the Aarhus Convention was adopted by the Meeting of the Parties at its third session through Decision III/8 on the Strategic Plan for 2009–2014.

International Processes

UN Environment Programme

The UNEP International Technical Guidelines for Safety in Biotechnology were enacted in 1995 to assist governments, intergovernmental organisations, private organisations, and others to strengthen capacities and exchange biosafety information. The guidelines are based on the principles of hazard identification, risk assessment (taking into account the probability of any hazards arising and the potential consequences of such hazards), and risk management (applying adequate management strategies, which includes developing procedures and methods to minimize risks and their consequences, or making decisions not to proceed). Such management strategies must be proportional to the risk assessment results.

REGIONAL AND DOMESTIC IMPLEMENTATION

Domestic experiences with the Cartagena Protocol detailed in previous chapters stress the need for common frameworks for regional and domestic implementation, and progress and challenges have resulted in Asia, Africa, the Americas, and Europe. For this implementation to be effective and not waste resources, it should complement and build on other law and good governance initiatives in countries and regional collaborations. Coherence and integration are not just good ideas; they are essential. This has been expressed in the examples of national biosafety regimes provided in Part III of this book and therefore is not reiterated here.

CONCLUSION

The Cartagena Protocol is embedded in a web of related international treaties and domestic laws, each reflecting the principles of sustainable development law to differing degrees and specific effects. There is a need to increase sensitivity to biosafety issues in other international legal regimes that share sustainable development objectives with the Protocol, however.[145] More be done to develop our understanding of the implications of these principles, and the linkages between the rules of the Protocol and the rules of other treaty regimes for domestic implementation. Ways must be found to incorporate the advances made in other international instruments in which sustainable development principles have been further recognised and implemented, to consider the lessons learned from these normative development experiences and practices that could be helpful to biosafety, and to analyse the meaning of these developments for domestic implementation of the Cartagena Protocol. Indeed, innovative tools have been designed and implemented to serve sustainable development as an objective in the context of several different regimes, and although these instruments relate to the specific area and scope of each regime, they might offer valuable ideas and lessons for countries seeking both to develop the international regime on biosafety and to implement their commitments on the domestic level. The example of the ITPGRFA, which applies a more common and multilateral approach to sustainable development objectives and principles, could be an option to investigate in the further implementation of regulatory aspects of the Cartagena Protocol.

[145] For example, human rights, trade, and investment, desertification, food, seeds, and oceans.

International law is still fragmented and diverse, although certain common principles are emerging in the field of sustainable development. As an objective, a commitment to sustainable development is central to the Cartagena Protocol regime, with related international principles clearly reflected in the treaty and domestic implementation as evidence of subsequent practice. Many questions remain unanswered, however, particularly the evolving status of some of the legal principles related to sustainable development and the overlap between the biosafety regime and other rules of international law. It is a rapidly changing and fascinating new field in which much more legal research and analysis are needed.

25

Trade and Investment Implications of Implementing the Cartagena Protocol

Marie-Claire Cordonier Segger and Markus Gehring

INTRODUCTION

A complex matrix of international rules has emerged to guide policies for a greener economy, including measures to encourage the sustainable development of genetic resources and biotechnologies. It is hoped that these policies support the development of more sustainable practices and technologies for global agriculture, forestry, health, and other systems. Much remains to be done, however.[1] Both developed and developing countries are considering new regulatory measures to promote biosafety, as part of a broader commitment to adopt new laws promoting more sustainable development.[2]

Evolving international economic regimes are an important piece of this puzzle. To date, multilateral efforts to liberalise trade and investment and to promote the sustainable use of biotechnologies have proceeded largely along separate tracks.[3] One system is defined by the treaties establishing the World Trade Organization (WTO) and its annexes, along with over 3,000 regional and bilateral trade and investment treaties. The other is shaped by the *Convention on Biological Diversity* (CBD),[4] *Cartagena Protocol on Biosafety*,[5] *Nagoya – Kuala Lumpur Supplementary Protocol*

[1] United Nations Environment Programme, 10 *Years After Rio: The UNEP Assessment* (UNEP: New York, 2002).

[2] Decision BS-III/5 – *Matters relating to the financial mechanism and resources*, UN Doc. UNEP/CBD/BS/COP-MOP/3/15. UNEP-GEF Biosafety Projects, online at http://www.unep.org/biosafety/.

[3] G Sampson, *The WTO and Sustainable Development* (UNU Press, 2005); M Gehring and MC Cordonier Segger, eds, *Sustainable Development in World Trade Law* (chapter on precaution / biosafety).

[4] *Convention on Biological Diversity*, 5 June 1992, 1760 UNTS 79, 31 ILM 818 (entered into force 29 December 1993) [CBD].

[5] *Cartagena Protocol on Biosafety to the Convention on Biological Diversity*, 29 January 2000, 2226 UNTS 208 (entered into force 11 September 2003), online: http://bch.cbd.int/protocol/text/.

The authors would like to thank Freedom-Kai Phillips for his substantive insights and invaluable assistance in preparing this chapter.

on Liability and Redress,[6] and several related international arrangements. Although both systems of international rules may share a common sustainable development objective, the relationship between the two is complex. As many have noted, these regimes are likely to come into closer contact as trade measures to secure biosafety lead to economic effects.[7] International trade and investment treaty rules also could affect the viability and effectiveness of new regulations to secure sustainable development of biotechnologies.[8] Just as trade and investment rules can and should shape biosafety policy, biosafety rules can and do shape trade and investment policy. Although biosafety laws and policies might restrict or constrain certain kinds of economic development, they also can provide incentives for new kinds of development. Indeed, certain biotechnology measures are already being characterised as trade and investment rules themselves.

New policies and regulations are being set in place to establish National Biosafety Frameworks (NBFs), which aim to reduce the risks of biotechnologies, especially the import and export of living modified organisms (LMOs), and to stimulate investment in more sustainable development of genetic resources. There are important differences between international disciplines that govern the use of trade- and investment-related barriers by States, international rules that encourage States to set trade measures in place to govern the import and export of LMOs, and domestic measures that encourage countries to conserve and develop genetic resources, or firms to transfer or develop new biotechnologies. The legal rules that govern the international movements of LMOs and investment in biotechnologies can – and should – be considered part of the wider body of evolving trade and investment law. Just as many trade provisions in the WTO Agreements contribute to the WTO's sustainable development objective, so do the *lex specialis* trade and investment provisions in the CBD and Cartegena Protocol contribute to the sustainable development objective of the global biodiversity and biosafety regime.

Based on recent legal analysis and scholarship in the field of sustainable development, this chapter examines the trade and investment implications of the bioeconomy. First considered is how trade regimes, investment rules, and biosafety measures can serve as tools for sustainable development. Then, we examine the trade and investment implications of new biosafety regulations, suggesting ways to

6 *Nagoya – Kuala Lumpur Supplementary Protocol on Liability and Redress to the Cartagena Protocol on Biosafety*, 15 October 2010.

7 Sabrina Safrin, "Treaties in Collision: The Biosafety Protocol and the World Trade Organization Agreements" (2002) 96(3) Am J Intl Law 606; Gilbert R Winham, "International regime conflict in trade and environment: the Biosafety Protocol and the WTO" (2003) 2 World Trade Review 131; Terence P Stewart and David S Johanson, "A Nexus of Trade and the Environment: The Relationship between the Cartagena Protocol on Biosafety and the SPS Agreement of the World Trade Organization" (2003) 14 Colo J Int'l Envtl L & Pol'y 1; Aarti Gupta, "Governing Trade in Genetically Modified Organisms: The Cartagena Protocol on Biosafety" 42(4) Environment: Science and Policy for Sustainable Development 22.

8 *Ibid.*

minimise overlap with trade and investment disciplines when drafting biosafety-related regulations, and addressing issues related to technology transfer. Next, we consider how trade and investment regimes might be refined to complement and support the bioeconomy. The chapter concludes with a call for integration and, at minimum, coherence among trade, investment, and biosafety regimes sharing common sustainable development objectives.

TRADE REGIMES, INVESTMENT RULES, AND BIOSAFETY MEASURES: TOOLS FOR SUSTAINABLE DEVELOPMENT?

International Trade Law Obligations

Similar to the rules of the international biosafety regime, the benefits of trade liberalization are not available to States and their citizens by right. Instead, governments negotiate economic treaties to secure access to foreign markets for their firms, and access to international goods and services for their consumers. In the WTO and many regional trade agreements, States have committed to lower tariffs progressively, to secure nondiscrimination and most-favoured-nation (MFN) status for goods and often services from trading partner countries, and to operate transparently. These four main obligations provide the foundations of world trade law,[9] aiming to secure stability and lower costs for firms from all State Parties to trade agreements. The rules might also affect a government's "room to move" in the design of new laws and measures to promote higher levels of biosafety and respond to perceptions of risk, in the interest of sustainable development of their bioeconomy.

First, States have committed to convert many nontariff barriers to trade, such as quotas, into more easily measured tariffs. Essentially, States have reciprocally agreed to be bound by schedules of specific tariff rates, and to continue negotiating reductions of these rates. The main commitments of WTO members are contained in individual detailed Country Schedules attached to the *General Agreement on Tariffs and Trade* (GATT)[10] for goods, and the *General Agreement on Trade in Services* (GATS),[11] which lists services. GATT rules apply to all goods unless they are reserved or excepted. GATS rules use a "positive-list" approach whereby only services listed by WTO Members are covered. These schedules provide comprehensive listings of all the products for which a WTO member has accepted a commitment to a binding tariff at a particular level. Each WTO member maintains a *schedule of concessions*, which is essentially a list of the tariff rates that it imposes on imported

9 M Gehring, J Hepburn & MC Cordonier Segger, *World Trade Law in Practice* (New York: Globe Law & Business, 2006) 17.

10 *General Agreement on Tariffs and Trade* (adopted 30 October 1947, provisionally entered into force 1 January 1948) 55 UNTS 194, CTS No 31 (1948) (GATT).

11 *General Agreement on Trade in Services* (adopted 15 April 1994, entered into force 1 January 1995) 1869 UNTS 183 (GATS).

goods. When members *bind* a certain tariff in their schedule, they guarantee that their customs tariff on the product in question will not be higher than the bound rate. Members have no obligation under WTO law to lower their bound tariff rates, but they do commit to continue negotiations to bind more product lines and to lower the tariffs on product lines already bound.[12] The WTO Committee on Trade in Goods and the WTO Committee on Trade in Services monitor implementation of these rules. This commitment to provide market access on a nondiscriminatory basis can curtail the type of rules that States adopt, affecting a State's ability to restrict certain imports selectively.[13] Article XI:1 of the GATT, the provision that prohibits quantitative restrictions, has been used to evaluate the GATT consistency of health- and environment-related trade bans, for example, in the *EC – Asbestos Case*,[14] the *US – Tuna Case*,[15] and *US – Shrimp Case*.[16] This, as is discussed later, is relevant for States seeking to ban trade in LMOs or other biotechnology products.

Second, Members commit to extend MFN status[17] to each other, precluding discrimination in the ways goods, services, and investments from different economic partners are treated. This can affect biosafety, should a State decide to limit LMOs or other products of biotechnology from one country, although still allowing them from another. The reason that this causes particular challenges for biosafety and for the Cartagena Protocol's trade measures springs from a third principle. Essentially, Members of the WTO and other trade agreements also commit to extend "national treatment" to other Members for goods, certain services, and other aspects of their trade policy.[18] This prevents WTO Members from treating the "like products" of other members differently[19] but is also meant to prevents like products from being distinguished on the basis of the process and production methods (PPMs) used to make them. The rules of the trade regime discipline the types of health, environmental, natural resource management, consumer safety, and other standards that

[12] Gehring, Hepburn, & Cordonier Segger, *supra* note 9 at 97.

[13] *Ibid.*; M. Trebilcock and R. Howse, *Regulation of International Trade* (New York: Routledge, 2005) at 336.

[14] *European Communities – Measures Affecting Asbestos and Products Containing Asbestos (Complaint by Canada)* (2001), WTO Doc. WT/DS135/AB/R at para. 99 (Appellate Body Report), online: http://www.wto.org/english/tratop_e/dispu_e/cases_e/ds135_e.htm (*EC – Asbestos*).

[15] *United States-Restrictions on the Import of Tuna* (1991) GATT BISD 39S/155, (1991) 30 ILM 1594.

[16] WTO, *United States: Import Prohibition of Certain Shrimp and Shrimp Products – Report of the Panel* (15 May 1998) WT/DS58/R; see also WTO, *United States: Import Prohibition of Certain Shrimp and Shrimp Products – Report of the Appellate Body* (6 November 1998) Doc WT/DS58/AB/R.

[17] GATT art I:1, GATS art II:2, *Agreement on Trade-Related Aspects of Intellectual Property Rights* (adopted 15 April 1994, entered into force 1 January 1995) 1869 UNTS 299, art 4 (TRIPS).

[18] GATT art III, GATS (n 14) art XVII, TRIPS (n 14 above) art 3.

[19] Likeness is determined by the dispute settlement mechanism in the WTO on a case-by-case basis, taking into account criteria such as "the product's end-uses in a given market; consumers' tastes and habits, which change from country to country; the product's properties, nature and quality" (Report of the Working Party on Border Tax Adjustments, BISD 18S/97, para 18).

WTO members can apply to products, if exceptions are not secured by the State implementing the measures in question.[20] MFN and national treatment commitments are implemented through the WTO *Agreement on Technical Barriers to Trade* (TBT Agreement),[21] which addresses technical regulations and standards, and the WTO *Agreement on the Application of Sanitary and Phytosanitary Measures* (SPS Agreement),[22] which addresses health and plant safety regulations and standards.

The TBT Agreement and the SPS Agreement define when certain restrictions on trade are allowed to limit protectionism (e.g., the use of regulations to unfairly privilege domestic firms vis-à-vis the firms of trading partners).[23] As such, for instance, the SPS Agreement essentially provides specific restrictions on the types of sanitary and phytosanitary standards governments should adopt, conditioning the relevant GATT rules and exceptions.[24] WTO TBT and WTO SPS Committees study and debate these issues and can grant time-limited exceptions to developing countries in light of their particular financial, trade, and development needs.[25]

WTO Members also commit to protect intellectual property rights through the WTO Agreement on Trade Related Aspects of Intellectual Property Rights[26] (TRIPs Agreement), to discipline subsidies in the WTO Agreement on Subsidies and Countervailing Measures[27] (Subsidies Agreement), to discipline government procurement through the WTO Agreement on Government Procurement,[28] and to discipline investment measures related to trade in goods in the Agreement on Trade-Related Investment Measures[29] (TRIMs). Such obligations might affect government attempts to regulate in relation to biosafety. The TRIPs Agreement obliges WTO Members to set laws in place to protect private intellectual property rights, potentially affecting technology transfer, including technologies to assess or reduce risks from

20 There is a growing opinion which argues that in the GATT the same rules as in the Agreement on Technical Barriers to Trade should be in force, permitting process and production methods to be taken into account under certain conditions.

21 *Agreement on Technical Barriers to Trade* (adopted 15 April 1994, entered into force 1 January 1995) 1868 UNTS 120 (TBT).

22 *Agreement on the Application of Sanitary and Phytosanitary Measures* (adopted 15 April 1994, entered into force 1 January 1995) 1867 UNTS 493 (SPS).

23 M. Echols, *Food Safety and the WTO: The Interplay of Science, Culture and Technology* (The Hague: Kluwer Law Int, 2001), C. Button, *The Power to Protect: Trade, Health and Uncertainty in the WTO* (Oxford: Hart Publishing, 2004) 43–90.

24 Button, *ibid.* at 10–11.

25 See WTO's Documents Online database (http://docsonline.wto.org) using document symbol G/SPS/GEN for all documents of the SPS Committee, including those related to exceptions for developing countries.

26 TRIPs, *supra* note 17.

27 *Agreement on Subsidies and Countervailing Measures* (adopted 15 April 1994, entered into force 1 January 1995) 1867 UNTS 14.

28 *Agreement on Government Procurement* (adopted 15 April 1994, entered into force 1 January 1995) 1867 UNTS 194.

29 *Agreement on Trade-Related Investment Measures* (adopted 15 April 1994, entered into force 1 January 1995) 1868 UNTS 186 (TRIMs).

unplanned releases of LMOs. The Subsidies Agreement disciplines the types of subsidies that WTO members can provide, potentially affecting incentives that governments could otherwise provide to promote sustainable agriculture, and subsidies that governments might provide for certain biosafety industries. The WTO Government Procurement Agreement and the TRIMs are minimalist accords, because governments were unwilling take on significant restrictions in these areas.[30] For instance, the TRIMs apply only to measures that affect trade in goods, committing to notify certain specific trade-related investment measures that discriminate against foreigners or foreign products.[31] If more stringent disciplines are adopted on government procurement or investment, however, they might constrain schemes for public purchasing of LMO-free products, or limit the adoption of biosafety regulations affecting foreign investors in agriculture or other industries.

Third, Members agreed to transparency and notification obligations, as well as binding, peaceful settlement of trade disputes.[32] The WTO's Dispute Settlement Understanding (DSU) aims to provide "security and predictability to the multilateral trading system."[33] WTO members settle disputes through consultation and other mechanisms, but if attempts fail to defuse a trade-related conflict, members can submit the disagreement to adjudication by a Panel, with appeals to an Appellate Body. If a decision is not implemented by the losing party, the Dispute Settlement Body can authorise retaliation (e.g., suspension of trade concessions like preferential tariffs).

The WTO Agreements are not without exceptions for measures related to sustainable development. First, Article XX of the GATT allows WTO members to violate WTO disciplines in certain circumstances, such as for the protection of health, the environment, or conservation of exhaustible natural resources. Article XX reads, in relevant part:

> Subject to the requirement that such measures are not applied in a manner which would constitute a means of arbitrary or unjustifiable discrimination between countries where the same conditions prevail, or a disguised restriction on international trade, nothing in this Agreement shall be construed to prevent the adoption or enforcement by any contracting party of measures: . . .
>
> (b) necessary to protect human, animal or plant life or health; . . .
>
> (g) relating to the conservation of exhaustible natural resources if such measures are made effective in conjunction with restrictions on domestic production or consumption; . . . .[34]

[30] Gehring, Hepburn, & Cordonier Segger, *supra* note 9 at 139.

[31] *Ibid.*

[32] GATT, *supra* note 10 at art X; GATS, *supra* note 11 at art III; TRIPs, *supra* note 17 at art 63; TBT, *supra* note 21 at art 10; SPS at art 7.

[33] *Understanding on Rules and Procedures Governing the Settlement of Disputes* (adopted 15 April 1994, entered into force 1 January 1995) 1869 UNTS 401 at art 3.2 (DSU).

[34] GATT, art XX.

Similar exceptions were agreed in the GATS and the TRIPs Agreement. As noted by the Retrospective Analysis of the 1994 Canadian Environmental Review of the WTO, GATT Article XX is an important safeguard for a State's ability to regulate for sustainable development.[35] Once a violation of trade law obligations has been established, however, the burden to defend environmental and social measures falls on the WTO Member State invoking the exception.[36]

Article XX exceptions have been tested in WTO disputes related to several topics highlighted in Agenda 21 and the Johannesburg Plan of Implementation.[37] Of particular relevance to this chapter, as discussed below, States have made claims related to the use of GMOs in the case of *EC – Approval and Marketing of Biotech Products*.[38] They have also used the dispute settlement mechanism to insist on the enforcement of domestic IP laws (*Denmark – Measures Affecting the Enforcement of Intellectual Property Rights*),[39] to challenge marine animal protection laws (*US – Shrimp/Turtle*),[40] and other domestic legislation (*US – Section 211 Appropriations Act*),[41] to question the regulation of carcinogenic asbestos in favour of asbestos substitutes (*European Communities – Measures Affecting Asbestos and Products Containing Asbestos*),[42] and to challenge measures for waste management to promote public health concerns, such as malaria avoidance (*Brazil – Measures Affecting Imports of Re-treaded Tyres*).[43] Such cases have been inconclusive, as not surprisingly, the trade dispute settlement body appears to place highest priority on trade law obligations.

A further systemic exception involves the recognition of nonreciprocal special and differential treatment for developing countries,[44] and at Article XXIV:5 of the GATT, WTO members also exclude customs unions and bilateral or regional free trade areas from compliance with WTO disciplines in certain circumstances.[45] These

35 DFAIT, *Retrospective Analysis of the 1994 Canadian Environmental Review – Uruguay Round of Multilateral Trade Negotiations* (DFAIT, Ottawa 1999).

36 V Lowe, International Law (OUP: Oxford, 2007) 219–20.

37 *Agenda 21*, online at http://www.un.org/esa/dsd/agenda21/res_agenda21_00.shtml, accessed 6 March 2011; *Johannesburg Plan of Implementation*, Report of the World Summit on Sustainable Development, Johannesburg UN Doc. A/CONF.199/20.

38 World Trade Organization, *European Communities – Measures Affecting the Approval and Marketing of Biotech Products: Panel Report* (2006), WT/DS291/R, 292/R, 293/R, 29 September 2006, Online: http://www.wto.org/english/tratop_e/dispu_e/cases_e/ds291_e.htm. (*EC – Biotech*)

39 WTO, *Denmark – Measures Affecting the Enforcement of Intellectual Property Rights* (21 May 1997) WT/DS83/1.

40 WTO, *United States – Import Prohibition of Certain Shrimp and Shrimp Products* (20 September 1999) WT/DS58/AB/R.

41 WTO, *United States – Section 211 Omnibus Appropriations Act of 1998* (1 February 2002) WT/DS17/AB/R.

42 *EC – Measures Affecting Asbestos and Asbestos-Containing Products* (12 March 2001) WT/DS135/AB/R.

43 WTO, *Brazil: Measures affecting Imports of Retreaded Tyres – Report of the Panel* (12 June 2007) WT/DS332/R.

44 1979 "Enabling Clause" decision of the GATT Contracting Parties, see *European Communities – Conditions for the Granting of Tariff Preferences to Developing Countries* (7 April 2004) WT/DS246/AB/R.

45 Lorand Bartels and Federico Ortino, *Regional Trade Agreements and the WTO Legal System* (Oxford: Oxford University Press, 2007) at 3.

regional agreements are important, establishing both disciplines that might affect the adoption of domestic rules on biosafety and measures to promote sustainable development and environmental cooperation.

In the field of biosafety, as mentioned earlier, the WTO dispute settlement body has had the occasion to examine the compatibility of precautionary regulations related to LMOs adopted by certain WTO Member States (i.e., the European Union). There is much to be learned from this dispute, although the *EC – Biotech Case* in the end did little to progressively address the risks posed by GMOs, and only highlighted the need to comprehensively address the issue of GMO integration through multilateral negotiation, rather than unilateral litigation.

Implications of the EC – Biotech *Dispute in the WTO*

The WTO's DSB faced a daunting and politically charged trade and biosafety law problem in the *EC – Biotech Case*: the fate of the EC's regulation of GMOs under WTO law. The various versions of the final judgments speak volumes to the complexity inherent in both the legal and scientific issues underlying the dispute.[46] This section details the facts and complainants of the dispute, and an outline of the principal legal arguments presented in the dispute follows, with a note on the role played by civil society and scientific experts through the submission of amicus curiae briefs, and a summary of the final decision of the WTO Panel, focusing on the Panel's interpretation and application of Articles 5.1 and 5.7 of the SPS Agreement.

The *EC – Biotech Case* took place in the context of a decades-long debate over the risks associated with the commercial use of GMOs, including LMOs. North American and European perceptions have differed greatly on this matter. Although North American WTO Members have permitted, indeed perhaps encouraged, the integration of GMOs for commercial use in the mid-1990s, European WTO Members have remained, on the whole, profoundly suspicious of the potential of these products to be safely integrated into agricultural systems and consumer markets.[47] In the summer of 1998, responding to growing public concern over the safety of GMOs, the EC implemented a moratorium on all GMO products pending legislation aimed at their commercial regulation.[48] Subsequently, the EC implemented a robust premarket approval regime.[49] Aimed at protecting human health and the

46 Robert Howse and Henrik Horn, "European Communities – Measures Affecting the Approval and Marketing of Biotech Products" (2009) World Trade Review, 8:1 49 at 82.

47 Gilbert Winham, "The GMO Panel: Applications of WTO Law to Trade in Agricultural Biotech Products" (2009) 31(3) European Integration 409.

48 *Ibid.*

49 Commission Regulation (EC) No. 258/97/EC of the European Parliament and of the Council of 27 January 1997 concerning novel foods and novel food ingredients, [1997] *Official Journal* L 043, of 14 February 1997, available: http://eur-lex.europa.eu/LexUriServ/LexUriServ.do?uri=CELEX:31997R0258:EN:HTML (retrieved 10 October 2010).

environment,[50] in practice the approval process resulted in indefinite if not undue delay.[51]

In 2001, the regime was clarified with the passage of the Deliberate Release Directive,[52] with a view to removing the moratorium.[53] In 2003, the EC amended the regime by putting forward legislation requiring both the labelling and tracing of all GMO products commercially sold.[54] These labelling requirements stigmatised the products, essentially deeming them commercially impractical. In May 2003, the conflict came to a head when the United States, Canada, and Argentina challenged these measures at the WTO. They collectively claimed that the regulatory controls imposed by Member States of the EC did not comply with their obligations under the covered agreements.

The complainants made three main assertions: (1) the existence of a moratorium from 1998 onward, which in effect fundamentally restricted the approval and subsequent importation of GMO products broadly; (2) the indefinite delay of twenty-seven specific product applications as a result of the moratorium; and (3) that the provisional "precautionary" measures adopted by six EC Member States (Austria, France, Germany, Greece, Italy, and Luxembourg), addressing previously EC approved GMO products, were discriminatory in nature.[55]

It was alleged that these EC actions breached various aspects of WTO law. First, it was alleged that the EC had, through the application of its premarket regulatory system, imposed an "undue delay" on the approval of GMO product applications contrary to numerous aspects of the SPS Agreement.[56] Second, this "undue delay" equated to "less favorable treatment" in contravention of the national-treatment

50 Mary Footer, "Post-Normal Science, in the Multilateral Trading System: Social Science Expertise and the *EC-Biotech* Panel" (2007) World Trade Review, 6:2 281 at 283.

51 Lawrence A Kogan, "World Trade Organization Biotech Decision Clarifies Central Role of Science in Evaluating Health and Environmental Risks for Regulation Purposes" Global Trade and Customs Journal, 2(3) (London: Kluwer Law International 2007) at 149, available: http://www.itssd.org/Publications/GTCJ_04-offprints_Kogan%5B2%5D.pdf.

52 Directive 2001/18/EC of the European Parliament and of the Council of 12 March 2001 on the deliberate release into the environment of genetically modified organisms and repealing Council Directive 90/220/EEC, [2001] *Official Journal* L 106/1 of 17 April 2001, online: http://www.bmwf.gv.at/fileadmin/user_upload/forschung/gentechnik/RL-2001-18-EG-_EN_.pdf (retrieved 10 October 2010).

53 *EC – Biotech*, *supra* note 38.

54 Commission Regulation (EC) No. 1829/2003 of the European Parliament and of the Council of 22 September 2003 on genetically modified food and feed, [2003] *Official Journal* L 268/1 of 18 October 2003, available: http://ec.europa.eu/food/food/animalnutrition/labelling/Reg_1829_2003_en.pdf (retrieved 10 October 2010); Commission Regulation (EC) No. 1830/2003 of the European Parliament and of the Council of 22 September 2003 concerning the traceability and labelling of genetically modified organisms and the traceability of food and feed products produced from genetically modified organisms and amending Directive 2001/18/EC [2003] *Official Journal* L 268/24 of 18 October 2003, available: http://www.food.gov.uk/multimedia/pdfs/reg18302003.pdf (retrieved 10 October 2010).

55 *Supra* note 2; Simon Lester, "WTO – *Sanitary and Phytosanitary Measures* Agreement-rules/exceptions-international law as interpretive tool" (2007) 101 Am. J. Intl L. 453 at 454.

56 *Supra* note 1 at 4–5; Articles 2.2, 2.3, 5.1, 5.2, 5.5, 5.6, 7 and 8, and Annexes B(1), B(2), B(5), C(1)(a), C(1)(b), and C(1)(e) of the *SPS Agreement*.

clause of the GATT, based on the assertion that GMO and non-GMO products were sufficiently "like."[57] Finally, the "precautionary" measures imposed by Member States were excessive in nature, lacked proper scientific evidence, and were not based on an "adequate" science-based risk assessment within the meaning of the SPS agreement.[58]

The EC responded by broadly (1) denying the existence of a moratorium, while arguing in the alternative that if one were found to exist, it was colloquial in nature and did not constitute an undue restriction on trade; (2) indicating that regulatory ineptitude was not a trade restrictive measure in and of itself and was merely a complaint on process; (3) proposing that the "precautionary" safeguards imposed by Member States were temporary in nature and were implemented because of the continued scientific uncertainty surrounding GMOs.[59] Underlying the EC's position was its reliance on the precautionary principle broadly, and its obligations under Cartagena Protocol Biosafety specifically.[60]

In response, the EC's legal position was one illustrating the need for caution regarding GMOs. First, the EU contended that owing to the potentially broad socioeconomic effects stemming from the integration of GMOs into the environment, it was more appropriate to apply provisions of the TBT rather than the narrowly focused SPS Agreement.[61] Second, the EC put forward that the stated objective of its regulatory regime was the protection of human health and the environment,[62] which was consistent with precautionary principle as a general principle of international law[63] and which could be considered by virtue of Article 31(3)(c) of the Vienna Convention on the Laws of Treaties. As evidence, the EC cited key international documents,[64] most notably the Cartagena Protocol, which it used to justify most of its regulatory safeguards.[65] Third, the EC specifically referenced the Protocol to rebut the assertion that GMO and non-GMO products were sufficiently "like."[66] Finally, the EC argued that Article 5.7 of the SPS Agreement is a specific illustration of the precautionary principle,[67] which the EC relied on to justify the implementation of its provisional safeguard measures.[68]

57 EC – *Biotech*, *supra* note 38; Articles I:1, III:4, X:1 and XI:1 of the GATT 1994.
58 EC – *Biotech*, *supra* note 38; Articles 2.2, 5.1 and 5.7 of the *SPS Agreement*.
59 EC – *Biotech*, *supra* note 38.
60 EC – *Biotech*, *supra* note 38 at 53.
61 EC – *Biotech*, *supra* note 38 at 415–17.
62 EC – *Biotech*, *supra* note 38 at para 2.
63 EC – *Biotech*, *supra* note 38 at 336.
64 *Ibid*; The EC relied on, *inter alia*, the *World Charter for Nature* of 1982, the *Rio Declaration* of 1992, the *United Nations Framework Convention on Climate Change* and the *Convention of Biological Diversity*.
65 EC – *Biotech*, *supra* note 38 at 54.
66 EC – *Biotech*, *supra* note 38 at 71, 342.
67 EC – *Biotech*, *supra* note 38 at 97.
68 EC – *Biotech*, *supra* note 38 at 73.

Because of the significant scientific debate over GMOs, this case raised a unique opportunity for the public to participate through two key channels. First, three independent groups aimed to contribute information for consideration by means of amicus briefs. These "unsolicited" briefs were submitted between May and June 2004, and the Parties were given the opportunity to comment on them.[69] The first submission came from a group of university professors,[70] whereas the second and third submissions came on behalf of two coalitions of NGOs represented by the Foundation for International Environmental Law and Development (FIELD) and the Center for International Environmental Law (CIEL), respectively.[71]

Cumulatively, the three briefs aimed to contextualise the Panel's interpretation of WTO jurisprudence in light of the substantial uncertainty around GMOs, particularly regarding their risks.[72] Although these briefs were not formally adopted by the Panel,[73] one point was raised. The Panel explicitly noted that two of the members of the NGO coalition represented by CIEL – namely Friends of the Earth and Institute for Agriculture and Trade Policy (IATP) – had publicly posted the Panel's confidential preliminary findings seemingly in contradiction to the term *amicus curiae* friends of the court.[74]

The second key channel for public participation came by way of expert consultations. Drawing on its powers pursuant to Article 11.2 of the SPS Agreement, Articles 14.2 and 14.3 of the TBT Agreement, and Article 13.1 of the DSU, the Panel requested "appropriate scientific and technical advice" because of the conflict's complex nature.[75] This choice echoed the formal request by the EC for the Panel to do so.[76] In August of 2004, the Panel identified experts from various international organisations,[77] four of whom were chosen by the Secretariat with two additional experts added in November.[78] Both the Panel and the Parties involved were granted the opportunity to question the experts. This consultation was essential to provide the Panel with the appropriate technical acumen to make its findings[79] and illustrates a landmark contribution by members of the public to WTO jurisprudence.

[69] EC – *Biotech, supra* note 38 at 284.

[70] EC – *Biotech*; specifically, Lawrence Busch (Michigan State University), Robin Grove-White (Lancaster University), Sheila Jasanoff (Harvard University), David Winickoff (Harvard University), and Brian Wynne (Lancaster University).

[71] *Ibid.*

[72] *World Trade Law in Practice, supra* note 9 at 290–95.

[73] EC – *Biotech, supra* note 38 at 285.

[74] EC – *Biotech, supra* note 38 at 282.

[75] EC – *Biotech, supra* note 38 at 285.

[76] *Ibid.*

[77] Specifically the CBD, Codex, FAO, IPPC, OIE, UNEP, and WHO.

[78] *Supra* note 1 at 287–88; Dr. David Andow (University of Minnesota), Dr. Marilia Regini Nutti (EMBRAPA), Dr. Allison Snow (Ohio State University), Dr. Geoff Squire (Scottish Crop Research Institute), Dr. Marion Healy (FSANZ), and Dr. John W Snape (John Innes Center).

[79] EC – *Biotech, supra* note 38 at 288.

In its final decision, the WTO Panel began by outlining the issues raised by the case that they would not address. First, the Panel did not find itself competent to determine the general safety of biotech products. Second, the Panel did not find that it should decide on whether GMO and non-GMO products were "like products," finding this question unnecessary to evaluate the complaint. Third, the Panel did not discuss the legality of the EC's labelling and tracing requirements as part of its regulatory regime. Finally, the Panel did not rely on the findings of the relevant EC scientific committees, choosing to undertake consultation with their own subject matter experts instead.[80]

To the factual issues presented, the Panel made the following determinations. (1) It was concluded that the EC had implemented a *de facto* moratorium on GMO-based products from June 1999 to August 2003. Subsequently, the EC was in violation of Annex C(1)(a) of the SPS Agreement prohibiting "undue delay."[81] (2) Following the same logic, the Panel found that the "product-specific" measures were equally in violation of Annex C(1)(a).[82] (3) The Panel concluded that the "precautionary measures" of all six EC Member States violated Articles 5.1 and 2.2 of the SPS Agreement,[83] because they were not based on an adequate scientific risk assessment, and that these measures were not saved by Article 5.7 of the SPS Agreement.[84] This final conclusion warrants further analysis.

To come to a final determination regarding the "precautionary measures" of WTO Member States, the Panel dissected all national measures in question to determine their "purpose" and "nature" as applied to Annex (A)(1) of the SPS Agreement.[85] Concluding that they were indeed measures under the SPS agreement, the Panel turned to the relationship of Articles 5.1 to 5.7.[86] The Panel characterised Article 5.7 "as a qualified right," which shifted the burden of proof to the complaining States to prove a violation of the SPS agreement prior to the application of Article 5.1.[87] Finally, based on a consideration of the scientific evidence provided by the Member States, the Panel concluded that the "precautionary measures" in question were not based on scientific risk assessment.[88] Fundamentally, the Panel had determined that Member States could not rely on NGO reports or peer-reviewed journal articles as the scientific basis of a trade-restrictive measure in the face of international applicable "risk assessment techniques," and that clearly the scientific evidence presented did not satisfy a determination of an "adequate risk assessment" as intended in Annex (A)(4).[89]

80 EC – *Biotech*, *supra* note 38 at 1067.
81 EC – *Biotech*, *supra* note 38 at 1070.
82 EC – *Biotech*, *supra* note 38 at 1071.
83 EC – *Biotech*, *supra* note 38 at 1073.
84 *Ibid.*
85 EC – *Biotech*, *supra* note 38 at 876.
86 EC – *Biotech*, *supra* note 38 at 937.
87 EC – *Biotech*, *supra* note 38 at 956.
88 EC – *Biotech*, *supra* note 38 at 1007–8.
89 *Supra* note 7 at 150.

Following a finding that none of the relevant "precautionary measures" satisfied the requirements of Article 5.1 of the SPS Agreement, the Panel moved to consider these measures could be *saved* under Article 5.7. First, relying on *Japan – Agricultural Products II*,[90] the Panel recognized that there were four "cumulative" requirements in Article 5.7 that must be satisfied to support the measures in question. The Panel noted that the first of these requirements was that the measure was imposed in a situation in which "relevant scientific information is insufficient," and that Article 5.7 reflected the precautionary principle.[91]

Second, the Panel moved to consider the measures in question and put forward the following conclusions: (1) the EC was not excused from conducting a scientific risk assessment owing to the insufficiency of scientific evidence,[92] and (2) both the EC approval procedures and the EC scientific committees' findings were risk assessments under the SPS Agreement, and as such the EC actually did have sufficient scientific evidence.[93] The Panel did not provide an opinion on the legal status of the Precautionary Principle in international law,[94] and although the Panel was able to take account of the Convention on Biological Diversity and the Cartagena Protocol, these agreements were not in force on all parties to the dispute and thus "need not be taken into account."[95] Cumulatively, the Panel found that the EC did not conduct a scientific risk assessment adequate enough to satisfy Article 5.1, and that the conditions necessary to invoke Article 5.7 were not present.[96]

Following the Panel's findings that the measures imposed by EC Member States were not in compliance with WTO, the EC needed to bring these measures into compliance. Negotiations on an appropriate timeline for compliance were completed in the fall of 2007.[97] The EC found it politically challenging to find consensus among Member States to remove the measures in question, however.[98] Over the next three years, the Parties continued prolonged negotiations, diligently working to find an appropriate date for the EC to comply. Ultimately, these negotiations resulted in a total of thirty-two addendums to the Panel's initial report. Although the EC was able to come to a suitable agreement with Canada[99] and Argentina[100] throughout that

90 *EC – Biotech, supra* note 38 at 940, 1008.

91 *EC – Biotech, supra* note 38 at 1008.

92 *EC – Biotech, supra* note 38 at 1023.

93 *EC – Biotech, supra* note 38 at 1023–4.

94 Oren Perez, "Anomalies at the Precautionary Kingdom: Reflections on the GMO Panel's Decision" (2007) World Trade Review, 6:2, 265 at 267.

95 Simon Lester, "WTO – *Sanitary and Phytosanitary Measures* Agreement-Rules/Exceptions-International Law as Interpretive Tool" (2007) 101 Am. J. Intl L. 453 at 457.

96 *Supra* note 3 at 422.

97 *Supra* note 3 at 424.

98 *Ibid.*

99 World Trade Organization, European Communities – Measures Affecting the Approval and Marketing of Biotech Products: Status Report of the EC (2009) WT/DS293/31/Add.20, 15 September 2009.

100 World Trade Organization, European Communities – Measures Affecting the Approval and Marketing of Biotech Products: Status Report of the EC (2009) WT/DS291/37/Add.27, 9 April 2010.

time, the United States and the EC did not resolve this issue. Perhaps complicating matters, on September 28, 2010, Austrian Health Minister Alois Stoger stated that he had signed three decrees banning otherwise EU-approved GMOs.[101]

EC – Biotech suggests that when negotiators fail to find consensus on contentious trade issues, consensus can be fashioned at the end of a gavel. Although the Panel in *EC – Biotech* did not put forward a profound affirmation of the status of the precautionary principle, or choose to accept the EC position (that its actions were justified under the Cartagena Biosafety Protocol), one positive remark can be made. Because the EC was unable to use the precautionary principle as an affirmative defense, the potential remains for another Panel to find that it does apply in different circumstances. Until then, the matter remains one of intense scholarly debate.

As noted previously, however, the world trading system is governed by a complex web of international treaties, and trade rules are found not only in the WTO but also in regional and bilateral trade agreements. These agreements are relevant to a discussion of the trade implications of implementation of the Cartagena Protocol on Biosafety in several ways.

Regional Trade and Investment Law Obligations Related to Biotechnology

Nearly all States today participate in bilateral or regional free trade arrangements (RTAs). These treaties are discriminatory by nature, designed to grant certain trade preferences to the Parties. They can be agreed on between developed States, developing States, or among both a developed and a developing State.[102] As of May 2011, 489 regional trade agreements had been notified to the WTO.[103] The WTO and others have raised concerns about this proliferation of regional accords.[104] Furthermore, depending on their specific provisions, regional trade agreements might encourage growth in unsustainable industries or promote unsustainable levels of resource exploitation, leading to related physical impacts, and also constrain the adoption or implementation of new standards. Small-scale regional trade agreements also could provide a useful testing ground for trade policy innovations, including for the sustainable development of the bioeconomy.[105]

[101] Government of Austria, "Stöger: Gen-Importverbote unterzeichnet," Press Release (Federal Ministry of Health: Vienna, Austria, 2010) available: http://www.ots.at/presseaussendung/OTS_20100928_OTS0017/stoeger-gen-importverbote-unterzeichnet.

[102] C.A. Cooper and B.F. Massell, "Toward a General Theory of Customs Unions for Developing Countries" (1965) Journal of Political Economy 461.

[103] WTO, "Regional Trade Agreements Gateway," http://www.wto.org/english/tratop_e/region_e/region_e.htm; WTO, "WTO launches new database on regional trade agreements," WTO Doc. Press/548, 14 January 2009.

[104] WTO, "Regional Trade Agreements: Scope of RTAs," http://www.wto.org/english/tratop_e/region_e/scope_rta_e.htm. Accessed 31 March 2009.

[105] M.C. Cordonier Segger, "The WTO, Regional Trade Agreements and Sustainable Development" in Bartels and Ortino, *supra* note 45.

Of further relevance to new laws and measures related to biosafety and the sustainable use of biotechnology, especially LMOs, in recent decades more than three thousand International Investment Agreements (IIAs) have been negotiated.[106] These have been supplemented internationally by rules and dispute settlement procedures developed through the United Nations Commission on International Trade Law (UNCITRAL) and the International Centre for Settlement of Investment Disputes (ICSID). Private sector investment could help to finance the adoption of biotechnologies.[107] Some even argue that "private finance [is] now the biggest show in town."[108] These IIAs seek to create favourable conditions and stable frameworks for the treatment of foreign investors and investments, encouraging private sector investment in developing countries. The IIAs' obligations normally guarantee a minimal standard of treatment, or "fair and equitable treatment" toward the foreign investor in "like circumstances." Some IIAs commit to "stabilization clauses," which can exclude IIA-covered investments from changes in the law of host States. Such clauses could be important to future attempts to develop domestic biosafety rules. The "legitimate expectations" of the investor regarding a regulatory framework could become grounds for a potential challenge by a foreign investor toward a possibly unfavourable change in circumstances owing to new climate change regulations, including LMO bans or risk management schemes that impose significant new costs on private firms.[109] By December 2007, there were 280 known IIA arbitrations, in which foreign investors challenged governments in confidential "investor-State" dispute settlement proceedings by invoking investment agreement clauses.[110] Potentially, these IIAs, measures, and rule-making bodies are more likely than trade law to constrain biosafety and related regulatory measures. It might be possible, however, to design biosafety regulations and measures carefully, to avoid becoming embroiled in disputes of this kind. It is also possible to design international investment agreements and trade agreements to ensure that legitimate other new measures for safe

[106] A. Newcombe and L. Paradell, *Law and Practice of Investment Treaties: Standards of Treatment* (Leiden: Kluwer Law International 2009) 57–64; UNCTAD Secretariat, International Investment Rulemaking, 22 May 2007 TD/B/COM.2/EM.21/2; UNCTAD, Latest Developments in Investor-State Dispute Settlement, IIAS Monitor No 1 (2008) UNCTAD/WEB/ITE.IIA/2008/3.

[107] UNEP, *Towards a Green Economy: Pathways to Sustainable Development and Poverty Eradication* (Nairobi: United Nations Environment Programme, 2011). General policies stating the need for increased private sector investment: www.fao.org/fileadmin/user_upload/abdc/documents/optpriore.pdf; www.ers.usda.gov/Features/AgRD/index.htm.

[108] M. Klein and T. Harford, "The Market for Aid" (Washington, DC: World Bank Publications, 2005) at 51.

[109] Fiona Marshall, *Fair and Equitable Treatment in International Investment Agreements* (International Institute for Sustainable Development, Issues in International Investment Law: *Background Papers for the Developing Country Investment Negotiators' Forum*, Singapore, October 1–2, 2007) at 10–12, available at http://www.iisd.org/pdf/2007/inv_fair_treatment.pdf.

[110] A. Newcombe & L. Paradell, *Law and Practice of Investment Treaties* (Austin, Texas : Wolters Kluwer Law & Business, 2009) 59.

and sustainable development of the bioeconomy are – at minimum – not frustrated, and perhaps even promoted.

DESIGNING BIOSAFETY POLICIES WITH TRADE AND INVESTMENT CONSIDERATIONS

Sustainable Development Objectives of the Global Biosafety Regime

In the 1992 CBD and its 2001 Cartagena Protocol, States committed to both health and environmental protection *and* sustainable development objectives. The treaties repeatedly emphasise the importance of sustainable economic development, especially for more than 140 developing country Parties. Trade and investment liberalisation is significant in this regard, providing both constraints and opportunities for developing countries. In the CBD, Parties agreed to "cooperate to promote a supportive and open international economic system," which, it was thought, would lead to sustainable economic growth and development in all Parties, particularly developing country Parties, thus enabling them better to address the sustainable use of biological diversity and genetic resources. It was essentially hoped that in accordance with the now-discredited Environmental Kuznets Curve theory,[111] trade liberalisation would lead to higher incomes, which would consequently lead to greater health and environment protections, and even sustainable development, in developing countries. By 1997, however, States had recognised the more complex series of linkages between the sustainable development of the bioeconomy, policies to address it, and international trade.[112] The Cartagena Protocol States in its Preamble that the Parties "Recognize that trade and environment agreements should be mutually supportive; Emphasize that the Protocol is not interpreted as implying a change in the rights and obligations under any existing agreements; and Understand that the above recital is not intended to subordinate the Protocol to other international agreements."[113]

This recognised that States might need to balance measures to address risks and potential adverse effects of LMOs with the adverse effects of measures on international trade, to minimise both.

[111] This theory proposed that as economic wealth increases, populations will demand higher environmental standards and environmental quality will automatically improve, without taking into account that certain environmental damage is irreversible, that many environmental problems are cumulative and global, or that the beneficiaries of economic growth do not necessarily live or vote where damage is done. See H. Nordstrom and S. Vaughan, "Trade and Environment" (1999) WTO Special Studies 4, Geneva.

[112] Steve Charnovitz, "A New WTO Paradigm for Trade and Environment" (2007) 11 S.Y.B.I.L 15 at 38–40.

[113] Cartagena Protocol on Biosafety to the Convention Biological Diversity, *preamble*, http://bch.cbd.int/protocol/text/article.shtml?a=cpb-00.

Trade and Investment Implications of Biosafety

Depending on regulatory instrument choices and how they are designed and implemented, biosafety measures and regulatory schemes can have important implications for trade and investment law. Complementarity and mutual supportiveness are not automatic. As we have seen, a foreign trading partner or investor that perceives itself to be economically disadvantaged by a biosafety regulatory scheme could encourage its State to launch a claim under the WTO, under a regional trade tribunal, or under an investment treaty investor-State tribunal, and the regulators would be required to spend significant resources defending their embryonic rules. Furthermore, domestic regulations to encourage the development of benign biotechnologies, such as drought-resistant plants, require both basic elements and accompanying measures, which, if not carefully designed, might discriminate against firms from foreign trading partners or foreign investors. The issues can be divided into two general categories. First, specific trade disciplines agreed in the WTO and other trade agreements might, if not properly understood, appear to disallow the use of certain domestic biosafety measures. Second, and perhaps most problematic, recently agreed-on investment treaty commitments, if not carefully interpreted, might be used to challenge biosafety measures, thereby constraining their adoption or limiting their effective implementation.

Trade Law Implications of Biosafety Measures

Disciplines agreed on in the WTO might inadvertently or otherwise be interpreted by regulators or dispute settlement bodies in a way that unduly constrains the effectiveness of biosafety regulations. There could be direct regulatory effects, in which a trade authority or dispute settlement mechanism questions or disallows certain measures, or indirect regulatory effects, in which environment and development authorities could be dissuaded from using biosafety measures with threats of trade challenges (so-called regulatory chill).

One trade issue that is often raised, for example, involves the use of government subsidies. If government support for the development of certain biotechnology products or government expenditures on risk guarantees or risk management were characterised as an unfair government subsidy for the purposes of a regional trade agreement, or more generally, WTO rules on subsidies and countervailing measures, the grants or subsidies might be the basis for a challenge in trade law. For instance, a blanket exemption from tax payments has been judged to be a subsidy.[114] Parallel to a domestic biosafety regulatory system, a regulator might provide incentives for firms to carefully assess impacts, separate or differentiate LMO products, or adopt other new control technologies. Such measures could be characterised as inappropriate

[114] See especially *US – Canada WTO Corn Trade Dispute*, WTO Doc. WT/DS357/11; discontinued.

border measures, and therefore probably inconsistent with WTO rules.[115] In many cases, however, trade rules are structured to accommodate such situations. Most trade rules on subsidies, including in the WTO, initially provided for "windows," or reservations for environmental measures, especially for subsidies meant to encourage the adoption of new technologies. Of course, as trade liberalisation continues and rules are refined through dispute settlement, such windows could become more limited. In WTO negotiations, some interests have proposed to set limits on the "green box" subsidies, the WTO-recognized category of subsidies that are permitted because of their environmental objectives, so that subsidies authorised in one State might not be recognised as legitimate by others.[116] In any case, many biosafety risk reduction subsidies could still conform owing to their lack of significant trade impact. In the case of support for research and development of biosafety procedures, such a trade impact might be assessed by a comparison with any previous, less effective rules. In most instances, biosafety regulations place an additional burden on the domestic companies affected, placing them generally at a disadvantage vis-à-vis nonregulated competitors. Viewed in this light, it would be difficult to challenge a biosafety regulatory regime using trade rules on subsidies. Furthermore, however, trade and investment issues can affect the political feasibility of new laws and policies to address biosafety. Were a State to attempt to introduce a new risk management tax for a LMO product or other biosafety regulations, these measures could jeopardize the international competitiveness of its domestic companies.[117] To address these concerns, States might seek to implement border tax adjustments (BTAs).[118] The use of BTAs might well be considered by regulators as a solution to the potential distortions created by biosafety regulatory requirements. A State seeking to implement new biosafety regulations could use BTAs to ensure continued competitiveness. To avoid challenges of discrimination, the proponents would argue that imports are being required to pay a biosafety tax at the border to equalise competition between actors, when the tax is then simply the extension to imported products of the tax or cost of reducing the risks of biotechnologies, which are imposed on domestic producers.[119]

[115] Jeffrey Frankel, "Environmental Effects of International Trade" (2009) Harvard Kennedy School Working Paper Series 39; online: http://www.regeringen.se/content/1/c6/11/88/18/9ba64bac.pdf, accessed 31 March 2009 (with thanks to Christina Voigt for drawing this article to the authors' attention).

[116] Canadian Federation of Agriculture, Trade Policy Statement; online: http://www.cfa-fca.ca/pages/index.php?main_id=61. Accessed 31 March 2009.

[117] It is also something that could raise questions of the "regressiveness" of any carbon tax, meaning that already poorer actors from developing countries would be penalised in the short term, although question of whether the long-term impact on the poor in the event of no carbon taxing would not be worse. See Stephen Tindale and Chris Hewitt, "Must the Poor Pay More? Sustainable Development, Social Justice and Environmental Taxation" in Andrew Dobson (ed), *Fairness and Futurity: Essays on Environmental Sustainability and Social Justice* (Oxford: Oxford University Press, 1999).

[118] Report of the Working Party on Border Tax Adjustments, BISD 18S/97, 2 December 1970.

[119] This has been argued with regards regulations on climate change, see, e.g., Joost Pauwelyn, "U.S. Federal Climate Policy and Competitiveness Concerns: The Limits and Options of International Trade Law." Working Paper (Nicholas Institute for Environmental Policy Solutions, Duke University, 2007) at 41.

The opposing argument would highlight that, in the context of biosafety, the costs of the biotechnology impact assessment and other risk reduction measures levied on imported products (to mirror their costs of production in a "nonregulated" State) are often provided free of charge to domestic actors by the health, agriculture, or other relevant authorities of the biosafety-promoting State, raising claims of "national treatment" violations.[120] Furthermore, it is unclear whether such BTAs would avoid challenges when the tax concerned an input such as LMO insects for pollination, which are fully consumed in the production process and not present in the final product itself. The decision of the WTO Panel in the *US–Superfund Case* offers some guidance.[121] In this case, the WTO Panel permitted BTAs for chemicals used during production – although only traces of these chemicals were still present in the final product.[122] Just as a biosafety regime could be seen in subsidies' terms, it could also be characterised as having the effect of a tax, permitting equalisations. A scheme characterised as a unilateral "carrot or stick" BTA could be a promising avenue for biosafety regulatory schemes within the framework of global trade rules.[123] It would be important to calculate the biosafety risk reduction measure equivalence of the BTA conservatively and to be prepared to address challenges in trade or investment tribunals.[124]

In another example, mentioned earlier, the WTO's Agreement on Technical Barriers to Trade (TBT) seeks to ensure that technical regulations are not "more trade restrictive than necessary to fulfil a legitimate objective" and do not discriminate between "like products."[125] Regulations that take into account process and production measures (PPMs), restricting the entry of goods that have been produced with the use of LMOs versus goods that have been produced organically (for example), could be open to trade law challenges based on nondiscrimination and prohibitions on quantitative restrictions. As noted earlier, the product itself is not different from its competing product – it is only the production method that differs. As such, the products *simpliciter* remain "like," and differing treatment would be caught by national treatment or nondiscriminatory provisions. For the biosafety regulation scheme, or a regulatory regime that sought to favour organic produce, this distinction means that States could have trouble giving significant trade advantages to products produced under the application of a national or even international organic label. If standards were perceived as being based on PPMs and appeared to discriminate between products from different countries, any preferential treatment (in terms of tariffs for

[120] With thanks to Christina Voigt for her insights.

[121] "United States – Taxes on Petroleum and Certain Imported Substances," Report of the Panel, Doc L/6175 – 34S/136, 1987.

[122] *Ibid*, at 5.2.

[123] *Ibid*.

[124] *Ethyl Corporation v. Canada*, Award on Jurisdiction, 24 June 1998, (1999) 38 ILM 708.

[125] Agreement on Technical Barriers to Trade (adopted 15 April 1994, entered into force 1 January 1998), 1186 UNTS 276, art 2.2.

those products) could be challenged in the WTO and other regional trade dispute settlement fora.[126] Few trade rules prevent general use of labels or certification schemes, however. Such "eco-labeling" allows the consumer to know that certain goods were produced in a more environmentally friendly (or, at least, less environmentally harmful) manner than the competing products.[127] As noted by Simon Baughen, in the context of sustainable development measures related to low-carbon products:

> Caution as regards PPMs is perhaps understandable, in that they can be seen as one [WTO] member's attempt to impose its environmental standards on other members. However, the issue of PPMs may, in future, come up in the rather different context of transboundary spill-overs, where the objection to the way in which a product is manufactured is based on adverse environmental consequences felt in the Member State imposing the measure.[128]

Taking this proposition one step further, the practice of "organic and sustainable agriculture" self-labelling could in theory fall within a prohibition on PPM-based measures, should a State use such voluntary declarations or self-labels to assign legal consequences. When the impact on sustainable development and the environment is "transboundary" in nature, however, such as for biosafety and biodiversity, then the measure could no longer be seen as extraterritorial. The issue then becomes one that WTO Panels, in the *US–Shrimp* dispute and others, have recognised as being within the competence of States.[129] Biosafety regulations and laws to promote organic agriculture could be characterised as addressing such transboundary issues.

Sustainable development has been recognised as an objective of world trade law under the WTO, both by the Panel and Appellate Body in the *US – Shrimp* and subsequent disputes, and in the 2001 Doha Declaration.[130] Two world trade law arguments could be used generally by regulators and trade dispute settlement bodies to address any such areas of overlap.

First, trade dispute settlement bodies can carefully consider the definition of *like products* when assessing the legality of measures designed to promote organic agriculture or reduce the risks of biotechnologies. Although the WTO Panel in the *EC – Biotech Case* did not pronounce on this issue, it is possible that States would be

[126] A. Cosbey, *Environment and Trade: A Handbook* (Winnipeg: UNEP/IISD, 2000).

[127] An interesting problem would be whether a carbon market "seller" – habitually selling credits rather than purchasing them – could voluntarily eco-label itself or its product.

[128] See, e.g., Simon Baughen, *International Trade and the Protection of the Environment* (New York: Cavendish, 2007) 4.

[129] United States – Import Prohibition of Certain Shrimp and Shrimp Products, AB-1998–4 at para 186.

[130] *Ibid.*, Doha Ministerial Declaration (14 November 2001) WT/MIN(01)/DEC/1 http://www.wto.org/english/thewto_e/minist_e/min01_e/mindecl_e.htm. Accessed 31 March 2009; M Gehring & MCCordonier Segger (eds.), *Sustainable Development in World Trade Law* (The Hague: Kluwer Law International, 2005) at Introduction.

permitted, under trade law, to take a product's biosafety implications into account in determining its "likeness" to another product.[131] The *EC – Asbestos* dispute remains indicative of the current stance on discrimination of like products, demonstrating that in certain instances, such as when a carcinogen is being substituted for a product with potentially fewer carcinogenic properties, the WTO Appellate Body will take minute physical differences into consideration, shifting the burden of proof onto the challenger to demonstrate that their goods are indeed "like" a less harmful substitute.[132]

In the alternative, any biosafety measures might still be justified under GATT XX(g) exceptions for measures to conserve exhaustible natural resources, or GATT XX(b) exceptions for measures necessary to protect human, plant, or animal life and health. The Panel in the *US – Shrimp* dispute explicitly reserved the right for environmental measures to be excused from WTO obligations through Article XX(g), provided that similar products from other States were not given preferential treatment through special side agreements. It would be hard to describe the Cartagena Protocol on Biosafety, an international agreement open to all WTO Member States, as setting discriminatory or exclusive standards.[133] To prove that a measure is necessary to protect health or the environment, as noted by the Panel in the *Brazil–Retreaded Tyres* dispute, it might be sufficient to demonstrate, on the balance of qualitative evidence, that it is *likely* to contribute to achieving the legitimate health or environmental objectives.[134]

Investment Law Implications of Biosafety Measures

Although biosafety-related regulations and measures to encourage organic non-LMO agriculture can raise certain specific trade issues, depending on the design of the regime, the most significant challenges are likely presented by recent developments in international investment law. Of course, international investment agreements (IIAs) could promote sustainable development by ensuring more stable investment climates and thus encouraging investors to provide private funds for sustainable agricultural development and important biotechnology research and crops, such as drought-resistant plants. In these accords, however, States have also agreed on disciplines that have been used recently to challenge regulatory measures involving

[131] Richard G. Tarasofsky and Kenneth P. Ewing, *The "trade and environment" agenda: survey of major issues and proposals. From Marrakesh to Singapore* (Bonn: IUCN, 1997) at 7; Richard G. Tarasofsky, *Towards a mutually supportive relationship between the Convention on biological diversity and World Trade Organization: an action guide* (Bonn: IUCN, 2002); Edward L. Miles, et al., *Environmental Regime Effectiveness: Confronting Theory with Evidence* (Cambridge, MA: MIT Press, 2002).

[132] Marie-Claire Cordonier Segger and Markus W. Gehring, "Precaution, Health and the World Trade Organisation: Moving Toward Sustainable Development" (2003) 29 Queens Law Journal 133.

[133] Committee on Trade and Environment – Committee on Trade and Environment Special Session – Matrix on Trade Measures Pursuant to Selected Multilateral Environmental Agreements (14 March 2007) WT/CTE/W/160/Rev.4.

[134] Brazil – Measures Affecting Imports of Retreaded Tyres (DS 332).

sustainable development.[135] As noted previously, in IIAs, Parties often grant foreign investors the right to challenge host States in investor-State arbitral tribunals under UNCITRAL or ICSID rules, particularly on claims related to performance requirements, fair and reasonable treatment, expropriation, and transparency. These privileges can be used to challenge biosafety measures, depending on how new domestic schemes are designed and how the IIAs are interpreted.[136] As with trade law, these challenges have both direct effects, in which a State is asked to compensate an investor or group of investors for the economic impact of new biosafety risk reduction or import restriction regulations, and indirect regulatory effects, in which environment and development regulators are "chilled" from adopting or implementing biosafety measures because of threats of investor-State litigation. Several examples illustrate these implications.

First, on an almost theoretical level, biosafety regulations, especially risk identification and management measures like impact assessments, are not market-based instruments,[137] absent of "command and control" origins. The very existence of a government-imposed restriction on imports that do not meet certain conditions, such as free prior informed consent, is evidently "control" in nature. The introduction of new performance standards could pose questions under certain investment treaties. In such treaties, the United States and Canada have sought to prevent or constrain the use of performance requirements or standards that were once popular in developing countries, as a way to enhance the value of an investment by mandating a certain way of producing a product, such as sourcing local services, labour, or content (*local content requirement*), or earning foreign exchange through export requirements.

Second, the scope and nature of the domestic biosafety regulatory regime are important. The design and legal consequences of biosafety regulations and their implementation potentially raise claims of "indirect expropriation."[138] States choose whether their scheme will be limited to a particular sector or economy-wide in application, and which jurisdictions will be subjected to (or allowed entry into) the scheme. Importers targeted by such regulations can include both those that plan to place LMO products directly on the markets, such as vendors of LMO seeds (or makers of jams with seeds), and also those "farther upstream" in the chain of LMO use, such as producers of consumer goods that use LMOs in production.

[135] Miles, *supra* note 131 at 26. See also M.C. Cordonier Segger, M. Gehring, and A. Newcombe, *Sustainable Development in World Investment Law* (The Hague: Kluwer Law International 2010).

[136] M.C. Cordonier Segger, M. Gehring, & A. Newcombe, eds., *Sustainable Development in World Investment Law* (chapter on biodiversity and biosafety); Rudolf Dolzer, "Indirect Expropriations: New Developments" (2002–2003) 11 N.Y.U. Envtl L.J. 64; Julie A. Soloway, "Environmental Regulation as Expropriation: The Case of Nafta's Chapter 11" (2000) 33 Can. Bus. L.J. 92; Simon Baughen, "Expropriation and Environmental Regulation: The Lessons of NAFTA Chapter Eleven" (2006) 18(2) J Environmental Law 207.

[137] See Chapter 4.

[138] Rudolf Dolzer, *supra* note 136, Julie A. Soloway, *supra* note 136, Simon Baughen, *supra* note 136.

Depending on the data collected and economic impact assessments, States decide whether their schemes will be comprehensive or simply sectoral, and whether partial coverage can achieve their objectives. Such choices can affect the competitiveness of national companies against each other and against foreign companies. If investments in foreign investor-dominated sectors were seen as being unjustly targeted by stringent and costly requirements "tantamount to expropriation," whereas other domestic investor-dominated biotechnology sectors were excluded from the regulatory scheme, challenges might be issued under IIAs. If a regulator places restrictions on the use of biotechnology in some sectors and not others, there is the potential for such measures to be characterised as indirect expropriation of that company's investment. Similarly, if the cost of risk assessment and risk control measures becomes high enough to threaten the economic viability of certain investments (e.g., the planting of thousands of hectares of LMO soya products), these biosafety measures could also be deemed tantamount to expropriation. The core debate focuses on who bears the risks of private investments in "high-risk biotechnology" sectors – host governments or investors. Essentially, as one commentator has explained, "[i]f a government measure is undertaken for a clear public welfare purpose (such as health and safety, environment, public morals or order, etc.), and is non-discriminatory, but has the effect of harming a . . . foreign investor, under what circumstances can that measure be held to be an indirect expropriation, for which the government must pay compensation?"[139]

As demonstrated in the *Ethyl v. Canada* case,[140] claims of indirect expropriation can be made when new government measures affect the value of a foreign investment in a specific or unique industry. Such issues could arise for governments implementing biosafety measures, because in some countries, biotechnology industries can be dominated by multinational agricultural enterprises (i.e., foreign investors) with the necessary know-how and capital for research, testing, and development. Moreover, the resulting repeal by Canada of the MMT ban clearly demonstrated the indirect effect of a foreign investor challenge on government policy directions.[141] Indeed, if a developed State such as Canada could be perceived to have "chilled" its regulatory decisions because of international investment law obligations, IIAs, and expropriation challenges, it seems possible that a developing country might face even higher pressure to avoid necessary regulatory changes. Whether the developing country could in fact afford to compensate for the expropriation is a particularly pressing issue in the case of biosafety measures.

139 A. Cosbey, "NAFTA's Chapter 11 and the Environment," Discussion Paper for CEC Public Workshop, Mexico City, 24 March 2003 at 3.

140 Ethyl, *supra* note 24.

141 A. Newcombe, "Regulatory Expropriation, Investment Protection and International Law: When Is Government Regulation Expropriatory and When Should Compensation be Paid?" (LLM Thesis, University of Toronto 1999) (unpublished); online: http://ita.law.uvic.ca/documents/RegulatoryExpropriation.pdf, accessed 31 March 2009.

Such problems are difficult but not impossible to address. For instance, certain biosafety regulations might incorporate provisions specifically ensuring no distinction between national and foreign-owned companies.

Another basic design element that triggers trade and investment issues involves the commitment to regulatory transparency, which might support the designers of biosafety regulatory schemes. It could be argued that by designing clear and updated new biosafety regulations to govern issues such as "safe release" of LMOs, States may also be more transparent and accessible than traditional command-and-control schemes. Anyone wishing to challenge the environmental effectiveness of the regulatory regime can question directly the new set of regulations, rather than having to unravel the often complex relationship between specific historically developed controls applied to an individual firm, farm, or product, and an arbitrary or outdated risk assessment and management by-law based on outdated science or practice.[142] The transparency of domestic law and policy-making process is important to any potential investor.[143] As mentioned previously, however, investors do try to seek investment treaty guarantees against changes in government policy (not just fiscal or tax policy, but also environment and development policy),[144] hoping to stabilise regulations for the lifetime of an investment.[145] Certain IIAs contain clauses agreeing to freeze the laws of a country to the time the investment was made, or not to apply new laws to the investment, or to bear the costs of all regulatory changes affecting an investment: so-called stabilisation clauses. These commitments could be problematic from the standpoint of biosafety regulations. The principal difficulty posed is that States are bound to continue treating the investment in a certain way, which might become no longer viable in light of the Cartagena Protocol objectives and the developing scientific discoveries that have driven the evolution of the precautionary biosafety regulatory regimes.

To address this additional issue, a regulator might need to make it clear to potential investors that postestablishment decisions, that is, governmental decisions influencing the investment after it has been made, will take State commitments under the CBD and the Cartagena Protocol into account. Moreover, transparency works in many ways. As discussed elsewhere, impact assessments and other such requirements can contribute to stability rather than detracting from it, by generating valuable

[142] R. A. Stewart, "Economic Incentives for Environmental Protection: Opportunities and Obstacles" in R. L Revesz, P. Sands, and R. A. Stewart (eds), *Environmental Law, the Economy and Sustainable Development* (Cambridge: Cambridge University Press, 2000).

[143] The Preamble to the Aarhus Convention expressly calls for "transparency in all branches of government" when implementing provisions related to the Aarhus Convention's aims: Convention on Access to Information, Public Participation in Decision-Making and Access to Justice in Environmental Matters (adopted 25 June 1998, entered into force 30 October 2001) 38 ILM 517.

[144] Pauwelyn, *supra* note 119 Andrea Ross-Robertson, "Is the Environment Getting Squeezed out of Sustainable Development?" (2003) Public Law 249.

[145] Baughen, *supra* note 136 at Chapter 7.

investment intelligence and creating a more level playing field for investors.[146] To that end, by securing transparent policy decisions, States might insulate their new biosafety policies from formal investor-State challenges, while also contributing to their clarity and legitimacy.[147]

Each of these aspects can influence the effectiveness of a biosafety regulatory scheme. Additionally, these basic design questions are accompanied by myriad technical and practical rules, and trade and investment rules apply to the *de facto* effect of measures, not just their letter or intention. These include, but are not limited to, assessment, measurement, and monitoring of risks of biotechnologies; calculation of risks and potential benefits of specific sectors and types of LMOs; and enforcement of sanctions for noncompliance. For instance, in many countries, deliberate or negligent release of untested LMOs into the environment might well trigger civil, common, or even criminal law pollution-related penalties.

As seen in the *Methanex v. USA* dispute,[148] not all investor-State tribunals are willing to interpret their jurisdiction over regulators so broadly. Although some IIAs can be used to question environmental protection and sustainable development regulations, others clearly cannot be extended so far.[149] Still, when States enact measures to favour LMO, traditional, or even certified organic projects, especially in developing countries, where the agricultural sectors are dominated by a few foreign investors, these interests could well frame investor-State challenges in terms of performance requirements, discrimination, fair and equitable treatment, or expropriation. When such allegations are raised, it will be important to secure an appropriate interpretation of the concept of "like circumstances." It is only when two Parties are in "like circumstances" and receive different treatment that sustained tribunals would find discrimination. For future biosafety measures, the inclusion of public interests in one proposed project as compared with another that does not take such issues into account might serve to distinguish hitherto "like circumstances" Parties from one another.

The recent ICSID arbitration between Parkerings and Lithuania demonstrates this principle.[150] The case concerned two competing firms from Norway and the Netherlands in their tenders to construct and operate traffic facilities in the Lithuanian capital, Vilnius. The Norwegian company (FET) argued discriminatory treatment in favour of the Dutch firm, and the case turned on the concept of "likeness." The Tribunal ultimately found that the State could take into account matters of a

146 M.C. Cordonier Segger, M. Gehring, and A. Newcombe (eds.), *Sustainable Development in World Investment Law* (The Hague: Kluwer Law International, 2010).

147 Tarasofsky, *supra* note 131.

148 *Methanex v. USA*, Award on Jurisdiction, 28 August 2002.

149 Jessica C. Lawrence, "Chicken Little Revisited: NAFTA Regulatory Expropriations After *Methanex*," (2006) 41 Georgia Law Review 261.

150 *Parkerings-Compagniet AS v. Republic of Lithuania*, ICSID Arbitration Case No ARB/05/8, Award, 11 September 2007, para 375, 392 at 2 October 2007; online: http://ita.law.uvic.ca/documents/Parkerings.pdf.

proposed project's impacts on the environment when deciding how to treat different projects, based on a basic right to legislate in a reasonable and fair manner and the requirement for investors to anticipate change, and thus structure their investment in order to adapt to a change in circumstances. The tribunal rejected the FET claim, concluding that Lithuania had not given any explicit or implicit promise that the legal framework of the investment would remain unchanged.[151] The tribunal's approach has been welcomed in leading legal scholarship on these issues:

> This decision points to the ecological impact of an investor's project as a determinative factor in the like circumstances test. If this approach is followed in future investor-state disputes, then the potential for non-discrimination requirements in international investment agreements to frustrate biosafety regulation will be significantly reduced.[152]

Learning from this experience, negotiators might have to recognise the importance of maintaining flexibility for biosafety measures in investment treaties, while regulators take care to design the rules for biotechnology and sustainable agriculture investments, to avoid discrimination between industries in "like circumstances." Further potential refinements to ensure that trade and investment laws, particularly in regional treaties, which advance beyond the globally agreed disciplines, do not unduly constrain domestic regulatory flexibility to address biosafety – and might even promote more sustainable development – are discussed in the next section.

Improving Trade and Investment Regimes for Sustainable Development

The obligations of States under international trade and investment law might intersect with certain elements of biosafety regulations, requiring careful work to design compatible measures to establish regulatory schemes, and could lead to constraints on policy and law-making flexibility.[153] As noted in the 1987 Report of the World Commission on Environment and Development,[154] the 1992 Rio Declaration and Agenda 21,[155] and the 2002 Johannesburg Declaration and Plan of Implementation,[156] trade and investment could also provide important contributions to sustainable development. Just as in the international biodiversity regime, in many trade agreements and IIAs, Parties explicitly highlight their shared commitment to sustainable development as part of the object or purpose of the treaty. For instance, the North American Free Trade Agreement (NAFTA) includes a reference to the need to

[151] UNCTAD, Latest Developments in Investor-State Dispute Settlements, (UNCTAD: Geneva, 2008) at 5; online: www.unctad.org/en/docs/iteiia20083_en.pdf.
[152] Miles, *supra* note 131 at 32.
[153] See, e.g., Miles, *supra* note 131.
[154] World Commission on Environment and Development, *Our Common Future* (Oxford: Oxford University Press, 1987).
[155] Rio Declaration on Environment and Development (1992) 31 ILM 874; Agenda 21 (n 35).
[156] JPOI, *supra* note 37.

"promote sustainable development" within its Preamble.[157] Both the Canada–Chile Free Trade Agreement and Chile–US Free Trade Agreement also recognise the importance of strengthening capacity to protect the environment and promote sustainable development.[158] The Canada–Peru Free Trade Agreement makes explicit reference, in the chapter titled "Investment," to corporate social responsibility and the need for Parties to encourage enterprises to incorporate such standards into their internal policies.[159] The EU–Chile Association Agreement goes further, committing to implement their accord in accordance with the "principle of sustainable development,"[160] and EU economic negotiations with Central America seek to "harness globalisation in support of sustainable development" and "ensure an appropriate balance between economic, social and environmental components in a sustainable development context."[161] Sustainable development is a key objective of the world community, not only in the abstract but also in the very arena that has most sought to encourage economic growth – investment and trade policy and law.[162]

In such regional trade and investment agreements, States have also begun to seek ways to promote sustainable development through the inclusion of innovative yet practical instruments. Several preventive provisions, cooperative mechanisms, and new trade liberalisation enhancement initiatives can be identified. There are also important procedural innovations that can support sustainable development. Many legal options are available to States seeking to deliver on a commitment to sustainable development in a regional trade and investment regime, either as a principle or an objective.

First, States can include introductory provisions that signal the Parties' commitment to sustainable development, such as preambular commitments to "promote sustainable development" as a "joint resolution" of the Parties to the accord, or other initial provisions that commit to engage in the accord in accordance with a "principle of sustainable development."

Second, States can include provisions that create *windows*, or exemptions from trade rules, in which trade obligations might otherwise constrain regulators and policy makers, mitigating their effects. For instance, in trade and investment agreements, many States adopt general exceptions for measures related to the conservation of exhaustible living and nonliving natural resources, and the use of measures,

[157] North American Free Trade Agreement, 32 ILM 289 (1993)[NAFTA].

[158] Canada-Chile Free Trade Agreement, 36 ILM 1079; US-Chile Free Trade Agreement, 42 ILM 1026.

[159] Canada-Peru FTA, signed 29 May 2008, Chapter 8, "Investment," Article 810.

[160] EU-Chile Association Agreement, 30 December 2002; online: www.ec.europa.eu/trade/issues/bilsteral/countries/chile/eu/euchlagr_en.htm.

[161] Draft EU-Central America Negotiating Directive (2007) at paras 3.4 and 3.7; the States involved are Costa Rica, El Salvador, Guatemala, Honduras, Nicaragua, and Panama. Online: www.bilaterals.org/article.php3?id_article=8336. Accessed 31 March 2009.

[162] M Gehring and MC Cordonier Segger (eds), *Sustainable Development in World Trade Law* (The Hague: Kluwer Law Int., 2005) at Introduction.

including environmental measures, necessary to protect human, animal, or plant life or health. States may further adopt specific exceptions in sections of the trade and investment treaty when it is clear that trade rules on *inter alia* sanitary and phytosanitary standards, technical barriers to trade, intellectual property rights, public procurement, services, or investment might constrain the use of environmental and social measures. States may insert explicit non-application clauses for socially or environmentally sensitive sectors (e.g., parks, land-use planning, energy policy, and other natural resources reserved from investment provisions, or health and education sectors from services disciplines), linking them to the findings of sustainability impact assessments or environmental assessments of the trade agreements. States can also include general interpretive statements to guide potential areas in which trade rules could otherwise constrain the use of measures agreed on in other international (or regional) agreements, specifically mentioning their commitments on biosafety and biotechnology found in the CBD and its Cartagena Protocol. These provisions seek to prevent trade and investment rules from directly frustrating new or existing regulations for biosafety.

Third, States also can negotiate mechanisms for "value-added" but parallel (non-integrated) social and environmental cooperation strategies, such as parallel agreements, chapters, or both, for cooperation on environmental and social matters; the development of institutions for social and environmental cooperation on research and development in the area of benign biotechnologies; the agreement to adopt and implement common work programmes on specific environmental or social projects, such as risk assessment or sustainable agricultural product labelling, particularly when accompanied by reliable capacity building, technology transfer, and financing commitments; and even factual report or complaint mechanisms to provide recourse when it appears that biosafety rules could be violated to gain trade- or investment-related advantages. These provisions seek to strengthen implementation of new or existing regulations for biosafety.

Fourth, States many include constructive "sustainable development"-oriented trade and investment rule enhancement initiatives, in which a positive "triple-win" situation might be achieved within the trade agreement. These include, for example, sanitary and phytosanitary provisions that promote scientific cooperation and risk assessment to improve levels of biosafety protection; government procurement provisions that make public purchasing of LMO-free goods or services more affordable; technical barriers to trade provisions to implement nondiscriminatory "organic or sustainable agriculture" certification processes and promote mutual recognition; intellectual property rights provisions that support biotechnology transfer or respect for traditional knowledge; investment provisions that privilege socially responsible corporations and investments into sustainable biotechnology; monitoring and inspection measures to promote reductions in illegal trade in LMOs; measures to secure additional liberalisation of environmental goods and services, including waste management or safe transportation of closed containers for LMO trans-shipment; or measures to secure reductions in unsustainable fossil subsidies for high-risk LMOs.

Finally, the Parties could undertake certain procedural innovations during the trade negotiations to promote sustainable development and secure the integration of environmental and social concerns into a trade and investment treaty. Such process changes also could assist Parties and others in identifying useful innovations to be included in a trade or investment agreement. For instance, States could undertake *ex ante* (or ongoing) environmental, developmental, human rights, or sustainability impact assessments and reviews of trade liberalisation policies and draft treaties. The outcomes of these assessments, as mentioned previously, can be used to identify areas in which preventive, cooperative, or enhancement initiatives could be useful in a trade or investment treaty. States also could host consultations among economic, environment and development authorities. They could agree on or strengthen diverse mechanisms to ensure transparency and public participation in trade negotiations, and they also could establish new mechanisms to inform tribunals about sustainable development issues, including amicus curiae, public participation, and expert consultation measures.

It is not yet clear which strategies or instruments will be most successful in integrating social and economic development and environmental protection. Probably no one single measure provides the solution to all sustainable development challenges; rather, many different provisions will needed throughout the treaty. Certain instruments, such as the normative or regulatory evaluation elements of *ex ante* sustainability impact assessments, are still underdeveloped. Others, such as the new system of certification to ensure that forestry products traded from Peru to the United States are not obtained through illegal logging, are simply new and relatively untested.[163] Such provisions alone will not necessarily ensure that sustainable development priorities are given more weight by the Parties in complying with their obligations, or by dispute settlement bodies in interpreting agreements, compared with the other relevant objectives of agreements. They probably will contribute to achieving a greater degree of integration in trade agreements, however, which is an important first step toward sustainable development.

STRENGTHENING INTERNATIONAL LEGAL COHERENCE FOR THE GREEN ECONOMY

Current intersections between global and regional rules on trade, investment, and biotechnology are far from being well understood.[164] How should countries proceed, especially developing countries, faced with these difficulties and overlapping legal regimes? On a global scale, a fast but effective regulatory action in the fields of trade

[163] United States – Peru Trade Promotion Agreement, signed on 12 April 2006, http://www.ustr.gov/Trade_Agreements/Bilateral/Peru_TPA/Final_Texts/Section_Index.html, accessed 31 March 2009; ICTSD, "US-Peru Bilateral to Address Illegal Logging, Boost MEA Implementation," Bridges Trade BioRes, 22 June 2007 http://ictsd.net/i/news/biores/9122. Accessed 31 March 2009.

[164] D. French, "Regulation of GMOs and International Law: A Call for Generality," in L. Bodiguel and M. Cardwell (eds.), *Regulation of GMOs: Comparative Approaches* (Oxford: Oxford University Press, 2010), 355.

and investment law is needed, and in biosafety law as well. Not only must the risks of biotechnologies be reduced at a much faster pace, but trade and investment in sustainable alternatives also must increase. Of equal importance is the improvement of technology transfer to ensure that developing countries are actually able to develop more sustainable systems for agriculture, forestry, and other sectors. In this light, mechanisms such as effective domestic and international sustainable agricultural development schemes and detailed sectoral agreements could be part of the way forward beyond the terms of the Cartagena Protocol, particularly for developing countries, where economy-wide participation is impractical.

The CBD, the Cartagena Protocol on Biosafety, and the new Nagoya – Kuala Lumpur Protocol on Liability and Redress establish a comprehensive system in which economy-wide biosafety rules are established, but this does not prevent future, more detailed, progress in specific sectors or scientific fields. If independent and unbiased scientific collaboration agreements are undertaken, this could increase the contribution of those States currently unwilling to commit to opening their economies and environments to new biotechnologies perceived as unsafe, or prone to risk. From a developing country's perspective, the feasibility of permitting specific biotechnology products to be imported, once sound scientific cooperation and effective regulations are in place, is significantly increased. Success in a few well-defined and monitored areas could lead to greater foreign investment in sustainable biotechnologies and could also focus private sector innovation efforts. The interpretation of international investment treaty law will become central in ensuring that there is coherence with sustainable development policies and avoiding litigation and conflict.[165] To that end, IIAs can include express conflict clauses, in which matters of sustainable development are the issue. Policies aimed at securing biosafety might also be introduced into the treaties and/or parallel sustainable development cooperation mechanisms by States.[166] Parallel to this, investment tribunals can recognise that protection and/or promotion of sustainable development values within a given project do affect the circumstances surrounding the investor, often placing superficially similar investments in different circumstances. In investment law, judicious interpretive statements can offer guidance to investors and regulators, helping States to understand under which conditions different treatment would not be found to be discriminatory. In that regard, the *Parkering* decision is welcome, and it is hoped that that tribunal's reasoning carries forward in other disputes.

At the same time, certain basic steps can be taken by regulators to minimise the likelihood of challenges based on indirect expropriation and (un)fair and (in)equitable treatment clauses. In the context of precautionary new biosafety regulations and transparency, clear decisions not to discriminate between imports because

[165] S.E. Gaines, "International Trade, Environmental Protection and Development as a Sustainable Development Triangle" (2002) 11 Rev. EC & Int Env Law 259.

[166] Steve Charnovitz, "Trade and Environment," (2007) J Int Economic Law, 10 at 12–13.

of the nationality of the firm or investor can be made. Careful regulation instead can address the actual or perceived biosafety risks of the product itself.

Several practical measures can secure better coherence between future biosafety policies and trade rules. There clearly are win-win-win trade, investment, and biosafety approaches, such as cooperation programmes for monitoring and reduction of emissions, or the transfer of sustainable biotechnologies, the liberalisation of trade in environmental goods and services related to safer trans-shipment of LMOs, or the careful design of government procurement rules to exempt green and sustainable procurement schemes. Future regional trade agreements can seek to incorporate provisions on biosafety and other sustainable development objectives explicitly, and to ensure transparency for future investors.

Guidance and analysis is still needed in this area, and lessons must be taken from actions taken to date, given the need to both implement measures to conserve exhaustible natural resources and still not unjustifiably restrict international trade or investment.[167] Cooperation between the CBD and the WTO is therefore not merely probable but essential for future coherence in this area. At the forefront of this must be a committed response to the CBD's call to accelerate innovation in the development, deployment, adoption, diffusion, and transfer of environmentally sound technologies among all Parties, and particularly from developed to developing countries.

[167] *Ibid.*

26

The Cartagena Protocol and the Regulation of Genetically Modified Food Aid

Martin Endicott

The donation of food aid containing genetically modified organisms (GMOs) has caused considerable international controversy in recent years. Some of the world's poorest countries, even when struck by famine or natural disasters, have refused to accept this food. Others have imposed special requirements before allowing it across their borders. This chapter considers the background to the issues in this area, analyses the standards set by the Cartagena Protocol for the regulation of food aid (including their interaction with the WTO agreements), and highlights key considerations for the regulation of food aid in domestic biosafety law.

WHAT IS FOOD AID?

The United Nations Development Programme (UNDP) has defined food aid as:

> [t]he provision of food for human consumption for development purposes, including grants and loans for the purchase of food. Associated costs such as transport, storage, distribution, etc., are included in this category as are donor-supplied food related items, such as animal food and agricultural inputs, related to food growing when these are part of food aid programmes.[1]

Food aid can therefore consist of "cash" or "in kind" donations. The latter is what most concerns us here, because whereas financial assistance enables countries to choose between sources of genetically modified (GM) and non-GM supply, in-kind donations do not. For this reason, and because financial donations tend to support

[1] United Nations Development Programme *Development Cooperation Report [Re: Nepal]* (UNDP 2000), online: United Nations Development Programme). www.undp.org.np. Accessed 19 October 2009.

This chapter is the result of ongoing CISDL research on biosafety decision making and international sustainable development law. Martin Endicott, Christine Frison, and Kathryn Garforth with Marie-Claire Cordonier Segger, Jorge Cabrera, and Sylvestre Manga, *Innovations in Biosafety Law: A CISDL Working Paper* (Montreal: Centre for International Sustainable Development Law, 2005).

local and regional producers and so enhance long-term food security, donations in cash are often considered to have the most positive development impact.[2] Accordingly, under the 1999 Food Aid Convention, one of the standards of best practice that donor countries agree to observe is to consider the possibility of making cash contributions for local food purchases.[3]

Around 10 million metric tons of food aid are delivered annually to countries facing food shortages.[4] Approximately 40 percent to 50 percent of these deliveries are made through the United Nations World Food Programme (WFP) and 80 percent to 90 percent of them consist of cereals.[5] The WFP is open about the fact that since 1996 its aid supplies have included GM food.[6] The United States is the largest contributor of food aid both through the WFP and independently. In 2003, it was responsible for 63 percent of donations, amounting to over $1.2 billion; 40 percent was in the form of loans or grants and the rest was in kind.[7] In the same year, the European Union (EU) donated food aid worth $161 million dollars; Japan, $119 million; and the United Kingdom, $100 million.[8]

The controversy that surrounds the donation of GM food aid can be understood in the context of two principal factors. First, as the value of the donations discussed previously makes clear, food aid is regularly shipped around the world in very substantial quantities. Secondly, food aid, in common with many other forms of aid, is often used not only in an attempt to alleviate food crises but also to promote the interests of donor countries. The U.S. Food for Peace (Public Law 480) programme[9] was established to handle the problems posed by increasing U.S. agricultural surpluses and decreasing commodity prices in the mid-1950s, while at the same time providing a tool with which to reward States that cooperated with the United States and create export markets for U.S. products.[10] Some commentators have suggested that United States has used food aid shipments to reward U.S. allies, to assist friendly

2 M Mayet, "GM Food Aid: Africa Denied Choice Once Again?" (Africa Center for Biosafety and others 2004) 7, online: BioSafety Africa, http://www.biosafetyafrica.net.

3 Article XII (a) (ii) and (ii). The Food Aid Committee (FAC) extended the Food Aid Convention by two years in June 2003; FAO *Economic & Social Department* Food Outlook No.1 (2004) (Food and Agriculture Organisation 2004), online: Food and Agriculture Organization, http://www.fao.org/documents. Accessed 19 October 2009.

4 World Food Programme Annual Report 2003, 50 ("Global Food Aid Profile"), online: World Food Programme, http://www.wfp.org/policies/Annual_Reports/. Accessed 19 October 2009.

5 *Ibid.*

6 F Pearce, "UN is slipping modified food into aid" (19 September 2002) 175 New Scientist 2361: "The WFP says it sees no need to warn about GM material in food aid. It says, 'We are just the middle man. If the food meets the national standards of the donors, we accept it.'"

7 M Mayet, "GM Food Aid: Africa Denied Choice Once Again?" (Africa Center for Biosafety and others 2004), 7, online: Biosafety Africa, http://www.biosafetyafrica.net. Accessed: 19 October 2009).

8 *Ibid.*

9 Established by the Agricultural Trade and Development Assistance Act, Public Law 480 of 1954.

10 N Zerbe, "Feeding the Famine? American Food Aid and the GMO Debate in Southern Africa" (2004) 29(6) Food Policy 593 at 6, online: http://www.humboldt.edu/~nrz3/research/index.html.

governments to raise funds to buy arms, and to encourage political and economic concessions, such as the opening of markets to U.S. fertilizer exporters.[11] Earl Butz, U.S. Secretary of Agriculture between 1971 and 1974, stated openly, "Food is a tool. It is a weapon in the U.S. negotiating kit."[12] Today, it is often suggested that the principal reason that GM food aid is being supplied to countries suffering food shortages is to promote the interests of the biotechnology industry and to expand the acceptance of GM food.[13] In a statement issued in response to the U.S. complaint about EU regulation of GMOs before the World Trade Organisation (WTO), the EU asserted:

> Food aid to starving populations should be about meeting the urgent humanitarian needs of those who are in need. It should not be about trying to advance the case for GM food abroad, or planting GM crops for export, or indeed finding outlets for domestic surplus, which is a regrettable aspect of the US food aid policy.[14]

If this allegation were substantiated, it would represent a breach of the Food Aid Convention, Article IX(i) of which obliges Convention Members to ensure that "the provision of food aid is not tied directly or indirectly, formally or informally, explicitly or implicitly, to commercial exports of agricultural products or other goods and services to recipient countries."

Risks Associated with GM Food Aid Distribution

The extent of concerns over the distribution of GM food aid is well illustrated by events in Southern Africa in 2002. That spring, more than 14 million people in Southern Africa were faced with critical food shortages. In response to appeals by the WFP and several NGOs, Western governments offered a range of cash and in-kind aid. The United States offered one half of the total food aid requirements for the region, but the governments of Malawi, Mozambique, Zimbabwe, and Zambia initially refused to accept deliveries of in-kind aid because of concerns over the inclusion of genetically modified maize. Malawi, Mozambique, and Zimbabwe eventually accepted the aid on condition that it was milled before distribution to prevent farmers replanting the seed in future years. Zambia maintained its refusal to accept GM maize, even if milled, for three principal reasons. First, it considered that the distribution of GM maize carried a high risk of eroding the genetic diversity of local maize varieties. Second, its view was that there was insufficient knowledge related to the toxicity, allergenicity, and antibiotic resistance of the maize. In line with this concern, the U.K.'s Chief Scientific Adviser, Professor David King, has criticised the donation of

[11] *Ibid* at 9–16.
[12] Idem at 12.
[13] Idem at 4.
[14] Idem at 33.

GM foods to African countries as "a massive human experiment."[15] Third, it was concerned that there was a risk that the cultivation of GM maize in Zambia would endanger future agricultural exports to the EU.[16]

It is beyond the scope of this chapter to analyse the merits of the concerns voiced about GM food aid, but to provide some context to the discussion, it is helpful to divide the rightly or wrongly perceived risks of GM food aid into three broad categories, which, as Garforth suggests, can be aligned with the "three pillars of sustainable development," in other words, risks to economic, environmental, and social interests.[17] One of the principal purposes of biosafety law is to manage these perceived risks. Indeed, the term *biosafety* refers to "the need to protect human health and the environment from the possible adverse effects of the products of modern biotechnology."[18] These concerns are discussed in the next section.

Economic Concerns – Loss of Export Opportunities

One of the key concerns of the Southern African states offered GM food aid in 2002 was the risk of GM contamination of domestic crops, leading to the loss of export opportunities in the EU, Japan, and Korea. A particular concern is that if GM grain were planted in Southern Africa, it would contaminate other grain varieties fed to livestock, which would then jeopardize the high value that African livestock enjoys on European markets as an organic product.[19]

Environmental Concerns – Unintended Contamination

This category of risk centres on the potential for the uncontrolled spread of transgenes from GMOs by cross-pollination with wild plant varieties. This can occur when GMO grain is introduced into the environment by planting or by accidental release (e.g., when seeds escape from vehicles while in transit). The spread of dominant genes might reduce biodiversity or, for example, pass increased pesticide resistance to target pests, making them more difficult to manage. In relation to the threats posed by the uncontrolled spread of GM varieties, there are frequent references to the discovery of StarLink (a variety of GM maize authorised in the United States for use only as animal feed because of potentially severe allergic reactions in humans)

15 Friends of the Earth, "Playing with Hunger" (Friends of the Earth International 2003), online: http://www.foei.org/en/publications/pdfs/food-sovereignty/2000-2007/playing_with_hunger2.pdf.

16 *Ibid* at 24.

17 K Garforth, "Background on Biosafety and the Cartagena Protocol" (CISDL Working Paper, June 2005) at 10.

18 Secretariat of the Convention on Biological Diversity, *Cartagena Protocol on Biosafety to the Convention on Biological Diversity* (Montreal, 2000), 1 (Introduction).

19 F Pearce, "UN is slipping modified food into aid" (19 September 2002) 175 New Scientist 2361.

in food products and aid in the United States, Central America, Japan, and Korea since 2000.[20]

Social Concerns – Human Health and Farming Practices

Although expert opinion is generally that the risks of consuming GM food are relatively small,[21] most experts also agree that additional research must be carried out and, as illustrated by the cases of contamination by StarLink maize, varieties that are not considered safe for human consumption could become mixed with those that are. There are also health concerns related to the use of antibiotic resistance marker genes, increases in the allergenicity and toxicity of food, changes in nutritional value, and unintended side effects resulting from the process of genetic modification. It has been suggested that these risks could increase in magnitude in the context of food aid because that aid probably will form much of the recipients' staple diet. There are also concerns about the way that the business practices of the major biotechnology corporations affect farming practices. For example, Monsanto, whose seeds are planted on more than 90 percent of the total area cultivated with GM crops, requires farmers who buy its seed to sign a user technology agreement. This agreement forbids certain traditional practices, such as seed saving, and Monsanto has aggressively enforced its rights under that agreement through litigation.

APPLICATION OF THE PROTOCOL TO CONSIGNMENTS OF FOOD AID

The scope of the Protocol is limited to "living modified organisms"[22] (LMOs), defined as "any biological entity capable of transferring or replicating genetic material, including sterile organisms, viruses and viroids."[23] This definition covers GM organisms that could survive if introduced into the environment (e.g. seeds) but is unlikely to extend to processed GMOs.

As a compromise between major GMO-exporting States and food-importing States, the Protocol sets up two parallel regimes. The Article 7 Advanced Informed Agreement (AIA) regime applies to LMOs "for intentional introduction into the environment," such as seeds for experimental or commercial planting or fish for release (referred to here as LMOs-FIE). The less restrictive Article 11 Biosafety Clearing-House (BCH) regime applies to LMOs "for direct use as food or feed or

[20] Environmental Rights Action/Friends of the Earth Nigeria, "Genetically Modified Crops: A Challenge for Africa" (Nigeria 2005) 7, online: Biosafety Africa, http://www.biosafetyafrica.net. Accessed 20 October 2009.

[21] H Baumueller, "Domestic Import Regulations for Genetically Modified Organisms and their Compatibility with WTO Rules" (IISD/ICTSD Trade Knowledge Network 2003) 19, citing findings by the U.S. Food and Drug Administration (1995), the Food and Agriculture Organisation (1996), the American Medical Association (2000), and the U.S. Department of State (2000).

[22] Cartagena Protocol, *supra* note 18 at Article 4.

[23] *Ibid* at Article 2(h).

for processing" (whole corn or soybeans, for example, known as LMOs-FFP). Most traded LMOs are LMOs-FFP, which is why exporters were keen to limit regulatory requirements.

There is no express mention of food aid in the Protocol but it is likely that aid shipments intended for direct use as food or for processing (e.g., grain shipments) fall within the Article 11 regime. Given the Protocol's definition of LMOs, however, processed GM foods such as soy flour, corn-soy blends, maize meal, and vegetable oils are unlikely to come within its scope. Whether to adopt national legislation that regulates processed GM foods is therefore left entirely to the discretion of individual States.

The approach of the World Food Programme (WFP) is to treat all LMOs that it distributes as LMOs-FFP. Its Operational Guidelines on the Donation of Foods Derived from Modern Biotechnology (2004) state:

> any commodity distributed by WFP to or transiting through a Party to the Protocol that meets the definition of an LMO (which at present would apply only to unprocessed maize and soybeans) should be considered as "intended for direct use as food or feed, or for processing" under terms of the Protocol.[24]

The WFP also undertakes to observe any definition of a regulated GM food in national laws, however, and requires that WFP country offices comply with any national regulations and policies on the transit and importation of food.

Publication of National LMO-FFP Regulations and Risk Assessments

By Article 11(1), a Party that makes a final decision regarding domestic use of an LMO-FFP (e.g., approving its use and commercial availability) agrees to publish that decision through the BCH within 15 days. Any such decision must be accompanied by, among other things, information about the identity of the LMO and a risk assessment report.[25] Article 15 requires that risk assessments be carried out in a scientifically sound and transparent manner, using recognised techniques to evaluate the potential adverse effects of LMOs on biodiversity in the likely receiving environment and taking into account risks to human health.[26]

The BCH is an information sharing system, consisting primarily of a Web site maintained by the Secretariat of the Convention on Biological Diversity.[27] The Secretariat uses the Web site to publish national biosafety laws, regulations, and

24 World Food Programme, "Operational Guidelines on the Donation of Foods Derived from Modern Biotechnology," 11 February 2004 WFP/EB.1/2004/10-C.

25 Annex II to the Cartagena Protocol has a list of the information required in notifications under Articles 8, 10, and 13.

26 Cartagena Protocol, Annex III at Articles 1–3.

27 Biosafety Clearing-House, online: http://bch.biodiv.org/. Accessed 19 October 2009.

guidelines, copies of which the Parties agree to supply for this purpose.[28] The idea of this transparency mechanism is for potential importing States to be notified by publication through the BCH that another State has decided to allow a particular GMO onto the market, and the State can then decide whether to introduce domestic regulatory controls on the LMO that has become available for international trade. Obviously, the efficacy of this system depends on careful monitoring of BCH publications. There are concerns that the information on the BCH is sometimes incomplete or inaccurate and that some States do not have the necessary capacity to access the available information.

Article 11(8) confirms that Parties are entitled to take decisions regarding the import of LMO-FFPs in accordance with the precautionary principle. Parties are entitled to impose regulations that they consider appropriate to avoid or minimise the potential adverse effects of importing LMOs-FFP even where there is a "lack of scientific certainty due to insufficient relevant scientific information and knowledge regarding the extent of the potential adverse effects" of an LMO on biodiversity and in relation to human health. This might include total bans or special requirements, such as that GMO grain must be milled before or on delivery to prevent planting. WTO Members must ensure that their precautionary approach is consistent with their obligations under the SPS Agreement. The key consideration in this regard, as is discussed later, is that precautionary restrictions should be regarded as provisional, in that the SPS Agreement imposes an obligation to seek further scientific information to justify the measure within a reasonable period.

Identifying Documentation

By Article 18(1), Parties agree to require that LMOs that are subject to intentional transboundary movement are "handled, packaged and transported under conditions of safety." LMO-FFPs must have accompanying documentation (e.g., commercial invoices) that clearly identifies that they "may contain" LMOs and are not intended for intentional introduction into the environment, as well as a contact point for further information.[29] At COP-MOP 1, Parties and other governments were encouraged to require the identifying documentation to state the common, scientific, and commercial names, as well as the transformation event code or unique identifier of the LMO-FFP in question, and in cases of intentional shipment, to state "contains LMOs" rather than "may contain LMOs." An expert group was set up to discuss documentation requirements.

[28] Cartagena Protocol at art 11(5). The requirements in relation to importing LMOs-FFP are significantly less demanding than in relation to LMOs-FIE. Under Article 7(1), prospective exporters of LMOs-FIE are required to notify the importing State of the intention to import LMOs prior to their first cross-border movement andawait the importing State's authorisation for the import (Article 10).

[29] Cartagena Protocol at art 18(2)(a).

At COP-MOP 3, Parties to the Protocol were requested to take measures to ensure that, in addition to the provisions laid out at COP-MOP 1, documentation accompanying LMOs-FFP clearly states that the shipment included LMOs, if they could be identified, or that the shipment might contain LMOs, if the identity were not known. In addition, it was recommended that the documentation include a stipulation that the LMOs are not intended for intentional introduction into the environment. The Parties also agreed to review the efficacy of the documentation requirements at COP-MOP 5.

The Implications of the WTO Agreements for the Regulation of Food Aid

As mentioned earlier, the Preamble to the Cartagena Protocol makes it clear that obligations under the WTO Agreements have to be considered alongside the Protocol's scheme. In other words, import and other trade-related restrictions implemented pursuant to the Protocol or national biosafety laws must comply with those agreements. As discussed in the previous chapter, this was tested before a WTO Panel in the dispute settlement proceedings brought by the United States, Canada, Argentina, and others against the European Communities (EC) in regard to an alleged moratorium on the approval of new GMOs and several import and marketing bans in some EC member States. The Panel found that the EC did in fact have a *de facto* moratorium on GMOs from the United States, Canada, and Argentina. Furthermore, it was characterised as a procedural decision to delay substantive approval decisions on new GMOs and not a measure to achieve the EC level of sanitary or phytosanitary protection, as the EC had claimed. As a result of this action, the Panel found that the EC had acted inconsistently with Annex C(1)(a) and, by implication, Article 8 of the SPS Agreement.[30]

Article XX of GATT sets out general exceptions allowing members to take measures that would otherwise violate GATT principles to, among other things, protect human, animal, or plant life or health. Even where such an exception applies, however, members are obliged to ensure that they adopt the least trade-restrictive measure possible and to ensure that they do not discriminate between like products. The issue of likeness is particularly controversial in this field. There are diverging views on whether GMOs and their products are "substantially equivalent" to their nonmodified counterparts, and there have yet to be any decisions on the issue. A good example is whether tuna packed in GM soya oil should be considered to be "like" the same tuna packed in non-GM oil. GATT practice indicates that four criteria should be taken into account in making this determination: (1) the product's physical properties, (2) its end uses, (3) its tariff classification, and (4) consumers' tastes and habits. It is the fourth of these that leaves the most room for argument.

30 *European Communities – Measures Affecting the Approval and Marketing of Biotech Products*, WT/DS291 WT/DS292, WT/DS293, 2006.

The SPS Agreement is intended to ensure that human and animal health (sanitary) and plant health (phytosanitary) measures do not amount to unnecessary, arbitrary, scientifically unjustifiable, or disguised restrictions on international trade.[31] It does this by requiring that such measures are based either on existing international standards or on a risk assessment.[32] Otherwise, when relevant scientific evidence is insufficient, members may adopt provisional measures but must seek to obtain additional information necessary for an objective risk assessment "within a reasonable period of time."[33] This suggests an important limitation on the right to restrict imports in accordance with the precautionary principle that is contained in the Cartagena Protocol: that precautionary measures can be only provisional.[34] Of course, much turns on the interpretation of terms such as *provisional* and a *reasonable length of time*. A Panel might be persuaded to take a liberal approach as to their interpretation taking into account the stronger precautionary rights adopted in the Cartagena Protocol, in addition to the concept that trade and environment agreements should be "mutually supportive."

The TBT Agreement provides that mandatory technical regulations, voluntary standards, and related assessment procedures (other than measures governed by the SPS Agreement) should not be more trade restrictive than necessary to achieve a legitimate objective. Article 2.2 of the TBT Agreement specifies that legitimate objectives include, among other things, "national security requirements; the prevention of deceptive practices; protection of human health or safety, animal or plant life or health, or the environment." It is not yet clear whether measures such as the labelling of GMO products to provide consumer information constitute a legitimate objective. Another issue arises from the fact that, like the GATT, the TBT Agreement does not permit discrimination among "like products" (Article 2.1 TBT). The meaning of *like products* is not defined, and there is considerable potential for disagreement over whether a GMO-FFP is like a non-GMO (i.e., traditionally bred) relative.[35]

KEY CONSIDERATIONS FOR THE REGULATION OF FOOD AID IN DOMESTIC LAW

National governments engaged in the process of drafting biosafety laws may well be inclined to consider the effect of those laws on the control of food aid. This section

[31] Sanitary and Phytosanitary Agreement at Preamble and Article 5.2.

[32] SPS, Articles 3(1), 3(3), and 5.

[33] *Ibid* at Article 5(7).

[34] In *EC – Measures Concerning Meat and Meat Products (Hormones)* AB-1997–3, WT/DS26/AB/R, 1998 the Appellate Body agreed with the Panel's finding that although the precautionary principle is reflected in Article 5.7 of the SPS Agreement, it "would not override the explicit wording of Articles 5.1 and 5.2."

[35] According to the concept of "substantial equivalence" adopted by the Organisation for Economic and Cooperative Development as a guiding principle for assessing the safety of GM foods, genetic modification "does not inherently lead to foods that are less safe than those develop by conventional techniques," and their safety evaluation does not require a different standard of safety.

provides a brief discussion of the factors most relevant to this issue. There is also a consideration of the approach adopted in the African Model Law on Safety in Biotechnology, approved by the Organisation for African Unity (now the African Union) in 2001.[36]

The intention behind the Model Law is to build on the structure of the Cartagena Protocol and to promote harmonised biosafety legislation in African countries. Although it is used here as an example of the potential form of biosafety regulations, it is also recognised that key considerations can vary across individual countries and that it is important that national law makers will want to consider their particular domestic legal background.

Scope

As explained above, the Protocol makes a significant distinction in the way that the movement of LMOs-FIE and LMOs-FFP is regulated. A key consideration is whether to maintain this distinction in national biosafety laws. Relevant factors in this context include the fact that the Protocol's distinction between the two categories is generally seen as a negotiating concession made to Parties that did not want to restrict the trade in GMOs; the acknowledged risk that LMOs-FFP could be introduced into the environment, perhaps by planting or by accidental escape from open-topped trucks;[37] and that there is no obligation to notify national authorities of LMO-FFP imports in international law. Indeed, the WFP has made clear that in the absence of national law requirements, it sees no need to provide notifications of GM material in food aid.[38] Accordingly, States may wish to make it a legal requirement that national authorities be notified of and have a reasonable period to decide whether to authorise the delivery of GM food aid consignments before they cross their borders (e.g., by introducing requirements similar to the AIA procedure that would apply to food shipments). National governments must also consider the resources that they can make available for the processing of notifications.

A further consideration is whether a national law should be drafted to apply to processed foods and other categories that are excluded from the scope of the Cartagena Protocol, such as foods in transit (a category excluded from the scope of the Protocol by Article 6). Article 2 of the African Model Biosafety Law provides that it shall also apply "to the import, export, transit, contained use, release or placing on the market of any genetically modified organism whether intended for release into the environment, for use as a pharmaceutical, for food, feed or processing or a product of a genetically modified organism." Thus it applies to several categories

36 OAU African Model Law on Safety in Biotechnology (2001), online: www.africabio.com.

37 "Recent research in Mexico found that GM maize imported from the US to make tortillas has been planted and is now contaminating fields far and wide"; F Pearce, "UN is slipping modified food into aid" (19 September 2002) 175 New Scientist 2361.

38 *Ibid.*

excluded from the Protocol's scope (e.g., GMOs in transit or destined for contained use) as well as to GM products (widely defined in Article 1). Article 4 goes on to require procedures very similar to those of the Protocol's AIA procedure in regard to all GMOs (rather than only LMOs-FIE, as in the case of the Protocol). One potential concern is the resources available to national authorities to review such an expansive range of notifications within a period that does not unnecessarily restrict international trade.

Approval Criteria and Risk Management

Two interesting additional features of the Model Law regime are the requirements imposed under Articles 6(9) and 6(10). Article 26 of the Protocol confirms that Parties may take into account socioeconomic considerations when making decisions on import approvals. Article 6(9) of the Model Law seeks to ensure that such considerations are taken into account by requiring that no approval for GMO imports or use is granted unless it is determined that granting it will benefit the country without causing any significant risk to the environment, biological diversity, or human health; contribute to sustainable development; not have adverse socioeconomic impacts; and accord with the ethical values and concerns of communities and not undermine community knowledge and technologies.

Article 6(10) requires as a condition for approval of import or use that the applicant "furnish evidence of insurance cover or some other arrangements sufficient to meet its obligations under this law." Insurance is an important factor to consider, given that Article 27 of the Protocol foresees the establishment of rules and procedures for liability and redress for damage resulting from transboundary movement of GMOs. GM contamination has the potential to lead to very substantial losses; for example, products might have to be recalled and export opportunities lost.

Labelling and Traceability

Given the substantial influx of unfamiliar GM products that might arrive in the country in the form of aid, national governments might consider the establishment of labelling and traceability requirements an important issue. The Protocol does not make any provision for them, but they are considered important elements of some existing regulatory regimes (the EU's regime is a prominent example). *Labelling requirements* ensure that GM products are fully identified and carry important information on their use. *Traceability* refers to the establishment of a system enabling the tracking of GM produce as it moves between distributors before it reaches the end user. Labelling and traceability requirements must not be more trade restrictive than necessary if they are to comply with the WTO TBT Agreement. They potentially have a trade restrictive effect because of the additional expense that can be incurred if imports have to undergo identification tests or if GM produce has to be segregated from non-GM produce. As to what is necessary, however, the Codex Alimentarius Commission, which sets international food safety standards, adopted standards for GM foods in 2003 that refer to the tracing of products and labelling as legitimate risk management tools.[39]

Transparency, Public Awareness, and Participation

Given that food aid imports often relate to emergency situations that call for a balancing between effective relief and the management of GMO-related risks, the public's involvement in the decision-making process about GMOs is considered especially important. Article 23 of the Protocol requires Parties to consult the public in the decision-making process and to make the results of such decisions available to the public. Article 5 of the African Model Law provides for the public to participate in the decision-making process through a notice and comment procedure and by public consultations, at the discretion of the competent authority. Article 5 also requires that the information provided by the applicant when making an application for import approval be made publicly available.

CONCLUSION

Given the concerns that have been widely expressed over imports of GM food, the fact that there is no requirement in international nor in many national laws that national authorities be notified of such imports is an important factor to be considered in the design of national biosafety laws. In the absence of such a requirement, countries that receive food aid should work on the assumption that that aid will contain GMOs and that they will not receive advance notification of this. This could be of concern, because past experience is that GMOs provided as food have been introduced into the environment by planting or by accidental release.

The Cartagena Protocol simply provides a framework that leaves much regulatory work to be done at the national level. National policy makers must consider whether they are satisfied with the protection offered by the BCH procedure in relation to imports of food. Key issues for national governments in the regulation of food aid, keeping in mind WTO obligations, therefore include first and foremost whether to extend the scope of AIA procedures to GMOs intended for food and whether to incorporate labelling and traceability requirements and a transparent decision-making process that allows public participation.

39 Online: http://www.codexalimentarius.net. Baumueller, "Domestic Import Regulations for Genetically Modified Organisms and their Compatibility with WTO Rules: Some Key Issues" (Trade Knowledge Network, 2003) at 5.

Conclusion

Marie-Claire Cordonier Segger, Frederic Perron-Welch, and Christine Frison

BIOSAFETY LAW: LOOKING FORWARD

This book has addressed the legal aspects of implementing the Cartagena Protocol on Biosafety by analysing the key issues that affect the design and implementation of national biosafety regulatory regimes, as well as the validity of existing regulations. First, the book has identified aspects of international law on biosafety that are reflective of sustainable development principles. Second, the book has analysed the national implementation of international law on biosafety; the merits and drawbacks of biosafety regulations from a sustainable development perspective; and the challenges encountered in the attempt to implement international biosafety obligations in contemporary settings. Last, the book discussed whether such obstacles are likely to prevent the successful implementation of biosafety regulations in developing countries, and how to support more coherent national and international rules, offering insights for international bodies that oversee international law on biosafety. The authors examined the practical, legal, political, and economic issues that have been encountered by various countries in the attempt to implement these regulations in contemporary settings. They focused on the different means of facilitating the implementation of biosafety regulations in developing countries, where there are further challenges because of limited scientific, human, technical, and financial resources.

Inevitably, when the process of creating, developing, and applying international legal norms in national biosafety legislation is discussed, the conversation leads to the scientific material. Legal and scientific links are an area characterised by obstacles in the development of national biosafety strategies. Indeed, one of the first concerns to arise during reviews of developing countries' National Biosafety Frameworks involves the "scientific" parts of the Protocol. References to scientific terms, techniques, or mechanisms that are included in the legislation on biosafety are often left unclear. Definitions can be taken out of context, techniques can be confused, and risk-assessment or decision-making processes can be incomplete. This in turn creates uncertainty in the national policy and/or law, which is exactly where precision and clarity are needed. Uncertainty threatens to hinder the implementation of a country's biosafety law and policy, which can lead to negative impacts on the safety, development, and well-being of the national population and environment, as well as a significant waste of very scarce resources and political will.

Difficulties in implementing national biosafety mechanisms, however, cannot be exclusively attributed to a lack of means or scientific capacity within a country. In fact, many countries have established highly competent scientific structures in very different fields (e.g., medical, agricultural, and botanical). Even though these structures and institutions are limited in the techniques available for their use and rely on widely available knowledge, a growing number of countries have the basic frameworks and competent people available to deal with biosafety issues, or have the ability to become part of subregional or regional cooperation frameworks that could complement one another.

A principal obstacle has been a lack of cooperation or teamwork among national bodies and competent officials working in both the legal and scientific domains addressed by the Protocol. Assistance with institutional and organizational collaboration (governance and regulatory reform) is an important priority and is as significantly and simultaneously essential as assistance in the understanding, development, and application of international norms. Bringing institutions and stakeholders together can increase the efficacy of regulatory regimes. Consulting scientific experts on their areas of competency during the drafting of a law to check the meaning and certainty of the facts or mechanisms described in the law will help to avoid confusion, contradictions, and gaps.

Additional weaknesses and concerns have been identified by developing country leaders, scientists, and civil society representatives, especially in Francophone Africa. These concerns relate mainly to the policy objectives set by a participatory process in the country in question. In some countries, policies in biosafety issues are very precautionary, including in small island developing States such as Cape Verde, the Seychelles, or Madagascar; Muslim countries like Algeria; countries with high dependence on tourism or agricultural trade with Europe; and fragile ecologies, such as Ecuador. In some Asian and Central Asian countries there is concern that the socioeconomic changes created by biotechnology could, if not properly understood and managed, have similar costs and benefits as the Green Revolution. These challenges require careful regulatory drafting to properly reflect these policies into law, and make compliance feasible.

In other countries, policies on biosafety are intermediately protective – simultaneously biosafety and development oriented. This is particularly true for countries with "porous" boundaries, such as Togo, Niger, and Mali, where public or governmental concerns are high and there is a strong dependence on subsistence agriculture or food trade, but there is also a need to learn to address products that enter the country regardless of controls. Regulations to implement these policies can be challenging to develop and more challenging to enforce.

In still other areas, policies have sought primarily to encourage development of biotechnology while ensuring an adequate level of biosafety. This is particularly true of those countries with a close relationship to the United States, such as Burkina Faso, Mozambique, Chile, or Argentina. Such countries might also serve as gateways for crops or aquaculture LMOs to the rest of their continents.

For most countries (including those that are WTO members, those seeking to eventually join, and those that receive significant levels of food aid), however, overlap and potential conflicts among different international instruments are of serious concern. Other important questions for future biosafety research related to legal and policy regimes include issues such as the need to balance the legal and scientific aspects of obligations and requirements under national and international law, the need to ensure the solid and informed participation of civil society in decision making, and the need to develop or learn the best possible risk assessment laws and techniques for effectiveness in different local contexts.

ENDURING QUESTIONS

Among the range of actors in the biosafety field, there is widespread agreement that many unanswered questions remain and much more work is needed on biosafety research that focuses on the concerns of developing countries. Several key questions, concerns, and areas of research that need to be developed to reach a better level of understanding and application of biosafety norms and issues are discussed here and have been grouped into categories for ease of reference.

The role of sustainable development cannot be overlooked in the implementation of the Cartagena Protocol. How do States, and in particular developing countries, see the link between the Cartagena Protocol and sustainable development? Do they consider sustainable development issues when complying with their obligations deriving from the Cartagena Protocol in national frameworks or legislation?

Biosafety, Integration and Interrelationship

The concept of risk management found in Article 16 of the Cartagena Protocol does not exist explicitly in the WTO Agreements. At best, one may consider that Article 5.3 of the Sanitary and Phytosanitary Agreement (SPS Agreement) represents an opening in that direction. Risk management and the Protocol's Article 16 might be *de facto* contested through the Doha Declaration's introduction of the term *Specific Trade Obligation* by being considered neither specific nor obligatory enough. What does this say about issues like compatibility, coherence, WTO-UN relations, or the environmental mandate, potential, or aspirations of the Doha Development agenda?

Can standard risk assessment procedures be developed that will be accepted by the international community both in the context of the Cartagena Protocol and beyond?

What are the legal stakes if the CBD Conference of the Parties serving as the meeting of the Parties to the Cartagena Protocol represents an "international standardizing body" in the sense of Article 1.1 of the WTO's Technical Barriers to Trade Agreement (TBT Agreement; assuming that it is about to create rules such as the ones on liability and redress), or if the Protocol itself is considered a standard in the sense of Annex 1, paragraph 2 of the TBT Agreement and Annex A, paragraph 3(d) of the SPS Agreement?

Can analysis, based on comprehensive case studies, be developed to demonstrate how a country's obligations under different international instruments related to biosafety (e.g., the Cartagena Protocol, the IPPC, Codex) can affect the national biosafety process?

How can developing countries design and use regional biosafety mechanisms for scientific and regulatory risk assessment, particularly when faced with porous borders or similar requests from exporters? What is the potential for cooperation between laboratories or scientific institutions and among regulatory bodies, and how can this cooperation be encouraged? Can joint systems be set in place to support processes such as monitoring and certification?

There are clear socioeconomic issues that must be taken into consideration when making decisions about biosafety. What is unclear, however, are the types of methodologies and procedures that can be used to effectively assess those considerations, which include the social, cultural, and economic dimensions of GMO risks and regulation, as well as the scientific dimensions of environmental health and safety effects.

Biosafety, Equity and Poverty Eradication

Of particular concern to the international community is the implementation of the Cartagena Protocol at the national level in developing countries. There must be a focus on the strategies that are being used by the biotechnology industry to break into markets where people are poor to ensure that the arrival of biotechnology in these markets helps people, rather than making them poorer.

Imported GM crops that have been grown efficiently with huge machinery on very large farms in industrialised countries, perhaps subsidised directly or indirectly by the country of export, could be relatively cheap in the country of import and interesting to urban markets. At the same time, these low-price imports could be devastating for domestic farmers, who often constitute most of the population and already tend to suffer most from poverty. An influx of low-cost GM crops might result in farmers being both unable to sell their products and without the ability to afford the patented seed/fertiliser/pesticide package. This in turn has the potential to drive rural populations into urban slums and create cycles of social exclusion. How can a developing country protect itself in cases of such occurrences or trends, under Article 26 on socioeconomic considerations?

What is the impact of the adventitious presence of LMOs in exports from developing countries on these countries' access to markets? Article 2.2 of the WTO's TBT Agreement on the preparation, adoption, and application of technical regulations allows certain objectives and elements to be considered when evaluating the legitimacy of a trade measure. How can an importing country invoke legitimate and relevant objectives and elements of consideration that are not based on biological risk assessment but rather on other approaches, such as socioeconomic risk assessment or other societal concerns and issues?

Which obstacles do developing countries face as they try to commercialise the products of local and domestic biotechnology research and development? What can be done to overcome these obstacles? What are other options besides biotechnology for solving the challenges faced by poor farmers?

What is the impact of a widespread switch to raising genetically modified (GM) organisms for export on domestic food consumption and nutrition? How can countries with high levels of public apprehension from religious or cultural factors address the dichotomy between international and domestic cultural standards?

Biosafety and Common but Differentiated Responsibilities

More research is necessary into the major scientific and regulatory capacity needs and limitations that presently hinder the effective implementation of the Cartagena Protocol in developing countries. An analysis is needed of the legal and institutional difficulties faced by developing countries in implementing a biosafety regulatory regime. This can be done by examining analogous fields of regulation, apart from the specifics of biosafety, that are also key for the effective functioning of a biosafety system (e.g. intellectual property, liability rules, consumer protection, other safety systems, access to information, access to justice, and competition).

How can developing countries best participate in and implement decisions of the COP-MOP in an informed manner? Which representation is most effective, and how can information be shared between one meeting and the next? How can developing countries best conduct regulatory assessments or use other new tools to identify the areas of their biosafety systems that need to be strengthened (e.g., scientific capacity, regulatory capacity, innovation), and how can these different areas, once identified, be prioritised? What are the scientific and technical challenges faced by developing countries in implementing a biosafety regulatory regime (e.g., understanding of and capacity to implement risk assessment and risk management, assess socioeconomic considerations, conduct decision making based on sound science)? How is it possible to handle scientific compliance issues in National Biosafety Frameworks and laws more comprehensively? How can issues such as technical or scientific definitions of terms, determination, limitation of rights, and obligations regarding certain scientific techniques and mechanisms, including/excluding certain scientific techniques from a legal mechanism be addressed in regard to each country's specific circumstances? Which types of tools have been used to build biosafety capacity to date (e.g., training workshops, manuals)? How effective have these tools been, and what are areas for improvement? How can developing countries optimise the use of their existing capacities and adapt them to cover the field of biosafety? How can existing legal and administrative frameworks in related areas (e.g., phytosanitary measures) be affected by and contribute to the biosafety policy of the country?

Biosafety and the Precautionary Approach

What constitutes sound science? How does the requirement for scientific risk assessments under the Protocol (in particular, Article 15) correspond to similar requirements in the WTO's SPS Agreement (in particular, Articles 2.2, 3.3, 5.1, and 5.2)? Can decision making include social science considerations, as provided in Article 26 of the Protocol? Furthermore, what is the importance of risk assessment in the research development and testing of organisms with novel traits to identify any hazards, and to assess risks and impacts that LMOs could have on human health, agricultural production, and the environment?

Biosafety, Public Participation and Access to Justice

What are the different views, interests, and roles played by nongovernmental institutions (including business firms) regarding LMO risks and regulations, and their implications for governance? How can full "participation" in a biosafety decision-making process be ensured while keeping this within the meagre resources of most countries (and civil society actors), and not allowing one sector to dominate a process? How can transparency and participation be ensured when there is not an adequate level of basic knowledge about biosafety issues, and biotechnology within a country's population? What is the best way to inform the population without propaganda? Within a process of democratic participation, decision making should necessarily involve and take into account the points of view, needs, and concerns of experts, people involved in each process, people potentially affected by the short- and long-term consequences of policies, and policy makers. How can this requirement be built into the requirements for scientific decision making under the Protocol? Which role do consumers play in decisions about access to public information and the use and labelling of GM products? How can the confidentiality of information be better understood and protected? If confidentiality is well-defined in a country's NBF, how can its application and implementation in law be improved? How can countries create public participation structures that do not overwhelm their other obligations under the Protocol? How can countries strike the right balance between the confidentiality of information and public participation? How can countries include public participation in their biosafety mechanisms? How much information does the biotechnology industry freely provide about the effects of biotechnology on land, water, food, the health and well-being of workers handling it, potential market effects on regional products and by-products, potential biodiversity loss, cost-benefit analyses that do not include negative externalities, industry liability for long-term effects of their products, and the burden of proof on industry? What should governments do to ensure that people are aware of the implications of biotechnology in these areas and which measures should be taken to provide people with full information about these issues when industry is not disposed to provide it willingly? Is this something that should be tackled by individual governments, or should it be done at the international level?

Good Governance for Biosafety

From a general governance perspective, this book evaluates the various areas of regulation that a country must take into account when attempting to implement effective national biosafety legislation and processes. What are the alternative governance options for addressing LMO regulatory issues and mechanisms for harmonising or resolving differences in existing approaches and methods? More research is required to examine the impediments that countries face in ratifying the Cartagena Protocol. Once a country does ratify the Protocol, there must be an examination of the hurdles that impede the creation and implementation of a regulatory framework on biosafety, either to comply with the Protocol or to go beyond its provisions, to inform best practices. Furthermore, domestic biosafety regimes must be evaluated within the context of existing or anticipated international obligations to ensure minimal contradictions and problems with dual implementation.

CONCLUSION

Although this book tackles many legal issues, new questions are arising that require answers. Implementing a complex international instrument such as the Cartagena Protocol is not a simple task for any country, let alone developing countries. This is particularly true in reference to the myriad obligations in the many international instruments that countries are bound to implement. Many fields and research areas are still to open to development. Many questions remain to be answered, but it is hoped that this book has met its objectives: to clarify the implementation process, identify the difficult issues, and provide the beginnings of answers to the many remaining questions. To those countries that still have to implement the Protocol or improve their regulations and process for managing LMOs, the editors hope that this contribution will aid them in their difficult task.

ANNEX I

The Components of a National Biosafety Framework (UNEP/GEF Toolkit)

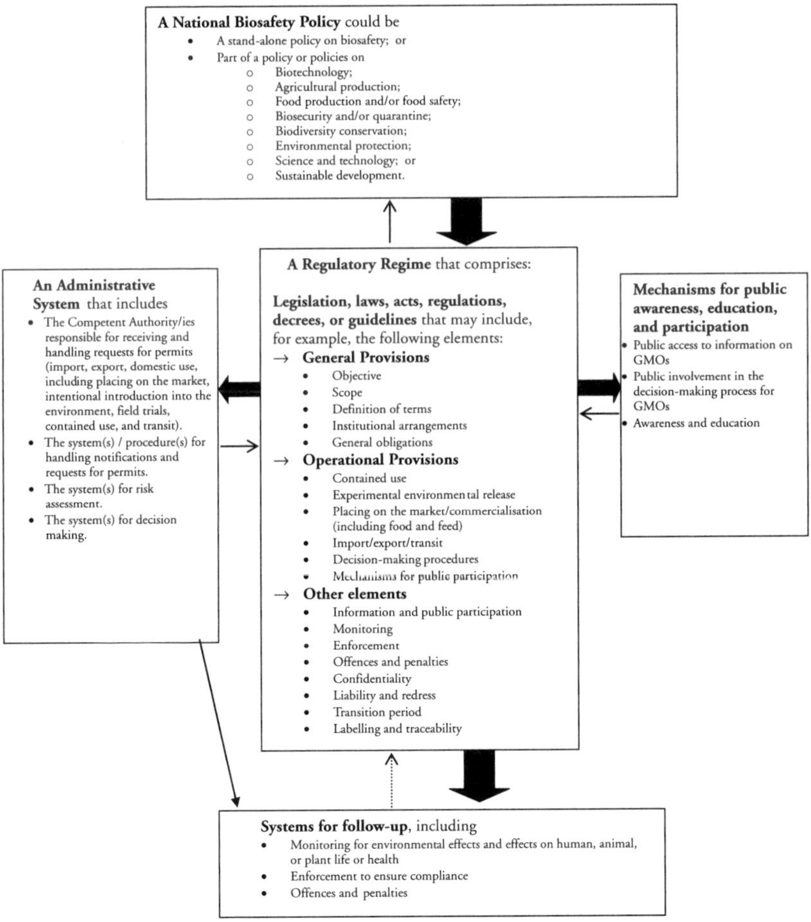

ANNEX II

Cartagena Protocol Implementation Tool Kit

1. *Administrative tasks*

	Tasks	Article	√
	Initial actions		
1.	Designate one national authority responsible for liaison with the Secretariat and provide name/address to Secretariat.	19(1),(2)	
2.	Designate one or more competent authorities responsible for performing administrative functions under the Protocol and provide name(s)/address(es) to the Secretariat. If more than one, indicate the types of LMOs for which each competent authority is responsible.	19(1),(2)	
3.	Provide to the Biosafety Clearing-House: any relevant existing laws, regulations or guidelines, including those applicable to the approval of LMOs-FFP; and any bilateral, regional or multilateral agreements or arrangements.	20(3)(a)-(b), 11(5), 14(2)	
4.	Specify to the Biosafety Clearing-House cases in which import may take place at the same time as the movement is notified.	13(1)(a)	
5.	Specify to the Biosafety Clearing-House imports of LMOs exempted from the AIA procedures.	13(1)(b)	
6.	Notify the Biosafety Clearing-House if domestic regulations shall apply with respect to specific imports.	14(4)	
7.	Provide the Biosafety Clearing-House with a point of contact for receiving information from other States on unintentional transboundary movements in accordance with Article 17.	17(2)	
8.	Notify the Secretariat if there is a lack of access to the Biosafety Clearing-House and hard copies of notifications to the Clearing House should be provided.	(e.g., 11(1))	

	Tasks	Article	√
	Follow-up actions		
9.	Provide to the Biosafety Clearing-House: Summaries of risk assessments or environmental reviews of LMOs generated by regulatory processes and conducted in accordance with Art. 15; Final decisions concerning the import or release of LMOs; and Article 33 reports.	20(3)(c)-(e)	
10.	Make available to the Biosafety Clearing-House information concerning cases of illegal transboundary movements.	25(3)	
11.	Monitor the implementation of obligations under the Protocol and submit to the Secretariat periodic reports at intervals to be determined.	33	
12.	Notify the Biosafety Clearing-House of any relevant changes to the information provided under part I above.		

II. *Legal requirements and/or undertakings*

	Tasks	Article	√
1.	Ensure that the development, handling, transport, use, transfer and release of LMOs are undertaken in a manner that prevents or reduces the risks to biological diversity, taking also into account risks to human health.	2(2)	
2.	Ensure that there is a legal requirement for the accuracy of information provided by domestic exporters for purposes of notifications for export to another country and by domestic applicants for domestic approvals for LMOs that may be exported as LMOs-FFP.	8(2) 11(2)	
3.	Ensure that any domestic regulatory framework used in place of the AIA procedures is consistent with the Protocol.	9(3)	
4.	Ensure that AIA decisions are taken in accordance with Article 15.	10(1)	
5.	Ensure that risk assessments are carried out for decisions taken under Article 10 and that they are carried out in a scientifically sound manner.	15(1),(2)	
6.	Establish and maintain appropriate mechanisms, measures and strategies to regulate, manage and control risks identified in risk assessments associated with the use, handling and transboundary movement of LMOs under the Protocol.	16(1)	
7.	Take appropriate measures to prevent the unintentional transboundary movements of LMOs, including measures such as requiring a risk assessment prior to the first release of an LMO.	16(3)	

(continued)

II *(continued)*

Tasks	Article	√
8. Endeavour to ensure that LMOs, whether imported or locally developed, have undergone an appropriate period of observation that is commensurate with its life cycle or generation time before it is put to its intended use.	16(4)	
9. Take appropriate measures to notify affected or potentially affected States, the Biosafety Clearing-House, and, where appropriate, relevant international organizations, when there is an occurrence within its jurisdiction that leads or may lead to an unintentional transboundary movement of and LMO that is likely to have significant adverse effects on the sustainable use and conservation of biodiversity, taking also into account risks to human health in such States.	17(1)	
10. Take necessary measures to require that LMOs that are subject to transboundary movement under the Protocol are handled, packaged and transported under conditions of safety, taking into account relevant international rules and standards.	18(1)	
11. Take measures to require that documentation accompanying LMOs-FFP • clearly identifies that they "may contain" LMOs and are not intended for intentional introduction into the environment; and • provides a contact point for further information.	18(2)(a)	
12. Take measures to require that documentation accompanying LMOs destined for contained use: • Clearly identifies them as LMOs; • Specifies any requirements for their safe handling, storage, transport and use; • Provides a contact point for further information; and • Provides the name and address of individuals or institutions to which they are consigned.	18(2)(b)	
13. Take measures to require that documentation accompanying LMOs that are intended for intentional introduction in the environment and any other LMOs within the scope of the Protocol: • Clearly identifies them as LMOs • Specifies the identify and relevant traits and/or characteristics; • Provides any requirements for the safe handling, storage, transport and use; • Provides a contact point for further information; • Provides, as appropriate, the name and address of the importer and exporter; and • Contains a declaration that the movement is in conformity with the requirements of the Protocol.	18(2)(c)	
14. Provide for the designation of confidential information by applicants, subject to the exclusions set forth in Article 21(6).	21(1),(6)	
15. Ensure consultation with applicants and review of decisions in the event of disagreement regarding claims of confidentiality.	21(2)	

	Tasks	Article	√
16.	Ensure the protection of agreed-upon confidential information and information claimed as confidential where a notification is withdrawn.	21(3),(5)	
17.	Ensure that confidential information is not used for commercial purposes without the written consent of the notifier.	21(4)	
18.	Promote and facilitate public awareness, education and participation concerning the safe transfer, handling and use of LMOs, taking also into account risks to human health.	23(1)(a)	
19.	Endeavour to ensure that public awareness and education encompass access to information on LMOs identified in accordance with the Protocol that may be imported.	23(1)(b)	
20.	In accordance with relevant domestic laws, consult with the public in decision making under the Protocol, while respecting confidential information.	23(2)	
21.	Endeavour to inform the public about the means of public access to the Biosafety Clearing-House.	23(3)	
22.	Adopt appropriate measures aimed a preventing and, if appropriate, penalizing transboundary movements in contravention of domestic measures to implement the Protocol.	25(1)	
23.	Dispose, at its expense, LMOs that have been the subject of an illegal transboundary movement through repatriation or destruction, as appropriate, upon request by an affected Party.	25(2)	

III. *Procedural requirements: advanced informed agreement*

	Tasks	Article	✓
1.	Notify, or require the exporter to ensure notification to, in writing, the competent national authority of the Party of import prior to the intentional transboundary movement of a living modified organism that falls within the scope of Article 7, paragraph 1	8(1)	
2.	Provide written acknowledgement of receipt of notification to notifier within 90 days, including:		
	– Date of receipt of notification;	9(2)(a)	
	– Whether notification meets requirements of Annex I;	9(2)(b)	
	– That the import may proceed only with written consent and whether to proceed in accordance with the domestic regulatory framework or in accordance with Article 10; **OR** - Whether the import may proceed after 90 days without further written consent.	10(2)(a), 9(2)(c) 10(2)(b)	
3.	Communicate in writing to the notifier, within 270 days of receipt of notification: • Approval of the import, with or without conditions; • Prohibition of the import; • A request for additional relevant information in accordance with domestic regulatory framework or Annex I; or • Extension of the 270 day period by a defined period of time; AND	10(3)(a)-(d)	
	Except where approval is unconditional, the reasons for the decision, including the reasons for the request for additional information or for an extension of time.	10(4)	
4.	Provide in writing to the Biosafety Clearing-House the decision communicated to the notifier.	10(3)	
5.	Respond in writing within 90 days to a request by an Exporting Party for a review of a decision under Article 10 where there has been a change in circumstances or additional relevant scientific or technical information has been made available, providing the reasons for the decision upon review.	12(2), (3)	

IV. *Procedural requirements: living modified organisms for direct use as food, feed or for processing*

	Tasks	Article	√
1.	Upon making a final decision regarding domestic use, including placing on the market, of LMOs that may be subject to transboundary movement for direct use as food or feed, or for processing, inform the Biosafety Clearing-House within 15 days of making that decision, including the information listed in Annex II.	11(1)	
2.	Except in the case of field trials, provide hard copies of the final decision to the National Focal Point of Parties that have notified the Secretariat in advance that they do not have access to the Biosafety Clearing-House.	11(1)	
3.	Provide additional information contained in paragraph (b) of Annex II about the decision to any Party that requests it.	11(3)	
4.	In response to the posting of a decision by another Party, a Party that decides to import may take a decision on the import of LMOs-FFP: either as approved under the domestic regulatory framework consistent with the Protocol; **OR** in the absence of a regulatory framework, on the basis of a risk assessment in accordance with Annex III within no more than 270 days. In this case, a declaration must be made to the Biosafety Clearing-House.	11(4), (67)	

Source: Reproduced from Cartagena Protocol, COP-MOP 1, Decision BS-I/5 “Capacity-building” at Annex III.

ANNEX III

Risk Management and Risk Assessment

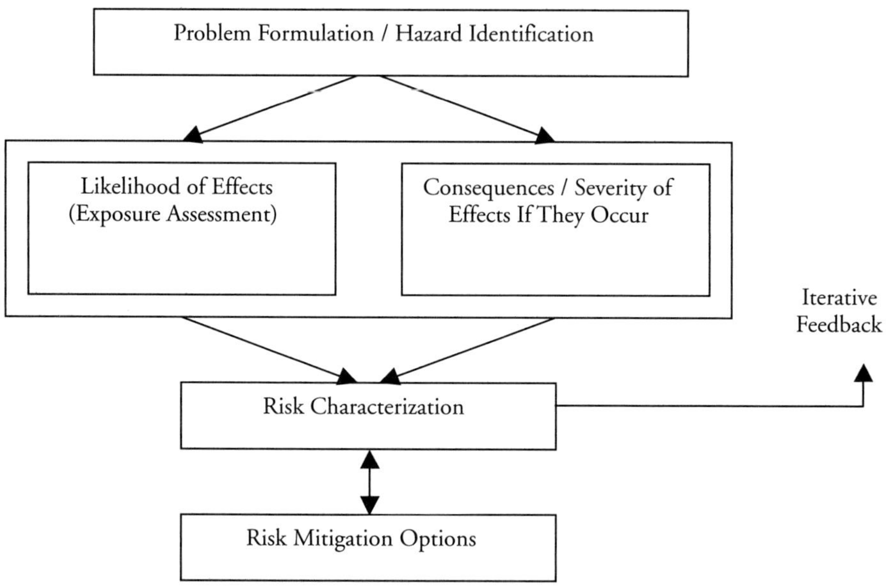

FIGURE 1. Generic Steps Common to Many Risk Assessment Frameworks (*adapted from* Hill 2005).

TABLE 1. *Selected principles and methodological components of the Cartagena Protocol reflected in other risk assessment frameworks*

Principles and methodological components	IPPC – ISPM#11[1]	Australia OGTR[2]	EC guidance notes
Risks should be considered in the context of risks posed by nonmodified recipients or parental organisms (i.e., in a comparative context) – Annex III, paragraph 5.	• "'LMO' refers to an LMO that is a potential quarantine pest due to new or altered characteristics or properties resulting from the genetic modification." • "For LMOs, the acceptable level of risk may also be expressed by comparison to the level of risk associated with similar or related organisms."	• "[R]isks posed by a particular GMO need to be considered in the context of the risks posed by the unmodified parental organism in the receiving environment"	"Identified characteristics of the GMO and its use which have the potential to cause adverse effects should be compared to those presented by the non-modified organism from which it is derived and its use under corresponding situations"
Risk assessment is carried out on a case-by-case basis, and the level of detail will vary depending on the case – Annex III, paragraph 6.	• The risk assessment should be carried out on a case-by-case basis.	"Risks . . . are assessed on a case by case basis."	"The ERA should be carried out on a case by case basis, meaning that the required information may vary."
Risk assessment begins with the *identification of LMO characteristics that may have adverse effects* (emphasis added) – Annex III, paragraph 8(a)	• "Initiation" precedes risk assessment and focuses on identifying a pest or pathway and the potential for consequences. • Pest categorization is then the first step of risk assessment.	• "Hazard identification" focuses on what could go wrong and the causal pathway leading to a potential adverse outcome.	Identification of the characteristics of a GMO that could cause adverse effects

(continued)

[1] IPPC Secretariat, "ISPM No. 11: Pest Risk Analysis for Quarantine Pests Including Analysis of Environmental Risks and Living Modified Organisms" (International Standard for Phytosanitary Measures, 2004).

[2] Australia, Commonwealth, Office of the Gene Technology Regulator, *Risk Analysis Framework for License Applications to the Office of the Gene Technology Regulator* (Handbook, 2005).

TABLE 1 *(continued)*

Principles and methodological components	IPPC – ISPM#11	Australia OGTR	EC guidance notes
Risks are *estimated* based on an evaluation of the *likelihood* and *consequences* of the identified adverse effects being realized – Annex III, paragraphs 8(b) to 8(d).	• The likelihood component is divided into entry, establishment, and spread. • Economic consequences include environmental and biodiversity impacts, either direct or indirect.	• Risk estimation is a combination of the likelihood ("how likely is it to happen?") and consequences ("will it be serious if it happens?") of adverse outcomes. Consequences are measured in terms of the severity of an adverse outcome.	An estimation of risks based on the likelihood and consequences of each potential adverse effect
Risk assessment includes identification of strategies to manage risks – Annex III, paragraph 8(e); also Articles 16.1 and 16.2.	• "Risk management options" are part of risk management, which is separate from risk assessment. Both risk assessment and risk management are interrelated in a risk-analysis framework.	• Risk management is separate from risk assessment, but both are interrelated in a risk analysis framework.	Step 5 of risk assessment consists in the "Application of management strategies." Step 6 of risk assessment consists in the determination of overall risk.
Uncertainty should be expected in risk assessment – see Articles 10.6, 11.8; also paragraphs 4 and 8(f) of Annex III.	• Overall risk estimates take uncertainty into account. "It is important to document the areas of uncertainty and the degree of uncertainty in the assessment."	• Dedicates a short chapter and an appendix solely to the consideration of uncertainty	"ERA has to take into account uncertainty at various levels." "The ERA may not always result in definitive answers."

ANNEX IV

International Liability and Redress Treaties

Instruments	Adoption date	Signatures	Ratified	Entry into force
Antarctic Treaty System				
Annex VI to the Protocol on Environmental Protection to the Antarctic Treaty (Liability Arising from Environmental Emergencies)	14 June 2005	N/A	4	Not in force
Convention on Biological Diversity (CBD)				
Nagoya – Kuala Lumpur Supplementary Protocol on Liability and Redress to the Cartagena Protocol on Biosafety	15 October 2010	0	0	Not in force
Council of Europe (COE)				
Lugano Convention on Civil Liability for Damage Resulting from Activities Dangerous to the Environment	21 June 1993	9	0	Not in force
International Civil Aviation Organization (ICAO)				
Convention on Damage Caused by	7 October 1952	25	49	4 February 1958
Foreign Aircraft to Third Parties on the Surface	23 October 1978	14	12	25 July 2002
• Amending Protocol				
Convention on Compensation for Damage Caused by Aircraft to Third Parties	2 May 2009	9	0	Not in force
Convention on Compensation for Damage to Third Parties, Resulting from Acts of Unlawful Interference Involving Aircraft	2 May 2009	7	0	Not in force

(continued)

(continued)

Instruments	Adoption date	Signatures	Ratified	Entry into force
Organization for Economic Cooperation and Development (OECD)				
Paris Convention on Third-Party	29 July 1960	18	16	1 April 1968
Liability in the Field of	28 January 1964	15	16	1 April 1968
Nuclear Energy	16 November 1982	15	16	1 August 1991
• Amending protocol	12 February 2004	16	1	Not in force
• Amending protocol	31 January 1963	13	12	4 December 1974
• Amending protocol	28 January 1964	13	12	4 December 1974
•Supplementary Convention	16 November 1982	14	12	7 October 1988
• Amending protocol	12 February 2004	13	2	Not in force
• Amending protocol				
• Amending protocol				
International Atomic Energy Agency (IAEA)				
Brussels Convention on the Liability of Operators of Nuclear Ships	25 May 1962	17	7	Not in force
Vienna Convention on Civil	21 May 1963	14	36	12 November 1977
Liability for Nuclear Damage	12 September 1997	15	5	4 October 2003
• Amending protocol	12 September 1997	13	4	Not in force
• Supplementary Convention	21 September 1988	22	26	27 April 1992
• Joint Protocol Relating to the Application of the Vienna Convention and the Paris Convention				
International Maritime Organization (IMO)				
International Convention on	27 November 1992	10	124	30 May 1996
Civil Liability for Oil	18 October 2000	N/A	N/A	1 November 2003
Pollution Damage (replaced 1969 Convention) Amendment				
Supplementary FUND	27 November 1992	10	105	30 May 1996
Convention (replaced 1971	18 October 2000	N/A	N/A	1 November 2003
Convention)				
• Amendment	16 May 2003	5	26	3 March 2005
• Protocol				
International Convention on Civil Liability for Bunker Oil Pollution Damage	23 March 2001	11	51	21 November 2008
International Convention on	3 May 1996	8	14	Not in force
Liability and Compensation in Connection with Carriage of Hazardous and Noxious Substances by Sea (HNS Convention)	30 April 2010	0	0	Not in force

Instruments	Adoption date	Signatures	Ratified	Entry into force
• Protocol				
Nairobi International Convention on the Removal of Wrecks	18 May 2007	6	1	Not in force
United Nations Environment Programme (UNEP)				
Basel Protocol on Liability and Compensation for Damage Resulting from Transboundary Movements of Hazardous Wastes and their Disposal	10 December 1999	13	10	Not in force
United Nations Economic Commission for Europe (UNECE)				
Convention on Civil Liability for Damage Caused during Carriage of Dangerous Goods by Road, Rail, and Inland Navigation Vessels	10 October 1989	2	1	Not in force
Protocol on Civil Liability and Compensation for Damage Caused by the Transboundary Effects of Industrial Accidents on Transboundary Waters	21 May 2003	24	1	Not in force
Miscellaneous				
Convention on International Liability for Damage Caused by Space Objects	29 November 1971	25	89	1 September 1972
Convention on Civil Liability for Oil Pollution Damage Resulting from the Exploration for and Exploitation of Seabed Mineral Resources	1 May 1977	6	0	Not in force

ANNEX V

Central and Eastern Europe

	What was in place before CPB ratification	What is in place now
Albania	No biosafety law at the time of CP ratification, but there were many laws related to biosafety and biotechnology that could be used and adapted, such as food safety laws, nature protection acts, etc.	Draft law prepared, not adopted yet.
Armenia	No biosafety law, although several related acts, including Laws on Principles of Legislation on Nature Protection, on Fauna, on Flora, Law on Selection Achievements, etc.	Armenia ratified CP in Spring 2004. It took three years to prepare draft law, which has not yet been adopted.
Azerbaijan	No biosafety law, several related acts: Law on Environmental Safety, on Environment Protection, on Food, etc.	Draft law under preparation.
Belarus	Comprehensive system was in place before CP ratification, although it did not explicitly regulate GMOs.	Law on Safety in Genetic Engineering Activities was adopted in 2006.
Bosnia and Herzegovina	No biosafety-related legal acts.	Food Act was adopted in 2004.
Bulgaria	Regulations on the Release of Genetically Modified Higher Plants were introduced as a temporary arrangement in anticipation of the adoption of a law on GMOs (1996).	Law on GMOs (2005), Food Act (2005), regulation on the contained use of GMOs (2005) and secondary legal acts.

	What was in place before CPB ratification	What is in place now
Croatia	Nature Protection Act (2003) had some provisions addressing the use of GMOs, but it was not fully in line with the requirements of the Cartagena Protocol. The Food Act (2003) was also relevant.	A new GMO Act was drafted and adopted in 2005, in addition to secondary regulations.
Czech Republic	Established a law in 2000, which entered into force in 2001 (Act 153/2000, which addresses the Use of Genetically Modified Organisms and Products in Addition to Amendments of Some Related Acts).	Amended Act 153/2000 after getting the member state of EU (on Genetically Modified Organisms and Genetic Products, 2004). Also amended other acts, including Act on Varieties, Seeds and Seedlings of Cultivated Plants; Act on Medicines; Act on Food and Tobacco Products; Act on Feed; Act on Phytosanitary Measures; Act on Organic Farming. Secondary legislation was also adopted.
Estonia	Comprehensive system has been in place since 2001.	New GMO Act was adopted in 2004 after having joined the EU. Many other acts were amended as well: Food Act, Fertilizers Acts, etc.
Georgia	No biosafety law, although various related acts, including the Environmental Protection Act, System of Protected Area Act, etc.	Draft law on "Genetically Modified Organisms" was prepared in 2004 but has yet to be adopted and is currently under revision. Secondary legislation is also being drafted.
Hungary	Act on Gene Technological Activity (1999)	No updated information available.
Latvia	Many legal acts that provide general guidance of biosafety, including the Law on Environmental Protection (adopted in 1991, amendments in 2004), the Law on Surveillance of the Food Circulation (adopted in 1998, amendments 2004), the Plant Protection Law (adopted in 1998, amendments in 2004). Regulation on the Monitoring Council of Genetically Modified Organisms and Novel Foods (adopted in 2000; amended in 2002).	A regulation on Contained Use, Deliberate Release into the Environment, and the Placing of Genetically Modified Organisms on the Market, as well as on their Monitoring (first version adopted in 2000, current regulation adopted in 2004).

(continued)

(continued)

	What was in place before CPB ratification	What is in place now
Lithuania	Framework Law on GMOs (2001), Law on Plant Seeds Growing (2001).	Law on GMOs, amended in 2003 The food safety sector is regulated by the Law on Food (2000, amended in 2004) and the ministerial order Novel Foods and Novel Food Ingredients (2001, amended in 2003).
Former Yugoslav Republic of Macedonia	No specific biosafety law. Although some laws might be relevant to biosafety, including the Law on Environment and Nature Protection and Promotion, Law on Seed Material, Planting Material and Material for Breeding, Recognizing, Approving and Protecting of Varieties (2000), etc.	Draft Law on GMOs established in 2005, no information available on its adoption.
Malta	The Contained Use of Genetically Modified Micro-organisms Regulations (2002), Deliberate Release into the Environment of Genetically Modified Organisms Regulations (2002), Biosafety Co-ordinating Committee Regulations (2002), Environment Protection Act (2001), etc.	The Contained Use of Genetically Modified Micro-organisms Regulations was amended in 2004. All other acts and regulations named in first column are still in force.
Moldova	National Biosafety Law (2002), and many acts under the Ministry of Agriculture and the Ministry of Health.	National Biosafety Law should be amended but no information available if this has been done. New amendments have not yet been adopted.
Montenegro	Had a common legal system with Serbia.	No information available.
Poland	GMO Act in force since 2001.	No updated information available.
Romania	Comprehensive system in place. Government Ordinance on obtaining, testing, and the use and commercialisation of genetically modified organisms obtained through modern biotechnology techniques, and of products resulting thereof (1999). In force since 2002.	Emergency Governmental Ordinance on the deliberate release and placing of genetically modified organisms on the market (2007).

	What was in place before CPB ratification	What is in place now
Serbia	Comprehensive system in place, Law on GMO (2001), Regulations on contained use, deliberate release, and placing of GMOs on the market (2002).	New GMO Act under discussion since 2005.
Slovakia	Act on GMOs (2002).	Act on GMOs amended in 2005.
Slovenia	Health and Hygiene Safety of Foods, and of Materials and Articles Intended to Come into Contact with Foods Act (2000, amended in 2002), Management of Genetically Modified Organisms Act (2002).	All acts in first column still in force.
Turkey	Communiqué on Field Trials of Cultivated Transgenic Plants (1998).	Law on Biosafety drafted in 2005. No information on its adoption available.
Ukraine	No biosafety law. Law on Quality and Safety of Food Products and Food Raw Materials (1997). Several secondary acts, including Temporary procedure of import, state testing, registration and use of transgenic plant varieties in Ukraine approved by the decree of the Cabinet of Ministers of Ukraine (1998).	Law on the National Biosafety System in Creation, Testing, Transportation and Use of GMOs adopted in 2007.

Cartagena Protocol *National Focal Points*

Ministry of environment	Ministry of agriculture	Others
Albania, Armenia, Azerbaijan, Bulgaria, Czech Republic, Estonia, Georgia, Hungary, Lithuania, Former Yugoslav Republic of Macedonia (FYR), Malta (Environment and Planning Authority), Moldova, Poland, Romania, Slovakia, Slovenia, Ukraine	Latvia, Montenegro, Serbia, Turkey	Belarus: Ministry of Foreign Affairs, (Secondary focal point Ministry of Natural Resources and Environmental Protection) Croatia: Ministry of Culture (Nature Protection Directorate) Bosnia and Herzegovina: Ministry of Foreign Trade and Economic Relations

***Ratification by** CEE countries of the Cartagena Protocol on Biosafety, the Aarhus Convention, and the WTO*

Country	Cartagena Protocol *ratification*	Aarhus Convention *ratification, approval (aa), accession (a)*	WTO membership
Albania	8 February 2005	27 June 2001	8 September 2000
Armenia	30 April 2004	1 August 2001	5 February 2003
Azerbaijan	1 April 2005	23 March 2000 (a)	Observer
Belarus	26 August 2002	9 March 2000 (AA)	Observer
Bosnia and Herzegovina	1 October 2009	1 October 2008 (a)	Observer
Bulgaria	13 October 2000	17 December 2003	1 December 1996
Croatia	29 August 2002	27 March 2007	30 November 2000
Czech Republic	8 October 2001	6 July 2004	1 January 1995
Estonia	24 March 2004	2 August 2001	13 November 1999
Georgia	2 February 2009	11 April 2000	14 June 2000
Hungary	13 January 2004	3 July 2001	1 January 1995
Latvia	13 February 2004	14 June 2002	10 February 1999
Lithuania	7 November 2003	28 January 2002	31 May 2001
Macedonia, Former Yugoslav Republic of (FYR)	14 June 2005	22 July 1999 (a)	4 April 2003
Malta	1 May 2007	23 April 2002	1 January 1995
Moldova, Republic of	4 March 2003	9 August 1999	26 July 2001
Montenegro	23 October 2006	2 October 2009 (a)	Observer
Poland	10 December 2003	15 February 2002	1 July 1995
Romania	30 June 2003	11 July 2002	1 January 1995
Serbia	8 February 2006	31 July 2009 (a)	Observer
Slovakia	24 November 2003	5 December 2005 (a)	1 January 1995
Slovenia	20 November 2002	29 July 2004	30 July 1995
Turkey	24 October 2003		26 March 1995
Ukraine	6 December 2002	18 November 1999	16 May 2008

***CEE Finalization** of NBFs*

Year	Country
2003	Estonia, Slovenia
2004	Armenia, Belarus, Czech Republic, Georgia, Latvia, Lithuania, Former Yugoslav Republic of Macedonia (FYR), Moldova (Republic of), Slovakia
2005	Croatia, Turkey, Romania
2006	Albania
2007	Serbia and Montenegro[1]
Under Review	Azerbaijan, Ukraine

[1] Montenegro became a party to the CPB by succession after it separated from Serbia on 3 June 2006. The NBF that was formulated for Serbia and Montenegro now only applies to Serbia.

***Biosafety databases** or Web sites in CEE countries*

CEE country	Biosafety or BCH national websites
Albania	http://al.biosafetyclearinghouse.net/
Armenia	http://am.biosafetyclearinghouse.net/
Azerbaijan	http://az.biosafetyclearinghouse.net/
Belarus	http://biosafety.org.by/
Bosnia and Herzegovina	
Croatia	http://en.gmo.hr/; http://hr.biosafetyclearinghouse.net
Czech Republic	http://www.cenia.cz/biosafety; http://gmo.vscht.cz
Estonia	http://ee.biosafetyclearinghouse.net/
Georgia	
Hungary	http://biodiv.kvvm.hu/
Latvia	http://lv.biosafetyclearinghouse.net/
Lithuania	
Macedonia	http://www.biosafety.gov.mk/
Malta	http://mt.biosafetyclearinghouse.net/
Moldova	http://www.biosafety.md/
Montenegro	
Poland	http://gmo.mos.gov.pl/english/index_en.html
Romania	
Russian Federation	
Serbia	
Slovakia	http://bch.gmo.sk
Slovenia	http://www.bch.bf.uni-lj.si/
Turkey	
Ukraine	

ANNEX VI

West Africa

Country	Cartagena biosafety protocol status	Completed national biosafety framework under UNEP-GEF development project (with date of completion)	Operational biosafety regulatory system in place (functional system that could receive applications)	Legally enforceable regulatory system in place (biosafety law, regulations, etc.)	Draft biosafety bill, regulations, or decree to establish biosafety regulatory system	Primary biosafety regulatory agency
Benin	Ratified	Yes (May 05, 2005)	No	No	Yes	Ministry of Environment
Burkina Faso	Ratified	Yes (May 8, 2005)	Yes	Yes	Yes (Biosafety Law passed in 2006; 8 draft decrees and 3 draft regulations to implement law)	National Biosafety Agency
Côte d'Ivoire	Not yet ratified	Yes (May 7, 2005)	No	No	Not Known[1]	National Biosafety Commission (within the Ministry of Environment)
Guinea- Bissau	Not yet ratified	Yes (March 2008)	No	No	No	Ministry of Environment
Mali	Ratified	Yes (May 10, 2005)	No	No	Yes (both draft bill for biosafety system and draft decree for contained research and confined field trial regulation)	Ministry of Environment (draft bill); Ministry of Agriculture (draft decree)
Niger	Accession	Yes (January 19, 2005)	No	No	Yes	National Biosafety Agency (within the Ministry of Environment)
Senegal	Ratified	Yes (March 30, 2005)	No	No	Yes	National Biosafety Authority
Togo	Ratified	Yes (December 12, 2004)	No	No	Yes	Ministry of Environment

[1] The contact from Côte D'Ivoire did not provide information on this issue.

TABLE 2. *Summary of key components of the biosafety regulatory systems in WAEMU countries*

Country	Government office legally in charge of biosafety	Legal scope of system: Legal authority to make a decision on an application	Legal scope of system: *What products are covered?*	Legal scope of system: *What activities are covered?*	*What risk issues are covered?*	Inclusion of decision-making standards	Proportionate risk based reviews of applications
Cartagena Biosafety Protocol	Any Agency, Ministry or Department can be designated	National Competent Authority	Living modified organisms	All Activities are covered but with different procedures	Biodiversity and Environmental issues	This is left to the discretion of the individual country	Yes; the Protocol sets up different procedures for different activities.
Benin	Ministry of Environment	Ministry of Environment	GMO or GMO products	Importation; exportation; confined use; placing on the market	Human health and the environment; conservation and sustainable use of biodiversity; socioeconomic considerations	None in NBF; some language in draft biosafety law to "avoid unfavorable effects for conservation and sustainable use of biodiversity and human health"	Neither the NBF nor the draft biosafety law identifies different procedures for different organisms or activities
Burkina Faso	National Biosafety Agency	National Biosafety Agency	GMOs and derived products	Laboratory activities; confined use; imports; voluntary and involuntary releases (commercial products)	Environmental; biodiversity; health and animal health (food safety); socioeconomic; ethical	Benefit without causing harm; contribute to sustainable development; not cause socioeconomic harm; and comply with ethical rules	No distinctions are made for different activities to be conducted

Côte D'Ivoire		The National Biosafety Commission, which is under the Ministry of Environment	Relevant Technical Ministries (e.g. Agriculture, Health, Animal Froduce, Industry, etc.)	GMOs and their products	Contained use; deliberate release for research purposes; deliberate release by placing on the market; and placing on the market genetically modified animal feed	Protection of population health; safeguarding the environment and biodiversity; compatibility with ethical values	None in NBF	Mentions the need for different procedures based on activities but states that every GMO action requires "advanced informed agreement"
Guinea-Bissau[2]		NA	NA	NA	NA	NA	NA	NA
Mali	Draft Law	Ministry of Environment	Ministry of Environment	GMOs and derived products	Contained use; release into environment; imports; research and development	Environment; biodiversity	Benefit without causing harm; contribute to sustainable development; not cause socioeconomic harm; and comply with ethical rules.	Some distinctions are made between deliberate release, food feed and processing; also allows for simplified procedure
	Draft Decree	Ministry of Agriculture	Inter-ministerial Committee	GM plants	Contained use and confined field trials	Not specified	No standard identified	No distinctions made between contained use and confined field trials
Niger		Ministry of Environment	Ministry of Environment in collaboration with other relevant ministries	GMOs and derived products	Laboratory activities; confined use (field trials); release as commercial products; importation; exportation; transit and transboundary movement	Human, animal, and vegetal health; biodiversity and environment	Avoid risks to human and animal health; save ethical and cultural values.	No proportionate treatment but law mentioned that they may enact different risk levels

(continued)

TABLE 2 *(continued)*

Country	Government office legally in charge of biosafety	Legal scope of system: Legal authority to make a decision on an application	*What products are covered?*	*What activities are covered?*	*What risk issues are covered?*	Inclusion of decision making standards	Proportionate risk based reviews of applications
Senegal	National Biosafety Authority (which has not yet been established)	National Biosafety Authority	GMOs and derived products	Importation; exportation; confined use; voluntary release; and placing on the market	Human and animal health; biodiversity; environment; welfare; socioeconomic considerations; ethical considerations	No standard specified in NBF or draft bill	No significant distinctions for different activities
Togo	Ministry of Environment	Ministry of Environment	GMOs and derived products	Confined use; importation; exportation; environmental releases placing on the market; and laboratory activities	Human and/or animal health; biodiversity; socioeconomic considerations; cultural values	Benefit without causing harm; contribute to sustainable development; no socioeconomic harm; no threat to traditional knowledge and values; comply with ethical rules	Identifies two methods of risk evaluation: one for pharmaceutical GMOs and one for non-pharmaceuticals

[2] Guinea-Bissau had not completed its NBF when this annex was written. Thus, there was no information available to complete any column of the table. "NA" means "not available." The NBF was completed in March 2008 but it is only available in Portuguese.

TABLE 3. *Summary of key components of the biosafety regulatory systems in WAEMU countries*

Country	Public participation: In regulatory procedures	Public participation: In decision on applications	Transparency	Type and Role of Advisory Committees	Special Procedures for Confined Field Trials	Involvement of Other Government Departments or Agencies	Legally Responsible Authority for Oversight and Enforcement	Other Issues Covered by System
Cartagena Biosafety Protocol	Not specifically addressed	Consultation should occur if consistent with national laws and regulations	Certain minimum information should be available to the public	No advisory committees are required	No specific reference to confined field trials; ambiguous as to their treatment			
Benin	Mentioned in NBF and draft law; details left to discretion of the competent authority	Mentioned in NBF and draft law; details left to the discretion of the competent authority	No specific provision identifying information to be released to the public	National Biosafety Committee in an advisory role. The competent authority may also set up public committees on biosafety	None specified in NBF or draft law	Different ministries are member of the National Biosafety Committee	Ministry of Environment	Labeling; civil liability and redress
Burkina Faso	Through the National Biosafety Observatory Committee	Chief surveyor collects public comments, conducts hearings, and interviews selected people near release area of GMO. Report summarizing public opinion is sent to NBA prior to final decision[3]	Provides public with portions of notifications and final decisions	Three different types of committees; all advisory	Not specified in the law	Each relevant ministry has an internal biosafety committee that can consult NBA on applications before it within their area of expertise	NBA for oversight; Ministry of Environment for Enforcement	Mandatory labeling; strict liability for any damages

(continued)

TABLE 3 *(continued)*

Country		Public Participation: In regulatory procedures	Public Participation: In decision on applications	Transparency	Type and Role of Advisory Committees	Special Procedures for Confined Field Trials	Involvement of Other Government Departments or Agencies	Legally Responsible Authority for Oversight and Enforcement	Other Issues Covered by System
Côte D'Ivoire		Yes, but not specified	Yes, through consultations with the public	Public will have access to technical dossier and results of risk analysis	Interministerial Monitoring Committee; Specialised Scientific Commissions	Not specified in the NBF	Different Ministries make decision depending on the type of GMO	Not specified in NBF	None
Guinea-Bissau[4]		NA	NA	NA	NA	NA	NA	NA	NA
Mali	Draft Law	Through the National Biosafety and Biotechnology Committee	Through the National Biosafety and Biotechnology Committee. Also, public can make comments and comments must be taken into account in decision making process	Provides the public with pertinent information from notification and final decisions	National Biosafety and Biotechnology Committee with broad membership in an advisory role; Institutional-Biosafety Committees	No special discussion in draft law; separate decree from Ministry of Agriculture to address this subject	Included as members of the National Biosafety and Biotechnology Committee	Ministry of the Environment	Labeling; liability and redress covered in draft law
	Draft Decree	None	None	Not specified	Subcommittee on Genetic Engineering implements decision and carries out risk assessment	Decree is set up to deal with this situation	Nine government ministries are represented on the Interministerial Committee	Not specified	None

Niger	None mentioned in documents	Not mentioned in draft bill, only in the NBF. Public hearings will be held before authorizing field trials. No other specified procedures for public participation for other GMO activities.	Provides the public with pertinent information from notification and final decision	Advice is given to the Ministry of Environment by the National Technical Biosafety Prevention Authority	No special provisions for this activity	Different ministries have members on the National Technical Biosafety Prevention Authority; bill allows Minister of Environment to consult with other ministers before making decision	To be determined in a decree issued under the biosafety law	Labeling; liability and redress
Senegal	Through the National Biosafety Committee and the Public Biosafety Committee	Public can make comments and the comments must be taken into account in the decision process	Public can ask the National Biosafety Authority for access to information about GMOs and derived products	NBC plays a advisory role; Institutional committees, National Scientific Committee and Public Committee on Biosafety are consultative bodies	No special procedures specified in NBF or draft bill	The NBC and the National Scientific Committee have members who represent the different ministries interested in biosafety	National Biosafety Authority	Liability and Redress is covered in the draft law
Togo	Ad hoc "Public Committees"	None specified in draft biosafety law	The National Competent Authority can provide information as long as it does not provide confidential information	National Biosafety Committee; Science and Technical Committee	No special procedures	Internal Biosafety Committees within each Ministry are consulted on relevant GMOs	Ministry of Environment	Management of accidents; labeling; regimes for liability and redress

3 Information is based on the draft Public Participation decree that implements the Burkina Faso Biosafety Law.

4 Guinea-Bissau had not completed its NBF when this study was completed. Thus, there was no information available to complete any column of the table. "NA" means "not available." The NBF was completed in March 2008, but it is only available in Portuguese.

ANNEX VII

East Africa

TABLE 1. *Comparison of countries and major issues*

Issues	Kenya	Uganda	Tanzania
Comprehensiveness	Addresses all GE organisms at different stages of development; addresses environmental issues; does not address food safety issues	Addresses all GE organisms at different stages of development; addresses environmental issues; does not address food safety issues	Addresses all GE organisms at different stages of development; addresses environmental issues; does not address food safety issues
Legal Authority	Recently passed Kenya Biosafety Act of 2009	Will use new legal authority established in biosafety-specific legislation to be passed by Parliament	Parliament passed comprehensive Environmental Management Act, which included legal authority to regulate biosafety
The Safety Standard	Not established in biosafety bill or guidelines	Not established in biosafety policy or regulations	Not established in Environmental Management Act, but National Biosafety Framework and EMA Regulations set forth some decision criteria
Socioeconomic Considerations	To be included in biosafety regulatory process but no details are provided	To be included in biosafety regulatory process but no details are provided	To be included in biosafety regulatory process as set forth in the EMA Regulations

Issues	Kenya	Uganda	Tanzania
Proportional Risk-Based Reviews	Kenyan Biosafety Act does differentiate procedures based on activity to be conducted with the GE organism	Some distinctions proposed in the draft Uganda Biosafety Bill	The National Biosafety Framework and Environmental Management Act make little distinction for different activities; Schedule 18, however, is a good example of a proportional risk-based approach for small-scale confined trials.
Public Participation	Good involvement of public in process establishing biosafety system; Kenya Biosafety Act of 2009 allows for public participation with certain types of individual applications	Good involvement of public in process establishing biosafety system; draft Uganda Biosafety Bill has one explicit requirement to obtain public comments before making decisions on individual applications	Good involvement of public in process establishing biosafety system; Environment Management Act regulations have some public participation provisions
Transparency	Biosafety system provides for giving public information about both process and individual applications	Biosafety system provides for giving public information about both process and individual applications	Biosafety system provides for giving public information about both process and individual applications
Inclusion of Products of GMOs	Biosafety system does not include such products in approval requirements	Biosafety system does not include such products in its approval requirements	The Environmental Management Act and the regulations include products of GMOs in their review and approval procedures

(continued)

TABLE 1 *(continued)*

Issues	Kenya	Uganda	Tanzania
Interim Measures	Kenyan National Science and Technology Council established NBC and regulations under its operating statute	Uganda Science and Technology Council established NBC and National Biosafety Guidelines under its operating statute	Established Schedule 18 to address small-scale confined field trials under the authority granted in the Plant Protection Act
Institutional Biosafety Committees	Established under Kenyan Guidelines to conduct activities such as reviewing risk assessments and monitoring release sites	IBCs in Uganda review and clear prospective applications before they go to NBC and conduct training and oversight.	IBCs in Tanzania review risk assessments, monitor release sites, and hold discussions on the social and economic impact of alternatives to the proposed GMO.

ANNEX VIII

Asia

TABLE 1. *CPB Status of the 23 Asian countries participating in the NBF development project*

CPB Status	Asian countries in NBF project
Countries that were Parties to the CPB before joining project	Bhutan and Maldives
Countries that were signatories to the CPB before joining project that 1. Have now ratified the CPB 2. Have not completed ratification process	1. Bangladesh, Indonesia, Iran, Jordan, Philippines, and Sri Lanka 2. Myanmar, Nepal, and Republic of Korea
Countries that joined the NBF project prior to ratification or signature 1. Have now ratified the CPB 2. Have not completed ratification process	1. Cambodia, Iran, Korea DPR, Kyrgyzstan, Laos PDR, Mongolia, Syria, Thailand, Vietnam, and Yemen 2. Kazakhstan and Lebanon

TABLE 2. *Status of the regulatory regime for biosafety in 23 Asian countries participating in the NBF development project*

Status of regulatory regime for biosafety	Asian countries in NBF project
Preexisting legal instruments for biosafety 1. Biosafety guidelines 2. Ministerial decree 3. Biosafety regulations 4. Biosafety law	1. Bangladesh, Sri Lanka, Syria and VietNam 2. Indonesia 3. Iran, Philippines, and Thailand 4. Republic of Korea
No preexisting legal instruments for biosafety	Bhutan, Cambodia, DPR Korea, Jordan, Kazakhstan, Kyrgyzstan, Laos, PDR, Lebanon, Maldives, Mongolia, Myanmar, Nepal, Tajikistan, and Yemen

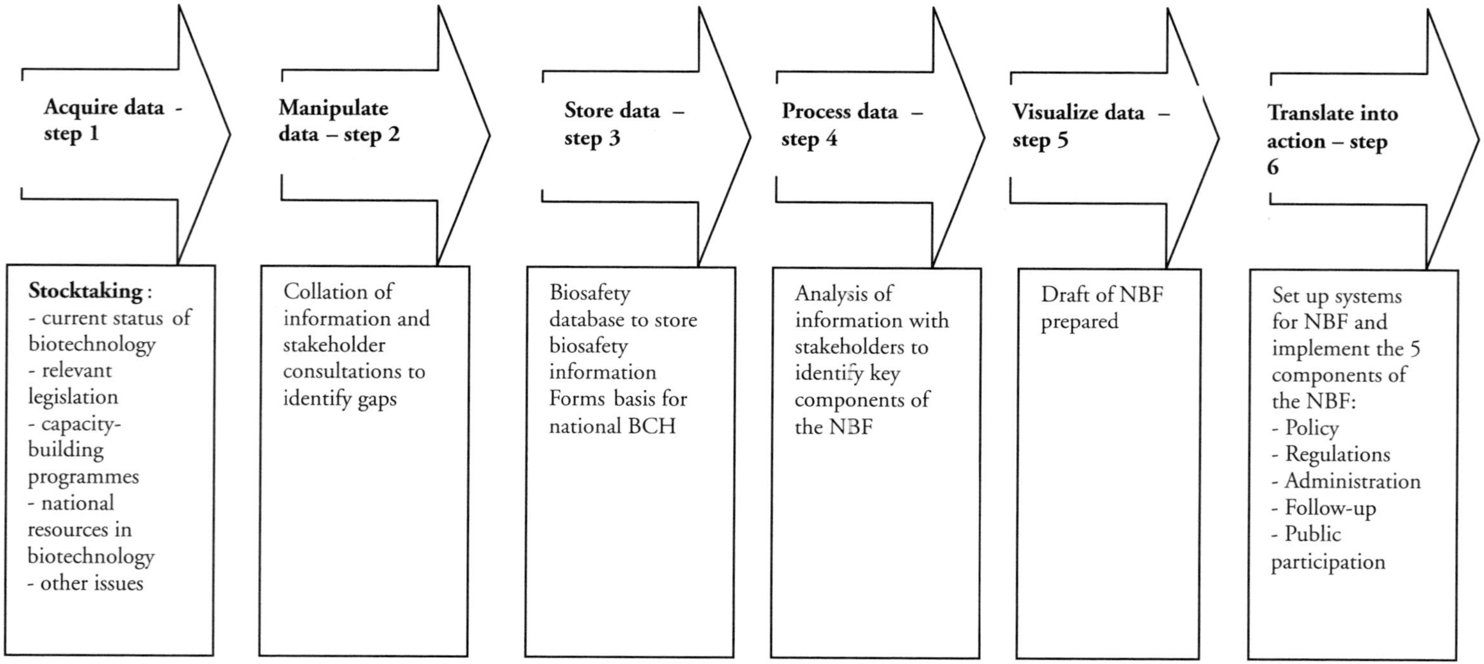

FIGURE 1. Mapping process for development of NBF

TABLE 3. *Composition of NCCs in 23 Asian countries participating in the NBF development project*

Stakeholder	# of countries	Examples
NCO representatives	13	Consumer associations (Jordan and Yemen), farmers' associations (Tajikistan), Women's organisations (DPR Korea), or environmental NGOs (Kazakhstan, Nepal, and Sri Lanka)
Private sector representatives	10	Chamber of Commerce (Nepal and Yemen) or similar commercial associations (Kyrgyzstan and Philippines) or private firms (Iran and Indonesia)
Public sector scientists	16	Universities (Jordan, Myanmar, and Syria) and public research institutes (Syria, Iran, and Republic of Korea)
Media representative	1	Yemen
Government agencies	23	Environment, Agriculture, Science and Technology, Trade, Commerce, Health, Education

TABLE 4. *Status of proposals for implementation of NBFs in the 12 Asian countries*

Status of NBF implementation	Countries
Implementation projects approved by GEF and started	Vietnam and Cambodia
Projects technically cleared by GEF but not yet approved for funding	DPR Korea and Tajikistan
Project proposals prepared but not submitted to GEF as yet	Bhutan, Indonesia, Iran, Laos PDR, Sri Lanka, and Yemen
NBF implementation through government funding	Republic of Korea (no longer eligible for GEF funding)

ANNEX IX

Americas

Country	Regulatory System	Agencies and Authorities	Institutional Framework	Labelling and Traceability	Scope
Canada[1]	– Seeds Act and Regulations – Feeds Act and Regulations – Plant Protection Act and Regulations – Health of Animals Act and Regulations – Fisheries Act and Fish Health Protection Regulations – Food and Drugs Act and Regulations – Canadian Environmental Protection Act – Fertilizers Act and Regulations – Pest Control Products Act and Regulations	The Canadian Food Inspection Agency (CFIA) regulates plants with novel traits (PNTs), novel fertilisers and supplements, novel livestock feeds, veterinary biologics in accordance with the Seeds Act, the Fertilizers Act, the Feeds Act, and the Health of Animals Act. The Pest Management Regulatory Agency (PMRA) and Health Canada regulate pest control products in accordance with the	The Canadian Regulatory System is based on the Federal Regulatory Framework for Biotechnology developed in 1993. The Canadian Environmental Protection Act 1999 (CEPA) mentions the precautionary principle. The act applies when no specific regulations are applicable, and essentially regulates microorganisms and certain types of genetically modified animals.	In consultation with the CFIA, Health Canada can require mandatory labelling if there is a health or safety concern related to allergens or a significant change in nutrition or composition. A national standard for the voluntary labelling of foods derived from biotechnology has been developed.	Activities related to LMOs and LMO products are regulated by different agencies according to their authority and the nature of the products and activities.

[1] *Sources:* The Canadian Agency for Food Inspection, *Regulation of Agricultural Biotechnology in Canada* (Her Majesty th Queen in Right of Canada, 2007), www.inspection.gc.ca/english/sci/biotech/gen/educ_postsece.pdf, accessed 20 January 200 The Canadian Agency for Food Inspection, "Regulating Agricultural Biotechnology," http://cbac-cccb.ca/epic/site/cbac-ccc nsf/en/Home, accessed 20 January 2008; Department of Justice Canada, "Consolidated Statutes and Regulations," http laws.justice.gc.ca/en/, accessed 20 January 2008; Government of Canada, "BioPortal," http://www.bioportal.gc.ca/englis BioPortalHome.asp?x=1, accessed 20 January 2008.

Country	Regulatory System	Agencies and Authorities	Institutional Framework	Labelling and Traceability	Scope
	– New Substance Notification Regulations (Organisms) – Masked Name Regulations – Draft Regulations on Living Modified Organisms – Regulatory Directive 2000-07: Release of Plants with Novel Traits in Contained Field Trials – Regulatory Directive 94-08: Assessment Criteria for Determining Environmental Safety of Plants with Novel Traits – Regulatory Directive 95-03: Guidelines for the Safety Assessment of Novel Feeds – Regulatory Directive 96-13: Import and Permit Requirements for Plants with Novel Traits and their Products – Guidelines for the Safety Assessment of Novel Foods – Guidelines for the Regulation of Veterinary Biology Produced by Biotechnology – Guidelines for the Notification and Testing of New Substances (Organisms)	Pest Control Products Act. Health Canada regulates novel foods in accordance with the Food and Drugs Act. Environment Canada and Health Canada regulate all animate products of biotechnology destined for uses not covered under other federal legislation in accordance with the Canadian Environmental Protection Act, 1999. The CFIA and Health Canada also have a role in the labelling of novel foods derived from genetic engineering. (Health Canada is responsible for setting labelling regulations with respect to safety and Health; the CFIA is responsible for establishing other labelling regulations.)			

(continued)

(continued)

Country	Regulatory System	Agencies and Authorities	Institutional Framework	Labelling and Traceability	Scope
United States[2]	– Federal Food, Drug and Cosmetic Act (FFDCA) Related regulations issued by the US Food and Drug Administration: – Statement of Policy: Foods Derived From New Plant Varieties (1992) – Consultation Procedures under FDA's 1992 Statement of Policy – Plant Protection Act (PPA) Related regulation issued by the US. Department of Agriculture: – 7 CFR Part 340: Introduction of Organisms and Products Altered or Produced Through Genetic Engineering Which are Plant Pests or Which There is Reason to Believe are Plant Pests – Federal Insecticide, Fungicide and Rodenticide Act (FIFRA)	The United States Department of Agriculture and the Animal and Plant Inspection Service (USDA – APHIS) regulates the import, handling, interstate movement, and release into the environment of LMOs that are known or suspected to be plant pests or to pose a plant pest risk. It also licenses animal vaccines that are products of biotechnology. The United States Environmental Protection Agency (EPA): – Through a registration process, regulates the sales, distribution, and use of pesticides that are produced by an organism derived through techniques of modern biotechnology – Issues Experimental Use Permits for field testing	According to the Coordinated Framework for Regulations of Biotechnology developed in 1986, agencies need to operate their programs in an integrated and coordinated fashion and together should cover the full range of plants, animals and microorganisms derived from genetic engineering techniques.	There are no mandatory regulations, only a Draft Guidance regarding voluntary labelling indicating whether or not a product was derived from bioengineering.	LMOs and LMO products are regulated by different agencies, according to their authority and the nature of the products.

[2] *Sources:* Office of Science and Technology Policy, Coordinated Framework for Regulation of Biotechnology (Announcement of Policy/Notice for Public Comment, 1986), http://usbiotechreg.nbii.gov/CoordinatedFrameworkForRegulatioOfBiotechnology1986.pdf, accessed 20 January 2008; United States Food and Drug Administration, Voluntary Labeling indicating whether foods have or have not been derived using bioengineering (Draft Guidance, 2001), http://www.cfsan.fda.gov/~dms/biolabgu.html, accessed 20 January 2008; The National Biological Information Infrastructure, United States Unified Regulatory Agencies website, http://usbiotechreg.nbii.gov/roles.asp, accessed 20 January 2008.

Country	Regulatory System	Agencies and Authorities	Institutional Framework	Labelling and Traceability	Scope
	Related regulations issued by the US. Environmental Protection Agency: – 40 CFR Parts 152 and 174: Pesticide Registration and Classification Procedures – 40 CFR Part 172: Experimental Use Permits – Toxic Substances Control Act (TSCA) Related regulation issued by the US. Environmental Protection Agency: – 40 CFR Part 725: Reporting Requirements and Review Processes for Micro-organisms – Virus, Serum, Toxin Act	– Sets tolerance limits for residues of pesticides produced by an organism derived through techniques of modern biotechnology on and in food and animal feed The United States Food and Drug Administration (FDA) ensures the safety and proper labelling of foods and feeds developed through biotechnology.			
Mexico[3]	– Biosafety Act enacted on March 18, 2005 and its regulation of 2008 (last amendment March 2009) – Federal Plant Health Act – Federal Seeds Production, Certification and Trade Act – Rural Sustainable Development Act (in the context of agrobiotechnology promotion)	The Ministry of Agriculture, Livestock, and Fishing handles agricultural issues related to agrobiotechnology. The Ministry of Environment and Natural Resources deals with environmental issues related to agrobiotechnology. The Ministry of Health handles health-related	A Commission on Biosafety and Genetically Modified Organisms (CIBIOGEM), which is composed of the National Science and Technology Council (CONACYT) and representatives from different Ministries, was created by agreement on	– General Health Law (art. 98 and 282 Bis, Bis-1 and Bis-2) – Regulation regarding the Sanitary Control of Products and Services (Art. 164 and 165): It is mandatory to notify the Ministry of Health of any biotechnological	The Biosafety Act applies to all LMOs and products obtained through modern biotechnology and destined to be used in: agriculture, livestock, fisheries, aquaculture, forestry, industrial, and commercial

(continued)

Sources: Comisión Intersecretarial de Bioseguridad de los Organismos Genéticamente Modificados (CIBIOGEM; Governmental Agency website), http://www.cibiogem.gob.mx/, accessed 20 January 2008; Mexican Government, Leyes federales de México (List of Mexican Federal Laws), http://www.gob.mx/wb/egobierno/egob_leyes_del_congreso, accessed 20 January 2008.

(continued)

Country	Regulatory System	Agencies and Authorities	Institutional Framework	Labelling and Traceability	Scope
	– Ecological Balance and Environmental Protection Act and related Environmental Impact Assessment Regulation – Sustainable Forest Development Act – Health Act and related regulations – Federal Criminal Code, art. 420 (offence related to nonauthorized LMO activities) – Law 056-FITO-1995: requirements for national movement, import, and field trials of genetically modified organisms (officially cancelled June 22, 2009). – Organic Products Act – Science and Technology Act	issues related to agrobiotechnology.	November 5, 1999, and legally established under the Biosafety Act. The Commission, which is assisted by a Technical Committee and a Consultative Advisory Group, coordinates and establishes policies. Technical aspects are managed by the Technical Committee. The Ministries of Agriculture, Health and Natural Resources are the competent authorities in their respective fields.	products or products derived from biotechnology and destined for human consumption. Official regulations will then set forth the specific process, and product features and labelling requirements for the above-mentioned products. Information regarding the safety and reliability of biotechnological products shall be provided to the Ministry of Health by the agent that has developed the products. Prior to allowing the trade in biotechnological products, the Ministry of Health can verify the accuracy of the information provided. According to ministerial requirements, labels shall contain information regarding health features and risks. The Health Act contains additional regulations on publicity. The Biosafety Act also provides some requirements regarding the labelling of LMOs.	applications etc.

Country	Regulatory System	Agencies and Authorities	Institutional Framework	Labelling and Traceability	Scope
Belize	– Belize Agricultural Health Authority Act (1999)[4] – Environmental Protection Act (1998)[5] – A Final Draft National Biosafety Framework is in the process of being developed as part of the UNEP-GEF Global Project on the Development of National Biosafety Frameworks (NBFs).[6]		The Agricultural Health Authority Act comprises biosafety and LMO-related aspects. A Biosafety National Committee was created in 2002, to handle LMO-related requests on a case-by-case basis.	There are no specific regulations.	Basically, plants.
Costa Rica	– Phytosanitation Law No. 7664 and related regulation No. 26921-MAG[7] – Biodiversity Law No. 7788[8] – General Health Regulation No. 5395[9]	– Ministry of Agriculture and Livestock (MAG)[10] – Ministry of Health[11] – Phytosanitary Protection Service[12]	There is a National Technical Commission on Biosafety[13] which is linked to the MAG, and composed of various ministries, including the	The Ministry of Health created an Institutional Commission for LMO-related issues. There are no labelling and traceability regulations.	Solely LMOs destined for agricultural use (plants and animals). Biotechnology is defined as genetic engineering.

(continued)

4 Attorney General's Ministry of Belize, "Making legal Information easy to find, use and share" (Legal Database), http://www.belizelaw.org, accessed 20 January 2008.

5 The Food and Agriculture Organization (FAO), FAO Country Profiles and Mapping Information System (Agricultural Profile of Belize), http://www.fao.org/countryprofiles/index.asp?lang=en&iso3=BLZ&subj=4, accessed 20 January 2008.

6 The United Nations Environment Program and the Global Environmental Facility (UNEP-GEF), UNEP-GEF Biosafety Projects – Country Pages: Belize, http://www.unep.ch/biosafety/partcountries/BZcountrypage.htm, accessed 20 January 2008.

7 Alex May Montero (Ministry of Agriculture and Livestock), Costa Rica: First Regular National Report on the Implementation of the Cartagena Protocol on Biosafety (Report) (11 September 2007) 4, http://www.cbd.int/doc/world/cr/cr-nr-cpb-01-es.pdf, accessed 21 January 2008.

8 *Ibid.*

9 *Ibid.*

10 Alex May Montero (Ministry of Agriculture and Livestock), Costa Rica: First Regular National Report on the Implementation of the Cartagena Protocol on Biosafety (Report) (11 September 2007) 5, http://www.cbd.int/doc/world/cr/cr-nr-cpb-01-es.pdf, accessed 21 January 2008.

11 *Ibid.*

12 Costa Rican Ministry of Agriculture and Livestock, Servicio Fitosanitario del Estado (official website for the Phytosanitary Protection Service), http://www.protecnet.go.cr/SFE/Organica.htm, accessed 21 January 2008.

13 La Gaceta N³ 83 (Gazette No. 83), Ley No 7.664 – Ley de protección fitosanitaria (Law which creates the National Technical Commission on Biosafety found in the ECOLEX Database, 1997) art. 40, http://faolex.fao.org/docs/pdf/cos9597.pdf, accessed 21 January 2008.

(continued)

Country	Regulatory System	Agencies and Authorities	Institutional Framework	Labelling and Traceability	Scope
	– Risk Assessment Guidelines (BIO-02)[14] – Regulation Concerning Agricultural Biosafety Audits (Decree No. 32486, 2005)[15] Animal Health Service Law No. 8495 of 2006 (for animal LMO)[16] Law for the Promotion of Organic Agriculture and Livestock No. 8591 of 2007 and its regulation Decree No. 35242-MAG-H-MEIC (4 de junio del 2009),[17] which regulates coexistence between LMO and organic agriculture and some liability rules for "contamination" are provided. – A Final Draft National Biosafety Framework was developed in 2005 as part of the UNEP-GEF Global Project on the Development of NBFs.[18]	– National Commission for Biodiversity Management (CONAGEBIO) of the Ministry of Environment and Energy (it has the authority to revoke MAG decisions in the field of biodiversity)[19]	Ministry of Environment, the National Sciences Academy, and as of recently, nongovernmental organizations. The National Seeds Office participates in the supervision of genetically modified seeds.	The Phytosanitary Protection Law establishes that authorised LMOs are not automatically approved as food. A Labelling and Traceability Bill has been submitted to the Assembly.	

[14] *Ibid.*

[15] The Costa Rican Justice Department (Procuraduría General de la República), Sistema Costarricense de Información Jurídic (Governmental Legal Database), http://www.pgr.go.cr/scij/index_pgr.asp, accessed 21 January 2008.

[16] *Ibid.*

[17] *Ibid.*

[18] Alex May Montero (National Project Coordinator), "Development of a National Biosafety Framework for Costa Rica" (Fin Report Provided as part of the UNEP-GEF Global Project on the Development of NBFs) (2005) http://www.unep.ch/biosafet development/Countryreports/CRNBFrepEN.pdf, accessed 21 January 2008.

[19] Costa Rican Legislative Assembly, Ley de Biodiversidad 7788 (Law found in the Ecolex Database, 1998), http://faolex.fao.or docs/pdf/cos13426S.pdf, accessed 21 January 2008 (law creating the commission).

Country	Regulatory System	Agencies and Authorities	Institutional Framework	Labelling and Traceability	Scope
El Salvador	– Article 68 of the Environmental Law establishes the Ministry of Environment and Natural Resources' responsibility to apply safety rules to products resulting from biotechnology to mitigate adverse effects on native biodiversity.[20] – Art. 21 of the Environmental Law requires that the production of LMOs be subjected to environmental risk assessment.[21] – General Environmental Regulation (Decree No. 17), which was amended in March 2007, regulates environmental impact assessment[22] – Article 30 of the 2001 Seeds Law No. 530 prohibits imports, research, production, and trade in transgenic seeds.[23]	Ministry of Environment and Natural Resources (MARN).[24] Ministerial authorities are supported by technical agencies.	The regulation of 2008 establishes a scientific committee with the mandate to provide technical advice to the Ministry of the Environment and Natural Resources.	No specific regulations	Products resulting from biotechnology

(continued)

20 Salvadorean Ministry of Environment and Natural Resources, Legislación Ambiental (list of laws that are related to the environment) http://www.marn.gob.sv/index.php?fath=208&categoria=208, accessed 21 January 2008.

21 *Ibid.*

22 *Ibid.*

23 Ministry of Environment and Natural Resources (MARN), *Draft of National Biosafety Framework for El Salvador (Final report provided as part of the UNEP-GEF Global Project on the Development of NBFs)*, 2006 at 14 http://www.unep.ch/biosafety/development/Countryreports/SVNBFrepEN.pdf, accessed 21 January 2008.

24 *Ibid*, 13.

(continued)

Country	Regulatory System	Agencies and Authorities	Institutional Framework	Labelling and Traceability	Scope
	– Special Regulation for the safe handling of genetically modified organisms of July 2008.[25] – A Final Draft National Biosafety Framework has been developed in 2006 as part of the UNEP-GEF Global Project on the Development of NBFs.[26]				
Guatemala	– Specific regulation on LMOs destined for agricultural use: Agreement 386-2006 -MAGA (July 2006)[27] – A Final Draft Biosafety Regulatory Framework was completed in 2004 as part of the UNEP-GEF Project on the Development of National Biosafety Frameworks.[28]	Ministry of Agriculture, Livestock and Nutrition (MAGA) Imports and field trials are subject to the approval of the Ministry of Agriculture, Livestock and Nutrition (MAGA). It is necessary to present a protocol of testing approved by the Institute for Agricultural Science and Technology (ICTA) to the MAGA.	It is optional at discretion of the MAGA to create a special advisory commission for the assessment of a product. The National Committee on Biosafety (CNCB) is composed of representatives from various sectors and is in charge of elaborating and promoting biosafety policies and regulations.[29]	No specific regulations	Only LMOs destined for agricultural use. Includes import, transport, handling within the country, field trials, and production of seeds for expor[t] purposes.

[25] www.csj.gov.sv.

[26] *Ibid.*, complete report.

[27] Ministry of Agriculture, Livestock and Nutrition (MAGA), Decreto Ministerial No. 386-2006 (Decree found in the ECOLE[X] Database, 2006) http://faolex.fao.org/docs/pdf/gua67557.pdf, accessed 21 January 2008.

[28] Terminal Report: Guatemala (final report presented as part of the UNEP-GEF Global Project on the Development of NBF[s] (31 August 2004), http://www.unep.ch/biosafety/development/Countryreports/GTNBFvol1.pdf, accessed 21 January 2008.

[29] Consejo Nacional de Áreas Protegidas (CONAP – National Council on Protected Areas), Interim National Report on th[e] Cartagena Protocol on Biosafety (Report; 2005) http://www.cbd.int/doc/world/gt/gt-nr-cpbi-es.doc, accessed 21 January 200[8]

Country	Regulatory System	Agencies and Authorities	Institutional Framework	Labelling and Traceability	Scope
Honduras	– Biosafety Regulation Focusing on Transgenic Plants (Agreement 1570-98)[30] – Plant and Animal Health Law (Decree No. 157-94) as modified and completed by the Reform of the Plant and Animal Health Law (Decree No. 344-05)[31] – Other references can be found in quarantine regulations No 1618 (September 26, 1997) – A Final Draft Biosafety Regulatory Framework was completed in 2007 as part of the UNEP-GEF Project on the Development of National	The Agriculture and Livestock Secretariat (SAG) is in charge of applying the Biosafety Regulation Focusing on Transgenic Plants and more generally, of the sanitary, phytosanitary, and quality control of products derived through biotechnology.[32] The import, export, investigation, movement, release into the environment, reproduction and commercialisation of LMOs is subject to the prior approval of the National Service on Seeds (SENASA), which is a division of the Agriculture and Livestock Secretariat (SAG).[33]	The National Biodiversity Commission (CONABIOH) is a consultative authority in the field of biosafety and biotechnology.[34] The National Committee on Biotechnology is a subdivision of the National Biodiversity Commission and helps the SAG enforce the Biosafety Regulation.[35] The Committee is comprised of public institutions, universities and producers.[36]	No specific regulations	The scope is limited to recombinant DNA techniques and techniques involving the direct incorporation into a hereditary cell of genetic material prepared outside of the cell. (Definition provided by Art. 2 of the Biosafety Regulation Focusing on Transgenic Plants)[37] Contained use, release into the environment, and plant trade are subject to the authorities' control.

(continued)

30 Dirección General de Biodiversidad (DiBio – General Biodiversity Service) and the Natural Resources and Environment Secretariat (SERNA), Borrador/Propuesta: Marco Nacional sobre Seguridad de la Biotecnología (Final report provided as part of the UNEP-GEF Global Project on National Biosafety Frameworks) (November 2007) 55, http://www.unep.ch/biosafety/development/Countryreports/HNNBFrepSP.pdf, accessed 21 January 2008.

31 *Ibid*, 56; La Gaceta No. 27.552 (Gazette No. 27.552), Decreto No. 157/94 – Ley Fitozoosanitaria (Law found in the Ecolex Database, 1994) http://faolex.fao.org/docs/pdf/hon42546.pdf, accessed 21 January 2008; The Honduran National Congress, DECRETO No. 344-05 (Decree found in the Ecolex Database, 2005) http://faolex.fao.org/docs/texts/hon64978.doc, accessed 21 January 2008.

32 *Ibid*, 32, 52–3; Honduran Ministry of Agriculture and Livestock, Secretaria de Agricultura y Ganaderia' (Governmental web site) http://www.sag.gob.hn/index.php?option=com_content&task=view&id=85&Itemid=41, accessed 21 January 2008.

33 *Ibid*, 53; The National Service on Seeds, Servicio Nacional de Sanidad Agropecuaria (Governmental web site) http://www.senasa-sag.gob.hn/index.php?option=com_frontpage&Itemid=1, accessed 21 January 2008.

34 *Ibid*, 32–3.

35 *Ibid*.

36 *Ibid*.

37 La Gaceta No. 28.715, Acuerdo N° 1.570/98 – Reglamento de bioseguridad con énfasis en plantas transgénicas (Agreement found in the Ecolex Database, 1998) art. 2, http://faolex.fao.org/docs/pdf/hon43010.pdf, accessed 21 January 2008.

(continued)

Country	Regulatory System	Agencies and Authorities	Institutional Framework	Labelling and Traceability	Scope
	Biosafety Frameworks.[38]	The SENASA can designate competent authorities in the matter. The Natural Resources and Environment Secretariat (SERNA) and the Health Secretariat also have functions that are related to biosafety.[39]			
Nicaragua	– The Animal and Plant Heath Law, No. 291 and regulations as amended by the Executive Decree 59-2003 issued on August 8, 2003, which introduced a new chapter entitled, "LMO risk assessment" (chapter XVI). – A draft biodiversity law exists and takes biosafety issues into consideration. – A National Biosafety Framework is in the process of being drafted as part of the UNEP-GEF Global Project on the	The Ministry of Agriculture and Forestry (MAGFOR) through the Agriculture and Livestock Health Office	The National Commission for LMO Risk Assessment (CONARGEM) is attached to MAGFOR and is comprised of public institutions, including environmental institutions and universities. The Commission provides non-binding technical reports to the Ministry.	A draft regulation on the handling, transport, and labelling of genetically engineered seeds, in accordance with the Technical and Quality Standardization Law, is expected in the near future.	Contained use, field testing and trials, crops assessment, reproduction of seeds, production or import for direc consumption, processing, or "other applications."

[38] Dirección General de Biodiversidad (DiBio – General Biodiversity Service) and the Natural Resources and Environmen Secretariat (SERNA), Borrador/Propuesta: Marco Nacional sobre Seguridad de la Biotecnología (final report provided a part of the UNEP-GEF Global Project on National Biosafety Frameworks; November 2007) http://www.unep.ch/biosafety development/Countryreports/HNNBFrepSP.pdf, accessed 21 January 2008.

[39] *Ibid*, 48–52.

Country	Regulatory System	Agencies and Authorities	Institutional Framework	Labelling and Traceability	Scope
	Development of National Biosafety Frameworks.[40]				
Panama	– Law No. 47 enacted on July 19, 1996 establishes phytosanitary protection measures and general regulations related to LMOs.[41] – Law No. 48 enacted on August 8, 2002, establishes the National Biosafety Commission for genetically modified organisms as well as other provisions.[42] – Although these regulations do not establish a regulatory system, they foresee its establishment by way of ministerial decision. The Ministry of Agriculture and Livestock Development (MIDA), in collaboration with the Ministry of Health, the	Permits research, contained use, and trials related to genetically modified organisms in greenhouses as well as authorisations of trial lots shall be issued by sector-based committees (Environment, Agriculture and Livestock, and Health committees). Technological development, in the context of research, also needs to be authorised. The sector-based committees shall send the proceedings of their meetings and the technical risk assessments to the Technical Secretariat of the Biosafety Commission (Article 9, Law No. 48).[43]	The National Commission is comprised of various ministries, including the Ministries of Environment and Foreign Affairs, civil society representatives, private sector corporations and universities. A Technical Secretariat and sector-based committees on biosafety (Agriculture and Livestock, Environment, and Health committees are created to regulate, perform risk assessments, and monitor LMO activities in their respective fields). Resolution 011 passed on January 23, 2002, by the Ministry of Health, creates	No specific regulations	Law No. 48 requires that measures be taken in order to ensure risk prevention and minimization of the negative impacts of LMO-related activities on the environment, biological diversity, human health, and agriculture and livestock production (Article 1). Law No. 47 requires that an authorization be obtained in order to import, research, test, transport, release, reproduce, and trade genetically modified plants and their products, biologic control agents and

(continued)

40 The United Nations Environment Program and the Global Environmental Facility (UNEP-GEF), UNEP-GEF Biosafety Projects – Country Pages: Nicaragua, http://www.unep.ch/biosafety/partcountries/NIcountrypage.htm, accessed 20 January 2008.

41 Panamanian Legislative Assembly, Ley No. 47 por la Cual se Dictan Medidas de Protección Fitosanitaria y se Adoptan Otras Disposiciones (Law found in the Ecolex Database, 1998), http://faolex.fao.org/docs/pdf/pan41712.pdf, accessed 22 January 2008.

42 Panamanian Legislative Assembly, Ley No. 48 que crea la Comisión Nacional de Bioseguridad para los Organismos Genéticamente Modificados y Dicta Otras Disposiciones (Law found in the Ecolex Database, 2002), http://faolex.fao.org/docs/pdf/pan41988.pdf, accessed 22 January 2008.

43 Panamanian Legislative Assembly, Ley No. 48 que crea la Comisión Nacional de Bioseguridad para los Organismos Genéticamente Modificados y Dicta Otras Disposiciones (Law found in the Ecolex Database, 2002) Art. 9, http://faolex.fao.org/docs/pdf/pan41988.pdf, accessed 22 January 2008.

(continued)

Country	Regulatory System	Agencies and Authorities	Institutional Framework	Labelling and Traceability	Scope
	National Environmental Authority (ANAM),[44] and the Science and Technology Secretariat, is in charge of coordinating the development of biosafety regulations and procedures. – New Decree No. 25 (2009) regulates the access to genetic and biological resources and creates the Unit for the Access to Genetic Resources.[45] – Resolution No. 208 (2007) details the Unit's responsibilities.[46] – General Environmental Law No. 41[47] – A National Biosafety Framework is in the process of being developed as part of the UNEP-GEF Global Project on the Development of NBFs.[48]	Law No. 48 establishes the responsibilities of each Ministry in the following manner: the Ministry of Agriculture and Livestock Development is responsible for agriculture and livestock production systems; the Ministry of Health is responsible for human health and human health protection issues related to LMOs; the Ministry of Commerce and Industry is responsible for coordinating elaborating policy on LMOs; the Ministry of Foreign Affairs is responsible for foreign policy and international agreements relating to LMOs; The Science and Technology Secretariat is responsible for the	the National Biosafety Committee. The Executive Decree No 137 issued on May 29, 2001,[49] creates the National Biosafety and Bioethics Commission, which is part of the Ministry of Agriculture and Livestock Development (MIDA).		other types of organisms destined for agricultural use.

44 Panamanian National Environmental Authority, Autoridad Nacional del Ambiente (Government Web site) http://www.anam.gob.pa/PATRIMONIO/ accessed 22 January 2008.

45 Panamanian Ministry of Economy and Finances, 'Decreto Ejecutivo No. 257 que Reglamenta el Artículo 71 de la Ley 41 d 1 julio 1998, General de Ambiente' (Decree found in the Ecolex Database, 2006), accessed January 19, 2008.

46 Panamanian National Environmental Authority (ANAM), 'Resolución No. AG-0208-2007' (Resolution found in the Ecole Database, 2007) http://faolex.fao.org/docs/pdf/pan72041.pdf, accessed 19 January 2008.

47 Panamanian Legislative Assembly, Ley 41 General de Ambiente de la República de Panamá (Law found in the Ecole Database, 1998) accessed 19 January 2008.

48 The United Nations Environment Program and the Global Environmental Facility (UNEP-GEF), UNEP-GEF Biosafet Projects – Country Pages: Panama http://www.unep.ch/biosafety/partcountries/PAcountrypage.htm, accessed 20 January 2008

49 Online: http://190.34.208.115/legis-agro/sanidad_animal/le39_01_059.html or http://www.asamblea.gob.pa/APPS/LEGISPAN PDF_GACETAS/2000/2001/24317_2001.PDF.

Country	Regulatory System	Agencies and Authorities	Institutional Framework	Labelling and Traceability	Scope
		promotion of research and transfer of biotechnology; the ANAM is responsible for the general regulation and control of the access to genetic resources and for ensuring compliance with international obligations in the field of biosafety (Articles 10 to 13).[50] The ANAM acts through the Unit for Access to Genetic Resources, which is a subdivision of the ANAM's Division on Protected Areas and Wildlife.	the National Biosafety Committee.		
Bolivia	Biosafety Regulation: Decree No. 24676 issued in June 1997.[51]	Ministry of Sustainable Development.	The National Biosafety Committee is comprised of several public institutions.	No specific regulations	LMOs in general
Colombia	– Decree No. 4525 issued in December 2005 regulates the transboundary movement of LMOs.[52] It generally complies with the CPB.	The Ministry of Agriculture and Rural Development (acting through the Colombian Agricultural Institute), the Ministry of Environment and the Ministry of Social Protection	There is a National Technical Advisory Council which advises the three competent national authorities.	No specific regulations. Decree No. 4525 enables the competent authorities to determine the information to be provided to consumers and users on	All LMOs that might have an adverse effect on the environment and biologicdiversity, taking into account risks for human health and agricultural productivity.

(continued)

50 *Ibid.* Art. 10-13.

51 Official Bolivian Gazette, Supreme Decree No. 24676 (Decree found in the Ecolex Database, 1997) http://faolex.fao.org/docs/texts/bol53033ENG.doc, accessed 22 January 2008.

52 Diario Oficial 46.115, Decreto No. 4525 (Decree found in the Ecolex Database, 2005) http://faolex.fao.org/docs/texts/col62480.doc, accessed 22 January 2008.

(continued)

Country	Regulatory System	Agencies and Authorities	Institutional Framework	Labelling and Traceability	Scope
	– Decree No. 3075 issued in 1997[53] – The General Colombian Environmental Law (No. 99) requires that an Environmental License be obtained to import, handle and commercialize LMOs.[54]	are the competent authorities in their respective fields.		labels and packaging.	
Ecuador	– Article 89 of the Constitution establishes that: "The State will adopt measures to regulate, according to strict biosafety regulations, the release into the environment, testing, use, trade and import of genetically modified organisms."[55] – The Environmental Management Law, enacted on July 30, 1999, namely Article 9.1, confers the power of regulating biosafety issues to the Ministry of Environment.[56]	The Ministry of Environment	The Biosafety National Commission, which is attached to the Ministry of Environment, is composed of public and private organizations, including consumers, NGOs, producers and farmers.	Articles 13 and 14 of the Ecuadorean Consumer Protection Law enacted on July 10, 2000, establishes that food labels should indicate if a product contains genetically modified organisms.[57]	The Environmental Management Law embraces a broad range of LMOs.

53 The Cartagena Protocol on Biosafety, "Colombia: First National Report on the Implementation of the Cartagena Protoco on Biosafety" (Report) (October 2002), http://www.cbd.int/doc/world/co/co-nr-cpb-01-es.pdf, accessed 22 January 2008.

54 *Ibid.*

55 National Government of the Republic of Ecuador, "La Constitución" (Ecuadorian Constitution), http://www.presidenci gov.ec/modulos.asp?id=109, accessed 23 January 2008.

56 Ecuadorian National Congress, Ley de Gestión Ambiental (law found in the Ecolex Database, 1999), http://faolex.fao.or docs/texts/ecu19248.doc, accessed 23 January 2008.

57 *Ibid*, 17.

Country	Regulatory System	Agencies and Authorities	Institutional Framework	Labelling and Traceability	Scope
	– Decree No. 2232, issued in January 2007, establishes the National Biodiversity Policy.[58] – A Final Draft Biosafety Framework was developed in 2006 as part of the UNEP-GEF Global Project on the Development of NBFs.[59] – A General Regulation on Biosafety has been proposed.[60] – A draft Law on Biodiversity has also been elaborated.				
Peru	– Law No. 27104 on Risks related to Biotechnology was enacted in 1999 and contains provisions on biosafety.[61] – Decree No. 108-2002-PCM (Regulation related to Law No. 27104).[62]	Competent sector-based authorities. The National Agricultural Research Institute (INIA) has powers of authorization and supervision in the field of agriculture. The Vice-Minister of Fisheries is	Each sector-based authority receives support from sector-based technical groups. The authorities are in charge of reviewing requests for the use of LMOs and issuing recommendations.	No specific regulations	Research, production, introduction, handling, transport, storage, conservation, exchange, trade, contained use, and release of LMOs as well as related activities.

(continued)

58 Constitutional President of the Republic of Ecuador, Decreto No. 2232 (decree found in the Ecolex Database, 2007), http://faolex.fao.org/docs/texts/ecu69279.doc, accessed 23 January 2008.

59 The United Nations Environment Program and the Global Environmental Facility (UNEP-GEF), Proyecto: Desarollo de un Marco Regulatorio de la Seguridad de la Biotecnología (Final Report completed as part of the UNEP-GEF Project on the Development of National Biosafety Frameworks, 2006), http://www.unep.ch/biosafety/development/Countryreports/ECNBFrep%20Informe%20Final.pdf, accessed 23 January 2008.

60 *Ibid.*, 14.

61 The Peruvian National Environmental Council (CONAM), Estrategia Nacional de Diversidad Biologica – Ley y Reglamento sobre Diversidad Biologica (Laws and Regulations related to the Peruvian National Diversity Policy) http://www.conam.gob.pe/endb/lyrdb.htm, accessed 23 January 2008.

62 El Peruano (The Peruvian), Decreto Supremo No. 108-2002-PCM (decree found in the Ecolex Database, 2002), http://faolex.fao.org/docs/texts/per35720.doc, accessed January 2008.

(continued)

Country	Regulatory System	Agencies and Authorities	Institutional Framework	Labelling and Traceability	Scope
	– The Framework Law on Agricultural Health, namely Article 30, refers to the measures enacted by the National Authority on health issues. – The Conservation and Sustainable Use of Biological Diversity Law (Law No. 26839) and its related regulation mention biotechnology and biosafety in their provisions.[63] – A Final Draft National Biosafety Framework was developed in 2005 as part of the UNEP-GEF Global Project on the Development of NBFs.[64]	competent in matters relating to fishing. The General Environmental Health Office of the Ministry of Health is competent in matters relating to health.	The National Environmental Council (CONAM) holds a Biotechnology and Biosafety Unit. The National Biological Diversity Commission is the consultative, advisory, and coordinative authority in the field of biosafety. Other relevant authorities include the National Service on Agrarian Health (SENASA) and the National institute on Natural Resources (INRENA).[65]		
Venezuela	– The Biodiversity Law, enacted in May 2000, contains biosafety regulations, particularly under Title VIII on	The Ministry of Environment and Natural Resources is in charge of applying the Biodiversity Law and constitutes	Decree No. 2.375, issued on April 24, 2003, created the National Biosafety Commission which is comprised of	No specific regulations	The research, development, production, use, release or introduction of genetically modified

[63] The Peruvian National Environmental Council (CONAM), Estrategia Nacional de Diversidad Biologica – Ley y Reglament sobre Diversidad Biologica (Laws and Regulations that are part of the Peruvian National Biodiversity Policy), http://www .conam.gob.pe/endb/lyrdb.htm, accessed 23 January 2008.

[64] The United Nations Environment Program and the Global Environmental Facility (UNEP-GEF), Marco Estructura Nacional de Bioseguridad Del Perú (Final Report completed as part of the UNEP-GEF Project on the Developmen of National Biosafety Frameworks, 2005), http://www.unep.ch/biosafety/development/Countryreports/PENBFrepSP.pdf accessed 23 January 2008.

[65] The Peruvian National Environmental Council (CONAM), Estrategia Nacional de Diversidad Biologica – La END en los Sectores' (National Authorities and Institutions in Charge of Implementing the Peruvian Biodiversity Policy http://www.conam.gob.pe/endb/lyrdb.htm, accessed 23 January 2008.

Country	Regulatory System	Agencies and Authorities	Institutional Framework	Labelling and Traceability	Scope
	"Development and Transfer of Biotechnology" and Title IX on "Biosafety and Ethics."[66] – A draft regulation on research, import, export, release into the environment, handling, confined use, production, distribution, movement, and storing of LMOs has been developed.[67] – A Final Draft National Biosafety Framework was developed in 2005 as part of the UNEP-GEF Global Project on the development of NBFs.[68]	the focal point of the CBD. The draft regulation establishes that the competent authorities in the matter are the ones relating to Health, Agriculture, Environment, Science and Technology.	public corporations, universities, producers, NGOs, and consumers.[69]		organisms or alien species requires an authorization.
Argentina[70]	– Law No. 656-92 on living modified microorganisms	The Secretariat of Agriculture, Livestock,	CONABIA, the National Agriculture	No specific regulations. Food safety	Genetically engineered plants, animals,

(continued)

66 The Official Gazette of the Bolivarian Republic of Venezuela, Ley De Diversidad Biológica (law found in the Governmental Legislation Database, 1999), http://www.gobiernoenlinea.ve/legislacion-view/sharedfiles/334.pdf, accessed 23 January 2008.

67 The Government of the Bolivarian Republic of Venezuela, Reglamento Parcial de la Ley de Diversidad Biológica, sobre el Registro, Control y Fiscalización de Organismos Modificados Genéticamente (OMG), Derivados y Productos que los Contengan (Draft Regulation found on the Govenmental Legislation Database), http://www.gobiernoenlinea.ve/legislacion-view/sharedfiles/reglamentoparcialleydiversidadbiologica.pdf, accessed 23 January 2008.

68 The United Nations Environment Program and the Global Environmental Facility (UNEP-GEF), Proyecto Marco Nacional De Seguridad De La Biotecnología en la República Bolivariana De Venezuela (Final Report completed as part of the UNEP-GEF Project on the Development of National Biosafety Frameworks, 2005), http://www.unep.ch/biosafety/development/Countryreports/VENBFrepSP.pdf, accessed 23 January 2008.

69 The Official Gazette of the Bolivarian Republic of Venezuela, Decreto No. 2.375, mediante el cual se crea, con carácter permanente y ad honorem, la Comisión Nacional de Bioseguridad como organismo técnico-científico asesor del Ejecutivo Nacional y de las autoridades nacionales competentes en las materias que en él se especifican. (decree found on the Official Gazette web site), http://www.tsj.gov.ve/gaceta/julio/160703/160703-37733-02.html, accessed 23 January 2008.

70 All of the pertinent legal norms that make up the Argentinian Regulatory Framework on Biotechnology can be found on the following web site: The Secretariat of Agriculture, Livestock, Fishing and Feeding (SAGyP), Marco Regulatorio de la Biotecnología Agropecuaria en La República Argentina, http://www.sagpya.mecon.gov.ar/, accessed 23 January 2008.

(continued)

Country	Regulatory System	Agencies and Authorities	Institutional Framework	Labelling and Traceability	Scope
	– Law No. 226-97 on isolation measures for genetically engineered plants – Law No. 39 enacted on July 11, 2003, on genetically engineered plants. – Law No. 57, enacted on July 18, 2003, on testing and release of genetically modified animals into the environment. – Law No. 644-03 on corn assessment – Law No. 46-04 on the LMO Register Biosafety and Labelling Bill – The Argentinian National Biosafety Framework was developed in 2004 as part of the UNEP-GEF Global Project on the Development of NBFs.[71]	Fishing and Feeding (SAGyP). Resolution No. 244-04, passed in 2004, creates the Biotechnology Office, which is attached to the SAGyP. Other institutions of the SAGyP also take part in the regulatory system (e.g. the National Service on Health, Agriculture and Livestock, the National Seeds Institute, and the National Agency on Food and Agriculture Markets).	and Livestock Biotechnology Advisory Commission (regulated by Resolution No. 124-91, and amended by Resolutions 669-93 and 328-97), is composed of public corporations, universities, producer associations, the Argentinian Ecology Society, etc. It also advises the Secretariat of Agriculture, Livestock, Fishing and Feeding. Once the release of an LMO into the environment is authorized, a request for regulatory "relaxation" can be made. In this case, subsequent requests for the release of this LMO into the environment shall be accompanied only by specific information related to the sowing surface, the date of sowing and harvest, and the location of the release.	assessment is based on the concept of "substantial equivalence" (Resolution No. 412-02).	*and microorganisms*

[71] The United Nations Environment Program and the Global Environmental Facility (UNEP-GEF), "Revision Of Argentina' National Biosafety Framework" (Final Report completed as part of the UNEP-GEF Project on the Development of Nationa Biosafety Frameworks, 2004), http://www.unep.ch/biosafety/development/Countryreports/ARNBFENVol1.pdf, accesse 23 January 2008.

Country	Regulatory System	Agencies and Authorities	Institutional Framework	Labelling and Traceability	Scope
Brazil	– Article 225(1) II, IV and V of the Federal Constitution, enacted in 1988, pertaining to the inspection of entities that research and handle genetic material. – National Biosafety Law No. 11.105, enacted in March 2005, and related regulation: Decree No. 5.591, issued in November 2005 – Decree No. 5.950, issued in 2006, on the restrictions regarding planting of genetically modified organisms near conservation areas. – A number of the National Technical Commission on Biosafety (CTNBio)'s internal acts (normative instructions) govern technical	The National Biosafety Council, which is linked to the President's Office, decides on the elaboration and implementation of national biosafety policy.[72] The National Technical Commission on Biosafety (CTNBio), which is linked to the Ministry of Science and Technology, is an entity that is destined for consultation and deliberation. It provides technical support to the government as well as mandatory technical safety norms regarding the authorization of LMO-related activities.[73] The ministries in charge of issues relating to the environment, agriculture, health and fisheries	The National Technical Commission on Biosafety (CTNBio), which is linked to the Ministry of Science and Technology, is composed of public sector and civil society representatives. The Commission is assisted by specific sector-based commissions (CIBios).[74]	Federal Decree No. 4.680 requires that genetically engineered food and genetically engineered food ingredients be labelled (ingredients that account for more than 1% of the product's total mass need to be identified in the label – Art. 2).[75] The National Biosafety Law establishes that all LMOs destined for food and feed need to be labelled in accordance with the pertinent regulations.	Extensive regulation of the production, cultivation, handling, transport, trade, release, and disposal of LMOs Exclusion of mutagenesis and other techniques

(continued)

72 See National Biosafety Law article 8.

73 The National Technical Commission on Biosafety (CTNBio), CTNBio (CTNBio main page), http://www.ctnbio.gov.br/index.php/content/view/2.html, accessed 24 January 2008.

74 The National Technical Commission on Biosafety (CTNBio), CIBio (Description of the Internal Biosafety Commissions), http://www.ctnbio.gov.br/index.php/content/view/143.html, accessed 24 January 2008.

75 The President of the Republic of Brasil, Decreto No. 4.680, de 24.04.2003 (Decree found on the CTNBio website, 2003), http://www.ctnbio.gov.br/index.php/content/view/3472.htm, accessed 24 January 2008.

(continued)

Country	Regulatory System	Agencies and Authorities	Institutional Framework	Labelling and Traceability	Scope
	aspects associated with LMO-related activities, such as biosafety quality certification, the internal functioning of biosafety commissions, the import of genetically modified plants destined for use in research, the release into the environment, the transport and confined use of genetically modified organisms, the classification of tests involving genetically modified plants, genetic manipulation and human cloning, genetic intervention in human beings, the import of genetically modified microorganisms destined for confined use, etc.[76]				
Chile	– Resolution No. 1.523, passed in 2001, regulates the introduction of living modified	Agricultural and Livestock Service (SAG)	There is a Biosafety Technical Comity on Genetically Modified	Health Regulation on Food No. 977-96, modified in	Genetically engineered plants produce[d] in the country [o]abroad, and

[76] The National Technical Commission on Biosafety (CTNBio), Legislação (Biosafety Legislation), http://www.ctnbio.gov.b[r/]index.php/content/view/55.html, accessed 24 January 2008. (This web site provides a complete list of pertinent legal biosafe[ty] norms.)

Country	Regulatory System	Agencies and Authorities	Institutional Framework	Labelling and Traceability	Scope
Chile	plants into the environment for dissemination purposes.[77] – Exemption Resolution n° 3.970, passed on December 30, 1997, authorizes transgenic corn for animal feed.[78] – Environmental Regulation No. 320 on aquaculture, enacted in 2001, is applied to aquatic activities involving LMOs. It was most recently updated in September 2009. – Environmental Law No. 19.300[79] – A Final Draft Biosafety Framework was developed in 2005 as part of the UNEP-GEF Global Project on the Development of NBFs.[80]		Organisms, which is mainly composed of representatives from the public sector (Resolution 6.996 passed in 2005).[81]	2006, establishes that the production, handling, and trade of transgenic foods and raw materials needs to comply with the regulations issued by the Ministry of Health to be authorised.[82] Genetic modification is prohibited in nursing products.	destined to be introduced into the environment for dissemination purposes

(continued)

77 Diario Oficial (the Official Journal), Resolución No. 1523 – Establece Normas Para la Internación e Introducción al Medio Ambiente De Organismos Vegetales Vivos Modificados De Propagación (Resolution listed on the SAG website, 2001), http://www.sag.gob.cl/pls/portal/docs/PAGE/PG_SAG_BIBLIOTECA/BIBL_IMPORTACIONES/BIBLIO_IMP_AGRICOLA/BIBLIO_IMP_AGRI_NORMATIVAS/RES_1523_2001.PDF, accessed 24 January 2008.

78 The National Agriculture and Livestock Service (SAG), Regulación No. 3970 – Establece Autorización Para Consumo Animal De Maíz Transgenico Con Modificación (BT) y Resistente a Glufosinato (Resolution listed on the SAG web site, 1997), http://www.sag.gob.cl/pls/portal/docs/PAGE/PG_SAG_BIBLIOTECA/BIBL_IMPORTACIONES/BIBLIO_IMP_AGRICOLA/BIBLIO_IMP_AGRI_NORMATIVAS/RES_3970_1997.PDF, accessed 24 January 2008.

79 Diario Oficial (the Official Journal) No. 34.810, Ley No. 19.300 de Bases Generales Del Medio Ambiente (law found in the Ecolex Database, 1994), http://faolex.fao.org/docs/texts/chi4873.doc, accessed 24 January 2008.

80 The United Nations Environment Program and the Global Environmental Facility (UNEP-GEF), Bases Para el Marco Nacional de Bioseguridad De Chile (Final Report completed as part of the UNEP-GEF Project on the Development of National Biosafety Frameworks, 2005), http://www.unep.ch/biosafety/development/Countryreports/CLNBFrepSP.pdf, accessed 24 January 2008.

81 The National Agriculture and Livestock Service (SAG), Resolución No. 6966 – Crea Comité Técnico De Organismos Genéticamente Modificados (OGMs). (Resolution listed on the SAG web site, 2005), http://www.sag.gob.cl/pls/portal/docs/PAGE/PG_SAG_BIBLIOTECA/BIBL_IMPORTACIONES/BIBLIO_IMP_AGRICOLA/BIBLIO_IMP_AGRI_NORMATIVAS/RES_6966_2005.PDF, accessed 24 January 2008.

82 Diario Oficial (Official Journal) No. 13.05.97, Reglamento Sanitario De Los Alimentos No. 977/96 (Regulation listed on the SAG website, 1996), http://www.sag.gob.cl/pls/portal/docs/PAGE/PG_SAG_BIBLIOTECA/BIBL_EXPORTACIONES/BIBLIO_EXP_PEC/BIBLIO_EXP_PEC_NORMAS/DECRETO_977_ALIMENTOS.PDF, accessed 24 January 2008.

(continued)

Country	Regulatory System	Agencies and Authorities	Institutional Framework	Labelling and Traceability	Scope
Paraguay	– Decree No. 18.481-97 establishes the National Biosafety Commission. The commission is in charge of analysing and issuing recommendations regarding the introduction, trial tests, research and release into the environment of transgenic plants.[83] – Ministerial Resolution No. 2 regulates "Requests and procedures to import cereal grains and oil products." – On October 20, 2004, by way of Resolution No. 1.261 of the Ministry of Agriculture and Livestock, four types of soy (AW7110, AW5581, MSOY 7878 and MSOY 8080) were entered into the National Register of Commercial Crops.	The National Biosafety Commission (ComBio) of the Ministry of Agriculture and Livestock (MAG)[84]	The Commission (ComBio) is comprised of public and scientific institutions, civil society organizations, and productive sector representatives.[85] The National Service on Quality and Sanitation in Plants and Seeds (SENAVE) constitutes the Technical Secretariat of the Biosafety Commission. In collaboration with the National Biosafety Commission, it authorizes requests for the release into the environment and the commercialisation of genetically modified vegetables, elaborates biotechnology policy and regulations, and raises public awareness.[86]	No specific regulations	Plants

83 The Secretary of Environment (SEAM), Development of the National Framework of Security of Biotechnology for Paragua (Final Report completed as part of the UNEP-GEF Project on the Development of NBFs; 30 June 2007) 29, http://www.une .ch/biosafety/development/Countryreports/PYNBFrep.pdf, accessed 24 January 2008.

84 *Ibid*.

85 *Ibid*, 20.

86 *Ibid*, 20, 29.

Country	Regulatory System	Agencies and Authorities	Institutional Framework	Labelling and Traceability	Scope
	– A Biosafety Bill has been presented to Congress by the MAG and is currently under study.[87] – A Final Draft National Biosafety Framework was completed in 2007 as part of the UNEP-GEF Global Project on the Development of NBFs.[88]		Other institutions that are linked to the MAG include the Department of Agriculture and Research (DIA), the Department of Research on Animal Production (DIPA) and the National Service on Animal Health and Quality (SENASCA).[89] Other institutions that are not related to the MAG include:[90] – The Secretariat of Environment (SEAM), which is the Focal Point of the CBD and monitors new developments in biotechnology. – The National Institute for Food and Nutrition (INAN), which sees to the quality and innocuousness of processed foods destined for human consumption (As for foods that are consumed fresh, they are regulated by SENAVE.) – The National University of Asuncion (UNA), which conducts research and develops biotechnology		

(continued)

[87] The Secretary of Environment (SEAM), Development of the National Framework of Security of Biotechnology for Paraguay (Final Report completed as part of the UNEP-GEF Project on the Development of NBFs) (30 June 2007), 17, 34 http://www.unep.ch/biosafety/development/Countryreports/PYNBFrep.pdf, accessed 24 January 2008.

[88] *Ibid.*

[89] *Ibid*, 20–1.

[90] *Ibid*, 20–2.

(continued)

Country	Regulatory System	Agencies and Authorities	Institutional Framework	Labelling and Traceability	Scope
Uruguay	– Article 23 of the Environmental Protection Law n° 17.283, enacted in 2000, establishes that the Ministry of Housing, Territorial Order, and Environment shall apply measures in order to avoid risks resulting from LMOs. In cases where there are no other competent authorities, the previously mentioned Ministry can issue permits.[91] – Decree No. 249, issued on September 4, 2000, establishes a Risk Assessment Commission for Genetically Modified Plants (CERV).[92] – Ministerial Resolutions No. 236A/003 and No. 276/003 establish conditions for the introduction, use and production, and trade of MON 810 Corn.[93]	Mainly the Ministry of Agriculture, Livestock and Fisheries, the National Seeds Institute, and the Ministry of Economy and Finance.	The Risk Assessment Commission for Genetically Modified Plants (CERV) is composed of the Ministry of Livestock, Agriculture and Fisheries (MGAP), the Ministry of Health (MSP), the Ministry of Housing, Territorial Order, and Environment (MVOTMA), the Ministry of Economy and Finance (MEF), the National Seeds Institute (INASE), and the National Institute for Agricultural Research (INIA). The Ministry of Livestock, Agriculture and Fisheries and the Ministry of Economy and Finance can authorise the import and release into the environment of genetically modified plants.	No specific regulations	The introduction, use, and handling of plants and their genetically modified components, can only be carried out with prior authorization from the competent authorities (regardless of th[e] purpose of the introduction, handling, and use).

91 Mariano Arana (Minister of Housing, Territorial Order and Environment), Jaime Igorra (Subsecretary of Housing, Territori[al] Order and Environment), and Alicia Torres (National Director of Environment), Propuesta de Marco Nacional de Biose[-]guridad Para Uruguay (Final Report completed as part of the UNEP-GEF Project on the Development of NBFs) (Septembe[r] 2007), 47 http://www.unep.ch/biosafety/development/Countryreports/UYNBFrepSP.pdf, accessed 24 January 2008.

92 *Ibid*, 52.

93 *Ibid*, 48.

Country	Regulatory System	Agencies and Authorities	Institutional Framework	Labelling and Traceability	Scope
	– Ministerial Resolution No. 290/004, passed by the Ministry of Economy and Finance (MEF) and the Ministry of Livestock, Agriculture and Fisheries (MGAP), authorizes conditions for the introduction, use, production and trade of Bt11 corn.[94] – Ministerial Resolution No. 292/004 establishes conditions for the introduction, use, production and trade of Bt11 corn.[95] – Ministerial Resolution No. 17/08/06, passed by the Ministry of Housing, Territorial Order and Environment (MVOTMA), suspends the use, production, and trade of genetically modified sweet corn.[96] – Decree 037/007 suspends new requests for the use of		The Ministry of Livestock, Agriculture and Fisheries and the Seeds Institute can issue authorisations relating to the reproduction of seeds, the National Registry of Agricultural Products and the conduct of experiments and trials under conditions of contained use, or under specific biosafety conditions. The Ministry of Health and the Ministry of Housing, Territorial Order and Environment can also intervene in the decision-making process according to Laws No. 9.292 (MSP), No. 16.112 and No. 17.283 (MVOTMA). [97]		

(continued)

[94] *Ibid.*
[95] *Ibid.*
[96] *Ibid*, 49.
[97] *Ibid*, 52. (The institutional order is summed up on this page.)

(continued)

Country	Regulatory System	Agencies and Authorities	Institutional Framework	Labelling and Traceability	Scope
	biotechnology in plants and creates a working group composed of various ministries, which it puts in charge of elaborating national policy in the field of biotechnology.[98]				
Cuba	Environmental Law No. 81, enacted in 1997, namely Article 12 h), designates the Ministry of Science and Technology as the authority in charge of establishing environmental policy in the field of biosafety, and coordinating its implementation. – Article 28 t) regulates biotechnological products in the context of a broader environmental impact assessment framework. – Finally, Article 88 l) establishes the minister's authority to regulate and control risks associated to the use and release of living modified	Through the National Centre on Biological Safety (CSB), the Ministry of Science and Technology regulates all issues relating to LMOs. (Resolution No. 67, passed on June 15, 1996)[99]	In collaboration with the Cuban Registry of Foods and the CSB, Technical Comities CTN 91 and CTN 94 are in charge of elaborating specific regulations regarding the technical aspects of biosafety in the field of genetically engineered foods.[100]	No specific regulations	Biological safety in general. It is not limited to plants or to modern biotechnology.

[98] *Ibid.*, 49.
[99] *Ibid.*
[100] *Ibid.*

Country	Regulatory System	Agencies and Authorities	Institutional Framework	Labelling and Traceability	Scope
	organisms resulting from biotechnology.[101] – Law No. 190 on Biosafety regulates the intentional introduction into the environment, the import, export and confined use of LMOs as well as dangerous the issue of biological waste, biological emergencies and the governing body's functions.[102] – Resolution No. 76 regulates the issuing of authorisations for the transboundary movement, release into the environment and productiton of LMOs as well as for the construction of facilities, investigations, and trials involving LMOs.[103] – Resolution No. 8 on Biosafety in facilities that manipulate biological agents and their by-products.[104] – Resolution No. 103 on the establishment of biosafety requirements and procedures in facilities using biological agents and their by-products.[105]				

(continued)

[101] Gaceta Oficial de la Republica de Cuba – Edicion Extraordinaria (the Official Cuban Gazette – Special Edition), Ley No. 81 DEL MEDIO AMBIENTE (Law listed on the Ministry of Science, Technology and Environment's Web site, 1997) http://www.medioambiente.cu/oregulatoria/cica/resoluciones/L-81.doc, accessed 25 January 2008.

[102] Jorge Luis Fernández Chamero (Ministry of Science, Technology and Environment), Cuba: First Regular National Report on the Implementation of the Cartagena Protocol on Biosafety (Report) (2007) 7, http://www.cbd.int/doc/world/cu/cu-nr-cpb-01-es.pdf, accessed 25 January 2008.

[103] *Ibid.*

[104] *Ibid.*

[105] *Ibid.*

(continued)

Country	Regulatory System	Agencies and Authorities	Institutional Framework	Labelling and Traceability	Scop
	– Resolution No. 112 on the establishment of requirements and procedures in facilities working with animals and plants, which present a biological risk.[106] – Resolution 38/2006 provides an official list of the biological agents which affect humans, animals and plants.[107] – Participating in the UNEP-GEF Project on the Implementation of National Biosafety Frameworks.[108]				
Antigua and Barbuda	No specific biosafety regulations,[109] but there is a Biosafety and Biotechnology Management Bill.[110]				
Barbados	No specific biosafety regulations[111] A Final Draft National Biosafety Framework was developed in 2005 as part of the UNEP-GEF	The Ministry of Agriculture and Rural Development and the Ministry of Energy and the Environment[112]	There is no formalised institutional structure, but a National Biosafety Committee was	No specific regulations	

[106] *Ibid.*

[107] *Ibid.*

[108] The United Nations Environment Program and the Global Environmental Facility (UNEP-GEF), UNEP-GEF Biosafe Projects – Country Pages: Cuba, http://www.unep.ch/biosafety/partcountries/ImpCuba.htm, accessed 25 January 2008.

[109] Antigua and Barbuda Environment Division, BioSafety Documents (Governmental web site), http://www environmentdivision.info/?q=node/47, accessed 25 January 2008.

[110] Antigua and Barbuda, "Biosafety and Biotechnology Management Bill" (Final Report provided as part of the UNE GEF Project on the Development of National Biosafety Frameworks; 2005), http://www.unep.ch/biosafety/developmen Countryreports/AGNBFrep.pdf, accessed 25 January 2008.

[111] Ministry of Housing, Lands and Environment, Draft National Biosafety Framework for Barbados (Final Report provided part of the UNEP-GEF Global Project on the Development of NBFs; November 2005) 9, http://www.unep.ch/biosafet development/Countryreports/BBNBFrep.pdf, accessed 25 January 2008; Ministry of Energy and the Environment, Barbado First Regular National Report on the Impementation of the Cartagena Protocol on Biosafety (Report) (2007) 2, 4; htt //www.cbd.int/doc/world/bb/bb-nr-cpb-01-en.pdf, accessed 25 January 2008.

[112] Ministry of Housing, Lands and Environment, Draft National Biosafety Framework for Barbados (Final Report provided part of the UNEP-GEF Global Project on the Development of NBFs) (November 2005) 17, http://www.unep.ch/biosafet development/Countryreports/BBNBFrep.pdf, accessed 25 January 2008; Government of Barbados, Ministries/Agencies (Go ernmental web site) http://www.barbados.gov.bb/ministries.htm, accessed 25 January 2008.

Country	Regulatory System	Agencies and Authorities	Institutional Framework	Labelling and Traceability	Scope
	Global Project on the Development of NBFs.[113]		created. The National Biosafety Committee (NBC) is the National Competent authority under the ambit of the Ministry of Energy and the Environment. The Ministry of Agriculture and Rural Development serves as the Contact Point for notifications on unintentional transboundary movements of LMOs and for emergency measures.[114]		
Guyana	The Environmental Protection Act, enacted in 1996, namely Part IV, sections 10 and 11, regulates the use control, release, and keeping of LMOs. The said provisions specifically require that impact assessments be performed in the case of projects involving LMOs.[115] Final Draft National Biosafety Framework completed in 2006 as part of the UNEP-GEF Global Project on the development of NBFs.[116]	The Environmental Protection Agency is in charge of biosafety issues. It is the leader of the National Coordinating Committee on Biosafety (NCC).[117]	The National Coordinating Committee on Biosafety (NCC) is as a subcommittee of the National Biodiversity Advisory Committee.[118]	No specific regulations	

(continued)

113 Ministry of Housing, Lands and Environment, Draft National Biosafety Framework for Barbados (Final Report provided as part of the UNEP-GEF Global Project on the Development of NBFs) (November 2005) http://www.unep.ch/biosafety/development/Countryreports/BBNBFrep.pdf, accessed 25 January 2008.

114 *Ibid*.

115 John Cartey Caesar (Environmental Protection Agency), Final Draft Biosafety National Framework for Guyana (Report provided as part ot the UNEP-GEF Global Project on the Development of NBFs) (5 February 2007) 36, http://www.unep.ch/biosafety/development/Countryreports/GYNBFrep.pdf, accessed 25 January 2008.

116 *Ibid* (entire report).

117 *Ibid* 10–13.

118 *Ibid* 8.

(continued)

Country	Regulatory System	Agencies and Authorities	Institutional Framework	Labelling and Traceability	Scope
Jamaica	No specific biosafety regulations. Plants (Importation) Control (Amendment) Regulations passed in 2005 regulate the import of genetically modified plants.[119] A Final Draft National Biosafety Framework was developed as part of the UNEP-GEF Global Project on the Development of NBFs.[120]		The National Biosafety Committee was formed in 1996. It is composed of public, private sector, and academic representatives. It is also Committee is in charge of developing procedural guidelines and regulations for the import of genetically modified plants for experimentation purposes and is responsible for granting permission for the import of plants, seeds, and cuttings for the purposes of the Regulation.[121]	No specific regulations	"Plant" is defined as a plant that has been genetically modified and imported into Jamaica for the purpose of experimentation under controlled conditions.[122]
Dominican Republic	There is currently no regulatory framework in place,[123] but it is in the process of developing a National Biosafety	The Secretariat of Environment and Natural Resources receives requests regarding LMOs and sends them to the Subsecretariat		No specific biosafety regulations	

[119] The Jamaica Gazette (Supplement), Plants (Importation) Control (Amendment) Regulations' Law found in the Ecole database, 2005), http://faolex.fao.org/docs/pdf/jam61362.pdf, accessed 26 January 2008.

[120] Legal and Policy Framework for the Implementation of the Cartagena Protocol on Biosafety (Final Report provided as part c the UNEP-GEF Global Project on the Development of NBFs), http://www.unep.ch/biosafety/development/Countryreports JMNBFrep.pdf, accessed 26 January 2008.

[121] *Ibid*, 34.

[122] *Ibid*; The Jamaica Gazette (Supplement), Plants (Importation) Control (Amendment) Regulations. Law found in the Ecole database, 2005), Part II. http://faolex.fao.org/docs/pdf/jam61362.pdf, accessed 26 January 2008.

[123] Marina Hernández (Department of Genetic Resources, Wildlife and Biodiversity), First Regular National Report on th Impementation of the Cartagena Protocol on Biosafety (Report) (20 October 2007) 2, http://www.cbd.int/doc/world/dc do-nr-cpb-01-es.pdf, accessed 26 January 2008.

Country	Regulatory System	Agencies and Authorities	Institutional Framework	Labelling and Traceability	Scope
	Framework as part of the UNEP-GEF Global Project on the Development of NBFs[124]	of Protected Areas and Biodiversity in order that they can be processed.[125] The Department of Genetic Resources, Wildlife and Biodiversity is the focal point of the CBD.[126]			
Saint Lucia[127]	No specific biosafety regulations but a Final Draft National Biosafety Framework was completed in 2006 as part of the UNEP-GEF Project on Development of Biosafety Frameworks.		National Biosafety Commission.	No specific regulations	
Trinidad and Tobago	No specific biosafety regulations. A National Biosafety Framework is in the process of being drafted as part of the UNEP-GEF Global Project on the Development of NBFs.[128]		There is a Committee for the Development of a National Policy and Regulations on Biosafety, which is integrated by representatives from various sectors. The Committee is comprised of universities, ministries, etc.[129]	No specific regulations	

[124] *Ibid.*

[125] *Ibid.*

[126] Secretariat of the Convention on Biological Diversity, Cartagena Protocol on Biosafety Competent National Authorities' (Document found on the CBD web site), http://bch.cbd.int/reports/authorities.pdf, accessed 26 January 2008.

[127] Gillian Vidal-Jules Drafter, Biosafety Act (Final Report provided as part of the UNEP-GEF Global Project on the Development of NBFs) (May 2006), http://www.unep.ch/biosafety/development/Countryreports/LCNBFrep.pdf, accessed 26 January 2008.

[128] The United Nations Environment Program and the Global Environmental Facility (UNEP-GEF), UNEP-GEF Biosafety Projects – Country Pages: Trinidad and Tobago, http://www.unep.ch/biosafety/partcountries/TTcountrypage.htm, accessed 26 January 2008.

[129] Trinidad and Tobago, Initial Project Report as at Project Starting Date (Report) (September 5 2008), http://www.unep.ch/biosafety/development/countryreports/TTprojectrepo.pdf, accessed 26 January 2008.

ANNEX X

Costa Rica

Projects that require EIA or other studies as a result of other laws or regulations[1]

Activity	Law or regulation
Importation of certain wildlife species under the criteria of the National System of Conservation Areas (SINAC)	Wildlife Conservation Law No. 7317, Article 26.
Projects that in the opinion of the Technical Office of the National Commission for Biodiversity Management (CONAGEBIO) might affect biodiversity	Biodiversity Law No. 7788, Article 92.

[1] *Source:* SETENA.

Number of transgenic events in Costa Rica allowed per year, by crop[2]

Year	Corn	Soybeans	Cotton	Banana	Rice	Cocoyam	Petunia	Total
1991–92		1						1
1992–93	1		6					7
1994–95		13						13
1995–96	1	23		1			1*	25
1996–97	1	18	8					27
1997–98		38	9					47
1998–99	4	42	14			1		61
1999–00	5	39	13		1			58
2000–01	3	61	12	2	1			79
2001–02		62	33		1			96
2002–03		41	81	2	1			125
2003–04		23	98	2	1			124
Total	15	361	274	7	5	1		663

* This event was not approved but is included for informational purposes.
For research approvals, only the year of approval or project initiation is shown, except for rice, for which each new sowing is counted.

[2] *Source:* Walter Quirós, National Seeds Office.

Number of events allowed in Costa Rica, according to the transgenic characteristics of the crop (1991–2004)[3]

Transgenic characteristic	Soybean	Cotton	Corn	Banana	Rice	Cocoyam
Glyphosate resistance (RR)[4]	356	105	9			
Lepidoptera resistance (Bt)[5]		42	4			
(RR + Bt)[6]		106	1			
Bromoxynil tolerance (BNX)		1				
(BNX + Bt)		1				
Resistance to Virus Hoja Blanca and PPT[7]					5	
Resistance to black sigatoka				5		
Fruit ripening				1		
Resistance to virus rayado fino and PPT			1			
Resistance to glufosinate ammonium	5	19				
Marker genes						1
Promotor genes				1		
Total	361	274	15	7	5	1

3 *Source:* Walter Quirós, National Seeds Office.

4 Includes RR Flex.

5 Includes different kinds of genes Bt: Cry1Ac (Bt1 or Bollgard I), Cry2Ab, Cry1Ab, Cry1F, Vip3A, and the combination of Bt2 or Bollgard II (Cry1Ac + Cry2Ab).

6 The following different combinations of genes are present: BR (RR + Bt1), DR (RR + Bt2), DG (RG + Bt1,2), PR (RR + Cry1F), VR (RR + VIP3A), FD (RR Flex + Bt2).

7 PPT (phosphinothricin) is the herbicide (glufosinate ammonium) commercially known as Liberty or Basta.

Table of Treaties

Year	Treaty
1945	*Statute of the International Court of Justice*, 26 June 1945, TS No 933, 59 Stat 1055, Bevans 1179.
1951	*International Plant Protection Convention*, 6 December 1951, 150 UNTS 67 (entered into force 3 April 1952).
1966	*International Covenant on Civil and Political Rights*, 19 December 1966, 999 UNTS 171, 6 ILM 368 (entered into force 23 March 1976).
1966	*International Covenant on Economic, Social and Cultural Rights*, 19 Dec. 1966, 993 UNTS 3 (entered into force 3 January 1976).
1969	*Vienna Convention on the Law of Treaties*, 23 May 1969, 1155 U.N.T.S. 331 (entered into force 27 January 1980).
1973	*Convention on International Trade in Endangered Species of Fauna and Flora*, 3 March 1973, 993 UNTS 243, 12 ILM 1085.
1987	*Montreal Protocol on Substances that Deplete the Ozone Layer*, 16 September 1987, 26 ILM 1550 (entered into force 1 January 1989).
1989	*Basel Convention on the Control of Transboundary Movements of Hazardous Wastes and their Disposal*, 22 March 1989, 1673 UNTS 126, 28 ILM 657 (entered into force 5 May 1992).
1991	*Convention concerning the Protection of the Alps*, 07 November 1991, B7 p. 991:83 (entered into force 6 March 1995).
1991	*Espoo Convention on Environmental Impact Assessment in a Transboundary Context*, 25 February 1991, B7 p. 991:15 (entered into force 10 September 1997).
1992	*Convention on Biological Diversity*, 5 June 1992, 1760 UNTS 79, 31 ILM 818 (entered into force 29 December 1993).
1992	*Treaty on European Union*, 31 ILM (1992) 247.
1992	*UN Framework Convention on Climate Change*, 31 ILM (1992) 849.

(continued)

(continued)

Year	Treaty
1994	*UN Convention to Combat Desertification in Those Countries Experiencing Drought and/or Desertification, Particularly in Africa*, 33 ILM (1994) 1328.
1994	*Agreement on the Application of Sanitary and Phytosanitary Measures*, being part of Annex IA to the *Agreement Establishing the World Trade Organization*, 15 April 1994, 33 ILM 1144, online: http://www.wto.org/english/tratop_e/sps_e/spsagr_e.htm.
1994	*Agreement on Technical Barriers to Trade*, being part of Annex IA to the *Agreement Establishing the World Trade Organization*, 15 April 1994, 33 ILM 1144.
1994	*General Agreement on Tariffs and Trade*, being part of Annex IA to the *Agreement Establishing the World Trade Organization*, 15 April 1994, 1867 UNTS 187, 33 ILM 1153.
1997	*International Plant Protection Convention (New Revised Text)*
1998	*Aarhus Convention on Access to Information, Public Participation in Decision-Making and Access to Justice in Environmental Matters*, 25 June 1998, 2161 UNTS 447 (entered into force 30 October 2001).
1998	*Kyoto Protocol to the Framework Convention on Climate Change*, 37 ILM (1998) 22.
1998	*Rotterdam Convention on Prior Informed Consent Procedure for Certain Hazardous Chemicals and Pesticides in International Trade*, 10 September 1998, 38 ILM 1 (entered into force 24 February 2004).
2000	*Cartagena Protocol on Biosafety to the Convention on Biological Diversity*, 29 January 2000, 2226 UNTS 208 (entered into force 11 September 2003), online: http://bch.cbd.int/protocol/text/.
2001	*International Treaty on Plant Genetic Resources for Food and Agriculture*, 3 November 2001, FAO Res 3/2001 (entered into force 29 June 2004), online: http://www.fao.org/nr/cgrfa/cgrfa-global/cgrfa-planttr/.
2001	*Stockholm Convention on Persistent Organic Pollutants*, 22 May 2001, 40 ILM 532 (entered into force 17 May 2004).
2003	*Traité de l'Union Économique et Monétaire Ouest Africaine*, online: http://www.uemoa.int/actes/2003/TraitReviseUEMOA.pdf.
2010	*Nagoya – Kuala Lumpur Supplementary Protocol on Liability and Redress to the Cartagena Protocol on Biosafety*, 15 October 2010.

Table of Declarations

Year	Declaration
1948	*Universal Declaration of Human Rights*, GA Res 217A, UN GAOR, 3rd. Sess., UN Doc A/810 (1948).
1972	*Stockholm Declaration on the Human Environment*, UN Doc. A/C. 48/14 (1972).
1982	*World Charter for Nature* UNGA Res. 7, UN GAOR 36th Sess., Supp. No. 51, UN Doc. A/51 (1982).
1986	*Declaration on the Right to Development*, 4 December 1986, GA res. 41/128, annex, 41 UN GAOR Supp. (No. 53) at 186, UN Doc. A/41/53 (1986).
1992	*Agenda 21: Programme of Action for Sustainable Development*, U.N. GAOR, 46th Sess., Agenda Item 21, UN Doc A/Conf.151/26 (1992).
1992	*Rio Declaration on Environment and Development*, UN Doc. A/CONF.151/26 (vol. I); 31 ILM 874 (1992).
1995	*Copenhagen Declaration on Social Development and Programme of Action of the World Summit for Social Development*, 12 March 1995, A/CONF.166/9 (1995).
2000	*Millennium Declaration*, GA Res. 55/2, 18 September 2000.
2002	*ILA New Delhi Declaration of Principles of International Law Relating to Sustainable Development*, A/CONF.199/8, Annex.
2002	*Johannesburg Declaration on Sustainable Development*, in *Report of the World Summit on Sustainable Development*, 26 August to 4 September 2002, UN Doc. A/CONF.199/20.
2002	*Johannesburg Plan of Implementation*, in *Report of the World Summit on Sustainable Development*, 4 September 2002, UN Doc. A/CONF.199/20.

Table of Decisions

Convention	Decision
Convention on Biodiversity	Consideration of the need for and modalities of a protocol for the safe transfer, handling, and use of living modified organisms, UNEP/CBD/COP/2/5 (1995).
	Adoption of the Cartagena Protocol and Interim Arrangements, UNEP/CBD/ExCOP/1/3 (2000).
Cartagena Protocol	UNEP/CBD/ICCP/3/10.
	COP-MOP 4, Decision BS-IV/11, "Risk Assessment and Risk Management" (2008).
	COP-MOP 4, Dec. BS-IV/17, "Public Awareness, Education and Participation" (2008).
	COP-MOP 3, Decision BS-III/10, "Handling, Transport, Packaging and Identification of Living Modified Organisms: Paragraph 2(a) of Article 18" (2006).
	UNEP/CBD/BS/COP-MOP/5/8, "Handling, Transport, Packaging and Identification of Living Modified Organisms – Synthesis of Information on Experience Gained with the Implementation of Requirements Related to Paragraph 2 (a) of Article 18" (13 May 2010).
	UNEP/CBD/BS/COP-MOP/4/9, "Handling, Transport, Packaging and Identification of Living Modified Organisms – Synthesis of Information on Experience Gained with the Use of Sampling of Living Modified Organisms and Detection Techniques and on the Need for and Modalities of Developing Criteria for Acceptability of, and Harmonizing, Sampling and Detection Techniques (Paragraph 2(a) of Article 18)" (7 April 2008).

Convention	Decision
European Commission	Commission Decision (EC) 2002/624 of 24 July 2002 establishing guidance notes supplementing Annex II to Directive 2001/18/EC of the European Parliament and of the Council on the deliberate release into the environment of genetically modified organisms and repealing Council Directive 90/220/EEC [2002] OJ L200/22.
Montreal Protocol	MOP Decision IV/5 (UNEP/OzL.Pro.4/15, 25 November 1992), subsequently amended by Decision X/10 (UNEP/OzL.Pro.10/9, 3 December 1998).
Espoo Convention	MOP Decision II/4 (ECE/MP.EIA/4, 2001), revised by MOP Decision III/2 (ECE/MP.EIA/6, 2004).
UNEP	UNEP Governing Council Decision 15/30 (1989).

Table of Legislation

Country	Title
Argentina	*Tratamiento diferenciado para el otorgamiento de la autorización de comercialización para aquellas acumulaciones provenientes del cruzamiento de parentales que contienen eventos que en forma individual ya tienen autorización para la comercialización en la República Argentina. Resolución* 60/2007. 2007. (BCH Record No. 40386).
	Liberación al Medio de Organismos Vegetales Genéticamente Modificados (OVGM) / Resolución No. 39. 2003. (BCH Record No. 5649).
	Solicitud para la concesión de permisos para Experimentación y/o Liberación al Medio de Microorganismos Genéticamente Modificados y/o sus productos para aplicaciones en animales. / Resolución No. 656. 2004. (BCH Record No. 11812).
	Experimentación y/o Liberación al Medio de Organismos Animales Genéticamente Modificados. / Resolución No. 57. 2004. (BCH Record No. 11911)
Republic of Kenya	*The Biosafety Bill* 2007 (Nairobi, Kenya: National Council for Science and Technology).
	The Biosafety Act 2009 (Nairobi, Kenya: National Council for Science and Technology).
Republic of South Africa	*Genetically Modified Organisms Act (No. 15 of 1997;* BCH Record No. 255) and *Regulations* (BCH Record No. 269).
	Genetically Modified Organisms Act Regulations (BCH Record No. 269).

Country	Title
Republic of Uganda	Republic of Uganda, *Statute* No. 1990 (Kampala, Uganda: Uganda National Council for Science and Technology, 1990).
United Republic of Tanzania	*The Environmental Management Act*, 2004 (Parliament of the United Republic of Tanzania, Dar es Salaam, Tanzania: United Republic of Tanzania, 2004).
Vietnam	*Regulation on Management of Biological Safety of Genetically Modified Organisms; Products and Goods Originating from Genetically Modified Organisms* (BCH Record No. 41474).

Table of Cases

Court	Case
ICJ CASES	*Corfu Channel Case (UK/Albania)*, [1949] I.C.J. Rep 4. *Nuclear Tests Case (Australia/France)*, [1974] I.C.J. Rep 253. *Advisory Opinion on the Legality of the Threat or Use of Nuclear Weapons*, (1996) ICJ Rep. 226. *Case Concerning the Gabèíkovo-Nagymaros Project (Hungary/Slovakia)* (1997), ICJ Rep. 7. *Pulp Mills on the River Uruguay (Argentina v. Uruguay)*, Judgment of 20 April 2010
ARBITRAL CASES	*Trail Smelter Arbitration (United States v. Canada)*, 3 R. Int'l Arb. Awards 1911 (1938), reprinted in 33 AJIL 182 (1939), 3 R. Int'l Arb. Awards 1938 (1941), reprinted in 35 AJIL 684 (1941). *Award in the Arbitration regarding the Iron Rhine ("Ijzeren Rijn") Railway between the Kingdom of Belgium and the Kingdom of the Netherlands, decision of* 24 May 2005 (2005), 27 RIAA 35 (Permanent Court of Arbitration).
WTO CASES	*European Communities – Measures Concerning Meat and Meat Products, (Hormones)* AB-1997–3, WT/DS26/AB/R, 1998. *European Communities – Measures Affecting the Approval and Marketing of Biotech Products*, WT/DS291 WT/DS292, WT/DS293, 2006.

Court	Case
CANADIAN CASES	114957 *Canada Ltée (Spraytech, Société d'arrosage) v Hudson (Town)*, 2001 SCC 40.
	Harvard College v. Canada (Commissioner of Patents), [2002] 4 SCR 45, 2002 SCC 76.
	Hoffman v Monsanto Canada Inc, 2005 SKQB 225, aff'd 2007 SKCA 47, leave to appeal to SCC denied.
	Monsanto Canada Inc V Schmeiser, [2004] 1 SCR 902, 2004 SCC 34.
U.K. CASES	*Rylands v Fletcher (1866)*, L.R. 1 Ex. 265; (1868), L.R. 3 H.L. 330.
	Anns v Merton London Borough Council, [1978] A.C. 728 (H.L.).
	Murphy v Brentwood District Council, [1991] 1 A.C. 398 (HL).
	R. v. Secretary of State for the Environment ex parte *Watson (1998)*, [1999] Env. L.R. 310 (C.A.).
U.S. CASES	*Diamond v Chakrabarty*, 206 U.S.P.Q 193 (U.S.S.C. 1980).
	Re Genetically Modified Rice Litigation, 2009 U.S. Dist LEXIS 98302 (October 9, 2009).

Bibliography

Abramson, Stanley H. & J. Thomas Carrato. "Crop Biotechnology: The Case for Product Stewardship," (2001) 20 Virginia Environmental L.J. 241.

Aerni, Philipp. "Assessing stakeholder attitudes to agricultural biotechnology in developing countries," Biotechnology and Development Monitor (2001), online: http://www.biotech-monitor.nl/4702.htm.

Aerni, Philipp. "Stakeholder attitudes towards the risks and benefits of genetically modified crops in South Africa" (2005) 8 Environmental Science & Policy 464.

Aerni, Philipp. "Public Policy Responses to Biotechnology" in *Our Fragile World: Challenges and Opportunities for Sustainable Development* (Paris: Encyclopaedia of Life Support Systems (ELOSS), UNESCO, 2001) 7.

African Union, Revised African Model Law on Biosafety (Draft). (African Union: Addis Ababa, Ethiopia, 2008) available at http://www.africa-union.org/root/au/AUC/Departments/HRST/biosafety/DOC/level2/DraftRevAMLBS_Jan08_EN.pdf.

AGBIOS, "MON 810 Environmental Risk Assessment Case Study," online: http://www.agbios.com/cstudies.php?book=ESA&ev=MON810&chapter=Preface, accessed 21 March 2008.

Alarcón, Enrique. "Panorama General de las Regulaciones en Bioseguridad en Las Américas: versión preliminary," San José, IICA, January 2002.

Alarcón, Enrique & Rodrigo Artunduaga, "América Latina y el Caribe: Nuevas agrobiotecnologías, desafíos, tendencias y consideraciones institucionales" (2006) 2 Comuniica Online, online: www.iica.org.uy/data/documentos/89876.pdf, accessed 18 January 2008.

Alarcón, Enrique & Rodrigo Artunduaga. "Nuevas Agrobiotecnologías: algunas consideraciones institucionales para su desarrollo, utilización y comercialización," February 2005.

Alblin, Eduardo. "Hacia la trazabilidad en el mercado mundial de soya: una nueva mirada a la Ley de la oferta y la Demanda," Cancillería Argentina, Buenos Aires, 2001.

Ali Mekour, Mohamed. "A global instrument on agrobiodiversity: the International Treaty on Plant Genetic Resources for Food and Agriculture," Environmental Policy and Law, FAO, Rome, 2002.

Andow, DA., et al. "Ecological Risk Assessment for Bt crops" (2006) 24 Nature Biotechnology 749.

Andrée, Peter & Lucy Sharratt, *Genetically Modified Organisms and Precaution: Is the Canadian Government Implementing the Royal Society of Canada's Recommendations?* (Ottawa: The Polaris Institute, 2004).

Ang, Marybelle. "A Slow Motion Tennis Match: GMO Debate Continues" (2010) January American Bar Association Section on Environment, Energy & Resources Agricultural Management Committee Newsletter, online: http://www.abanet.org/environ/committees/agricult/newsletter/archiveslist.html (last accessed October 3, 2010).

Argentina, Draft National Biosafety Framework, 2004, (BCH Record No. 43463).

Australia, Commonwealth, Office of the Gene Technology Regulator. Risk Analysis Framework for License Applications to the Office of the Gene Technology Regulator (Handbook, 2005).

Azmi & Associates, "Biotech: Natural Resources, New Economy," Asia Law (October 2003) online: http://www.asialaw.com/Article/1972413/Channel/16961/Biotech-Natural-Resources-New-Economy.html.

Bahamas, Draft National Biosafety Framework, 2004, (BCH Record No. 43464).

Baker, G.A. & T.A Burnham, "Consumer response to genetically modified foods: market segment analysis and implications for producers and policy makers" (2001) 26 J. Agric. Resour. Econ 387.

Bangladesh, Biosafety Guidelines of Bangladesh (Ministry of Environment and Forests, Government of the Peoples Republic of Bangladesh. 2005).

Bangladesh, Draft National Biosafety Framework, 2006 (BCH Record No. 43436).

Barnum, Susan R. *Biotechnology: An Introduction* (Scarborough, UK: Wadsworth Publishing Company, 1998).

Baumueller, H., "Domestic Import Regulations for Genetically Modified Organisms and their Compatibility with WTO Rules" (IISD/ICTSD Trade Knowledge Network 2003).

Becker, Barbara, *Sustainability Assessment: A Review of Values, Concepts and Methodological Approaches* (Washington, DC: Consultative Group on International Agricultural Research, 1997).

Berg, P., et al., "Potential biohazards of recombinant DNA molecules" (1974) 185 Science 303. Online: http://www.sciencemag.org/cgi/issue_pdf/toc_pdf/185/4148.pdf.

Bergmans, H., "Basic Framework for Risk Assessment for Transgenic Plants Developed by the OECD: 20 Years After the OECD Blue Book" (2006) 5 Environmental Biosafety Research 213.

Binnie, Patricia W., Alan E. Boyle, & Catherine Redgwell. *International Law and the Environment*, 3rd ed. (Oxford: Oxford University Press, 2009).

Birner, Regina, et al. "Biofortified Foods and Crops in West Africa: Mali and Burkina Faso" (2007) 10(3) AgBioForum 192.

Biosafety: Building further consensus for action (CBD Secretariat, 2007).

Boisson de Chazournes, L. & U.P. Thomas, "WTO Law, Science and Risk Communication" (2007) Special Edition 2006 of Journal of Trade and Environment, online: http://www.ecolomics-international.org/.

Bouët, Antoine, Guillaume Gruère, & Laetitia Leroy, "From 'May Contain' to 'Does Contain': The price and trade effects of strict information requirements for GM maize under the Cartagena Protocol on Biosafety," paper for Agricultural & Applied Economics Association 2010 (25–27 July, 2010).

Brañes, Raún & Orlando Rey, "Política, Derecho y Administración de la Bioseguridad en América Latina y el Caribe," CEPAL, Chile, 2001.

Brundtland, G. *Our Common Future: World Commission on Environment and Development* (Oxford: Oxford University Press, 1987), 8.

Burachik, Moisés & Patricia L. Traynor, "An analysis of a national biosafety system: regulatory policies and procedures in Argentina," ISNAR Country Report No. 63, April 2002.

Burgiel, Stanley W., "The Cartagena Protocol on Biosafety: Taking the Steps from Negotiation to Implementation" (2002) 11(1) Review of European Community and International Environmental Law 53.

Burton, G.A., et al., "Weight-of-Evidence Approaches for Assessing Ecosystem Impairment" (2002) 8 Journal of Human and Ecological Risk Assessment 1657.

Cabrera, Jorge, "A comparative analysis on ABS laws and policies: critical issues for implementation and interpretation" (Bonn: IUCN Environmental Law Centre 2004).

Cabrera Medaglia, Jorge, "Comercio Internacional Agrícola de Organismos Genéticamente Modificados: entre la bioseguridad y el libre intercambio," in Justicia Agraria y Ambiental en América, CADA, San José, 1998.

Cabrera Medaglia, Jorge & Kathryn Garforth, "Sustainable International Biodiversity Law" at 349 in Marie-Claire Cordonier Segger & Ashfaq Khalfan, eds. *Sustainable Development Law: Principles, Practices and Prospects* (Oxford: Oxford University Press, 2004).

Cambodia, National Biosafety Framework, 2004, (BCH Record No. 43438).

Canada, "Canada's Agriculture, Food and Beverage Industry; Canada's Seed Industry" (23 December 2003), online: Agriculture and Agri-Food Canada, http://www.sea.agr.gc.ca/supply/3319_e.htm.

Canada, Canadian Agency for Food Inspection, "Regulating Agricultural Biotechnology," online: http://cbac-cccb.ca/epic/site/cbac-cccb.nsf/en/Home, accessed 20 January 2008.

Canada, Canadian Agency for Food Inspection, Regulation of Agricultural Biotechnology in Canada (Her Majesty the Queen in Right of Canada, 2007) online: www.inspection.gc.ca/english/sci/biotech/gen/educ_postsece.pdf, accessed 20 January 2008.

Canada, Government of Canada, "BioPortal" online: http://www.bioportal.gc.ca/english/BioPortalHome.asp?x=1, accessed 20 January 2008.

Canada, Government of Canada, "A Framework for the Application of Precaution in Science-Based Decision Making about Risk," online: Privy Council Office, http://www.bcp-pco.gc.ca/docs/information/publications/precaution/precaution_e.pdf.

Carter, Colin & Guillaume Gruere, "International Approaches to the Labeling of Genetically Modified Food" (March 2003), online: http://www.agmrc.org/media/cms/cartergruere_16D28F0207B6C.pdf (accessed December 24, 2009).

CGIAR ISPC. Product Stewardship and Liability in the Context of IPR: Report of a Study (Rome, Italy: ISPC Secretariat, 2010).

Chaffin, Joshua. "Farmers face animal feed shortage," Financial Times (September 7, 2009), online: http://www.ft.com/cms/s/0/6983a01a-9bd7-11de-b214-00144feabdc0,dwp_uuid=70662e7c-3027-11da-ba9f-00000e2511c8.html?nclick_check=1.

Chapman, P.M., et al., "Weight-of-Evidence Issues and Frameworks for Sediment Quality (and Other) Assessments" (2002) 8 Journal of Human and Ecological Risk Assessment 148.

COCERAL Press Release, "Concern Over Viability of EU Livestock Sector" (13 October 2008).

Cohen, Joel, "Poorer nations turn to public developed GM Crops" (2005) 23(1) Nature Biotechnology."

Cohen, J., J. Komen, & P. Zambrano, "Public GM Innovations, Safety Regulation, and... Results?" (Presentation delivered to The Globalization of Agricultural Biotechnology: Multi-disciplinary Views from the South, University of Warwick, Coventry, U.K., 11–13 March 2005).

Conference of the Parties Serving as the Meeting of the Parties of the Cartagena Protocol on Biosafety (COP-MOP-1), Report of COP MOP-1, (UNEP/CBD/BS/COP-MOP/1/15, 27 February 2004).

Consent Stakeholder Consultative and Awareness Initiative, First Biotechnology and Biosafety Stakeholders Dialogue Report (Report of 18 June 2003) (Kampala, Uganda: Consumer Education Trust: 2003).

Consumers Union of Japan, "Getting Ready for Nagoya" (15 September, 2010), online: http://www.nishoren.org/en/.

Cordonier Segger, Marie-Claire and Ashfaq Khalfan, eds. *Sustainable Development Law: Principles, Practices and Prospects* (Oxford: Oxford University Press, 2004)

Cordonier Segger, Marie-Claire, "Sustainable Development in International Law," in David Armstrong, ed. *Routledge Handbook of International Law* (New York: Routledge, 2008), 255.

Cordonier Segger, Marie-Claire, "The Role of International Forums in the Advancement of Sustainable Development" (2009) 10(1) Sustainable Development Law & Policy.

Cordonier Segger, Marie-Claire, Ashfaq Khalfan, & Salim A. Nakhjavani, *Weaving the Rules for Our Common Future: Principles, Practices and Prospects for International Sustainable Development Law* (Montreal: CISDL, 2002).

deBeer, Jeremy, "Reconciling Property Rights in Plants" (2005) 8 Journal of World Intellectual Property 5.

deBeer, Jeremy, "Rights and Responsibilities in the Agricultural Biotechnology Industry" (2006) [unpublished].

deBeer, Jeremy, "The Rights and Responsibilities of Biotech Patent Owners" (2007) 40 University of British Columbia Law Review 343.

Delmer, Deborah P., "Agriculture in the Developing World: Connecting Innovations in Plant Research to Downstream Applications" (2005) 103(44) Proceedings of the National Academy of Sciences 15739 at 15745.

Dominican Republic, Decisisión para la introducción de pez cebra genéticamente modificado (BCH Record No 43828).

Drahos, Peter & John Braithwaite, *Information Feudalism: Who Owns the Knowledge Economy?* (New York: New Press, 2002).

Dunlop, Claire, "GMOs and Regulatory Styles" (2000) 9(2) Environmental Politics 149.

The East African Community, Environmental Assessment Guidelines for Shared Ecosystems in East Africa (Arusha, Tanzania: East African Community, 2005), online: http://www.eac.int/environment/index.php?option=com_docman&task=cat_view&gid=66&Itemid=106.

The East African Community, Harmonized Sanitary and Phytosanitary Standards, Measures, and Procedures (Arusha, Tanzania: The East African Community, 2004).

Ebener S., et al., "Knowledge mapping as a technique to support knowledge translation" (2006) 84(8) Bulletin of the World Health Organization 636.

Egziabher, Tewolde, "Ethiopia" in C. Bail, R. Falkner, & H. Marquard (eds.), *The Cartagena Protocol on Biosafety: Reconciling Trade in Biotechnology with Environment and Development?* (London: Earthscan, 2002) 115.

Environmental Rights Action/Friends of the Earth Nigeria. "Genetically Modified Crops: A Challenge for Africa" (Nigeria 2005), online: Biosafety Africa http://www.biosafetyafrica.net, accessed 20 October 2009.

Ethiopia, National Biosafety Framework 2007 (BCH Record No. 43387).

Endicott, Martin, Christine Frison, and Kathryn Garforth with Marie-Claire Cordonier Segger, Jorge Cabrera, and Sylvestre Manga, *Innovations in Biosafety Law: A CISDL Working Paper* (Montreal: Centre for International Sustainable Development Law, 2005).

European Mycotoxin Awareness Network, Fact Sheets on HACCP – Prevention and Control, available at services.leatherheadfood.com/mycotoxins/index.asp (accessed October 3, 2010).

Executive Secretary of the Convention on Biological Diversity, "Public Awareness and Participation: Interim report on the status of implementation of public awareness, education and participation concerning the safe transfer, handling and use of living modified organisms (Article 23, paragraph 1[a])," UN Doc. UNEP/CBD/BS/COP-MOP/4/16 (3 March 2008).

Expert Group Workshop on International Environmental Law, Final Report of the Expert Group Workshop on International Environmental Law aiming at Sustainable Development, UN doc. UNEP/IEL/WS/3/2 (1996).

Falck-Zepeda, José B., "Socio-economic Considerations, Article 26.1 of the Cartagena Protocol on Biosafety: What Are the Issues and What Is at Stake?" 12(1) AgBioForum 90.

Falkner, Robert, "Negotiating the Cartagena Protocol: The International Process" in C. Bail, R. Falkner, & H. Marquard (eds.), *The Cartagena Protocol on Bioafety: Reconciling Trade in Biotechnology with Environment and Development?* (London: Earthscan, 2002) 3.

FEFAC Press Release, "EU feed and food price inflation on the rise" (3 August 2010), http://www.fefac.org/file.pdf?FileID=30415.

FIELD, *The Cartagena Protocol on Biosafety: A Negotiating History* (Montreal: Secretariat of the Convention on Biological Diversity, 2003).

Final Report of the Expert Group Workshop on International Environmental Law Aiming at Sustainable Development, UN Doc. UNEP/IEL/WS/3/2 (1996).

Footer, Mary, "Post-Normal Science, in the Multilateral Trading System: Social Science Expertise and the EC-Biotech Panel," (2007) 6:2 World Trade Review 281.

Fransen, L et al., *Integrating Socio-Economic Considerations into Biosafety Decisions* (Washington, DC: World Resources Institute 2005).

Friends of the Earth, "Playing with Hunger" (Friends of the Earth International 2003), online: http://www.foei.org/en/publications/pdfs/food-sovereignty/2000–2007/playing_with_hunger2.pdf, accessed 20 October 2009.

Frison, Christine, "The Principles of Sustainable Development in the Context of the Implementation of the International Treaty on Plant Genetic Resources for Food and Agriculture" (2006) 2 JSDLP.

Frison, Christine, et al., "Intellectual Property and Facilitated Access to Genetic Resources under the International Treaty on Plant Genetic Resources for Food and Agriculture" (2010) 1 European Intellectual Property Review, online: http://papers.ssrn.com/sol3/papers.cfm?abstract_id=1576823.

Fukuda-Parr, Sakiko, *The Gene Revolution: GM Crops and Unequal Development* (London: Earthscan, 2007).

Garcia-Alonso, M., et al., "A Tiered System for Assessing the Risk of Genetically Modified Plants to Non-target Organisms" (2006) 5 Environmental Biosafety Research 57.

Garforth, K., "Background on Biosafety and the Cartagena Protocol" (CISDL Working Paper June 2005).

Garforth, Kathryn, "Balancing Industry Confidentiality with the Public Right of Access: The Case of Biotechnology in Canada" (CISDL Working Paper, September 2005), online: Forum on Privatization and the Public Domain http://www.forumonpublicdomain.ca/files/CISDL_WP%20Access%20to%20Info.pdf.

Garforth, Kathryn, "When Biosafety Becomes Binding" in Marie-Claire Cordonier Segger and C.G. Weeramantry (eds.), *Sustainable Justice: Reconciling Economic, Social and Environmental Law* (Boston: Martinus Nijhoff Publishers, 2005).

Garforth, Kathryn, et al., "An overview of the national and regional implementation of measures on access to genetic resources and benefit sharing" (Montreal: Centre for International Sustainable Development Law, 2005).

Garforth, Kathryn & Christine Frison, "Occasional Paper 2, Key Issues for the relationship between the Convention on Biological Diversity & the International Treaty on Plant Genetic Resources for Food and Agriculture" (2007) July, online: QIAP website, http://www.qiap.ca/pages/documents/OP2-Final_000.pdf.

Gaskell, George, et al., "Europeans and Biotechnology in 2005: Patterns and Trends" (Eurobarometer 64.3, May 2006) 3.

GEF Council, Initial Strategy for Assisting Countries to Prepare for the Entry into Force of the Cartagena Protocol on Biosafety, November 1–3, 2000, GEF/C.16/4/Rev.1. Online: http://www.unep.org/biosafety/Documents/GEF_strategy.pdf.

Glover, Dominic, et al., "Public Participation and the Cartagena Protocol on Biosafety," Part I: The Main Report, Institute of Development Studies 30. Online: http://www.ids.ac.uk/UserFiles/File/knots_team/NBFreport_Main.pdf.

"GM canola spreading: Agriculture Canada report" CBC News (28 June 2002), online: CBC News: http://www.cbc.ca/story/science/national/2002/06/27/gncanola020627.html.

Government of Austria, "Stöger: Gen-Importverbote unterzeichnet," press release (Federal Ministry of Health: Vienna Austria, 2010) online: http://www.ots.at/presseaussendung/OTS_20100928_OTS0017/stoeger-gen-importverbote-unterzeichnet.

Gouvernement du Burkina Faso, Cadre National Pour la Prevention Des Riques Biotechnologiques au Burkina Faso, (Ouagadougou, Burkina Faso: Comité National de Biosécurité, 2005) online: http://www.unep.org/biosafety/files/BFNBFrep.pdf.

Greenpeace, "Genetically Engineered Canola Contamination across Japan" (May 24, 2005), online: http://www.greenpeace.or.jp/campaign/gm/imgindex_e_html, accessed December 11, 2006.

Hallman, William K., et al., "Public Perceptions of Genetically Modified Foods Americans Know Not What They Eat" (Food Policy Institute, Rutgers, The State University of New Jersey, RR-0302–001, 2002) available online at http://www.foodpolicyinstitute.org.

Hallman, W. K., et al., Public Perceptions of Genetically Modified Foods: A National Study of American Knowledge and Opinion (Publication number RR-1003–004) (Food Policy Institute, 2003).

Harsh, M., "Formal and Informal Governance of Agricultural Biotechnology in Kenya: Participation and Accountability in Controversy Surrounding the Draft Biosafety Bill" (2005) 17 J Int. Dev. 661.

Hårstad, J., T. Young, J. McNeely, D. MacKenzie, D. Brann, J. Morris, & H.R. Macauley, *Evaluation of GEF Support for Biosafety* (Washington, DC: GEF Evaluation Office, 2006), online: http://www.thegef.org/gef/sites/thegef.org/files/documents/Biodiversity-Biosafety-2006.pdf.

Heim, Kristi, "Gates Foundation's Agriculture Aid a Hard Sell" Seattle Times (20 Jan. 2008) online: www.seattletimes.nwsource.com/html/local.

Heinzerling, Lisa, "Climate Change, Human Health, and the Post-Cautionary Principle" (2007) Georgetown Law Journal Research Paper No. 4/2007.

Hervé, Dominique, "Un análisis del marco regulatorio de la bioseguridad en Chile," Centro de Derecho Ambiental de la Universidad de Chile, Santiago, 2003.

Hilbeck, A. and D.A. Andow (eds.), *Environmental Risk Assessment of Genetically Modified Organisms vol. 1: A Case Study of Bt Maize in Kenya* (Wallingford: CABI Publishing, 2004).

Hilbeck, A., et al. (eds.), *Environmental Risk Assessment of Genetically Modified Organisms vol. 2: A Case Study of Bt Cotton in Brazil* (Wallingford: CABI Publishing, 2006).

Hill, Ryan A., "Conceptualizing Risk Assessment Methodology for Genetically Modified Organisms" (2005) 4 Environmental Biosafety Research 67.

Hill, Ryan A. & Cyrie Sendashonga, "General Principles for Risk Assessment of Living Modified Organisms: Lessons from Chemical Risk Assessment" (2003) 2 Environmental Biosafety Research 81.
Hill, Ryan A. & Cyrie Sendashonga, Guidelines for Ecological Risk Assessment (United States Environmental Protection Agency, EPA/630/R-95/002F, April 1998), available online: http://www.epa.gov/lawsregs/guidance/bytopic-e.html#ecosystems, accessed 21 March 2008.
Hill, R. et al., "A Weight-of-Evidence Approach for Evaluating Ecological Risks: Report of the Massachusetts Weight-of-Evidence Workgroup" (1996) 2 Journal of Human and Ecological Risk Assessment 277.
Hill, R. et al., "Level of Detail in Ecological Risk Assessment" (2000) 40 Marine Pollution Bulletin 471.
Hill, Ryan, et al., "Risk Assessment and Precaution in the Cartagena Protocol" (2004) 13 RECIEL 263.
Howse, Robert & Henrik Horn, "European Communities – Measures Affecting the Approval and Marketing of Biotech Products" (2009) 8:1 World Trade Review 49.
Huang J., D. Zhang, J. Yang, S. Rozelle, & N. Kalaitzandonakes, "Will the Cartagena Protocol hinder or protect the developing world: Learning from China's experience" (2008) 33 (1) Food Policy 1.
Hull, R.N. and S. Swanson, "Sequential Analysis of Lines of Evidence – an Advanced Weight-of-Evidence Approach for Ecological Risk Assessment" (2006) 2 Integrated Environmental Assessment and Management 302.
ICTSD, "Between Trade and Sustainable Development," Bridges, No. 5, June 1999.
ICTSD, "Between Trade and Sustainable Development," Bridges No. 9, November-December 2000.
ICTSD, "Malaysian Biosafety Act in Force," Crop Biotech Update, 4 December 2009, online: http://ictsd.org/i/news/biores/65569/.
ICTSD, "News, events and resources at the intersection of trade and biodiversity", 6:6 Bridges Trade BioRes (3 April 2006), online: http://www.ictsd.org/biores/index.com.
Intergovernmental Committee for the Cartagena Protocol on Biosafety, Plenary Report of ICCP 1 (UNEP/CBD/ICCP/1/9, 19 December 2000).
Intergovernmental Committee for the Cartagena Protocol on Biosafety, Note of the Executive Secretary. Synthesis of Views Regarding Elements and Options for a Compliance Regime, (UNEP/CBD/ICCP/2/13, 17 July 2001).
Intergovernmental Committee for the Cartagena Protocol on Biosafety, Report of the Open-Ended Meeting of Experts on a Compliance Regime (UNEP/CBD/ICCP/2/13 Add 1, 28 September 2001).
Intergovernmental Committee for the Cartagena Protocol on Biosafety, Report of ICCP 2 (UNEP/CBD/ICCP/2/15, 15 October 2001).
Intergovernmental Committee for the Cartagena Protocol on Biosafety, Report of ICCP 3 (UNEP/CBD/ICCP/3/10, 26 April 2002).
International Institute for Sustainable Development (IISD), Earth Negotiations Bulletin, IISD, ENB Vol 9 No. 298, online: http://www.iisd.ca/vol09/enb09298e.html.
International Institute for Sustainable Development (IISD), Earth Negotiations Bulletin, IISD, ENB No. 9, Vol. 320. Online: http://www.iisd.ca/vol09/enb09320e.html.
International Institute for Sustainable Development (IISD), Earth Negotiations Bulletin, IISD, ENB No. 9, Vol. 351. Online: http://www.iisd.ca/vol09/enb09351e.html.
International Law Association, "International Law on Sustainable Development" in *Report of the Seventy-Second Conference: Toronto* (London: International Law Association, 2006) 467.

International Law Association, "International Law on Sustainable Development" in *Report of the Seventy-Third Conference: Rio de Janeiro* (London: International Law Association, 2008) 895.

International Monetary Fund, Burkina Faso: Poverty Reduction Strategy Paper – Annual Progress Report (IMF Country Report No. 08/212) online: http://www.imf.org/external/pubs/ft/scr/2008/cr08212.pdf.

IPPC Secretariat, "ISPM No. 11: Pest Risk Analysis for Quarantine Pests Including Analysis of Environmental Risks and Living Modified Organisms" (International Standard for Phytosanitary Measures, 2004), online: https://www.ippc.int/servlet/CDSServlet?status=ND0xMzM5OSY2PWVuJjMzPSomMzc9a29z, accessed 21 March 2008.

Jacobs, M., "Sustainable Development as a Contested Concept" in A. Dobson (ed.), *Fairness and Futurity: Essays on Environmental Sustainability and Social Justice* (Oxford: Oxford University Press, 1999) 35.

Jaffe, Gregory, "Implementing the Cartagena Biosafety Protocol through National Biosafety Regulatory Systems: An Analysis of Key Unresolved Issues" (2005) 5 Journal of Public Affairs 299.

Jaffe, Gregory, "Regulating Transgenic Crops: A Comparative Analysis of Different Regulatory Processes" (2004) 13 Transgenic Research 5.

James, Clive, ISAAA Global status of commercialized Biotech/GM crops 2008 (International Service for the Acquisition of Agri-biotech Applications, 2009).

Kalaitzandokes, Nicholas, "The Potential Impacts of the Biosafety Protocol on Agricultural Commodity Trade," International Food and Agricultural Trade Policy Council, Issue Brief, December 2004.

Kalibwani, F., et al. "African Policy Dialogues on Biotechnology – Southern Africa" (Harare, Zimbabwe, September 20–21, 2004).

Kalibwani, F., et al. *Governing Biotechnology in Africa: Toward Consensus on Key Issues in Biosafety NEPAD*, (Johannesburg, South Africa; Harare, Zimbabwe; and Washington, DC: FANRPAN/IFPRI, 2004).

Kenya Council for Science and Technology. Implementation of the National Biosafety Framework for Kenya, Proceedings of the UNEP-GEP Workshop, 14th-18th April 2003 (Nairobi, Kenya: UNEP-GEF, 2003).

Kershen, Drew, "Health and Food Safety: The Benefits of Bt-corn", (2006) 61 Food Drug L. J. 197.

Kilman, Scott, "McDonald's, Other Fast-Food Chains Pull Monsanto's Bio-Engineered Potato," Wall Street Journal, April 28, 2000, p. B4.

Kimani, Virginia & Guillaume Gruere, Implications of Import Regulations Information Requirements under the Cartagena Protocol on Biosafety for GM Commodities in Kenya (April 2010).

Kinderlerer, Julian, "Regulation of Biotechnology: needs and burdens for developing countries" (UNEP-GEF Biosafety Projects: 2002), online: UNEP-GEF http://www.unep.org/biosafety/Documents/BTregulationJK.pdf.

Kinderlerer, Julian & M. Adcock, "Agricultural Biotechnology, Politics, Ethics, and Policy" in Steven Were Omamo and Klaus von Grebmer (eds.), *Biotechnology, Agriculture, and Food Security in Southern Africa* (Washington, DC: Food, Agriculture, and Natural Resources Policy Network, 2005).

Kiss, A. & D. Shelton, *International Environmental Law*, 2nd ed. (New York: Transnational Publishers, 1994).

Kloppenburg Jr., Jack Ralph, *First the Seed: The Political Economy of Plant Biotechnology*, 2d ed. (Madison: University of Wisconsin Press, 2004).

Koester, Veit, "Review of Compliance under The Aarhus Convention: A Rather Unique Compliance Mechanism" (2005) 2(1) Journal for European Environmental and Planning Law 31.

Koester, V., "The Convention on Access to Information Public Participation in Decision-Making and Access to Justice in Environmental Matters (Aarhus Convention)," in G. Ulfstein et al. (eds.), *Making Treaties Work: Human Rights, Environmental and Arms Control* (Cambridge University Press, 2007) 179.

Koester, V., "The Compliance Committee of the Aarhus Convention: An Overview of Procedures and Jurisprudence" (2007) 37:2–3 Environmental Policy and Law 83.

Koester, V., "The Cartagena Protocol: A New Hotspot in the Trade–Environment Conflict" (2001) 31(2) Environmental Policy and Law 82.

Koester, V., "Some Insider Reflections on the Compliance Mechanism under the Cartagena Protocol on Biosafety" (2008) 3:5 Cartagena Protocol News 5, online: http://www.cbd.int/doc/newsletters.

Kogan, Lawrence A., "World Trade Organization Biotech Decision Clarifies Central Role of Science in Evaluating Health and Environmental Risks for Regulation Purposes" (2007) 2:3 Global Trade and Customs Journal 149, online: http://www.itssd.org/Publications/GTCJ_04-offprints_Kogan%5B2%5D.pdf.

Kwan, Chris, "Malaysia's Biosafety Bill – Throwing Precaution to the Wind?" (2007) 5 Current Law Journal, online: http://papers.ssrn.com/sol3/papers.cfm?abstract_id=1016008, accessed September 17, 2010.

La Viña, A., and L. Fransen, *Integrating Socio-Economic Considerations into Biosafety Decisions: The Challenge for Asia* (Washington DC: World Resources Institute, 2004).

Lang, A. et al., "Early-tier Tests Insufficient for GMO Risk Assessment" (2007) 25 Nature Biotechnology 35.

Lapeña, Isabel, "Genéticamente modificado: principio precautorio y derechos del consumidor en el Perú," Lima, September 2004.

Lean, M.A. et al., A conceptual framework for implementing biosafety: linking policy, capacity and regulation (ISNAR), www.isnar.cgiar.org, accessed 3 February 2008.

Lefeber, René, *Transboundary Environmental Interference and the Origin of State Liability* (Cambridge, MA: Kluwer Law International, 1996).

Lester, Simon. "WTO – Sanitary and Phytosanitary Measures Agreement-rules/exceptions-international law as interpretive tool" (2007) 101 Am. J. Intl L. 453.

Chee Yoke Ling & Lim Li Lin, "Public Participation in the Implementation of the Cartagena Protocol" (2005) Third World Network. Online: http://www.biosafety-info.net/pubart.php?pid=41&PHPSESSID=2046f6e8832cd06ce7f93bf68cfe1767.

Macdonald, P., and S. Yarrow, "Current Harmonization Activities Related to Risk/Safety Assessment by the OECD Working Group on Harmonization of Regulatory Oversight in Biotechnology" (2006) 5 Environmental Biosafety Research 223.

Macharia, H.K., "East Africa PBS Roundtable" (Presentation at PBS Roundtable in Entebbe, Uganda 18 April 2005).

Mackenzie, Ruth, et al., *An Explanatory Guide to the Cartagena Protocol on Biosafety* (Gland, Switzerland and Cambridge, UK: IUCN, 2003).

MacKenzie, Ruth, "The Cartagena Protocol after the First Meeting of the Parties" (2004) 13(3) Review of European Community and International Environmental Law 270.

Marris, Claire et al., *Report on Public Perceptions of Agricultural Biotechnologies in Europe* (Lancaster: Centre for the Study of Environmental Change, Lancaster University, 2001).

Mayet, M., "GM Food Aid: Africa Denied Choice Once Again?" (2004) Africa Center for Biosafety and others 7, online: BioSafety Africa http://www.biosafetyafrica.net, accessed 19 October 2009.

McCaffrey, Stephen, "Biotechnology: Some Issues of General International Law" (2001) 14 Transnational Lawyer 91.

McLean, M.A. et al., *A conceptual framework for implementing biosafety: Linking policy, capacity, and regulation* (Netherlands: ISNAR, 2002), online at www.isnar.cgiar.org.

McLeod-Kilmurray, Heather. "Hoffman v. Monsanto: Courts, Class Actions, and Perceptions of the Problem of GM Drift" (2007) 27 Bulletin of Science, Technology & Society 188.

Mehta-Bhatt, P., R.V. Ebora, J.I. Cohen, J.R. Zepeda, and P. Zambrano, "An overview of Regulation, Perceptions and Priorities for GM Crops in Asia" (2005) Asian Biotechnology and Development Review 7(3).

Minde, I.J., *Facilitating Regional Agricultural Trade for Increased Food Security and Incomes: The Case of Rationalization and Harmonization of Seed Policies, Laws, Rules and Procedures in Eastern Africa* (Entebbe, Uganda: ASARECA-ECAPAPA, 2004).

Missmer, Stacey A., et al., "Exposure to Fumonisins and the Occurrence of Neural Tube Defects along the Texas-Mexico Border" (2006) 114 Environ Health Perspect 237. Available at ehp.niehs.nih.gov/docs/2005/8221/abstract.html, accessed March 2, 2006.

Monsanto, "2006 Monsanto Technology/Stewardship Agreement," online: Farmsource: Monsanto US Ag Resource, http://www.farmsource.com/images/pdf/2006%20EMTA%20Rev3.pdf.

Morgan, David & Gavin Goh, "Genetically Modified Food Labelling and the WTO Agreements," (2004) 13(3) Review of European Community & International Environmental Law 306.

Nampala, P., C. Mugoya, & T. Sengooba, "Biosafety Regulatory System in Uganda" (presentation at PBS Roundtable in Entebbe, Uganda, 2005).

National Foreign Trade Council Inc., "EU regulation, standardization and the precautionary principle," Washington, DC, August 2003.

National Research Council, *Committee on Risk Assessment Methodology, Understanding Risk – Informing Decisions in a Democratic Society* (Washington, DC: National Academy Press, 1996).

Nickson, T.E. and M.J. McGee, "Ecological Assessment of Crops Derived through Biotechnology" in John A. Thomas and Roy L. Fuchs, (eds.), *Biotechnology and Safety Assessment*, 3rd ed. (Oxford: Elsevier, 2002).

Nuffield Council on Bioethics, *Genetically modified crops: the ethical and social issues* (London: Nuffield Council on Bioethics, 1999).

OAPF, Letter from the OAPF Committee (13 December 2007), online: Saskatchewan Organic Directorate – Organic Agriculture Protection Fund http://www.saskorganic.com/oapf/pdf/13Dec07-letter.pdf.

Oberthür, Sebastian and Thomas Gehring, "Institutional interaction in global environmental governance: The case of the Cartagena Protocol and the World Trade Organization" (2006) 6(2) Global Environmental Politics 1.

OECD, Environmental Directorate, Guidance Document on the Use of Taxonomy in Risk Assessment of Micro-organisms: Bacteria Doc. No. ENV/JM/MONO(2003)13 (2003), online: http://www.olis.oecd.org/olis/2003doc.nsf/LinkTo/NT000042BE/$FILE/JT00148130.PDF, accessed 21 March 2008.

Oliva, María Julia, "Science and Precaution in the GMO dispute: a brief analysis of the first US Submission," in Bridges, No. 5, May 2004.

Paarlberg, Robert, "Principios para dirigir la revolución causada por los cultivos GM: opciones de política para los países en desarrollo," Visión 2020, Resumen, No. 68, December 2000.

Paarlberg, Robert L., "Governing the GM Crop Revolution: Policy Choices for Developing Countries" (Food Agriculture and the Environment Discussion Paper 33, IFPGRI, 2000) 251.

Pearce, F., "UN is slipping modified food into aid" (2002) New Scientist 175: 2361.

Perez, Oren, "Anomalies at the Precautionary Kingdom: Reflections on the GMO Panel's Decision" (2007) 6:2 World Trade Review 265.

Peterson, R.K.D., et al., "Genetically Engineered Plants, Endangered Species, and Risk: a Temporal and Spatial Exposure Assessment for Karner Blue Butterfly Larvae and Bt Maize Pollen" (2006) 26 Risk Analysis 845.

Pew Initiative on Food and Biotechnology, "Issues in the regulation of genetically engineered plants and animals" (Washington, DC: Pew Initiative on Food and Biotechnology, 2004).

Phillipson, Martin, "Giving Away the Farm? The Rights and Obligations of Biotechnology Multinationals: Canadian Developments" (2005) 16 The King's College Law Journal 362.

PNUMA, "Manual Preliminar sobre el cumplimiento y la observancia de los Acuerdos Multilaterales Ambientales" (Nairobi: PNUMA, 2004).

Policy Planning Division, Department of Food Safety, Pharmaceutical and Food Safety Bureau, MHLW, "Mandatory Labeling of Genetically Modified Foods and Foods Containing Allergens," online: http://www.mhlw.go.jp/english/topics/qa/gm-food/gm4.html.

Public Research and Regulation Initiative, "Guide for Notifications and Risk Assessments for Releases into the Environment of Genetically Modified Organisms" (Draft, 9 November 2005), online: http://pubresreg.org/index.php?option=com_content&task=blogcategory&Id=29&Itemid=40, accessed 21 March 2008.

Rao, S.R., "Issues in Capacity Building: an Overview" (2005) 7(3) Asian Biotechnology and Development Review 1.

Raybould, Alan, "Problem Formulation and Hypothesis Testing for Environmental Risk Assessments of Genetically Modified Crops" (2006) 5 Environmental Biosafety Research 119.

Raybould, A. & I. Cooper, "Tiered Tests to Assess the Environmental Risk of Fitness Changes in Hybrids Between Transgenic Crops and Wild Relatives: the Example of Virus Resistant Brassica Napus" (2005) 4 Environmental Biosafety Research 127.

Redgwell, Catherine, "Biotechnology, Biodiverisity and Sustainable Development" in Francesco Francioni and Tullio Scovazzi (eds.), *Biotechnology and International Law* (Oxford: Hart Publishing Ltd., 2006), 61.

Rey, Orlando, "Régimen Legal para la Seguridad de la Biotecnología en América Latina," December 2004.

Report of the Executive Secretary of the Convention on Biological Diversity, Implementation of United Nations environmental conventions (advance unedited copy), online: http://www.un.org/esa/dsd/resources/res_docugaecos_65.shtml.

Report of the Fifth Session of the Codex Ad Hoc Intergovernmental Task Force on Foods Derived from Biotechnology (July 2006), online: www.codexalimentarius.net/download/report/653/al29_34e.pdf.

Report of the United Nations Conference on Environment and Development, UN Doc. A/CONF.151/26/Rev.1 (Vol.1), (1992) 31 I.L.M. 874.

Republic of Benin, National Biosafety Framework and Action Plan for Benin (Cotonou, Benin: Ministry of the Environment, Housing and Urban Planning, 2005).

Republic of Côte d'Ivoire, Biosafety Framework (Abidjan, Côte d'Ivoire: Ministry of State/Ministry of the Environment: 2005), available at: http://www.unep.org/biosafety/files/CINBFrepEN.pdf.

Republic of Guinea-Bissau, National Biotechnology and Biosafety Framework of Guinea-Bissau (Bissau, Guinea-Bissau: Ministry of Natural Resources and Environment, 2008), online: http://www.unep.org/biosafety/Documents/NBFs/GB_NBF_En.pdf.

Republic of Kenya, National Guidelines for the Release of Genetically Modified Organisms (GMOs) into the Environment (Nairobi, Kenya: National Council for Science and Technology, 1998).

Republic of Korea, Draft National Biosafety Framework, 2004 (BCH Record No. 43444).

Republic of Mali, National Biosafety Framework Project (Bamako, Mali: Ministry of Environment and Sanitation, 2005), online: http://www.unep.org/biosafety/files/MLNBFrepEN.pdf.

Republique du Niger, Cadre National de Biosecurite (Niamey, Niger: Ministre de L'Hydraulique de L'Environnement et de la Lutte Contre la Desertification, 2005), online: http://www.unep.org/biosafety/files/NENBFrep.pdf.

Republique du Senegal, Cadre National de Biosecurite (Dakar Hann, Senegal: Ministre de L'Environnement et de la Protection de la Nature, 2005), online: http://www.unep.org/biosafety/files/SNNBFrep.pdf.

Republic of South Africa, Department of Agriculture, Forestry & Fisheries, Briefing of Parliamentary Committee on GMOs, online: www.pmg.org.za/files/docs/100504gmo.doc (last visited September 28, 2010).

Republic of Togo, National Biosafety Framework, (Lomé, Togo: Ministry of the Environment and Forest Resources, 2004), online: http://www.unep.org/biosafety/files/TGNBFrepEN.pdf.

Republic of Uganda, Guidelines on Biosafety in Biotechnology for Uganda 2002 (Kampala, Uganda: Uganda National Council for Science and Technology, 2002).

Republic of Uganda, Draft National Biotechnology and Biosafety Policy 2004 (Kampala, Uganda: Uganda National Council for Science and Technology, 2004).

Republic of Uganda, National Biotechnology and Biosafety Policy 2006 (Kampala, Uganda: Uganda National Council for Science and Technology, 2006).

Risk Assessment and Risk Management, CBD Dec., CBD COP-MOP/3, 2006, Doc. MOP BS-III/11, online: http://www.cbd.int/biosafety/cop-mop/result.aspx?id=11067, accessed 21 March 2008 [2]–[6].

Roberts, Martin, "U.S. soya shipments resume but more EU woes feared," Reuters Newswire (April 19, 2010).

Rodgers, Christopher P. "Liability for the Release of GMOs into the Environment: Exploring the Boundaries of Nuisance" (2003) 62 Cambridge Law Journal 371.

Romeis, J., et al.,"Reply to Early-tier Tests Insufficient for GMO Risk Assessment" (2007) 25 Nature Biotechnology 36.

Romeis, J., et al., "Reply to Ecological Risk Assessment for Bt Crops" (2006) 24 Nature Biotechnology 751.

Romeis, J., et al., "Transgenic Crops Expressing Bacillus Thuringiensis Toxins and Biological Control" (2006) 24 Nature Biotechnology 63.

The Royal Society of Canada, Elements of Precaution: Recommendations for the Regulation of Food Biotechnology in Canada (Ottawa: The Royal Society of Canada, 2001), online: Royal Society of Canada http://www.rsc.ca//files/publications/expert_panels/foodbiotechnology/GMreportEN.pdf.

Russo, E. "Reconsidering Asilomar – scientists see a much more complex modern day environment" (2000) 14(7) The Scientist 14, online: http://www.the-scientist.com/article/display/11789/.

Sahel and West African Club, Agricultural Biotechnology and the Transformation of West African Agriculture: Synthesis of the Regional Consultation with West African actors (Paris: Sahel and West African Club Secretariat, 2006).

Safyan, Susan, "Old MacDonald Had a Seed" (November 2005) 277 Alive Magazine, online: http://www.alive.com/3858a6a2.php?subject_bread_cramb=636.

Sands, Philippe, *Principles of International Environmental Law*, 2d ed. (Cambridge: Cambridge University Press, 2003).

Saskatchewan Organic Directorate – Organic Agriculture Protection Fund, Media release, "Individual action not the way to go: organic farmers" (16 April 2008), online: http://www.saskorganic.com/oapf/.

Saskatchewan Organic Directorate – Organic Agriculture Protection Fund, "Memorandum of argument" in the Supreme Court of Canada, file no. 32135, online: http://www.saskorganic.com/oapf/pdf/Argument-Jul-25–07.pdf.

Schaper, Marianne y Parada, Soledad, "Organismos Genéticamente Modificados: su impacto socioeconómico en la agricultura de los países de la Comunidad Andina," MERCOSUR and Chile.

Schoonejans, Eric, "Advanced Informed Agreement Procedures," in C. Bail, R. Falkner, & H. Marquard (eds.), *The Cartagena Protocol on Bioafety: Reconciling Trade in Biotechnology with Environment and Development?* (London: Earthscan, 2002) 299.

Sears, M.K. et al., "Impact of Bt Corn Pollen on Monarch Butterfly Populations: A Risk Assessment" (2001) 98 Proceedings of the National Academy of Sciences 11937.

Secretariat of the Convention on Biological Diversity, "Biosafety Information Resource Centre" (Database of Guidance materials), online: http://bch.cbd.int/database/resources/, accessed 21 March 2008.

Secretariat of the Convention on Biological Diversity, "Global Biosafety from concepts to action. Decisions from the First Meeting of the COP-MOP" (Montreal: SCBD, 2004).

Secretariat of the Convention on Biological Diversity, "Facing the Biosafety Challenge: Towards Effective Implementation of the Protocol: Decisions of the Second Meeting of the Conference of the Parties to the Convention on Biological Diversity Serving as the Meeting of the Parties to the Cartagena Protocol on Biosafety" (Montreal: Secretariat of the Convention on Biological Diversity, 2005).

Secretariat of the Convention on Biological Diversity, "Biosafety: Building Further Consensus for Action: Decisions of the Third Meeting of the Conference of the Parties to the Convention on Biological Diversity Serving as the Meeting of the Parties to the Cartagena Protocol on Biosafety" (Montreal: Secretariat of the Convention on Biological Diversity, 2007).

Secretariat of the Convention on Biological Diversity, "Biosafety: Taking Further Steps Towards Effective Implementation of the Protocol. Decisions of the Fourth Meeting of the Conference of the Parties to the Convention on Biological Diversity Serving as the Meeting of the Parties to the Cartagena Protocol on Biosafety" (Montreal: Secretariat of the Convention on Biological Diversity, 2008).

Secretariat of the Convention on Biological Diversity, "The Cartagena Protocol on Biosafety: A Record of the Negotiations" (Montreal: Secretariat of the Convention on Biological Diversity, 2003) at 30, online: http://bch.cbd.int/protocol/publications/a_record_of_negotiation_en.pdf, accessed 5 September 2010).

Seed Regional Working Group, Harmonization of Seed Policies and Regulations in Eastern Africa (Entebbe, Uganda: ASARECA-ECAPAPA, 2003).

Shaw, Malcolm N., *International Law*, 5th ed. (Cambridge: Cambridge University Press, 2003).

Sindico, Francesco, "The GMO dispute before the WTO: legal implications for trade and environment debate," Fondaziones Enrico Mattei, Note di Lavoro, 2005.

Sithole-Niang, I., J. Cohen, & P. Zambrano. "Putting GM technologies to work: public research pipelines in selected African countries" (2004) 3 African Journal of Biotechnology 564.

Stoll, Peter-Tobias, "Controlling the Risks of Genetically Modified Organisms: The Cartagena Protocol on Biosafety and the SPS Agreement" (1999) 10 Yearbook of International Environmental Law 82.

Tamale, Erie, "Building Capacities for the Effective Implementation of the Biosafety Protocol" Biosafety Protocol News (16 October 2006). Available online: Convention on Biological Diversity http://www.cbd.int/doc/newsletters/bpn/bpn-issue01h.pdf.

Tansey, G. & T. Rajotte, eds., *The Future Control of Food* (London: Earthscan, 2008).

Ten Kate, Kerry, et al., "The Undertaking Revisited: a commentary on the revision of the international undertaking on plant genetic resources for food and agriculture" (1997) 6(3) Review of European Community and International Environmental Law.

Thomas, U.P., "The CBD, the WTO, and the FAO: The Emergence of Phytogenetic Governance," in Philippe G. LePrestre, ed., *Governing Global Biodiversity: The Evolution and Implementation of the Convention on Biological Diversity* (Burlington, VT: Ashgate Publishing Co., 2002).

Tran, M., "Branded" (2005) Guardian, UK, online: http://blogs.guardian.co.uk/business insight/archives/2005/09/01/branded.html.

Traynor, P. & H.K. Macharia, *Analysis of the Biosafety System for Biotechnology in Kenya: Application of a Conceptual Framework*, ISNAR Country Report 6 (The Hague, The Netherlands: International Service for National Agricultural Research, 2003).

Traxler, Greg, "The GMO experience in North and South América – Where to from here," prepared for the Symposium on Biotechnology for a better World, Brisbane, Queensland, Australia, September 26 to October 1, 2004.

Traynor, P., and H.K. Macharia, Analysis of the Biosafety System for Biotechnology in Kenya: Application of a Conceptual Framework, ISNAR Country Report 6 (The Hague, The Netherlands: International Service for National Agricultural Research, 2003).

Treves, T. et al. (eds.), *Non-Compliance Procedures and Mechanisms, and the Effectiveness of International Environmental Agreements* (The Hague, Netherlands: T.M.C. Asser Press, 2009).

Trigo, Eduardo, et al. "Agricultural Biotechnology and rural development in Latin America and the Caribbean," Inter-American Development Bank, Sustainable Development Department, Technical Series, Washington, DC, 2000.

Trinidad and Tobago, "Initial Project Report as at Project Starting Date" (Report), online: http://www.unep.ch/biosafety/old_site/partcountries/TTcountrypage.htm, accessed 26 January 2008.

UNEP-GEF, Phase 3: tool kit module, Part (1), Developing the regulatory regimen, 2002.

UNEP, UNEP-GEF Biosafety Projects, What Is a National Biosafety Framework?, UNEP-Geneva.

UNEP, UNEP-GEF Biosafety Projects – Country Pages: Trinidad and Tobago, online: http://www.unep.ch/biosafety/old_site/partcountries/TTcountrypage.htm, accessed 26 January 2008.

UNEP-GEF Guide for the implementation of national biosafety frameworks, online: http://www.unep.org/biosafety/Documents/Impl.Guide-RegReg.pdf.

UNEP, Second Meeting of the Intergovernmental Committee for the Cartagena Protocol on Biosafety Nairobi, 1–5 October 2001, capacity building (articles 22, article 28), UNEP/CBD/ICCP/2/10, Annex II, 31 July 2001.

UNEP-GEF. Workshop on European experiences with LMOs, for developing country delegates (Bonn, Germany, 2004).

UNEP (Division of Global Environment Facility Coordination (GEF) – Biosafety Unit), "A Comparative Analysis of Experiences and Lessons" (December 2006).

UNEP, "International Technical Guidelines for Safety in Biotechnology" (1995), online: http://www.unep.org/biosafety/Documents/Techguidelines.pdf, accessed 5 May 2010).

United Nations, Report of the World Summit on Sustainable Development (New York: UN, 2002), online: http://www.johannesburgsummit.org (accessed 6 May 2010).

United Republic of Tanzania, The National Biosafety Framework for Tanzania, 2005 (Dar es Salaam, United Republic of Tanzania: Division of Environment, Vice President's Office, 2005).

United Republic of Tanzania, Schedule 18: Requirements for the Conduct of Confined Field Trials of Genetically Engineered Plants in Tanzania (Ministry of Agriculture and Food Security, Dar es Salaam, Tanzania: United Republic of Tanzania).

United Republic of Tanzania, Government Notice 265 published on July 24, 2009.

U.S. Department of Agriculture, Global Agriculture Information Network (GAIN) report VM8005, (17 January 2008), online: http://www.fas.usda.gov/gainfiles/200801/146293488.pdf.

U.S. Environmental Protection Agency, Biopesticides Registration Document – Bacillis thuringiensis Plant-Incorporated Protectants (Washington, DC: USEPA, 2001).

U.S. Food and Drug Administration, "Voluntary Labeling indicating whether foods have or have not been derived using bioengineering" (Draft Guidance, 2001), online: http://www.cfsan.fda.gov/~dms/biolabgu.html, accessed 20 January 2008.

U.S. National Biological Information Infrastructure, "United States Unified Regulatory Agencies Website," online: http://usbiotechreg.nbii.gov/roles.asp, accessed 20 January 2008.

U.S. Office of Science and Technology Policy, "Coordinated Framework for Regulation of Biotechnology" (Announcement of policy/Notice for Public Comment, 1986), online: http://usbiotechreg.nbii.gov/CoordinatedFrameworkForRegulationOfBiotechnology1986.pdf, accessed 20 January 2008.

Vail, E.F., "Knowledge Mapping: Getting Started with Knowledge Management" (1999) 16(4) Information Systems Management 16.

Vidal-Jules, Cillian, "Biosafety Act" (Final Report provided as part of the UNEP-GEF Global Project on the Development of NBFs) (May 2006), online: http://www.unep.ch/biosafety/development/Countryreports/LCNBFrep.pdf, accessed 26 January 2008.

Von Grebmer, Klaus, & Steven Were Omamo, "Lessons and Recommendations," in S. Were Omamo & K. von Grebmer (eds.), *Biotechnology, Agriculture, and Food Security in Southern Africa* (IFPRI/FANRPAN: Washington, DC/Harare, Zimbabwe, 2005).

Wafula, D., & N. Clark, "Science and Governance of Modern Biotechnology in Sub-Saharan Africa-the Case of Uganda" (2005) 17 J. Int. Dev. 679.

Watanabe, K.N., M. Taeb, and H. Okusu, "Putting Cartagena into Practice" (2004) 22 Nature Biotechnology 1207.

Winham, Gilbert, "The GMO Panel: Applications of WTO Law to Trade in Agricultural Biotech Products" (2009) 31:3 European Integration 409.

Wirth, David A., "Hazardous Substances and Activities" in Daniel Bodansky, Jutta Brunnée, & Ellen Hey, *The Oxford Handbook of International Environmental Law* (Oxford: Oxford University Press, 2007), 394.

Wolf, Amanda, "The Emergence and Implementation of the Advance Informed Agreement Procedure," in Philippe G. LePrestre, ed., *Governing Global Biodiversity: The Evolution and Implementation of the Convention on Biological Diversity* (Burlington, VT: Ashgate Publishing Co., 2002) 127.

Wolt, Jeffrey D. et al., "An Ecological Risk Assessment of Cry1F Maize Pollen Impact to Pale Grass Blue Butterfly" (2005) 4 Environmental Biosafety Research 243.

Wolt, J.D. & S.S. Olin, "Problem Formulation for Environmental Risk Assessments of Genetically Modified Crops" (Abstract of a Presentation for the Annual Meeting of the Society for Risk Analysis in San Antonio, December 2007), online: http://birenheide.com/sra/2007AM/program/singlesession.php3?sessid=W4-B, accessed 21 March 2008.

World Commission on Environment and Development, *Our Common Future* (Oxford: Oxford University Press, 1987).

World Food Programme, Annual Report 2003, 50 ("Global Food Aid Profile"), online: World Food Programme, online: http://www.wfp.org/policies/Annual_Reports/, accessed 19 October 2009.

World Food Programme, "Operational Guidelines on the Donation of Foods Derived from Modern Biotechnology," 11 February 2004 WFP/EB.1/2004/10-C.

World Summit on Sustainable Development, Plan of Implementation of the World Summit on Sustainable Development, in United Nations, Report of the World Summit on Sustainable Development (New York: UN, 2002), online: http://www.johannesburgsummit.org.

World Trade Organization, European Communities – Measures Affecting the Approval and Marketing of Biotech Products: Panel Report (2006), WT/DS291/R, 292/R, 293/R, 29 September 2006, online: http://www.wto.org/english/tratop_e/dispu_e/cases_e/ds291_e.htm.

World Trade Organization, European Communities – Measures Affecting the Approval and Marketing of Biotech Products: Status Report of the EC (2009) WT/DS293/31/Add.20, 15 September 2009.

World Trade Organization, European Communities – Measures Affecting the Approval and Marketing of Biotech Products: Status Report of the EC (2009) WT/DS291/37/Add.27, 9 April 2010.

Wouter, S. & Nick F. Pidgeon, "Trust in Risk Regulation: Cause or Consequence of the Acceptability of GM Food?" (2005) 25 Risk Analysis 199.

Young, Tomme, "Genetically Modified Organisms and Biosafety: a background paper for decision-makers and others to assist in consideration of GMO issues" (IUCN, Gland 2004).

Zarrilli, S., International Trade in GMOs and GM Products: National and Multilateral Legal Frameworks (New York: United Nations Conference on Trade and Development, 2005).

Zerbe, N., "Feeding the Famine? American Food Aid and the GMO Debate in Southern Africa" (December, 2004) 29(6) Food Policy 593, online: http://www.humboldt.edu/~nrz3/research/index.html, accessed 20 October 2009.

Ziff, Bruce, "Travels with my Plant: Monsanto v. Schmeiser Revisited" (2005) 2 University of Ottawa Law and Technology Journal 493, online: http://www.uoltj.cA/articles/vol2.2/2005.2.2.uoltj.Ziff.493-509.pdf.

Index